Handbook *of* Agricultural Geophysics

Handbook *of* Agricultural Geophysics

Edited by

Barry J. Allred
Jeffery J. Daniels
M. Reza Ehsani

CRC Press
Taylor & Francis Group
Boca Raton London New York

CRC Press is an imprint of the
Taylor & Francis Group, an **informa** business

CRC Press
Taylor & Francis Group
6000 Broken Sound Parkway NW, Suite 300
Boca Raton, FL 33487-2742

First issued in paperback 2019

© 2008 by Taylor & Francis Group, LLC
CRC Press is an imprint of Taylor & Francis Group, an Informa business

No claim to original U.S. Government works

ISBN-13: 978-0-8493-3728-4 (hbk)
ISBN-13: 978-0-367-38721-1 (pbk)

Library of Congress Card Number 2006004732

Library of Congress Cataloging-in-Publication Data

Handbook of agricultural geophysics / editors: Barry Allred, Jeffrey J. Daniels, and Mohammad Reza Ehsani.
    p. cm.
  Includes bibliographical references and index.
  ISBN 978-0-8493-3728-4 (alk. paper)
  1. Geophysics. 2. Agricultural innovations. I. Allred, Barry. II. Daniels, Jeffrey J. III. Ehsani, M. R. (Mohammad Reza) IV. Title.

QC806.H35 2008
550--dc22   2008011648

Visit the Taylor & Francis Web site at
http://www.taylorandfrancis.com

and the CRC Press Web site at
http://www.crcpress.com

Contents

SECTION I   Agricultural Geophysics Overview

SECTION II   Agricultural Geophysics Measurements and Methods

SECTION III   The Global Positioning System and Geographic Information Systems

SECTION IV   Resistivity and Electromagnetic Induction Case Histories

SECTION V Ground-Penetrating Radar Case Histories

Preface

Geophysical methods continue to show great promise for use in agriculture. The term "agricultural geophysics" denotes a subdiscipline of geophysics that is focused only on agricultural applications. As this geophysics subdiscipline becomes better established, there may come a time when a contracted name becomes appropriate, such as "agrigeophysics;" however, for this book, the more recognized term, "agricultural geophysics" will be used predominantly. Potential agricultural applications for geophysical methods are widespread and include precision agriculture, site infrastructure assessment, hydrologic monitoring, and environmental investigations. For example, changes in soil electrical conductivity measured across a farm field using geophysical methods may reflect spatial variability in soil properties, and this information can in turn be used along with precision agriculture techniques to ensure that the right quantity of fertilizer is applied to different parts of the field. In cases where placement records have been lost, ground-penetrating radar can be employed to locate and map buried agricultural drainage pipes. Geophysical measurements can be affected by the amount of water within soil, thereby allowing assessment of shallow hydrologic conditions. Geophysical methods could also be used in an environmental investigation to determine if there are leaks present in animal waste storage ponds or treatment lagoons.

The *Handbook of Agricultural Geophysics* was compiled to include a concise overview of the geophysical methods that can be utilized in agriculture and provides detailed descriptions of situations in which these techniques have been employed. The book is divided into four sections, with the first section devoted to both a general introduction of agricultural geophysics (Chapter 1) and a summary of past applications of geophysical methods to agriculture (Chapters 2 and 3). The second section systematically describes the three geophysical methods now most commonly employed: resistivity (Chapter 5), electromagnetic induction (Chapter 6), and ground-penetrating radar (Chapter 7). The second section also presents some theoretical insight on soil electrical conductivity measurement (Chapter 4) and describes, although in limited detail, three geophysical methods not typically used for agriculture but possibly having more widespread future application: magnetometry, self-potential, and seismic (Chapter 8). The Global Positioning System (GPS) and geographic information systems (GISs) are revolutionizing the way geophysical data are acquired and analyzed, and therefore warrant separate discussion in the third section of the book (Chapters 9 and 10). Agricultural geophysics case histories comprise roughly half the book. The resistivity and electromagnetic induction method case histories are included in Section IV (Chapters 11–22). The ground-penetrating radar method case histories are found in Section V (Chapters 23–30). The value of these case histories is that they document a wide range of scenarios in which geophysical methods have been successfully employed, thereby giving the reader an indication as to the potential effectiveness of using agricultural geophysics for their particular purpose.

Geophysicists will undoubtedly be very knowledgeable regarding most of the material presented within the first two, and probably, the third sections of this book. Although not the intended audience, geophysicists who are unfamiliar with geophysical applications to agriculture may have an interest in the case histories presented within the last two sections of the book. The primary audience for the *Handbook of Agricultural Geophysics* is expected to be quite diverse and include government agency personnel, university agricultural researchers, and agricultural and environmental consultants. To meet the growing demands of the intended audience, the book has been written specifically for those working in agriculture who need to use geophysical tools for gathering valuable information to solve problems. The reader of this book is therefore not expected to have a strong background in geophysics, and as such, when theoretical aspects are introduced, they are described, as much as possible, in an easy to understand manner. Furthermore, throughout the book, emphasis is placed on practical considerations regarding the application of geophysical methods to agriculture.*

Barry J. Allred
M. Reza Ehsani
Jeffrey J. Daniels

---

* Note: The use of product names throughout the book is for informational purposes only and does not imply endorsement by the editors or the organizations they represent.

Editors

Barry Allred is an agricultural engineer with the U.S. Department of Agriculture – Agricultural Research Service – Soil Drainage Research Unit, Columbus, Ohio. He is also an adjunct assistant professor in the Food, Agricultural, and Biological Engineering Department at Ohio State University. He has a B.S. degree in geology and two M.S. degrees, the first M.S. with emphasis in geophysics and the second M.S. with emphasis in hydrogeology. He earned his Ph.D. in Biosystems Engineering from Oklahoma State University. Prior employment experience includes working as a geophysicist in the petroleum industry (ConocoPhillips Inc.) and as a hydrogeologist with both an environmental consulting firm (ERM-Southwest, Inc.) and a state government agency (New Mexico Bureau of Geology and Mineral Resources). His research interests involve application of near-surface geophysical methods to agriculture; agricultural drainage and drainage water recycling systems; and laboratory hydraulic/mechanical property, solute transport, and environmental remediation studies. His agricultural geophysics investigations have focused largely on drainage pipe detection and golf course infrastructure assessment using ground-penetrating radar. He has also studied aspects of soil electrical conductivity mapping using both resistivity and electromagnetic induction methods. He is the lead author or coauthor of over 50 journal, technical, or book chapter publications.

Jeffrey Daniels is a professor in the School of Earth Sciences at The Ohio State University. He received his Ph.D. degree from the Colorado School of Mines and M.S. and B.S. degrees from Michigan State University. Daniels is an applied geophysicist, with a broad base of experience in surface and borehole geophysical methods applied to subsurface science. Prior to 1985, he was a research scientist for the U.S. Geological Survey in Denver, Colorado, working on surface electrical methods and borehole geophysical methods (electrical, nuclear, and acoustic) for characterizing physical properties near the borehole in the uranium (exploration and nuclear waste disposal) and coal exploration and development programs. He is an experienced computer programmer, and his personal research focuses on

forward and inverse computer modeling applied to geophysical imaging. In addition to his current work on agricultural problems, he has research projects and interests in the areas of subsurface void detection and environmental problems related to energy development and sustainability.

Reza Ehsani is an assistant professor of Agricultural and Biological Engineering at the University of Florida/Institute of Food and Agricultural Sciences (UF/IFAS) Citrus Research and Education Center (CREC). His current areas of research include precision technology for fruit and vegetable production, application of Global Positioning System/Geographic Information System (GPS/GIS) for orchard management, development of soil and plant sensors, and machine enhancement for citrus mechanical harvesters. He also develops educational materials on the application of GPS/GIS and sensor technology for fruit and vegetable growers and organizes grower conferences on precision agriculture and mechanical harvesting for citrus. Ehsani has a Ph.D. in Biological and Agricultural Engineering from the University of California, Davis. He also obtained a B.S. and M.S. in Agricultural Engineering from Tehran University, Iran. He was an assistant professor and a precision agriculture specialist at the Department of Food, Agricultural, and Biological Engineering at The Ohio State University before joining the University of Florida. He has authored or coauthored more than 65 journal and technical papers.

Contributors

James E. Ayars
Water Management Research Laboratory
U.S. Department of Agriculture–Agricultural
  Research Service (USDA-ARS)
Parlier, California

R. Boniak
Department of Plant, Soil and Agricultural
  Systems
Southern Illinois University
Carbondale, Illinois

Mary K. Brodahl
U.S. Department of Agriculture–Agricultural
  Research Service (USDA-ARS)
Water Management Unit
Fort Collins, Colorado

Gerald W. Buchleiter
U.S. Department of Agriculture–Agricultural
  Research Service (USDA-ARS)
Water Management Unit
Fort Collins, Colorado

John R. Butnor
U.S. Department of Agriculture (USDA) Forest
  Service
Southern Research Station
Research Triangle Park, North Carolina

S. -K. Chong
Department of Plant, Soil, and Agricultural
  Systems
Southern Illinois University
Carbondale, Illinois

David A. Claypool
Department of Plant Sciences
University of Wyoming
Laramie, Wyoming

William P. Clement
Center for Geophysical Investigation of the
  Shallow Subsurface
Boise State University
Boise, Idaho

M. E. Collins
Soil and Water Science Department
University of Florida
Gainesville, Florida

Dennis L. Corwin
U.S. Department of Agriculture–Agricultural
  Research Service (USDA/ARS)
U.S. Salinity Laboratory
Riverside, California

Frank P. Day
Department of Biological Sciences
Old Dominion University
Norfolk, Virginia

J. A. Doolittle
U.S. Department of Agriculture (USDA)
Natural Resources Conservation Service
Newtown Square, Pennsylvania

John W. Doran
U.S. Department of Agriculture–Agricultural
  Research Service (USDA-ARS) (Retired)
Agroecosystem Management Research Unit
University of Nebraska
Lincoln, Nebraska

Rhae A. Drijber
Department of Agronomy and Horticulture
University of Nebraska
Lincoln, Nebraska

Roger A. Eigenberg
U.S. Department of Agriculture–Agricultural
  Research Service (USDA-ARS)
U.S. Meat Animal Research Center
Clay Center, Nebraska

Hamid J. Farahani
International Center for Agricultural Research
  in the Dry Areas (ICARDA)
Aleppo, Syria

Richard B. Ferguson
Department of Agronomy and Horticulture
University of Nebraska
Lincoln, Nebraska

Robert A. Flynn
Department of Soil and Crop Sciences
Colorado State University
Fort Collins, Colorado

R. S. Freeland
Biosystems Engineering and Soil Science
University of Tennessee
Knoxville, Tennessee

Robin Gebbers
Department of Engineering for Crop
  Production
Leibniz–Institute for Agricultural Engineering
  Bornim
Potsdam, Germany

Dorota A. Grejner-Brzezinska
Civil and Environmental Engineering and
  Geodetic Sciences
Ohio State University
Columbus, Ohio

Douglas Groom
GeoElectric Sales
Geometric, Inc.
San Jose, California

S. J. Indorante
U.S. Department of Agriculture (USDA)
Natural Resources Conservation Service
Carbondale, Illinois

Dan B. Jaynes
U.S. Department of Agriculture–Agricultural
  Research Service (USDA-ARS)
National Soil Tilth Laboratory
Ames, Iowa

Kurt H. Johnsen
U.S. Department of Agriculture (USDA) Forest
  Service
Southern Research Station
Research Triangle Park, North Carolina

Cinthia K. Johnson
Plainview Farms, Inc.
Sterling, Colorado

R. P. Kelli Belden
Department of Renewable Resources
University of Wyoming
Laramie, Wyoming

Audun Korsaeth
Arable Crops Division
Norwegian Institute for Agricultural and
  Environmental Research
Kapp, Norway

A. N. Kravchenko
Department of Crop and Soil Sciences
Michigan State University
East Lansing, Michigan

Sigrun H. Kværnø
Soil and Environment Division
Norwegian Institute for Agricultural and
  Environmental Research
Ås, Norway

Bernd Lennartz
Institute for Land Use
University Rostock
Rostock, Germany

Scott M. Lesch
Department of Environmental Sciences
University of California
Riverside, California

Erika Lück
Institute of Geosciences
University of Potsdam
Potsdam, Germany

Edward L. McCoy
School of Natural Resources
Ohio State University
Columbus, Ohio

Daniel McInnis
U.S. Department of Agriculture (USDA) Forest
  Service
Southern Research Station
Research Triangle Park, North Carolina

Carolyn J. Merry
Civil and Environmental Engineering and
  Geodetic Sciences
The Ohio State University
Columbus, Ohio

Larry C. Munn
Department of Renewable Resources
University of Wyoming
Laramie, Wyoming

John A. Nienaber
U.S. Department of Agriculture–Agricultural
  Research Service (USDA-ARS)
U.S. Meat Animal Research Center
Clay Center, Nebraska

J. David Redman
Sensors & Software Inc.
Mississauga, Ontario
Canada

Hugh Riley
Arable Crops Division
Norwegian Institute for Agricultural and
  Environmental Research
Kapp, Norway

Michael Rogers
Department of Physics
Ithaca College
Ithaca, New York

Brian E. Roth
Forest Biology Research Cooperative
School of Forest Resources and Conservation
University of Florida
Gainesville, Florida

Britta Schmalz
Department of Hydrology and Water Resources
  Management
Ecology Centre
Kiel University
Germany

Dale L. Shaner
U.S. Department of Agriculture–Agricultural
  Research Service (USDA-ARS)
Water Management Unit
Fort Collins, Colorado

Peter J. Shouse
U.S. Department of Agriculture–Agricultural
  Research Service (USDA-ARS)
George E. Brown Jr. Salinity Laboratory
Riverside, California

Richard Soppe
Wageningen UR
Wageningen, The Netherlands

Heiner Stoffregen
Institute of Ecology
Technical University Berlin
Berlin, Germany

Daniel B. Stover
Department of Marine and Ecological Sciences
Florida University
Smithsonian CO_2 Gulf Coast Research Site
Kennedy Space Center, Florida

Richard S. Taylor
Dualem Inc.
Milton, Ontario
Canada

Mark Vendl
U.S. Environmental Protection Agency
Region V
Chicago, Illinois

Live S. Vestgarden
Soil and Environment Division
Norwegian Institute for Agricultural and
  Environmental Research
Ås, Norway

Derk Wachsmuth
Max-Planck-Institute for Limnology
Ploen, Germany

Andy L. Ward
Natural Resources Division
Pacific Northwest National Laboratory
Richland, Washington

Brian J. Wienhold
U.S. Department of Agriculture–Agricultural
  Research Service (USDA-ARS)
Agroecosystem Management Research Unit
University of Nebraska
Lincoln, Nebraska

Bryan L. Woodbury
U.S. Department of Agriculture–Agricultural
  Research Service (USDA-ARS)
U.S. Meat Animal Research Center
Clay Center, Nebraska

List of Important Agricultural and Geophysical Quantities

Quantity	Acronym or Symbol	Commonly Reported Measurement Units	Abbreviated Measurement Units
Apparent soil electrical conductivity	EC_a, σ_a	millisiemens/meter	mS/m
Apparent soil electrical resistivity	ER_a, ρ_a	ohm-meter	Ωm
Bulk modulus	k	gigapascals	GPa
Cation exchange capacity	CEC	milliequivalents/100 grams	meq/100 g
Crop yield	Y	kilograms/hectare	kg/ha
Density	ρ	grams/cubic centimeter	g/cm^3
Dielectric constant (or relative permittivity)	κ, ε_r	dimensionless ratio value	dimensionless
Electric current	I	amperes	A
Electric potential	V	volts	V
Electric potential difference	ΔV	volts	V
Electric Field electric potential gradient	E	volts/meter	V/m
Electromagnetic wave skin depth	δ, $\frac{1}{\alpha}$	meters	m
Hydraulic conductivity	K	centimeters/second	cm/s
Frequency (wave)	f	hertz	Hz
Magnetic field strength	B	nanoteslas	nT
Magnetic permeability	μ	henrys/meter	H/m
Magnetic susceptibility	κ	dimensionless quantity	dimensionless
Pesticide partition coefficient	K_d	liters/kilogram	L/kg
Porosity	n, ϕ	fraction by volume	dimensionless
Rigidity (shear) modulus	μ	gigapascals	GPa
Seismic adsorption coefficient	α	1/kilometers	1/km
Degree of saturation (soil)	S	fraction by volume	dimensionless
Salinity (electrical conductivity of saturated soil paste)	EC_e	decisiemens/meter	dS/m
Sodium adsorption ratio	SAR	fraction based on concentration	dimensionless
Soil dry bulk density	ρ_b	grams/cubic centimeter	g/cm^3
Soil particle size distribution	sand, silt, and clay	percentage by weight	%
Soil gravimetric water content	w	fraction by weight	dimensionless
Soil organic matter content	SOM	fraction by weight	dimensionless
Soil organic carbon content	SOC	fraction by weight	dimensionless
Soil pH	pH	dimensionless quantity	dimensionless
Soil volumetric water content	θ	fraction by volume	dimensionless
Soil solution concentration	Cc	moles/liter, milligrams/liter, or parts/million	mol/L, mg/L, or ppm
Pore water pressure potential	H_p, Ψ	centimeters of water	cm
Specific surface	S_s	meters squared/gram	m^2/g
Temperature	T	degrees celsius	°C

(continued on next page)

Quantity	Acronym or Symbol	Commonly Reported Measurement Units	Abbreviated Measurement Units
Velocity (seismic and radar)	seismic P-wave (V_P), seismic S-wave (V_S), radar (v)	V_P: kilometers/second V_S: kilometers/second v: meters/nanosecond	V_P: km/s V_S: km/s v: m/ns
Wavelength	λ	meters	m

Note: The same symbol is often utilized for more than one quantity.

Section I

Agricultural Geophysics Overview

1 General Considerations for Geophysical Methods Applied to Agriculture

Barry J. Allred, M. Reza Ehsani, and Jeffrey J. Daniels

CONTENTS

1.1 INTRODUCTION: GEOPHYSICS DEFINITIONS, DEVELOPMENT CHRONOLOGY, INVESTIGATION SCALE

Geophysics can be defined several ways. In the broadest sense, geophysics is the application of physical principles to studies of the Earth (Sheriff, 2002). This general definition of geophysics encompasses a wide range of disciplines, such as hydrology, meteorology, physical oceanography, seismology, tectonophysics, etc. Geophysics, as it is used in this book, has a much more focused definition. Specifically, geophysics is the application of physical quantity measurement techniques to provide information on conditions or features beneath the Earth's surface. With the exception of borehole geophysical methods and soil probes like a cone penetrometer, these techniques are generally noninvasive, with physical quantities determined from measurements made mostly at or near the ground surface. (Note: Some large-scale airborne surveys are carried out with geophysical measurements collected by airplanes and helicopters positioned well above the surface, but these types of surveys are not within the scope of this book.) The geophysical methods employed to obtain subsurface information from surface-based measurements include resistivity, electromagnetic induction, ground-penetrating radar, magnetometry, self-potential, seismic, gravity, radioactivity, nuclear magnetic resonance, induced polarization, etc.

One of the first known instruments for geophysical measurement is a seismoscope invented in A.D. 132 by the Chinese philosopher, Chang Hêng (Needham, 1959). This seismoscope reportedly had the capability to not only detect earthquakes, but could also determine the direction from which the earthquake originated. Many of the geophysical methods employed today originated or were more fully developed based on the needs of the mining and petroleum industries. In fact, present levels of worldwide production for minerals, oil, and natural gas could not have been achieved without the use of geophysics as an exploration tool.

The magnetic compass was used to find iron ore as early as 1640 (Dobrin and Savit, 1988). Robert Fox devised the self-potential method using copper-plated electrodes and a galvanometer to find copper sulfide ore bodies in Cornwall, England, during 1830 (Reynolds, 1997). Robert Thalén wrote *On the Examination of Iron Ore Deposits by Magnetic Methods* in 1879 and contributed to the invention of some of the first magnetometers (Telford et al., 1976). These Thalén–Tiberg and Thomson–Thalén magnetometers proved very successful for mineral prospecting in Sweden during the late 1800s. Initial development of resistivity and electromagnetic induction methods for the mining industry occurred between 1910 and 1930. Airborne magnetometers refined for submarine detection during the Second World War were employed shortly afterward to quickly prospect for minerals over large areas (Dobrin and Savit, 1988). The introduction of airborne electromagnetic surveys for mineral exploration also occurred shortly after the Second World War ended.

Dobrin and Savit (1988) and Lawyer et al. (2001) detail some of the early history involving initial applications of geophysical methods for the petroleum industry. A torsion balance field device for measuring anomalies in the Earth's gravitational field was refined by Baron Roland von Eötvös of Hungary in the late 1800s. Crude seismic methods were developed by the French, British, Germans, and Americans during the First World War as a means to locate enemy artillery positions. Torsion balance gravity measurements and fan-pattern seismic refraction surveys were then used to find oil fields associated with Texas Gulf Coast salt domes in the 1920s. Conventional seismic refraction methods introduced in 1928 to the Middle East were soon found to be particularly effective within Iran for locating limestone structures containing substantial oil reserves. J. C. Karcher conducted the first seismic reflection experiments from 1919 to 1921 and then demonstrated the potential of this geophysical method for oil exploration by mapping a shallow rock unit in central Oklahoma during 1921. The first oil discovery attributed to seismic reflection occurred during 1927 with the Maud Field in Oklahoma. Seismic reflection is the predominant geophysical method used for petroleum exploration today.

Although radar technologies were introduced during the Second World War, it was not until the early 1960s that ground-penetrating radar was first employed as a geophysical tool, initially to investigate the subsurface characteristics of polar ice sheets (Bailey et al., 1964). Archeological, environmental, geotechnical engineering, and hydrological geophysical surveys became more and more common in the latter half of the past century. There was some agricultural research activity in the 1930s and 1940s related to soil moisture measurement with resistivity methods (Edlefsen and Anderson, 1941; Kirkham and Taylor, 1949; McCorkle, 1931), but for the most part, the application of geophysical methods to agriculture did not gain momentum until the 1960s, and to a greater extent in the 1970s, with the use of resistivity methods for soil salinity assessment (Halvorson and Rhodes, 1974; Rhoades and Ingvalson, 1971; Rhoades et al., 1976; Shea and Luthin, 1961). Greater historical detail on the application of geophysical methods to agriculture is provided in Chapters 2 and 3 of this book.

Geophysical surveys conducted for petroleum, mining, hydrological, environmental, geotechnical engineering, archeological, and agricultural applications vary dramatically in scale with respect to the investigation depth of interest. Petroleum industry oil and gas wells have been drilled to levels 8 km beneath the surface based on information obtained from seismic reflection surveys. Most geophysical surveys conducted in the mining industry have an investigation depth of interest that is less than 1 km. There are, however, some deep mining operations extending more than 3 km below ground, and therefore mining geophysical surveys can occasionally require greater investigation depths down to 3 or 4 km. A geophysical survey conducted as part of a hydrological investigation to determine groundwater resources usually has an investigation depth no greater than 300 m. Geophysical investigation depths for environmental, geotechnical engineering, and archeological applications typically do not exceed 30 m. Agricultural geophysics tends to be heavily focused on a 2 m zone directly beneath the ground surface, which includes the crop root zone and all, or at least most, of the soil profile.

With regard to the application of geophysics to agriculture, this extremely shallow 2 m depth of interest is certainly an advantage, in one sense because most geophysical methods have investigation depth capabilities that far exceed 2 m. However, there are complexities associated with agriculture geophysics not typically encountered with the application of geophysical methods to other industries or disciplines. One such complexity involves transient soil temperature and moisture conditions that can appreciably alter the values of measured geophysical quantities over a period of days or even hours. Additionally, physical quantities measured in the soil environment with geophysical methods often exhibit substantial variability over very short horizontal and vertical distances.

1.2 GEOPHYSICAL METHODS APPLICABLE TO AGRICULTURE

Geophysical methods can be classified as passive or active. There is no artificial application of energy with passive geophysical methods. On the other hand, active geophysical methods do require the artificial application of some form of energy. The three geophysical methods predominantly used for agricultural purposes are resistivity, electromagnetic induction, and ground-penetrating radar. All three of these predominantly employed methods are active, and each is summarized within this book; resistivity in Chapter 5, electromagnetic induction in Chapter 6, and ground-penetrating radar in Chapter 7. Chapter 8 provides shorter descriptions of three additional geophysical methods: magnetometry (passive), self-potential (passive), and seismic (active), all of which have the potential for substantial future use in agriculture, but at present are being employed sparingly or not at all for agricultural purposes. To provide an introduction, the six geophysical methods— resistivity, electromagnetic induction, ground-penetrating radar, magnetometry, self-potential, and seismic—are all concisely defined as follows.

1.2.1 RESISTIVITY METHODS

Resistivity methods measure the electrical resistivity, or its inverse, electrical conductivity, for a bulk volume of soil directly beneath the surface. Resistivity methods basically gather data on the subsurface electric field produced by the artificial application of electric current into the ground. With the conventional resistivity method, an electrical current is supplied between two metal electrode stakes partially inserted at the ground surface, while voltage is concurrently measured between a separate pair of metal electrode stakes also inserted at the surface. The current, voltage, electrode spacing, and electrode configuration are then used to calculate a bulk soil electrical resistivity (or conductivity) value.

1.2.2 ELECTROMAGNETIC INDUCTION METHODS

Electromagnetic induction methods also measure the electrical conductivity (or resistivity) for a bulk volume of soil directly beneath the surface. An instrument called a ground conductivity meter is commonly employed for relatively shallow electromagnetic induction investigations. In operation, an alternating electrical current is passed through one of two small electric wire coils spaced a set distance apart and housed within the ground conductivity meter that is positioned at, or a short distance above, the ground surface. The applied current produces an electromagnetic field around the "transmitting" coil, with a portion of the electromagnetic field extending into the subsurface. This electromagnetic field, called the primary field, induces an alternating electrical current within the ground, in turn producing a secondary electromagnetic field. Part of the secondary field spreads back to the surface and the air above. The second wire coil acts as a receiver measuring the resultant amplitude and phase components of both the primary and secondary fields. The amplitude and phase differences between the primary and resultant fields are then used, along with the intercoil spacing, to calculate an "apparent" value for soil electrical conductivity (or resistivity).

1.2.3 GROUND-PENETRATING RADAR METHODS

With the ground-penetrating radar (GPR) method, an electromagnetic radio energy (radar) pulse is directed into the subsurface, followed by measurement of the elapsed time taken by the radar signal as it travels downward from the transmitting antenna, partially reflects off a buried feature, and eventually returns to the surface, where it is picked up by a receiving antenna. Reflections from different depths produce a signal trace, which is a function of radar wave amplitude versus time. Radar waves that travel along direct and refracted paths through both air and ground from the transmitting antenna to the receiving antenna are also included as part of the signal trace. Antenna frequency, soil moisture conditions, clay content, salinity, and the amount of iron oxide present have a substantial influence on the distance beneath the surface to which the radar signal penetrates. The dielectric constant of a material governs the velocity for the radar signal traveling through that material. Differences in the dielectric constant across a subsurface discontinuity feature control the amount of reflected radar energy, and hence radar wave amplitude, returning to the surface. As an end product, radar signal amplitude data are plotted on depth sections or areal maps to gain insight on below-ground conditions or to provide information on the position and character of a subsurface feature.

1.2.4 MAGNETOMETRY METHODS

This geophysical method employs a sensor, called a magnetometer, to measure the strength of the Earth's magnetic field. Anomalies in the Earth's magnetic field indicate the presence of subsurface features. An anomaly is produced when a subsurface feature has a remanent magnetism or magnetic susceptibility that is different from its surroundings. A gradiometer is an instrument setup composed of two magnetometer sensors mounted a set distance apart. Gradiometers are typically used to measure the vertical gradient of the magnetic field, which is not affected by transient magnetic field changes. In comparison to a single magnetometer sensor, the gradiometer has the additional advantage of being better adapted for emphasizing magnetic field anomalies from shallow sources.

1.2.5 SELF-POTENTIAL METHODS

Self-potential methods collect information on a naturally occurring electric field associated with nonartificial electric currents moving through the ground. Unlike resistivity methods, no electric power source is required. Naturally occurring electric potential gradients can arise a number of different ways, including the subsurface flow of water containing dissolved ions, spatial concentration differences of dissolved ions present in subsurface waters, and electrochemical interactions between mineral ore bodies and dissolved ions in subsurface waters. Self-potential methods are fairly simple operationally. All that is required to obtain information on a natural electric field below ground is the voltage measurement between two nonpolarizing electrodes placed or inserted at the ground surface.

1.2.6 SEISMIC METHODS

Seismic methods employ explosive, impact, vibratory, and acoustic energy sources to introduce elastic (or seismic) waves into the ground. These seismic waves are essentially elastic vibrations that propagate through soil and rock materials. The seismic waves are timed as they travel through the subsurface from the source to the sensors, called "geophones." The energy source is positioned at the surface or at a shallow depth. Geophones are typically inserted at the ground surface. Seismic waves move through the subsurface from source to geophone along a variety of direct, refracted, and reflected travel paths. The velocity of a seismic wave as it travels through a material is determined by the density and elastic properties for that particular material. Differences in the density and elastic properties across a subsurface discontinuity feature control the amount of reflected or refracted seismic energy, and hence the seismic wave amplitudes returning to the surface. Information on the timed arrivals and amplitudes of the direct, refracted, and reflected seismic waves measured by the geophones are then used to gain insight on below-ground conditions or to locate and characterize subsurface features.

1.3 ASPECTS OF AGRICULTURAL GEOPHYSICS DATA COLLECTION AND ANALYSIS

1.3.1 SELECTING THE PROPER GEOPHYSICAL METHOD

A clear goal must be defined in the initial planning stage of a geophysical survey regarding the soil condition or subsurface feature information that needs to be acquired. In order to choose the proper geophysical method for monitoring changing soil conditions, consideration must first be given to the different physical properties responded to by the various geophysical methods and then whether any of these physical properties are influenced by the soil condition of interest. Delineating a subsurface feature with geophysics requires there to be a contrast between the feature and its surroundings with respect to some physical property responded to by a geophysical method. To summarize, the geophysical method selected must respond to a physical property that is in turn affected by temporal changes in soil conditions or the spatial patterns of subsurface features; otherwise, useful information cannot be obtained on these soil conditions or subsurface features of interest. For example, soil cation exchange capacity (CEC) will often have a substantial impact on soil electrical conductivity (or resistivity); therefore, resistivity or electromagnetic induction methods that measure soil electrical conductivity may be useful for delineating spatial patterns in CEC. On the other hand, magnetometry methods respond to anomalies in remanent magnetism or magnetic susceptibility, properties that are not likely to be affected by CEC, and consequently, magnetometry methods would not be a good choice for delineating spatial patterns in CEC.

1.3.2 INVESTIGATION DEPTH AND FEATURE RESOLUTION ISSUES

Once a geophysical method is chosen, there are usually options with respect to the equipment and its setup. The investigation depth required and the size of the feature to be detected are two important issues that should be taken into account when deciding on the equipment to use and its setup. There is normally a trade-off between the investigation depth and the minimum size a feature must have to be detected. Finding a large, deeply buried object or a small, shallow object with geophysical methods is much easier than locating a small, deeply buried object. One potential example is the use of GPR to locate buried plastic or clay tile agricultural drainage pipe. The radar signal penetration depth and minimum size at which an object can be detected are both inversely related to GPR antenna frequency. Low-frequency GPR antennas are better for locating larger deeply buried objects, and high-frequency GPR antennas are more applicable for small, shallow objects. Therefore, a GPR unit with 100 MHz transmitting and receiving antennas might work well at finding a 30 cm diameter drainage pipe 2 m beneath the surface in a clay soil, and a GPR unit with 250 MHz transmitting and receiving antennas is likely capable of finding a 10 cm diameter drainage pipe 0.5 m beneath the surface in a clay soil. But, finding a 10 cm diameter drainage pipe 2 m beneath the surface in a clay soil is probably an extremely difficult undertaking regardless of the GPR antenna frequency employed.

An important implication with respect to the issues of investigation depth and feature resolution (detection) is to use equipment with the proper setup that provides an investigation depth similar to the investigation depth of interest. Using an equipment setup with an investigation depth substantially greater than the investigation depth of interest results in the minimal size for feature resolution being increased over what would be the case if the equipment investigation depth coincided with the investigation depth of interest. Additionally, by using an equipment setup with an investigation depth substantially greater than the investigation depth of interest, a problem could arise of not being able to determine whether a detected feature is located within the investigation depth of interest or at a deeper level. An equipment setup investigation depth substantially less than the investigation depth of interest means that features positioned between the equipment setup investigation depth and the depth of interest will not be detected. For example, when a resistivity survey is employed to map lateral changes within a well-developed soil profile, a specific electrode array length might be chosen to provide an approximate 2 m investigation depth. Significantly shorter or longer electrode array lengths than that selected for a 2 m investigation depth would respectively produce investigation depths much less or much greater than 2 m, thereby producing information that does not include the entire soil profile (short electrode array length problem) or information where it is difficult to determine whether resistivity changes occurred within the soil profile or at a greater depth (long electrode array problem). Finally, there are instances where small, deeply buried features are unlikely to be detected, and therefore, time and expense should not be wasted conducting a geophysical survey.

1.3.3 Field Operations: Station Interval, Stacking, Survey Line/Grid Setup, and Global Positioning System (GPS) Integration

The distance is usually fixed or at least fairly consistent from one geophysical measurement location to the next along a transect, and this distance between measurement locations is referred to as the station interval. A short station interval provides a better chance for finding the smaller features that are capable of being resolved with the geophysical equipment used. Reducing the station interval has the downside of increasing the time needed to conduct a geophysical survey. Consequently, it makes sense to use the shortest station interval possible that still allows the geophysical survey to be carried out in the time allotted.

Often, several measurements are collected at each measurement location and then are added or averaged. This overall process is called stacking. Unwanted signals referred to as noise tend to be random and can thus be cancelled out by adding or averaging multiple geophysical measurements obtained at the same location. Although data quality is improved, increased stacking can slow the geophysical survey. Data collection procedures should be optimized to provide the greatest amount of stacking possible within the time frame during which the geophysical survey needs to be conducted.

For a larger subsurface feature where the general directional trend is known, a sufficient number of geophysical measurement transects should be oriented perpendicular to the feature's trend so as to better delineate the feature. A measurement transect parallel to a linear subsurface feature, but offset from it by sufficient distance, will in all probability not detect the feature. A geophysical survey grid covering a study area is commonly composed of either one set of parallel measurement transects or two sets of parallel measurement transects oriented perpendicular to one another. Setting up a geophysical survey grid composed of two sets of parallel measurement transects oriented perpendicular to one another reduces the risk of not finding long, narrow subsurface features, such as agricultural drainage pipes, whose trend prior to the survey is unknown. The spacing distance between adjacent transects is usually fixed at some constant or fairly consistent value for a particular set of parallel transects. This spacing distance should be set small enough, within reasonable limits, to avoid missing important features.

The integration of Global Positioning System (GPS) receivers with geophysical equipment is becoming more and more common, particularly with regard to agricultural applications. GPSs can provide accurate determinations of measurement locations while the geophysical survey is in progress. As a result of GPS integration, marking off a well-defined grid in the field is no longer required, thereby allowing rapid geophysical data collection over large areas, especially in regard to horizontal soil electrical conductivity mapping with resistivity or electromagnetic induction methods. The importance of GPS to agricultural geophysics will undoubtedly continue to experience growth in the near future; therefore, a detailed discussion on aspects related to GPS is certainly warranted and can be found in Chapter 9.

1.3.4 ANALYSIS OF GEOPHYSICAL DATA

Depth sections and contour maps are two of the most common geophysical data analysis end products. Two-dimensional depth sections characterize the distribution of some geophysically measured property beneath a measurement transect along the surface. Different geophysical methods employ different computer processing steps to produce these depth sections. Contour maps are typically used to show the horizontal spatial pattern of some geophysically measured property. Various spatial interpolation algorithms are employed by the computer software used to generate these contour maps. Where there is a choice, careful consideration is needed in selecting the interpolation algorithm so as not to introduce features on the contour map that do not truly exist or to remove features that are actually present.

Rather than focusing just on a single geophysical data set at a time, the integration of several geophysical data sets along with other spatial information is an approach that can potentially improve agricultural data interpretation for a particular farm site. Integration of multiple geophysical and nongeophysical spatial data sets is accomplished using a geographic information system (GIS). A GIS is a powerful data analysis tool that is just beginning to find widespread use in agricultural geophysics. Because GIS is expected to become essential to agricultural geophysics in the future, a detailed discussion on some important GIS elements is definitely relevant and is presented in Chapter 10.

1.4 POTENTIAL AGRICULTURAL USES FOR GEOPHYSICAL METHODS

Past research indicates a wide range of potential uses for the three geophysical methods predominantly employed in agriculture (resistivity, electromagnetic induction, and ground-penetrating radar). Table 1.1 serves to emphasize the variety of possible applications by listing just a few of the numerous ways that these three geophysical methods can provide valuable information for agriculture purposes. The resistivity and electromagnetic induction case histories in Chapter 11 and the ground-penetrating radar case histories in Chapter 12 provide in-depth descriptions for many of the agricultural geophysics applications listed in Table 1.1. However, some aspects regarding the last four agricultural geophysics applications listed in Table 1.1 warrant further mention at this juncture.

Figure 1.1 provides two examples of GPR drainage pipe detection. Figure 1.1a and Figure 1.1b are GPR time-slice amplitude maps. Each map represents the reflected radar amplitudes (and radar energy) returning to the surface from a particular depth interval. Lighter shaded elements on gray-scale GPR time-slice amplitude maps typically denote subsurface features that reflect significant amounts of radar energy. The lighter shaded elements with linear trends found in Figure 1.1 are indicative of buried drain lines. Shown in Figure 1.1a is the subsurface drainage pipe system in a northwest Ohio agricultural field, and depicted in Figure 1.1b is the subsurface drainage pipe system for a central Ohio golf course green. In addition to GPR, magnetometry methods have exhibited some success in locating buried drainage pipes (Rogers et al., 2005, 2006). An example regarding the application of magnetometry methods to locate subsurface drain lines at a dairy operation in Oregon is included in Chapter 8.

TABLE 1.1

Potential Agricultural Applications for Resistivity, Electromagnetic Induction, and Ground-Penetrating Radar Methods

Geophysical Method	Agricultural Application	Literature Source
Resistivity	Soil drainage class mapping	Kravchenko et al., 2002
Electromagnetic induction	Determining clay-pan depth	Doolittle et al., 1994
Electromagnetic induction	Estimation of herbicide partition coefficients in soil	Jaynes et al., 1995a
Electromagnetic induction	Mapping of flood deposited sand depths on farmland adjacent to the Missouri River	Kitchen et al., 1996
Electromagnetic induction	Soil nutrient monitoring from manure applications	Eigenberg and Nienaber, 1998
Ground-penetrating radar	Quality/efficiency improvement and updating of U.S. Department of Agriculture/Natural Resources Conservation Service (USDA/NRCS) soil surveys	Doolittle, 1987; Schellentrager et al., 1988
Ground-penetrating radar	Measurement of microvariability in soil profile horizon depths	Collins and Doolittle, 1987
Ground-penetrating radar	Bedrock depth determination in glaciated landscape with thin soil cover	Collins et al., 1989
Ground-penetrating radar	Plant root biomass surveying	Butnor et al., 2003; Konstantinovic et al., 2007; Wöckel, et al., 2006
Ground-penetrating radar	Identification of subsurface flow pathways	Freeland et al., 2006; Gish et al., 2002
Ground-penetrating radar	Farm field and golf course drainage pipe detection	Allred et al., 2005a; Boniak et al., 2002; Chow and Rees, 1989
Resistivity and electromagnetic induction	Soil salinity assessment	Doolittle et al., 2001; Hendrickx et al., 1992; Rhoades and Ingvalson, 1971; Rhoades et al., 1989; Shea and Luthin, 1961
Resistivity and electromagnetic induction	Delineation of spatial changes in soil properties	Allred et al., 2005b; Banton et al., 1997; Carroll and Oliver, 2005; Johnson et al., 2001; Lund et al., 1999
Resistivity, electromagnetic induction, and ground-penetrating radar	Soil water content determination	Grote at al., 2003; Huisman et al., 2003; Kirkham and Taylor, 1949; Lunt et al., 2005; McCorkle, 1931; Sheets and Hendrickx, 1995

Substantial efforts have been devoted toward evaluating the capabilities of resistivity and electromagnetic induction methods for soil salinity assessment. The standard laboratory technique for determining salinity involves measuring the electrical conductivity of water extracted from a soil sample saturated paste (Smedema et al., 2004). The soil salinity obtained by the electrical conductivity of a saturated paste extract is designated EC_e. Resistivity and electromagnetic induction methods are used in the field to measure an "apparent" electrical conductivity for a bulk volume of soil beneath the surface, and this apparent electrical conductivity is designated EC_a. By incorporating various soil moisture, soil density, and soil textural parameters, EC_e can be calculated from EC_a (Rhoades et al., 1989, 1990). Discussions regarding the impact of soil conditions and soil properties on EC_a can be found in Chapters 2, 4, and 5. With the protocols now available for calculating EC_e from EC_a, resistivity and electromagnetic induction methods are indeed valuable tools for monitoring soil salinity levels in an agricultural field.

Precision agriculture is a growing trend combining geospatial data sets, state-of-the-art farm equipment technology, GIS, and GPS receivers to support spatially variable field application of

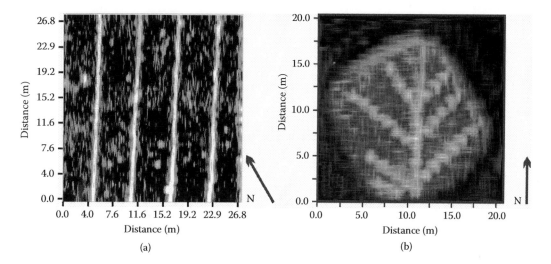

FIGURE 1.1   Ground-penetrating radar drainage pipe detection examples: (a) agricultural test plot in north-west Ohio and (b) golf course green in central Ohio.

fertilizer, soil amendments, pesticides, and even tillage effort (Morgan and Ess, 1997; National Research Council, 1997). The benefits of precision agriculture to farmers are maximized crop yields and reduced input costs. There is an important environmental benefit as well. Overapplication of agrochemicals and soil tillage is fairly common. Because precision agriculture operations result in optimal amounts of fertilizer, soil amendments, pesticides, and tillage applied on different parts of the field, there are potentially less agrochemicals and sediment released offsite via subsurface drainage and surface runoff. With reduced offsite discharge of agrochemicals and sediment, adverse environmental impacts on local waterways are diminished. In essence, precision agriculture techniques allow a farm field to be divided into different management zones for the overall purpose of optimizing economic benefits and environmental protection.

Geophysical surveys can have an important role in precision agriculture. The apparent soil electrical conductivity (EC_a) map of a farm field, obtained from resistivity or electromagnetic induction measurements, is often significantly correlated with the crop yield map for the same field. Presented in Figure 1.2 is an example comparing EC_a and soybean yield for a 3 ha field in northwest Ohio. The spatial correlation coefficient (r) between EC_a and soybean yield is −0.51 for the field shown in Figure 1.2. Over a two-year period for three different fields, Jaynes et al. (1995b) found r values between EC_a and corn/soybean yield of −0.73, −0.63, −0.55, −0.50, −0.09, and 0.45, so in five out of six cases for this study, there was substantial correlation.

Furthermore, the mapped horizontal EC_a patterns for a farm field often tend to remain consistent over time, which implies that the horizontal EC_a pattern is governed by lateral variations in soil properties (Allred et al., 2005b, 2006; Lund et al., 1999). The EC_a can be affected in a complex manner by a number of different soil properties; therefore, a limited soil sampling and analysis program is typically required to determine which soil property or properties have the greatest influence on the horizontal EC_a field pattern. Again, discussions regarding the effect of soil conditions and soil properties on soil electrical conductivity can be found in Chapters 2, 4, and 5. Soil property information based on EC_a measurement is useful for formulating management practices (strategies for fertilizer, soil amendment, and pesticide application along with tillage effort) that will improve crop yields while limiting the offsite release of agrochemicals and sediment. Consequently, because the spatial pattern for crop yield commonly exhibits a strong correlation with the horizontal EC_a pattern, which is in turn governed by lateral changes in soil properties, it becomes apparent that EC_a maps generated from resistivity and electromagnetic induction surveys can be a valuable precision agriculture tool providing insight on how to best divide a field into zones based on soil property

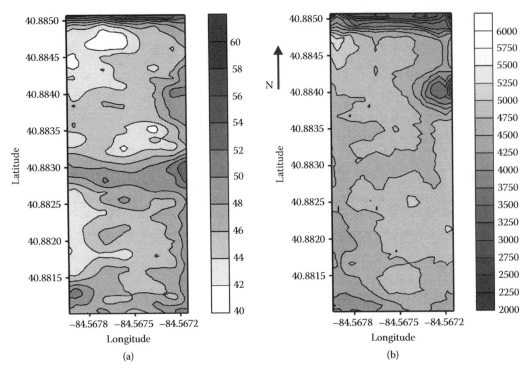

FIGURE 1.2 Soil electrical conductivity and crop yield comparison for a 3 ha field in northwest Ohio: (a) apparent soil electrical conductivity map obtained with a Veris 3100 Soil EC Mapping System (values are in millisiemens/meter) and (b) soybean yield map (values are in kilograms/hectare).

differences. These field zones can then be separately managed in an effective and efficient manner so as to maximize economic benefits while protecting the environment.

McCorkle (1931) conducted one of the first agricultural geophysics studies that focused on the use of resistivity methods to determine the soil gravimetric water content. There have been a number of research investigations since that have quantified the bulk soil volumetric water content (θ) level in soil using EC_a measured with resistivity or electromagnetic induction methods. There are, however, certain drawbacks regarding the use of resistivity and electromagnetic induction methods to determine θ. One drawback is that soil properties in which affect EC_a differ from one location to the next, and as a consequence, the EC_a versus θ relationship must be developed at each particular field site. In addition, temperature effects on EC_a values need to be taken into account before accurate θ estimates can be obtained.

GPR has recently proven to be an effective tool for rapidly measuring soil volumetric water content (θ) over large areas (Grote et al., 2003; Huisman et al., 2003; Lunt et al., 2005). The GPR methods employed for θ measurement are all based on the determination of a soil's dielectric constant, also called the relative permittivity (ε_r). The value of ε_r is strongly correlated to θ. For mapping θ, a couple of approaches are used to determine a soil's ε_r value. One approach is to calculate ε_r directly from the radar signal velocity in a soil. The soil radar signal velocity is easily computed by dividing the length of the direct or reflected signal travel path through the soil by the elapsed time taken by the signal to travel along the path from the transmitting antenna to the receiving antenna (velocity equals distance divided by time). The second approach involves positioning the transmitting and receiving antennas a short distance above the ground surface and then measuring the reflection coefficient at the ground surface. The reflection coefficient in this case equals the ratio of the radar signal amplitude reflected from the ground surface to the radar signal amplitude incident at the ground surface. With the ε_r of air known (equal to 1), the ε_r at the soil surface can be calculated from the ground surface reflection coefficient in a straightforward manner. Theoretically based site-specific calibrated equations or empirically derived relationships are used in the final step to calculate θ values from soil ε_r values. The empirical relationship most commonly used to calculate θ from ε_r for soils was developed by Topp et al. (1980) and is given by:

$$\theta = -0.053 + 0.0292\varepsilon_r - 0.00055\varepsilon_r^2 + 0.0000043\varepsilon_r^3 \qquad (1.1)$$

Additional information on the application of GPR to soil volumetric water content determination can be found in three of the Chapter 12 case studies.

1.5 AGRICULTURAL GEOPHYSICS OUTLOOK

New developments in the overall discipline of geophysics are ongoing, with innovative methods, equipment, and field procedures continuing to be introduced. The same is particularly true for agricultural geophysics. Many concepts being tested and initiated at present will eventually become commonplace for agricultural geophysics. In this regard, the following is a list summarizing the probable future trends (some previously mentioned) for agricultural geophysics.

1. New agricultural applications will continue to be discovered for the geophysical methods already used in agriculture (resistivity, electromagnetic induction, and GPR).
2. Geophysical methods not traditionally employed in the past for agricultural purposes will find significant use in the future. The geophysical methods likely to make inroads into agriculture include magnetometry, self-potential, and seismic. Agricultural opportunities for other geophysical methods, such as nuclear magnetic resonance, induced polarization, and seismoelectric, may also exist.
3. The incorporation of GPS receivers will become the norm, especially with regard to real-time kinematic (RTK) GPS, which will allow geophysical measurement positions to be determined with horizontal and vertical accuracies of a few centimeters or less. Guidance devices, video display tracking systems, or even simple on-the-go guesstimates of the spacing distance between transects, when used with an accurate GPS, can provide the capability of efficiently conducting geophysical surveys over large agricultural field areas without the need to mark out a well-defined grid at the ground surface. For some geophysical methods, the computer processing procedures used for horizontal mapping of measurements may require some modification for input of data collected along a set of transects with somewhat irregular orientations and spacing distances.
4. Geophysical surveying with more than one sensor will become a standard approach because of the variety of field information required to make correct agricultural management decisions. Multisensor systems based on a single geophysical technique have already been produced, and these systems are certainly beneficial to agriculture. Examples include GPR systems having more than one transmitter and receiver antenna pair (the individual transmitter and receiver antenna pairs can have the same frequency or different frequencies), or continuously pulled resistivity electrode arrangements containing more than one four-electrode array. However, multisensor systems based on more than one geophysical technique still need to be developed for agricultural purposes, something likely to happen in the near future. For reference, the physical properties responded to by the different geophysical methods are reviewed in Table 1.2.
5. Multiple geophysical data sets integrated and analyzed together along with other geospatial information can provide agricultural insight not available when analyzing each geophysical data set separately. Geostatistical analysis techniques can be especially useful in this regard. GISs are particularly well adapted for integration and geostatistical analysis of

TABLE 1.2

Physical Properties Responded to by Geophysical Methods

Geophysical Method	Physical Property
Resistivity	Electrical resistivity (or electrical conductivity)
Electromagnetic induction	Electrical conductivity (or electrical resistivity)
Ground-penetrating radar	Dielectric constant and electrical conductivity
Magnetometry	Magnetic susceptibility and remanent magnetism
Self-potential	Electric potential gradient
Seismic	Density and elastic moduli (bulk modulus, shear modulus, etc.)

multiple geophysical and nongeophysical spatial data sets. Consequently, GIS will play a greater role in the analysis of geophysical data collected in agricultural settings. Furthermore, as the practice of precision agriculture continues to grow, there is expected to be an increasing need to input geophysical data into the GIS used to make proper management decisions in regard to different areas of a farm field.

6. Expert system computer software and learning-capable computer software incorporating neural networks will be developed for specific agricultural applications to automatically analyze and interpret geophysical data.

7. Tomographic procedures will be employed to obtain geophysical data in agricultural settings when the situation is warranted. It is usually not possible to conduct geophysical surveys in an agricultural field during the growing season, once the crop emerges and begins to develop. Tomographic data collection and analysis procedures are a potential solution to this field access problem, allowing the within field horizontal spatial pattern of a physical property to be determined without actually having to obtain geophysical measurements inside the field. Tomographic data collection and analysis procedures can also provide valuable geophysical information even for circumstances when field access is not a problem. For the geophysical field measurement tomographic approach, geophysical energy source and sensor locations are moved along the perimeter of the field. They will typically involve multiple source and sensor positionings in which the geophysical sensor locations are always on opposite or adjacent sides of the field with respect to the side of the field where the geophysical energy source is located. A map of the horizontal spatial pattern for some physical property within an agricultural field is then generated with measurement data from a sufficient number of geophysical source and sensor positionings used as input for image reconstruction computer software employing inversion techniques.

8. The application of geophysical methods to agriculture will eventually become a well-recognized subdiscipline of geophysics, at which time it may become appropriate to use the contracted term "agrigeophysics" instead of the longer term "agricultural geophysics."

REFERENCES

Allred, B. J., J. J. Daniels, N. R. Fausey, C. Chen, L. Peters, Jr., and H. Youn. 2005a. Important considerations for locating buried agricultural drainage pipe using ground penetrating radar. *Appl. Eng. Agric.* v. 21, pp. 71–87.

Allred, B. J., M. R. Ehsani, and D. Saraswat. 2005b. The impact of temperature and shallow hydrologic conditions on the magnitude and spatial pattern consistency of electromagnetic induction measured soil electrical conductivity. *Trans. ASAE.* v. 48, pp. 2123–2135.

Allred, B. J., M. R. Ehsani, and D. Saraswat. 2006. Comparison of electromagnetic induction, capacitively coupled resistivity, and galvanic contact resistivity methods for soil electrical conductivity measurement. *Appl. Eng. Agric.* v. 22, pp. 215–230.

Bailey, J. T., S. Evans, and G. de Q. Robin. 1964. Radio echo sounding in polar ice sheets. *Nature*. v. 204, pp. 420–421.

Banton, O., M. K. Seguin, and M. A. Cimon. 1997. Mapping field-scale physical properties of soil with electrical resistivity. *Soil Sci. Soc. Am. J.* v. 61, pp. 1010–1017.

Boniak, R., Chong, S.K., Indorante, S.J., and J.A. Doolittle. 2002. Mapping golf course green drainage systems and subsurface features using ground penetrating radar. In *Proceedings of SPIE, Vol. 4758, Ninth International Conference on Ground Penetrating Radar*. pp. 477–481. S. K. Koppenjan and H. Lee, editors. April 29–May 2, 2002. Santa Barbara, CA. SPIE. Bellingham, WA.

Butnor, J. R., J. A. Doolittle, K. H. Johnson, L. Samuelson, T. Stokes, and L. Kress. 2003. Utility of ground-penetrating radar as a root biomass survey tool in forest systems. *Soil Sci. Soc. Am. J.* v. 67, pp. 1607–1615.

Carroll, Z. L., and M. A. Oliver. 2005. Exploring the spatial relations between soil physical properties and apparent electrical conductivity. *Geoderma*. v. 128, pp. 354–374.

Chow, T. L., and H. W. Rees. 1989. Identification of subsurface drain locations with ground-penetrating radar. *Can. J. Soil Sci.* v. 69, pp. 223–234.

Collins, M. E., and J. A. Doolittle. 1987. Using ground-penetrating radar to study soil microvariability. *Soil Sci. Soc. Am. J.* v. 51, pp. 491–493.

Collins, M. E., J. A. Doolittle, and R. V. Rourke. 1989. Mapping depth to bedrock on a glaciated landscape with ground-penetrating radar. *Soil Sci. Soc. Am. J.* v. 53, pp. 1806–1812.

Dobrin, M. B., and C. H. Savit. 1988. *Introduction of Geophysical Prospecting,* 4th Edition. McGraw-Hill. New York.

Doolittle, J. A. 1987. Using ground-penetrating radar to increase the quality and efficiency of soil surveys. In *Soil Survey Techniques*. pp. 11–32. W. U. Reybold and G. W. Petersen, editors. SSSA Special Publication Number 20. Soil Science Society of America. Madison, WI.

Doolittle, J. A., K. A. Sudduth, N. R. Kitchen, and S. J. Indorante. 1994. Estimating depths to claypans using electromagnetic induction methods. *J. Soil and Water Cons.* v. 49, pp. 572–575.

Doolittle, J., M. Petersen, and T. Wheeler. 2001. Comparison of two electromagnetic induction tools in salinity appraisals. *J. Soil and Water Cons.* v. 56, pp. 257–262.

Edlefsen, N. E., and A. B. C. Anderson. 1941. The four-electrode resistance method for measuring soil-moisture content under field conditions. *Soil Sci.* v. 51, pp. 367–376.

Eigenberg, R. A., and J. A. Nienaber. 1998. Electromagnetic survey of cornfield with repeated manure applications. *J. Environ. Qual.* v. 27, pp. 1511–1515.

Freeland, R. S., L. O. Odhiambo, J. S. Tyner, J. T. Ammons, and W. C. Wright. 2006. Nonintrusive mapping of near-surface preferential flow. *Appl. Eng. Agric.* v. 22, pp. 315–319.

Gish, T. J., W. P. Dulaney, K.-J. S. Kung, C. S. T. Daughtry, J. A. Doolittle, and P. T. Miller. 2002. Evaluating use of ground-penetrating radar for identifying subsurface flow pathways. *Soil Sci. Soc. Am. J.* v. 66, pp. 1620–1629.

Grote, K., S. Hubbard, and Y. Rubin. 2003. Field-scale estimation of volumetric water content using ground-penetrating radar ground wave techniques. *Water Resour. Res.* v. 39, pp. 1321–1333.

Halvorson, A. D., and J. D. Rhoades. 1974. Assessing soil salinity and identifying potential saline-seep areas with field soil resistance measurements. *Soil Sci. Soc. Am. Proc.* v. 38, pp. 576–581.

Hendrickx, J. M. H., B. Baerends, Z. I. Rasa, M. Sadig, and M. A. Chaudhry. 1992. Soil salinity assessment by electromagnetic induction of irrigated land. *Soil Sci. Soc. Am. J.* v. 56, pp. 1933–1941.

Huisman, J. A., S. S. Hubbard, J. D. Redman, and A. P. Annan. 2003. Measuring soil water content with ground penetrating radar: A review. *Vadose Zone J.* v. 2, pp. 476–490.

Jaynes, D. B., J. M. Novak, T. B. Moorman, and C. A. Cambardella. 1995a. Estimating herbicide partition coefficients from electromagnetic induction measurements. *J. Environ. Qual.* v. 24, pp. 36–41.

Jaynes, D. B., T. S. Colvin, and J. Ambuel. 1995b. Yield mapping by electromagnetic induction. In *Proceedings of Site-Specific Management for Agricultural Systems: Second International Conference*. pp. 383–394. P. C. Robert, R. H. Rust, and W. E. Larson, editors. March 27–30, 1994. St. Paul, MN. ASA, CSSA, and SSSA. Madison, WI.

Johnson, C. K., J. W. Doran, H. R. Duke, B. J. Wienhold, K. M. Eskridge, and J. F. Shanahan. 2001. Field-scale electrical conductivity mapping for delineating soil condition. *Soil Sci. Soc. Am. J.* v. 65, pp. 1829–1837.

Kirkham, D., and G. S. Taylor. 1949. Some tests of a four-electrode probe for soil moisture measurement. *Soil Sci. Soc. Am. Proc.* v. 14, pp. 42–46.

Kitchen, N. R., K. A. Sudduth, and S. T. Drummond. 1996. Mapping of sand deposition from 1993 Midwest floods with electromagnetic induction measurements. *J. Soil and Water Cons.* v. 51, pp. 336–340.

Konstantinovic, M., S. Wöckel, P. Schulze Lammers, J. Sachs, and M. Martinov. 2007. Detection of root biomass using ultra wideband radar—An approach to potato nest positioning. *Agr. Eng. Intl.* v. 9, Manuscript IT 06 003. cigr-ejournal.tamu.edu.

Kravchenko, A. N., G. A. Bollero, R. A. Omonode, and D. G. Bullock. 2002. Quantitative mapping of soil drainage classes using topographical data and soil electrical conductivity. *Soil Sci. Soc. Am. J.* v. 66, pp. 235–243.

Lawyer, L. C., C. C. Bates, and R. B. Rice. 2001. *Geophysics in the Affairs of Mankind, A Personalized History of Exploration Geophysics*, 2nd Edition. Society of Exploration Geophysics. Tulsa, OK.

Lund, E. D., P. E. Colin, D. Christy, and P. E. Drummond. 1999. Applying soil electrical conductivity technology to precision agriculture. In *Proceedings 4th Int. Conf. Precision Agric.* pp. 1089–1100. P. C. Robert, R. H. Rust, and W. E. Larson, editors. July 19–22, 1998. St. Paul, MN. ASA, CSSA, and SSSA. Madison, WI.

Lunt, I. A., S. S. Hubbard, and Y. Rubin. 2005. Soil moisture content estimation using ground-penetrating radar reflection data. *J. Hydrology.* v. 307, pp. 254–269.

McCorkle, W. H. 1931. *Determination of Soil Moisture by the Method of Multiple Electrodes.* Texas Agricultural Experiment Station Bulletin 426. Texas A & M University. College Station, TX.

Morgan, M., and D. Ess. 1997. *The Precision-Farming Guide for Agriculturists.* John Deere Publishing. Moline, IL.

National Research Council. 1997. *Precision Agriculture in the 21st Century.* National Academy Press. Washington, DC.

Needham, J. 1959. *Science and Civilization in China, Volume 3, Mathematics and the Sciences of the Heavens and Earth.* Cambridge University Press. Cambridge, UK.

Reynolds, J. M. 1997. *An Introduction to Applied and Environmental Geophysics.* John Wiley & Sons. Chichester, UK.

Rhoades, J. D., and R. D. Ingvalson. 1971. Determining salinity in field soils with soil resistance measurements. *Soil Sci. Soc. Am. Proc.* v. 35, pp. 54–60.

Rhoades, J. D., P. A. C. Raats, and R. J. Prather. 1976. Effects of liquid-phase electrical conductivity, water content, and surface conductivity on bulk soil electrical conductivity. *Soil Sci. Soc. Am. J.* v. 40, pp. 651–655.

Rhoades, J. D., N. A. Manteghi, P. J. Shouse, and W. J. Alves. 1989. Soil electrical conductivity and soil salinity: New formulations and calibrations. *Soil Sci. Soc. Am. J.* v. 53, pp. 433–439.

Rhoades, J. D., P. J. Shouse, W. J. Alves, N. A. Manteghi, and S. M. Lesch. 1990. Determining soil salinity from soil electrical conductivity using different models and estimates. *Soil Sci. Soc. Am. J.* v. 54, pp. 46–54.

Rogers, M. B., J. R. Cassidy, and M. I. Dragila. 2005. Ground-based magnetic surveys as a new technique to locate subsurface drainage pipes: A case study. *Appl. Eng. in Agric.* v. 21, pp. 421–426.

Rogers, M. B., J. E. Baham, and M. I. Dragila. 2006. Soil iron content effects on the ability of magnetometer surveying to locate buried agricultural drainage pipes. *Appl. Eng. in Agric.* v. 22, pp. 701–704.

Schellentrager, G. W., J. A. Doolittle, T. E. Calhoun, and C. A. Wettstein. 1988. Using ground-penetrating radar to update soil survey information. *Soil Sci. Soc. Am. J.* v. 52, pp. 746–752.

Shea, P. F. and J. N. Luthin. 1961. An investigation of the use of the four-electrode probe for measuring soil salinity in situ. *Soil Sci.* v. 92, pp. 331–339.

Sheets, K. R., and J. M. H. Hendrickx. 1995. Noninvasive soil water content measurement using electromagnetic induction. *Water Resour. Res.* v. 31, pp. 2401–2409.

Sheriff, R. E. 2002. *Encyclopedic Dictionary of Applied Geophysics,* 4th Edition. Society of Exploration Geophysics. Tulsa, OK.

Smedema, L. K., W. F. Vlotman, and D. W. Rycroft. 2004. *Modern Land Drainage: Planning, Design and Management of Agricultural Drainage Systems.* A.A. Balkema. Leiden, Netherlands

Telford, W. M., L. P. Geldart, R. E. Sheriff, and D. A. Keys. 1976. *Applied Geophysics.* Cambridge University Press. Cambridge, UK.

Topp, G. C., J. L. Davis, and A. P. Annan. 1980. Electromagnetic determination of soil water content: Measurements in coaxial transmission lines. *Water Resour. Res.* v. 16, pp. 574–582.

Wöckel, S., M. Konstantantinovic, J. Sachs, P. Schulze Lammers, and M. Kmec. 2006. Application of ultra-wideband M-sequence-radar to detect sugar beets in agricultural soils. In *Proceeedings of the 11th International Conference on Ground Penetrating Radar.* June 19–22, 2006. Columbus, OH.

2 Past, Present, and Future Trends of Soil Electrical Conductivity Measurement Using Geophysical Methods

Dennis L. Corwin

CONTENTS

2.1 INTRODUCTION

Arguably, the beginnings of geophysics can be traced to Gilbert's discovery that the world behaves like a massive magnet and Newton's theory of gravitation. Since that time, researchers in geophysics have developed a broad array of measurement tools involving magnetic, seismic, electromagnetic, resistivity, induced polarization, radioactivity, and gravity methods. Although at times a formidable technological feat, the adaptation of geophysical techniques from the measurement of geologic strata to the measurement of surface and near-surface soils for agricultural applications was the next logical step.

Geophysical techniques currently used in agricultural research include electrical resistivity (ER), time domain reflectometry (TDR), ground-penetrating radar (GPR), capacitance probes (CPs), radar scatterometry or active microwaves (AM), passive microwaves (PM), electromagnetic induction (EMI), neutron thermalization, nuclear magnetic resonance (NMR), gamma ray attenuation, and near-surface seismic reflection. Several of the geophysical techniques fall into the category of electromagnetic (EM) methods because they rely on an EM source, including TDR, GPR, CP, AM, PM, and EMI. Each varies from the other in a subtle way. For TDR, the applied electromagnetic pulse is guided along a transmission line embedded in the soil. The time delay between the reflections of the pulse from the beginning and the end of the transmission line is used to determine the velocity of propagation through soil, which is controlled by the relative dielectric permittivity or dielectric constant. Both TDR and GPR are based on the fact that electrical properties of soils are primarily determined by the water content (θ) in the frequency range from 10 to 1000 MHz (Topp et al., 1980). For GPR, however, radio frequency signals are radiated from an antenna at the soil surface into the ground, while a separate antenna receives both reflected and transmitted signals. Signals arriving at the receiving antenna come from three pathways: (1) through the air, (2) through the near surface soil, and (3) reflected from objects or layers below the soil surface. Signal velocity and attenuation are used, like TDR, to infer both θ and soil apparent electrical conductivity (EC_a), which is the electrical conductivity through the bulk soil. Capacitance probes for measuring θ are placed in the soil so that the soil acts like the dielectric of a capacitor in a capacitive-inductive resonant circuit, where the inductance is fixed. Active microwaves or radar scatterometry are similar to GPR, except that the antennae are located above the soil surface. The signal penetrates to a shallow depth, generally <100 mm below the soil surface, for the transmitted frequencies used. Analysis of the reflected signal results in a measure of θ and electrical conductivity at the near surface. Passive microwaves are unique in that no signal is applied, rather the surface soil is the EM source and a sensitive receiver located at the soil surface measures temperature and dielectric properties of the surface soil from which θ and EC_a are inferred. Finally, EMI, unlike GPR, employs lower-frequency signals and primarily measures the signal loss to determine EC_a. The common operating frequency ranges of instrumentation for these electromagnetic techniques are EMI (0.4 to 40 kHz), CP (38 to 150 MHz), GPR (1 to 2,000 MHz), TDR (50 to 5,000 MHz), AM (0.2 to 300 GHz), and PM (0.3 to 30 GHz).

Of these geophysical techniques, the agricultural application of geospatial measurements of EC_a, as measured by EMI, ER, and TDR, has had tremendous impact over the past two decades. Currently, EC_a is recognized as the most valuable geophysical measurement in agriculture for characterizing soil spatial variability at field and landscape spatial extents (Corwin, 2005, Corwin and Lesch, 2003, 2005a). It is the objective of this chapter to present a historical perspective of the adaptation of geophysical techniques for use in agriculture with a primary focus on trends in the adaptation of EC_a to agriculture, as well as the practical and theoretical factors that have forged these trends.

2.2 HISTORICAL PERSPECTIVE OF APPARENT SOIL ELECTRICAL CONDUCTIVITY (EC_a) TECHNIQUES IN AGRICULTURE—THE PAST

The adaptation of geophysical EC_a measurement techniques to agriculture was largely motivated by the need for reliable, quick, and easy measurements of soil salinity at field and landscape spatial extents. However, it became quickly apparent that EC_a was influenced not only by salinity, but also by a variety of other soil properties that influenced electrical conductivity in the bulk soil, including θ, clay content and mineralogy, organic matter, bulk density (ρ_b), and temperature. The EC_a measurement is a complex physicochemical property resulting from the interrelationship and interaction of these soil properties. Researchers subsequently realized that geospatial measurements of EC_a can potentially provide spatial distributions of any or all of these properties. This realization resulted in the evolution of EC_a in agriculture from a tool for measuring, profiling, and mapping soil salinity into a present-day tool for characterizing the spatial variability of any soil property that correlates with EC_a.

The impetus behind the evolution of EC_a in agriculture stems from several factors that make it well suited for characterizing spatial variability at field and larger spatial extents. Most importantly, measurements of EC_a are reliable, quick, and easy to take. This factor was instrumental in the initial adaptation of EC_a for agricultural use. Historically, considerable research was conducted using

EC_a measurements of soils. Consequently, there is a reasonable understanding of what is being measured, even though the measurement is complicated by the interaction of several soil properties that influence the conductive pathways through the bulk soil. Another factor is that the mobilization of EC_a measurement equipment is comparatively easy and can be accomplished at a reasonable cost. Tractor- and all-terrain vehicle (ATV)-mounted platforms have made intensive field-scale measurements commonplace (Cannon et al., 1994; Carter et al., 1993; Freeland et al., 2002; Jaynes et al., 1993; Kitchen et al., 1996; McNeill, 1992; Rhoades, 1993). Basin- and landscape-scale assessments are possible with airborne electromagnetic (AEM) systems (Cook and Kilty, 1992; George and Woodgate, 2002; George et al., 1998; Munday, 2004; Spies and Woodgate, 2004; Williams and Baker, 1982). However, AEM applications in agriculture have been primarily used to identify geological sources of salinity, because AEM penetrates well below the root zone to depths of tens of meters, whereas surface EMI for agricultural applications, such as the Geonics EM38* or DUALEM-2† electrical conductivity meters, generally penetrates to depths confined mainly to the root zone (i.e., 1.5 to 2 m). Mobilization made it possible to create maps of EC_a variation at field scales, making EC_a a practical field measurement. Finally, because EC_a is influenced by a variety of soil properties, the spatial variability of these properties can be potentially established, providing a wealth of spatial soil-related information.

2.2.1 MEASUREMENT OF SOIL SALINITY WITH EC_a

The measurement of soil salinity has a long history prior to its measurement with EC_a. Soil salinity refers to the presence of major dissolved inorganic solutes in the soil aqueous phase, which consist of soluble and readily dissolvable salts including charged species (e.g., Na^+, K^+, Mg^{+2}, Ca^{+2}, Cl^-, HCO_3^-, NO_3^-, SO_4^{-2}, and CO_3^{-2}), nonionic solutes, and ions that combine to form ion pairs. The need to measure soil salinity stems from its detrimental impact on plant growth. Effects of soil salinity are manifested in loss of stand, reduced plant growth, reduced yields, and, in severe cases, crop failure. Salinity limits water uptake by plants by reducing the osmotic potential making it more difficult for the plant to extract water. Salinity may also cause specific-ion toxicity or upset the nutritional balance of plants. In addition, the salt composition of the soil water influences the composition of cations on the exchange complex of soil particles, which influences soil permeability and tilth.

Six methods have been developed for determining soil salinity at field scales: (1) visual crop observations, (2) the electrical conductance of soil solution extracts or extracts at higher than normal water contents, (3) in situ measurement of ER, (4) noninvasive measurement of electrical conductance with EMI, (5) in situ measurement of electrical conductance with TDR, and (6) multi- and hyperspectral imagery.

Visual crop observation is the oldest method of determining the presence of soil salinity. It is a quick method, but it has the disadvantage that salinity development is detected after crop damage has occurred. For obvious reasons, the least desirable method is visual observation because crop yields are reduced to obtain soil salinity information. However, remote imagery is increasingly becoming a part of agriculture and represents a quantitative approach to the antiquated method of visual observation that may offer a potential for early detection of the onset of salinity damage to plants. Even so, multi- and hyperspectral remote imagery are still in their infancy with an inability at the present time to differentiate osmotic from matric or other stresses, which is key to the successful application of remote imagery as a tool to map salinity and water content.

---

* Geonics Limited, Inc., Mississaugua, Ontario, Canada. Product identification is provided solely for the benefit of the reader and does not imply the endorsement of the USDA.

† DUALEM, Inc., Milton, Ontario, Canada. Product identification is provided solely for the benefit of the reader and does not imply the endorsement of the USDA.

The determination of salinity through the measurement of electrical conductance has been well established for decades (U.S. Salinity Laboratory Staff, 1954). It is known that the electrical conductivity of water is a function of its chemical composition. McNeal et al. (1970) were among the first to establish the relationship between electrical conductivity and molar concentrations of ions in the soil solution. Soil salinity is quantified in terms of the total concentration of the soluble salts as measured by the electrical conductivity (EC) of the solution in dS m^{-1}. To determine EC, the soil solution is placed between two electrodes of constant geometry and distance of separation (Bohn et al., 1979). At constant potential, the current is inversely proportional to the solution's resistance. The measured conductance is a consequence of the solution's salt concentration and the electrode geometry whose effects are embodied in a cell constant. The electrical conductance is a reciprocal of the resistance as shown in Equation (2.1):

$$EC_T = k/R_T \qquad\qquad (2.1)$$

where EC_T is the electrical conductivity of the solution in dS m^{-1} at temperature T (°C), k is the cell constant, and R_T is the measured resistance at temperature T.

Electrolytic conductivity increases at a rate of approximately 1.9 percent per degree centigrade increase in temperature. Customarily, EC is expressed at a reference temperature of 25°C for purposes of comparison. The EC measured at a particular temperature T (°C), EC_T, can be adjusted to a reference EC at 25°C, EC_{25}, using the below equations from Handbook 60 (U.S. Salinity Laboratory staff, 1954):

$$EC_{25} = f_T \cdot EC_T \qquad\qquad (2.2)$$

where f_T is a temperature conversion factor. Approximations for the temperature conversion factor are available in polynomial form (Rhoades et al., 1999a; Stogryn, 1971; Wraith and Or, 1999) or other equations can be used such as Equation (2.3) by Sheets and Hendrickx (1995):

$$f_T = 0.4470 + 1.4034 e^{-T/26.815} \qquad\qquad (2.3)$$

Customarily, soil salinity is defined in terms of laboratory measurements of the EC of the saturation extract (EC_e) because it is impractical for routine purposes to extract soil water from samples at typical field water contents. Partitioning of solutes over the three soil phases (i.e., gas, liquid, solid) is influenced by the soil:water ratio at which the extract is made, so the ratio must be standardized to obtain results that can be applied and interpreted universally. Commonly used extract ratios other than a saturated soil paste are 1:1, 1:2, and 1:5 soil:water mixtures.

Soil salinity can also be determined from the measurement of the EC of a soil solution (EC_w). Theoretically, EC_w is the best index of soil salinity because this is the salinity actually experienced by the plant root. Nevertheless, EC_w has not been widely used to express soil salinity for two reasons: (1) it varies over the irrigation cycle as θ changes, and (2) methods for obtaining soil solution samples are too labor and cost intensive at typical field water contents to be practical for field-scale applications (Rhoades et al., 1999a). For disturbed samples, soil solution can be obtained in the laboratory by displacement, compaction, centrifugation, molecular adsorption, and vacuum- or pressure-extraction methods. For undisturbed samples, EC_w can be determined either in the laboratory on a soil solution sample collected with a soil-solution extractor or directly in the field using in situ, imbibing-type porous-matrix salinity sensors. Briggs and McCall (1904) devised the first extractor system. Kohnke et al. (1940) provide a review of early extractor construction and performance.

The ability of soil solution extractors and porous-matrix salinity sensors (also known as soil salinity sensors) to provide representative soil water samples is doubtful (England, 1974; Raulund- Rasmussen, 1989; Smith et al., 1990). Because of their small sphere of measurement, neither extractors nor salt sensors adequately integrate spatial variability (Amoozegar-Fard et al., 1982; Haines et al., 1982; Hart and Lowery, 1997); consequently, Biggar and Nielsen (1976) suggested that soil solution samples are qualitative point-sample measurements of soil solutions that are not representative quantitative measurements because of the effect of local-scale variability on small sample volumes. Furthermore, salinity sensors demonstrate a lag in response time that is dependent upon the diffusion of ions between the soil solution and solution in the porous ceramic, which is affected by (1) the thickness of the ceramic conductivity cell, (2) the diffusion coefficients in soil and ceramic, and (3) the fraction of the ceramic surface in contact with soil (Wesseling and Oster, 1973). The salinity sensor is generally considered the least desirable method for measuring EC_w because of its low sample volume, unstable calibration over time, and slow response time (Corwin, 2002).

Developments in the measurement of soil EC to determine soil salinity shifted away from extractions to the measurement of EC_a because the time and cost of obtaining soil solution extracts prohibited their practical use at field scales, and the high local-scale variability of soil rendered salinity sensors and small volume soil core samples of limited quantitative value. Rhoades and colleagues at the U.S. Salinity Laboratory led the shift in the early 1970s to the use of EC_a as a measure of soil salinity (Rhoades and Ingvalson, 1971). The use of EC_a to measure salinity has the advantage of increased volume of measurement and quickness of measurement, but suffers from the complexity of measuring EC for the bulk soil rather than restricted to the solution phase. Furthermore, EC_a measurement techniques, such as ER and EMI, are easily mobilized and are well suited for field-scale applications because of the ease and low cost of measurement with a volume of measurement that is sufficiently large (>1 m^3) to reduce the influence of local-scale variability. Developments in agricultural applications of ER and EMI have occurred along parallel paths with each filling a needed niche based upon inherent strengths and limitations.

2.2.1.1 Electrical Resistivity

Electrical resistivity was developed in the second decade of the 1900s by Conrad Schlumberger in France and Frank Wenner in the United States for the evaluation of ground ER (Telford et al., 1990; Burger, 1992). The earliest application of ER in agriculture was to measure θ (Edlefsen and Anderson, 1941; Kirkham and Taylor, 1950). This adaptation was later eclipsed by the use of ER to measure soil salinity (Rhoades and Ingvalson, 1971). Electrical resistivity has been most widely used in agriculture as a means of measuring soil salinity. A review of this early body of salinity research can be found in Rhoades et al. (1999). Arguably, the early salinity research with ER provided the initial momentum to the subdiscipline of agricultural geophysics.

Electrical resistivity methods involve the measurement of the resistance to current flow across four electrodes inserted in a line on the soil surface at a specified distance between the electrodes (Figure 2.1). The resistance to current flow is measured between a pair of inner electrodes while electrical current is caused to flow through the soil between a pair of outer electrodes. Although two electrodes (i.e., a single current electrode and a single potential electrode) can also be used, this configuration is highly unstable, and the introduction of four electrodes helped to stabilize the resistance measurement. According to Ohm's Law, the measured resistance is directly proportional to the voltage (V) and inversely proportional to the electrical current (i):

$$R = \frac{V}{i} \tag{2.4}$$

where resistance (R) is defined as one ohm (ω) of resistance that allows a current of one ampere to flow when one volt of electromotive force is applied. The resistance of a given volume of soil depends on its length (l, m), its cross-sectional area (a, m^2), and a fundamental soil property called resistivity (ρ, ω m^{-1}):

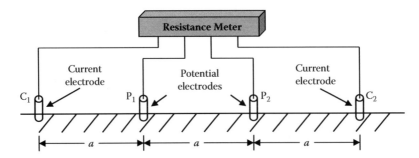

FIGURE 2.1 Wenner array electrodes: C_1 and C_2 represent the current electrodes, P_1 and P_2 represent the potential electrodes, and a represents the interelectrode spacing. (From Rhoades, J.D., and Halvorson, A.D., Electrical conductivity methods for detecting and delineating saline seeps and measuring salinity in Northern Great Plains soils, ARS W-42, USDA-ARS Western Region, Berkeley, CA, pp. 1–45, 1977. With permission.)

$$R = \rho \left(\frac{l}{a} \right) \tag{2.5}$$

The conductance (C, ω^{-1} or Siemens, S) is the inverse of resistance, and the EC_a (dS m^{-1}) is the inverse of the resistivity:

$$EC_a = \frac{1}{\rho} = \frac{1}{R}\left(\frac{l}{a} \right) = C\left(\frac{l}{a} \right) \tag{2.6}$$

When the four electrodes are equidistantly spaced in a straight line at the soil surface, the electrode configuration is referred to as the Wenner array (Figure 2.1). The resistivity measured with the Wenner array is shown in Equation (2.7):

$$\rho = \frac{2\pi a \Delta V}{i} = 2\pi a R \tag{2.7}$$

and the measured EC_a is as shown in Equation (2.8):

$$EC_a = \frac{1}{2\pi a R} \tag{2.8}$$

where a is the interelectrode spacing (m). The equations for other electrode configurations can be found in Dobrin (1960), Telford et al. (1990), and Burger (1992).

The volume of measurement of the Wenner array is relatively large and includes all the soil between the inner pair of electrodes from the soil surface to a depth equal to roughly the interelectrode spacing. Figure 2.2 illustrates the volume of measurement. For a homogeneous soil, the volume of measurement is approximately πa^3. The depth of penetration of the electrical current and the volume of measurement increase as the interelectrode spacing, a, increases.

Apparent soil electrical conductivity for a discrete depth interval of soil, EC_x, can be obtained with the Wenner array by measuring the EC_a of successive layers by increasing the interelectrode spacings (a_{i-1} and a_i) and using Equation (2.9) for parallel resistors (Barnes, 1952):

$$EC_x = EC_{a_i} - EC_{a_{i-1}} = \frac{\left(EC_{a_i} a_i - EC_{a_{i-1}} a_{i-1} \right)}{a_i - a_{i-1}} \tag{2.9}$$

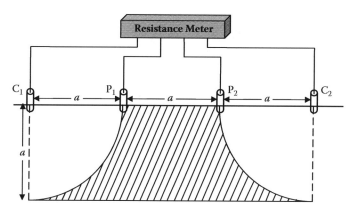

FIGURE 2.2 The volume of measurement for a Wenner-array electrode configuration. The shaded area represents measurement volume. C_1 and C_2 represent the current electrodes, P_1 and P_2 represent the potential electrodes, and a represents the interelectrode spacing. (From Rhoades, J.D., and Halvorson, A.D., Electrical conductivity methods for detecting and delineating saline seeps and measuring salinity in Northern Great Plains soils, ARS W-42, USDA-ARS Western Region, Berkeley, CA, pp. 1–45, 1977. With permission.)

where a_i is the interelectrode spacing, which equals the depth of sampling; a_{i-1} is the previous interelectrode spacing, which equals the depth of previous sampling; and EC_x is the conductivity for a specific depth interval. This is often referred to as vertical profiling.

Electrical resistivity is an invasive technique that requires good contact between the soil and electrodes inserted into the soil; consequently, it produces less reliable measurements in frozen, dry, or stony soils than noninvasive EMI measurement. Furthermore, depending upon the manner in which the ER electrodes are mounted onto the mobile fixed-array platform, microtopography, such as a bed-furrow surface, may cause contact problems between the electrodes and soil. Even so, ER is widely used in agriculture and has been adapted for commercial field-scale applications primarily because the ease of calibration is appealing and the linear relationship of EC_a with depth, which makes the application of Equation (2.9) possible, is simple and readily understood.

2.2.1.2 Electromagnetic Induction

In the late 1970s and early 1980s, de Jong et al. (1979), Rhoades and Corwin (1981), and Williams and Baker (1982) began investigating the use of EMI to measure soil salinity. de Jong et al. (1979) published the first use of EMI for measuring soil salinity. The early studies with EMI by Rhoades and Corwin were efforts to profile soil salinity through the root zone (Corwin and Rhoades, 1982; Rhoades and Corwin, 1981). Unlike ER, vertical profiling with EMI is not a trivial task, because a relatively simple linear model can be used for low conductivity media, but for higher conductivity values, a nonlinear model is required. Williams and Baker (1982) sought to use EMI as a means of surveying soil salinity at landscape scales and larger with the first use of AEM to map geologic sources of salinity having agricultural impacts.

Through the 1980s and early 1990s, the focus of EMI work in agriculture was on vertical profiling (Cook and Walker, 1992; Corwin and Rhoades, 1982, 1990; Rhoades and Corwin, 1981; Rhoades et al., 1989; Slavich, 1990; Wollenhaupt et al., 1986). Vertical profiling of soil salinity with EMI involves raising the EMI conductivity meter to various heights at or above the soil surface (i.e., 0, 30, 60, 90, 120, and 150 cm) to measure the EC_a corresponding to incremental depths below the soil surface (i.e., 0 to 150, 0 to 120, 0 to 90, 0 to 60, and 0 to 30, respectively). Site-specific empirical relationships were developed, which were not widely used because they could not be extrapolated to other sites without calibration. It was not until the work of Borchers and colleagues (1997) that inverse procedures for the linear and nonlinear models (Hendrickx et al., 2002) were developed to profile soil salinity with above-ground EMI measurements. Vertical profiling of EC_a

with EMI is mathematically complex and a difficult quantitative undertaking (Borchers et al., 1997; Hendrickx et al., 2002; McBratney et al., 2000). As a result, qualitative evaluations of EC_a at shallow and deep depths with EMI are generally used by positioning the EMI instrument at the soil surface in the vertical (EM_v) and then the horizontal (EM_h) dipole mode (i.e., receiver and transmitter coils are oriented perpendicular or parallel with the soil surface, respectively), which measures to depths of 0.75 and 1.5 m, respectively. This provides measurements of EC_a at shallow and deeper depths, which enables the qualitative determination of whether an EC_a profile is uniform with depth ($EM_h \approx EM_v$), inverted ($EM_h > EM_v$), or normal ($EM_h < EM_v$).

The depth-weighted nonlinear response of EMI is shown in Equation (2.10) and Equation (2.11) from McNeill (1980) for the vertical and horizontal dipoles, respectively:

$$R_v(z) = \frac{1}{(4z^2 + 1)^{1/2}} \tag{2.10}$$

$$R_h(z) = (4z^2 + 1)^{1/2} - 2z \tag{2.11}$$

where $R_v(z)$ and $R_h(z)$ are the cumulative relative contributions of all soil electrical conductivity with the vertical and horizontal EMI dipoles, respectively, for a homogeneously conductive media below a depth of z (m).

At low conductivity values ($EC_a < 100$ mS m^{-1}), McNeill (1980) showed that the measured EC_a when the EMI instrument is located at the soil surface is given by Equation (2.12):

$$EC_a = \left[\frac{4}{2\pi\mu_0 f_s^2} \right] \left(\frac{H_s}{H_p} \right) \tag{2.12}$$

where EC_a is measured in S m^{-1}; H_p and H_s are the intensities of the primary and secondary magnetic fields at the receiver coil (A m^{-1}), respectively; f is the frequency of the current (Hz); μ_0 is the magnetic permeability of air ($4\pi10^{-7}$ H m^{-1}); and s is the intercoil spacing (m).

The calibration of EMI equipment (e.g., Geonics EM38[1]), which can be difficult and time consuming, is another dissimilarity with ER. However, the DUALEM-2 does not appear to suffer from the same calibration difficulties as the EM38 due to the increased distance between the transmitter and receiver coils. Complexity of the EMI measurement and difficulties in calibration are distinct disadvantages of the EMI approach that have reduced its use in agriculture. These limitations are the most likely reasons that there are no commercially available EMI mobile platforms. This has caused the use of EMI in agriculture, even today, to be principally as a research tool.

Following the early vertical profiling efforts, research with EMI, and concomitantly with ER, drifted away from salinity and concentrated more on observed associations between ER and EMI measurements of EC_a and other soil properties. This research trend significantly contributed to the base of knowledge compiled in Table 2.1.

2.2.2 MEASUREMENT OF WATER CONTENT WITH EC_a

Several geophysical techniques have been adapted for agriculture to measure θ within the root zone including TDR, GPR, CP, AM, PM, EMI, neutron thermalization, NMR, gamma ray attenuation, and ER. Aside from ER and EMI, neutron thermalization, CP, TDR, and GPR have received the greatest use for laboratory and field-scale agricultural applications. The history of the agricultural application of CP and neutron thermalization predates all other geophysical-based approaches for measuring θ except ER. Gamma ray attenuation has been in use in agriculture since the 1950s, but it

TABLE 2.1

Compilation of Literature Measuring EC$_a$ with Geophysical Techniques (ER or EMI) Categorized According to the Soil-Related Properties Directly or Indirectly Measured by EC$_a$

Soil Property	References
Directly Measured Soil Properties	
Salinity (and nutrients, e.g., NO$_3^-$)	Bennett and George (1995); Cameron et al. (1981); Cannon et al. (1994); Corwin and Rhoades (1982, 1984, 1990); de Jong et al. (1979); Diaz and Herrero (1992); Drommerhausen et al. (1995); Eigenberg and Nienaber (1998, 1999, 2001); Eigenberg et al. (1998, 2002); Greenhouse and Slaine (1983); Halvorson and Rhoades (1976); Hanson and Kaita (1997); Hendrickx et al. (1992); Herrero et al. (2003); Johnston et al. (1997); Kaffka et al. (2005); Lesch et al. (1992, 1995a, 1995b,1998); Mankin and Karthikeyan (2002); Mankin et al. (1997); Nettleton et al. (1994); Paine (2003); Ranjan et al. (1995); Rhoades (1992, 1993); Rhoades and Corwin (1981, 1990); Rhoades and Halvorson (1977); Rhoades et al. (1976, 1989, 1990, 1999a, 1999b); Slavich and Petterson (1990); van der Lelij (1983); Williams and Baker (1982); Williams and Hoey (1987); Wollenhaupt et al. (1986)
Water content	Brevik and Fenton (2002); Farahani et al. (2005); Fitterman and Stewart (1986); Freeland et al. (2001); Hanson and Kaita (1997); Kachanoski et al. (1988, 1990); Kaffka et al. (2005); Kean et al. (1987); Khakural et al. (1998); Morgan et al. (2000); Sheets and Hendrickx (1995); Vaughan et al. (1995); Wilson et al. (2002)
Texture related (e.g., sand, clay, depth to claypans or sand layers)	Anderson-Cook et al. (2002); Banton et al. (1997); Boettinger et al. (1997); Brevik and Fenton (2002); Brus et al. (1992); Doolittle et al. (1994, 2002); Inman et al. (2001); Jaynes et al. (1993); Kitchen et al. (1996); Rhoades et al. (1999a); Scanlon et al. (1999); Stroh et al. (1993); Sudduth and Kitchen (1993); Triantafilis et al. (2001); Williams and Hoey (1987)
Bulk density related (e.g., compaction)	Gorucu et al. (2001); Rhoades et al. (1999a)
Indirectly Measured Soil Properties	
Organic matter related (including soil, organic carbon, and organic chemical plumes)	Benson et al. (1997); Bowling et al. (1997); Brune and Doolittle (1990); Brune et al. (1999); Farahani et al. (2005); Greenhouse and Slaine (1983, 1986); Jaynes (1996); Nobes et al. (2000); Nyquist and Blair (1991)
Cation exchange capacity	Farahani et al. (2005); McBride et al. (1990); Triantafilis et al. (2002)
Leaching	Corwin et al. (1999); Rhoades et al. (1999a); Slavich and Yang (1990)
Groundwater recharge	Cook and Kilty (1992), Cook et al. (1992); Salama et al. (1994)
Herbicide partition coefficients	Jaynes et al. (1995a)
Soil map unit boundaries	Fenton and Lauterbach (1999); Stroh et al. (2001)
Corn rootworm distributions	Ellsbury et al. (1999)
Soil drainage classes	Kravchenko et al. (2002)

Source: From Corwin, D.L., and Lesch, S.M., *Comput. Electron. Agric.*, 46, 11–43, 2005a. With permission.

relies on disturbed soil samples rather than an in situ measurement or noninvasive surface measurement like the other geophysical techniques. The remaining techniques for measuring θ (i.e., TDR, GPR, AM, PM, EMI, NMR, and near seismic reflection) are geophysical adaptations to agriculture that have principally developed since the 1980s and 1990s.

Unlike the electromagnetic approaches (e.g., CP, ER, TDR, and GPR), neutron thermalization relies on a radioactive source of high-energy, epithermal neutrons that collide with H nuclei in soil as a means of inferring volumetric water content. Because the H nucleus is similar in mass to a neutron, H atoms will thermalize the neutrons upon collision. The thermalized neutrons returning to a detector give a measure of the saturation of H atoms in the soil, which is related to volumetric water content as most H atoms in common soils are associated with water. Neutron thermalization first appeared in agriculture in the early 1950s.

Even though capacitance was introduced in the 1930s as a means of measuring θ (Smith-Rose, 1933), its greatest development occurred in the 1990s as a result of advances in microelectronics. Numerous papers and reviews are available in the literature that detail historical developments in capacitance probes (Chernyak, 1964; Dean et al., 1987; Gardner et al., 1991, 1998; Kuràz et al., 1970; Paltineanu and Starr, 1997; Robinson et al., 1998; Schmugge et al., 1980; Thomas, 1966; Wobschall, 1978).

Ground penetrating radar is an area of geophysical instrumentation where electromagnetic signals that propagate as waves are used to map subsurface structure. It has great potential for rapid, noninvasive soil water content measurements over large areas (Basson et al., 1993; Chanzy et al., 1996; van Overmeeren et al., 1997). Agricultural applications in GPR began in the early 1990s and are still in their infancy. Nevertheless, GPR technology has rapidly advanced due to (1) tremendous reduction in the cost of GPR instrumentation over the past decade, (2) advances in instrumentation, and (3) advances in data processing that make it practical and reliable for non-GPR experts to operate and use the instrumentation (Annan and Davis, 1997; Annan et al., 1997).

The underlying principles of GPR and TDR are identical (Davis and Annan, 2002; Weiler et al., 1998). As pointed out by Davis and Annan (2002), TDR is effectively a two-dimensional radar system where radio frequency signals are guided along a transmission line formed from metal conductors embedded in the soil, while GPR radiates the signals from a transmitting antenna through soil to the receiver antenna, which makes it better for rapid bulk measurements over large areas because the signal path is less constrained. Both methods measure the travel time and amplitude of electromagnetic wave fields to determine θ.

Spatial θ information is of particular value in light of the fact that distributions of soil moisture are often the single most important factor influencing within-field variation in crop yield, particularly in irrigated agriculture (Corwin et al., 2003a). Reliable noninvasive techniques that can be mobilized, such as GPR and EMI, where θ within the root zone can be quickly measured, offer a tremendous source of spatial data at field extents and larger, regardless of the dryness or condition of the field (e.g., frozen, rocky). In contrast, invasive techniques, such as ER and TDR, need good contact between the sensor and the soil. Nevertheless, invasive techniques such as TDR still have their place in agriculture outside the controlled conditions of the laboratory.

2.2.2.1   Time Domain Reflectometry

Time domain reflectometry was initially adapted in the early 1980s by Topp and colleagues as a point source technique for measuring θ in the laboratory and for obtaining field θ profiles (Topp and Davis, 1981; Topp et al., 1980, 1982a, 1982b). Over the past 25 years, TDR has become a standard method for measuring θ, which is second only to thermogravimetric methods. Great strides have been taken in the past decade to mobilize TDR and improve its use at field extents (Wraith et al., 2005).

Dalton et al. (1984) demonstrated the utility of TDR to also measure EC_a, based on the attenuation of the applied signal voltage as it traverses through soil. The ability to measure both θ and EC_a makes TDR a versatile geophysical technique in agriculture. The monitoring of the dynamics and spatial patterns of θ and EC_a with TDR was bolstered with the advent of automating and multiplexing capability (Baker and Allmaras, 1990; Heimovaara and Bouten, 1990; Herkelrath et al., 1991). Noborio (2001) provides a review of TDR with a thorough discussion of the theory for the measurement of θ and EC_a; probe configuration, construction, and installation; and strengths and limitations. Wraith (2002) provides an excellent overview of the principles, equipment, procedures, range and precision of measurement, and calibration of TDR. More recently, Wraith et al. (2005) provided an excellent review of TDR with emphasis given to the spatial characterization of θ and EC_a with TDR.

The TDR technique is based on the time for a voltage pulse to travel down a soil probe and back, which is a function of the dielectric constant (ε) of the porous media being measured. By measuring ε, θ can be determined through calibration (Dalton, 1992). The ε is calculated with Equation (2.13) from Topp et al. (1980):

$$\varepsilon = \left(\frac{ct}{2l_p}\right)^2 = \left(\frac{l_a}{l_p v_p}\right)^2 \tag{2.13}$$

where c is the propagation velocity of an electromagnetic wave in free space (2.997×10^8 m s^{-1}), t is the travel time (s), l_p is the real length of the soil probe (m), l_a is the apparent length (m) as measured by a cable tester, and v_p is the relative velocity setting of the instrument. The relationship between θ and ε is approximately linear and is influenced by soil type, ρ_b, clay content, and OM (Jacobsen and Schjønning, 1993).

By measuring the resistive load impedance across the probe (Z_L), EC$_a$ can be calculated with Equation (2.14) from Giese and Tiemann (1975):

$$EC_a = \frac{\varepsilon_0 c}{l} \frac{Z_0}{Z_L} \tag{2.14}$$

where ε_0 is the permittivity of free space (8.854×10^{-12} F m^{-1}), Z_0 is the probe impedance (Ω), and $Z_L = Z_u [2V_0/V_f - 1]^{-1}$, where Z_u is the characteristic impedance of the cable tester, V_0 is the voltage of the pulse generator or zero-reference voltage, and V_f is the final reflected voltage at a very long time. To reference EC$_a$ to 25°C, Equation (2.15) is used:

$$EC_a = K_c f_t Z_L^{-1} \tag{2.15}$$

where K_c is the TDR probe cell constant (K_c [m^{-1}] = $\varepsilon_0 c Z_0/l$), which is determined empirically.

The advantages of TDR for measuring EC$_a$ include (1) a relatively noninvasive nature, (2) an ability to measure both θ and EC$_a$, (3) an ability to detect small changes in EC$_a$ under representative soil conditions, (4) the capability of obtaining continuous unattended measurements, and (5) a lack of a calibration requirement for θ measurements in many cases (Wraith, 2002). However, because TDR is a stationary instrument with which measurements are taken from point-to-point, thereby preventing it from mapping at the spatial resolution of ER and EM approaches, it is currently impractical for developing detailed geo-referenced EC$_a$ maps for large areas.

Although TDR has been demonstrated to compare closely with other accepted methods of EC$_a$ measurement (Heimovaara et al., 1995; Mallants et al., 1996; Reece, 1998; Spaans and Baker, 1993), it is still not sufficiently simple, robust, and fast enough for the general needs of field-scale soil salinity assessment (Rhoades et al., 1999a, 1999b). Currently, the use of TDR for field-scale spatial characterization of θ and EC$_a$ distributions is largely limited. Even though TDR has been adapted to fit on mobile platforms such as ATVs, tractors, and spray rigs (Inoue et al., 2001; Long et al., 2002; Western et al., 1998), vehicle-based TDR monitoring is in its infancy, and only ER and EMI have been widely adapted for detailed spatial surveys consisting of intensive geo-referenced measurements of EC$_a$ at field extents and larger (Rhoades et al., 1999a, 1999b).

2.2.3 FROM OBSERVED ASSOCIATIONS TO EC$_a$-DIRECTED SOIL SAMPLING

Much of the early observational work with EC$_a$ correlated EC$_a$ to soil properties measured from soil samples taken on a grid, which required considerable time and effort. This early work noted the spatial correlation between EC$_a$ and soil properties and subsequently between EC$_a$ and crop yield. However, some of these observational studies were not solidly based on an understanding of the principles and theories encompassing EC_a measurements, which led to presentations and even pub-lications with misinterpretations. To ground researchers in the basic theories and principles of EC_a, guidelines for EC_a surveys and their interpretation were developed by Corwin and Lesch (2003).

After the research associating EC_a to soil properties and to crop yield, the direction of research gradually shifted to mapping within-field variation of EC_a as a means of directing soil sampling to characterize the spatial distribution and variability of properties that statistically correlate with EC_a. The early observational studies compiled in Table 2.1 served as a precursor to the mapping of edaphic (e.g., salinity, clay content, organic matter, etc.) and anthropogenic (e.g., leaching fraction, compaction, etc.) properties using EC_a-directed soil sampling.

Soil sampling directed by geospatial EC_a measurements is the current trend and direction for characterizing spatial variability. The use of EC_a-directed sampling has significantly reduced intensive grid sampling from tens of samples or even a hundred or more samples to eight to twelve sample locations for the characterization of spatial variability in a given field. The earliest work in the soil science literature for the application of geospatial EC_a measurements to direct soil sampling for the purpose of characterizing the spatial variability of a soil property (i.e., salinity) was by Lesch et al. (1992).

2.3 CURRENT STATE-OF-THE-SCIENCE OF EC_a APPLICATIONS IN AGRICULTURE—THE PRESENT

The current status of geophysical techniques in agriculture is reflected in ongoing research of the U.S. Department of Agriculture–Agricultural Research Service (USDA-ARS) laboratories at Ames, IA; Columbia, MO (Kitchen, Lund, and Sudduth); Columbus, OH (Allred); Fort Collins, CO (Buchleiter and Farahani); and Riverside; CA (Corwin and Lesch). Researchers at these facilities have been instrumental in organizing and contributing to symposia and special issues of journals that demonstrate the current role of geophysical techniques, particularly the measurement of EC_a, in agriculture: Soil Electrical Conductivity in Precision Agriculture Symposium at the 2000 American Society of Agronomy-Crop Science Society of America-Soil Science Society of America Annual Meetings, Applications of Geophysical Methods in Agriculture Symposium at the 2003 Annual American Society of Agricultural Engineers International Meeting, special symposium issue of *Agronomy Journal* (2003, vol. 95, number 3) on Soil Electrical Conductivity in Precision Agricul-ture, and special issue of *Computers and Electronics in Agriculture* (Corwin and Plant, 2005) on Applications of Apparent Soil Electrical Conductivity in Precision Agriculture. The most up-to-date review of EC_a measurements in agriculture is provided by Corwin and Lesch (2005a).

2.3.1 FACTORS DRIVING EC_a-DIRECTED SOIL SAMPLING

Three essential factors have driven the development of EC_a-directed soil sampling as a tool to char-acterize the spatial variability of soil properties: (1) the mobilization of EC_a measurement equip-ment, (2) the commercialization and widespread availability of a Global Positioning System (GPS), and (3) the development or adaptation of a statistical sampling approach to select sample sites from spatial EC_a data. All of these came to fruition in the 1990s.

The development of mobile EC_a measurement equipment coupled to a GPS (Cannon et al., 1994; Carter et al., 1993; Freeland et al., 2002; Jaynes et al., 1993; Kitchen et al., 1996; McNeill, 1992; Rhoades, 1993) has made it possible to produce EC_a maps with measurements taken every few meters. Mobile EC_a measurement equipment has been developed for both ER and EMI geophysi-cal approaches. In the case of ER, by mounting the electrodes to "fix" their spacing, considerable time for a measurement is saved. Veris Technologies* developed a commercial mobile system for

---

* Veris Technologies, Salinas, KS. Product identification is provided solely for the benefit of the reader and does not imply the endorsement of the USDA.

FIGURE 2.3   Veris 3100 mobile electrical resistivity equipment. (From Corwin, D.L., and Lesch, S.M., *Comput. Electron. Agric.*, 46, 11–43, 2005a. With permission.

measuring EC_a using the principles of ER (Figure 2.3). In the case of EMI, the EMI conductivity meter is carried on a sled or nonmetallic cart pulled by a pickup, ATV, or four-wheel-drive spray rig (Cannon et al., 1994; Carter et al., 1993; Corwin and Lesch, 2005a; Freeland et al., 2002; Jaynes et al., 1993; Kitchen et al., 1996; Rhoades, 1992, 1993). Both mobile ER and EMI platforms permit the logging of continuous EC_a measurements with associated GPS locations at time intervals of just a few seconds between readings, which results in readings every few meters. The mobile EMI platform permits simultaneous EC_a measurements in both the horizontal (EM_h) and vertical (EM_v) dipole configurations, and the mobile ER platform (i.e., Veris 3100) permits simultaneous measurements of EC_a at 0 to 30 and 0 to 90 cm depths. No commercial mobile system has been developed for EMI, but several fabricated mobile EMI rigs have been developed (e.g., see Figure 2.4).

To establish where soil sample sites are to be located based on the spatial EC_a data, the third essential component of EC_a-directed sampling is needed (i.e., statistical sample design). Currently, two EC_a-directed soil sampling designs are used: (1) design-based sampling and (2) model-based

FIGURE 2.4   Mobile dual-dipole electromagnetic induction equipment developed at the United States Salinity Laboratory. (From Corwin, D.L., and Lesch, S.M., *Comput. Electron. Agric.*, 46, 11–43, 2005a. With permission.)

sampling. Design-based sampling primarily consists of the use of unsupervised classification (Johnson et al., 2001), whereas model-based sampling typically relies on optimized spatial response surface sampling (SRSS) design (Corwin and Lesch, 2005b). Design-based sampling also includes simple random and stratified random sampling. Lesch and colleagues (Lesch, 2005; Lesch et al., 1995a, 1995b, 2000) developed a model-based SRSS software package (ESAP) that is specifically designed for use with ground-based soil EC_a data. The ESAP software package identifies the optimal locations for soil sample sites from the EC_a survey data. These sites are selected based on spatial statistics to reflect the observed spatial variability in EC_a survey measurements. Generally, eight to twelve sites are selected depending on the level of variability of the EC_a measurements for a site. The optimal locations of a minimal subset of EC_a survey sites are identified to obtain soil samples. Protocols are currently available to maintain reliability, consistency, accuracy, and compatibility of EC_a surveys and their interpretation for characterizing spatial variability of soil physical and chemical properties (Corwin and Lesch, 2005b).

There are two main advantages to the response-surface approach. First, a substantial reduction in the number of samples required for effectively estimating a calibration function can be achieved in comparison to more traditional design-based sampling schemes. Second, this approach lends itself naturally to the analysis of remotely sensed EC_a data. Many types of ground-, airborne-, and satellite-based remotely sensed data are often collected specifically because one expects this data to correlate strongly with some parameter of interest (e.g., crop stress, soil type, soil salinity, etc.), but the exact parameter estimates (associated with the calibration model) may still need to be determined via some type of site-specific sampling design. The response-surface approach explicitly optimizes this site-selection process.

2.3.2 Characterization of Soil Spatial Variability with EC_a

The shift in the emphasis of field-related EC_a research from observed associations to directed-sampling design has gained momentum, resulting in the accepted use of geospatial measurements of EC_a as a reliable directed-sampling tool for characterizing spatial variability at field and landscape extents (Corwin and Lesch, 2003, 2005a, 2005b). At present, no other measurement provides a greater level of spatial soil information than that of geospatial measurements of EC_a when used to direct soil sampling to characterize spatial variability (Corwin and Lesch, 2005a). The characterization of spatial variability using EC_a measurements is based on the hypothesis that spatial EC_a information can be used to develop a directed soil sampling plan that identifies sites that adequately reflect the range and variability of soil salinity and other soil properties correlated with EC_a. This hypothesis has repeatedly held true for a variety of agricultural applications (Corwin, 2005; Corwin and Lesch, 2003, 2005a, 2005c, 2005d; Corwin et al., 2003a, 2003b; Johnson et al., 2001; Lesch et al., 1992, 2005).

The EC_a measurement is particularly well suited for establishing within-field spatial variability of soil properties because it is a quick and dependable measurement that integrates within its measurement the influence of several soil properties that contribute to the electrical conductance of the bulk soil. The EC_a measurement serves as a means of defining spatial patterns that indicate differences in electrical conductance due to the combined conductance influences of salinity, θ, texture, and ρ_b. Therefore, maps of the variability of EC_a provide the spatial information to direct the selection of soil sample sites to characterize the spatial variability of those soil properties correlating, either for direct or indirect reasons, to EC_a.

The characterization of the spatial variability of various soil properties with EC_a is a consequence of the physicochemical nature of the EC_a measurement. Three pathways of current flow contribute to the EC_a of a soil: (1) a liquid phase pathway via dissolved solids contained in the soil water occupying the large pores, (2) a solid–liquid phase pathway primarily via exchangeable cations associated with clay minerals, and (3) a solid pathway via soil particles that are in direct and continuous contact with one another (Rhoades et al., 1989, 1999a). These three pathways of current

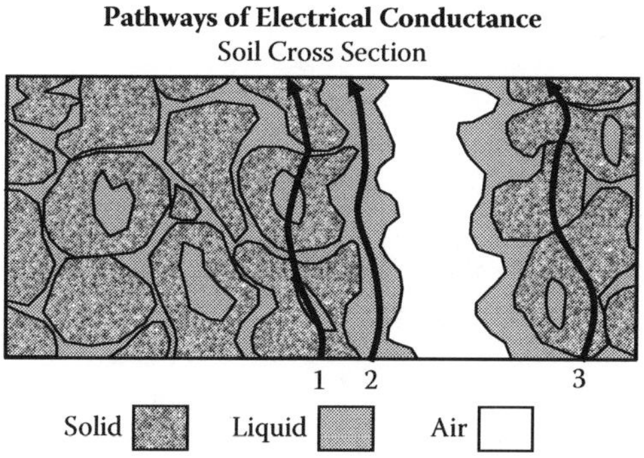

Pathways of Electrical Conductance
Soil Cross Section

Solid ▨   Liquid ▨   Air ☐

FIGURE 2.5   The three conductance pathways for the ECa measurement. (Modified from Rhoades, J.D., Manteghi, N.A., Shouse, P.J., and Alves, W.J., *Soil Sci. Soc. Am. J.*, 53, 433–439, 1989. With permission.

flow are illustrated in Figure 2.5. Rhoades et al. (1989) formulated an electrical conductance model that describes the three conductance pathways of EC_a:

$$EC_a = \left[\frac{(\theta_{ss} + \theta_{ws})^2 \cdot EC_{ws} \cdot EC_{ss}}{\theta_{ss} \cdot EC_{ws} + \theta_{ws} \cdot EC_s} \right] + (\theta_{sc} \cdot EC_{sc}) + (\theta_{wc} \cdot EC_{wc}) \qquad (2.16)$$

where θ_{ws} and θ_{wc} are the volumetric soil water contents in the soil–water pathway (cm^3 cm^{-3}) and in the continuous liquid pathway (cm^3 cm^{-3}), respectively; θ_{ss} and θ_{sc} are the volumetric contents of the surface-conductance (cm^3 cm^{-3}) and indurated solid phases of the soil (cm^3 cm^{-3}), respectively; EC_{ws} and EC_{wc} are the specific electrical conductivities of the soil–water pathway (dS m^{-1}) and continuous-liquid pathway (dS m^{-1}); and EC_{ss} and EC_{sc} are the electrical conductivities of the surface-conductance (dS m^{-1}) and indurated solid phases (dS m^{-1}), respectively. Equation (2.16) was reformulated by Rhoades et al. (1989) into Equation (2.17):

$$EC_a = \left[\frac{(\theta_{ss} + \theta_{ws})^2 \cdot EC_{ws} \cdot EC_{ss}}{(\theta_{ss} \cdot EC_{ws}) + (\theta_{ws} \cdot EC_s)} \right] + (\theta_w - \theta_{ws}) \cdot EC_{wc} \qquad (2.17)$$

where $\theta_w = \theta_{ws} + \theta_{wc}$ = total volumetric water content (cm^3 cm^{-3}), and $\theta_{sc} \cdot EC_{sc}$ was assumed to be negligible. The following simplifying approximations are also known:

$$\theta_w = \frac{(PW \cdot \rho_b)}{100} \qquad (2.18)$$

$$\theta_{ws} = 0.639\theta_w + 0.011 \qquad (2.19)$$

$$\theta_{ss} = \frac{\rho_b}{2.65} \qquad (2.20)$$

$$EC_{ss} = 0.019(SP) - 0.434 \qquad (2.21)$$

$$EC_w = \left[\frac{EC_e \cdot \rho_b \cdot SP}{100 \cdot \theta_w} \right] \qquad (2.22)$$

where PW is the percent water on a gravimetric basis, ρ_b is the bulk density (Mg m^{-3}), SP is the saturation percentage, EC_w is average electrical conductivity of the soil water assuming equilibrium (i.e., $EC_w = EC_{sw} = EC_{wc}$), and EC_e is the electrical conductivity of the saturation extract (dS m^{-1}).

The reliability of Equation (2.17) through Equation (2.22) has been evaluated by Corwin and Lesch (2003). These equations are reliable except under extremely dry soil conditions. However, Lesch and Corwin (2003) developed a means of extending equations for extremely dry soil conditions by dynamically adjusting the assumed water content function.

Because of the three pathways of electrical conductance, EC_a is influenced by several soil physical and chemical properties: (1) soil salinity, (2) saturation percentage, (3) water content, (4) bulk density, and (5) temperature. The quantitative influence of each factor is reflected in Equation (2.17) through Equation (2.22). The SP and ρ_b are both directly influenced by clay content and organic matter (OM). Furthermore, the exchange surfaces on clays and OM provide a solid–liquid phase pathway primarily via exchangeable cations; consequently, clay content and mineralogy, cation exchange capacity (CEC), and OM are recognized as additional factors influencing EC_a measurements. Apparent soil electrical conductivity is a complex property that must be interpreted with these influencing factors in mind.

Field measurements of EC_a are the product of both static and dynamic factors, which include soil salinity, clay content and mineralogy, θ, ρ_b, and temperature. Johnson et al. (2003) described the observed dynamics of the general interaction of these factors. In general, the magnitude and spatial heterogeneity of EC_a in a field are dominated by one or two of these factors, which will vary from one field to the next, making the interpretation of EC_a measurements highly site specific. In instances where dynamic soil properties (e.g., salinity) dominate the EC_a measurement, temporal changes in spatial patterns exhibit more fluidity than systems that are dominated by static factors (e.g., texture). In texture-driven systems, spatial patterns remain consistent because variations in dynamic soil properties affect only the magnitude of measured EC_a (Johnson et al., 2003). For this reason, Johnson et al. (2003) warn that EC_a maps of static-driven systems convey very different information from those of less-stable dynamic-driven systems.

Numerous EC_a studies have been conducted that revealed the site specificity and complexity of spatial EC_a measurements with respect to the particular property influencing the EC_a measurement at that study site. Table 2.1 is a compilation of various laboratory and field studies and the associated dominant soil property measured.

The complex nature of EC_a has a positive benefit. Because of its complexity, geospatial measurements of EC_a provide a means of potentially characterizing the spatial variability of those soil properties influencing EC_a or even soil properties correlated to EC_a without a direct cause-and-effect relationship. The characterization of spatial variability of soil properties correlated with EC_a at a specific field has been achieved through EC_a-directed soil sampling (Corwin and Lesch, 2005c; Lesch et al., 1995b).

2.3.3 AGRICULTURAL APPLICATIONS OF EC_a-DIRECTED SOIL SAMPLING

The characterization of soil spatial variability using EC_a-directed soil sampling has been applied to a variety of landscape-scale agricultural applications: (1) spatial input for solute transport models of the vadose zone, (2) mapping edaphic and anthropogenic properties, (3) characterizing and assessing soil quality, (4) delineating site-specific management units (SSMUs) and productivity zones, and (5) monitoring management-induced spatiotemporal change in soil condition.

To date, the only study to use EC_a-directed soil sampling to characterize soil variability for use in the modeling of solute transport in the vadose zone is by Corwin et al. (1999). In a landscape-scale study modeling salt loading to tile drains in California's San Joaquin Valley, Corwin et al. (1999) used EC_a-directed soil sampling to define spatial domains of similar solute transport capacity in the vadose zone. These spatial domains, referred to as stream tubes, were volumes of soil that are assumed to be isolated from adjacent stream tubes in the field (i.e., no solute exchange) so that a one-dimensional, vertical solute transport model can be applied to each stream tube without concern for lateral flow of water and transport of solute. The application of a functional, tipping-bucket, layer-equilibrium model to each stream tube resulted in the prediction of salt loading to within 30% over a five-year study period.

Mapping soil properties with EC_a-directed soil sampling has been conducted by a limited number of researchers because this approach is comparatively new. The earliest work was conducted by Lesch et al. (1995b) mapping soil salinity. Johnson et al. (2001, 2004) used an EC_a-directed stratified sampling approach to delineate within-field variability of physical, chemical, and biological properties and to relate observations made at different experimental scales. Corwin and Lesch (2005c) used EC_a-directed soil sampling to map a variety of properties for a saline-sodic soil, including salinity, clay content, and sodium adsorption ratio. Triantafilis and Lesch (2005) mapped clay content over a 300 km^2 area. Lesch et al. (2005) used EC_a-directed soil sampling (1) to map and monitor salinity during the reclamation of a field by leaching, (2) to map soil texture and soil type classification, and (3) to identify and locate buried tile lines of a drainage system. Sudduth et al. (2005) provide the most comprehensive compilation relating EC_a to soil properties covering the north-central United States.

An extension of the ability to map individual soil properties is the ability to characterize and assess soil condition based on a compilation of spatial data for individual soil properties influencing the intended function of a soil. The application of EC_a-directed soil sampling to characterize and assess soil condition has been largely found in the Great Plains area and the southwestern United States. Using EC_a maps to direct soil sampling, Johnson et al. (2001) and Corwin et al. (2003b) spatially characterized the overall soil quality of physical and chemical properties thought to affect yield potential. To characterize the soil quality, Johnson et al. (2001) used a stratified soil sampling design with allocation into four geo-referenced EC_a ranges. Correlations were performed between EC_a and the minimum data set of physical, chemical, and biological soil attributes proposed by Doran and Parkin (1996). Their results showed a positive correlation of EC_a with percentage clay, ρ_b, pH, and $EC_{1:1}$ over a soil depth of 0 to 30 cm, and a negative correlation with soil moisture, total and particulate organic matter, total C and N, microbial biomass C, and microbial biomass N. No relationship of the soil properties to crop yield was determined. Corwin et al. (2003b) characterized the soil quality of a saline-sodic soil using a SRSS design. A positive correlation was found between EC_a and the properties of volumetric water content; electrical conductivity of the saturation extract (EC_e); Cl^-, NO_3^-, SO_4^-, Na^+, K^+, and Mg^{+2} in the saturation extract; SAR (sodium adsorption ratio), exchangeable sodium percentage (ESP); B; Se; Mo; $CaCO_3$; and inorganic and organic C. The positive correlation indicated that the spatial variability of soil properties would be accurately characterized. Most of these properties are associated with soil quality for arid zone soils. A number of soil properties (i.e., ρ_b; percentage clay; pH_e; SP; HCO_3^- and Ca^{+2} in the saturation extract; exchangeable Na^+, K^+, and Mg^{+2}; As; CEC; gypsum; and total N) did not correlate well with EC_a measurements, indicating that the SRSS sample design would not accurately characterize the spatial variability of these particular properties. Johnson et al. (2001) and Corwin et al. (2003b) did not actually relate the spatial variation in the measured soil physical and chemical properties to crop yield variations.

To a varying extent from one field to the next, crop patterns are influenced by the spatial variability of edaphic properties. Conventional farming does not address these variations because it manages a field uniformly; as a result, within-field variations in soil properties cause less than optimal crop yields. Site-specific crop management (SSCM) seeks to address variations in crop yield by managing edaphic, anthropogenic, biological, meteorological, and topographic factors to optimize yield and economic return. Bullock and Bullock (2000) point out the importance to SSCM of developing efficient methods for accurately measuring within-field variations in soil physical and chemical properties that influence spatial variation in crop yield. The geospatial measurement of EC_a is a technology that has become an invaluable tool for identifying the soil physical and chemical properties influencing crop yield patterns and for establishing the spatial variation of these soil properties (Corwin et al., 2003a).

The application of EC_a to the SSCM arena is largely due to the past and current research efforts of Kitchen and colleagues (2003, 2005), Lund and colleagues (1999, 2001), and Jaynes and colleagues (1995b, 2003, 2005) in the Midwest using EC_a to delineate productivity zones. Productivity zones refer to areas of similar productivity potential and are of interest to producers, because some key management decisions depend upon reliable estimates of expected yield. Productivity zones associate productivity with a soil property or condition but do not provide the producer with site-specific information for optimizing yield in low-yield portions of a field. For instance, the productivity zones of dryland agriculture have been primarily related to available water as affected by soil and topography (Jaynes et al., 2003; MacMillan et al., 1998). In contrast, SSMUs are units of soil that can be managed similarly to optimize yield.

Corwin et al. (2003a) carried the EC_a-directed soil sampling approach to the next level in SSCM by integrating crop yield to delineate SSMUs with associated recommendations. This work was based on the hypothesis that in the field where yield spatially correlates with EC_a, then geospatial measurements of EC_a can be used to identify edaphic and anthropogenic properties that influence yield. Through spatial statistical analysis, Corwin et al. (2003a) were able to show the influence of salinity, leaching fraction, θ, and pH on the spatial variation of cotton yield for a 32.4 ha field in the Broadview Water District of central California. With this information, a crop yield response model was developed and management recommendations were made that spatially prescribed what could be done to increase cotton yield at those locations with less than optimal yield. Subsequently, Corwin and Lesch (2005a) delineated SSMUs. Highly leached zones were delineated where the leaching fraction (LF) needed to be reduced to <0.5; high salinity areas were defined where the salinity needed to be reduced below the salinity threshold for cotton, which was established at $EC_e = 7.17$ dS m^{-1} for this field; areas of coarse texture were defined that needed more frequent irrigations; and areas were pinpointed where the pH needed to be lowered below a pH of 8 with a soil amendment such as OM. This work brought an added dimension because it delineated within-field units where associated site-specific management recommendations would optimize the yield, but it still falls short of integrating biological, meteorological, economic, and environmental impacts on within-field crop-yield variation. However, prior to the work by Corwin and colleagues, SSCM applications of EC_a had been restricted to the identification of productivity zones (Boydell and McBratney, 1999; Jaynes et al., 2003, 2005; Kitchen et al., 2005; Ping and Dobermann, 2003) rather than management zones that vary in some management input or practice.

Because of its ability to spatially characterize soil properties, EC_a-directed soil sampling easily transitions into a means of monitoring management-induced spatiotemporal changes through the interjection of a temporal component (Corwin et al., 2006). However, even though EC_a-directed soil sampling is far more efficient and less costly than conventional grid sampling, it is still limited in the frequency with which spatio-temporal changes can be studied. Highly dynamic changes, such as those occurring between irrigation or precipitation events or within a crop growing season, are probably too dynamic to monitor effectively. Gradual changes that occur during the course of soil reclamation (Lesch et al., 2005) or due to changes in management, such as drainage water reuse (Corwin et al., 2006), are well suited for EC_a-directed soil sampling. These typically require monitoring at annual intervals or longer.

2.4 PROGNOSIS OF GEOPHYSICAL TECHNIQUES IN AGRICULTURE— FUTURE TRENDS AND NEEDS

The use of geospatial measurements of EC_a for directing soil sampling to characterize soil spatial variability will continue to be a useful approach for field and larger spatial extents. There is considerable potential impact because the characterization of spatial variability is a fundamental component of a variety of field- and landscape-scale issues, including soil quality assessment, solute transport modeling in the vadose zone, SSCM, assessing management-induced changes, and mapping and inventorying soil properties.

When geospatial measurements of EC_a are spatially correlated with geo-referenced yield data, their combined use provides an excellent tool for identifying edaphic and anthropogenic factors that influence yield, which can be used to delineate SSMUs (Corwin and Lesch, 2005a; Corwin et al., 2003a). The delineation of productivity zones from geospatial measurements of EC_a provides another approach to SSCM (Jaynes et al., 2005; Kitchen et al., 2005). Even so, an understanding of the soil-related factors influencing yield or the identification of productivity zones does not provide the whole picture for SSCM because crop systems are affected by a complex interaction of edaphic, biological, meteorological, anthropogenic, and topographic factors. Moreover, the precise manner in which these factors influence the dynamic process of plant growth and reproduction is not always well understood. Geo-referenced EC_a will only help to provide a spatial understanding of edaphic and anthropogenic influences. To be able to manage within-field variation in yield, it is necessary to have an understanding within a spatial context of the relationship of all dominant factors causing the variation.

Current applications of geophysical techniques in agriculture have made it evident that the temporal and spatial complexity of soil–plant systems at field and larger spatial extents will require a combined use of multiple geophysical sensors to obtain the full spectrum of spatial data necessary to identify and characterize the factors influencing yield. Of these, the use of hyperspectral imagery, EMI, real-time kinematic GPS, and GPR probably have the greatest potential from a cost-benefit perspective for providing the greatest information impact. The fruition of EC_a in SSCM will likely come from future plant indicator approaches where combinations of geo-referenced data are used (Corwin and Lesch, 2003). These geo-referenced data will likely include airborne multi- and hyperspectral imagery, EMI, GPR, and real-time kinetic GPS. Plant and soil sampling with model- or design-based sampling strategies will be based on the combined data inputs. Manipulation, organization, and display of these inputs and outputs will be performed with a geographic information system, image analysis, and spatial statistical analysis.

Remotely sensed imagery and EMI measurements of EC_a provide complementary information. Remotely sensed imagery is generally best suited for spatially characterizing dynamic properties associated directly with plant vegetative development, and EC_a measurements are best suited for spatially characterizing static soil properties such as texture, water table depth, and steady-state salinity. Remotely sensed imagery is particularly well suited for obtaining spatial crop information during the maturation of a crop. Furthermore, hyperspectral imagery may hold the key for identifying the spatial effects of nonedaphic factors (e.g., disease, climate, humankind, etc.) on crops. Geospatial measurements of EC_a are most reliable for measuring static soil properties that may influence crop yield because of the associated soil sampling required for ground truth to establish what soil property or properties are influencing EC_a at a given point of measurement. Soil sampling and analysis is time and labor intensive, making the measurement of dynamic soil properties using EC_a generally untenable. Ground truth for remotely sensed imagery is also necessary, but (1) wide-coverage real-time remote images are generally easier to obtain than spatially comparable real-time EC_a data unless EC_a is measured from an airborne platform and (2) calibrations are often faster because soil sampling for EC_a can involve several depth increments and numerous soil properties. Conventional mobilized ground-based EC_a platforms cannot begin to compete with satellite or airborne imagery from the perspective of extent of coverage of real-time data. Nonetheless, ground-based EC_a surveys at field scales have their place because they allow greater control and potentially increased spatial resolution.

There is no question that geospatial measurements of EC_a have found a niche in agricultural research and will likely continue to serve a significant role in the future. However, additional spatial information is needed to fill gaps in the database necessary for SSCM, including (1) the need for integrated spatial data of topographical, meteorological, biological, anthropogenic, and edaphic factors influencing yield; (2) the need for real-time data and rapid processing and analysis to enable temporal as well as spatial management decisions; and (3) the need for sensors that can measure dynamic soil properties and crop responses to those properties. Furthermore, no single study has been conducted that evaluates SSCM from a holistic perspective of environmental, productivity, and economic impacts. This task remains as a future goal for agronomists and soil and environmental scientists. Geophysical techniques will play a crucial role in any future holistic evaluations.

The integrated use of multiple remote and ground-based sensors is the future direction that agriculture will likely take to obtain the extensive spatial data that will be needed to direct variable-rate technologies. Variable-rate technologies driven by a network-centric system of multiple sensors will ultimately take SSCM from a drawing board concept to a reality.

REFERENCES

Amoozegar-Fard, A., Nielsen, D.R., and Warrick, A.W., Soil solute concentration distributions for spatially varying pore water velocities and apparent diffusion coefficients, *Soil Sci. Soc. Am. J.*, 46, 3–9, 1982.

Anderson-Cook, C.M., Alley, M.M., Roygard, J.K.F., Khosia, R., Noble, R.B., and Doolittle, J.A., Differentiating soil types using electromagnetic conductivity and crop yield maps, *Soil Sci. Soc. Am. J.*, 66, 1562–1570, 2002.

Annan, A.P., and Davis, J.L., Ground penetrating radar—Coming of age at last, in *Proc. of the Fourth Decennial International Conf. on Mineral Exploration*, Gubins, AG., Ed., Toronto, ON, Canada, pp. 515–522, 1997.

Annan, A.P., Redman, J.D., Pilon, J.A., Gilson, E.W., and Johnston, G.B., Cross hole GPR for engineering and environmental applications, in *Proc. of the High Resolution Geophysics Workshop* [CD-ROM], Univ. of Arizona, Tucson, AZ, 6–9 January 1997.

Baker, J.M., and Allmaras, R.R., System for automating and multiplexing soil moisture measurement by time-domain reflectometry, *Soil Sci. Soc. Am. J.*, 54, 1–6, 1990.

Banton, O., Seguin, M.K., and Cimon, M.A., Mapping field-scale physical properties of soil with electrical resistivity, *Soil Sci. Soc. Am. J.*, 61, 1010–1017, 1997.

Barnes, H.E., Soil investigation employing a new method of layer-value determination for earth resistivity interpretation, *Highway Res. Board Bull.*, 65, 26–36, 1952.

Basson, U., Gev, I., and Ben-Avrahsm, Z., Ground penetrating radar as a tool for mapping of moisture content and stratigraphy of sand dunes, *Water Technol.*, 19, 11–13, 1993.

Bennett, D.L., and George, R.J., Using the EM38 to measure the effect of soil salinity on *Eucalyptus globulus* in south-western Australia, *Agric. Water Manage.*, 27, 69–86, 1995.

Benson, A.K., Payne, K.L., and Stubben, M.A., Mapping groundwater contamination using DC resistivity and VLF geophysical methods—A case study, *Geophysics* 62, 80–86, 1997.

Biggar, J.W., and Nielsen, D.R., Spatial variability of the leaching characteristics of a field soil, *Water Resour. Res.*, 12, 78–84, 1976.

Boettinger, J.L., Doolittle, J.A., West, N.E., Bork, E.W., and Schupp, E.W., Nondestructive assessment of rangeland soil depth to petrocalcic horizon using electromagnetic induction, *Arid Soil Res. Rehabil.*, 11, 372–390, 1997.

Bohn, H.L., McNeal, B.L., and O'Connor, G. A., *Soil Chemistry*, John Wiley & Sons, New York, 1979.

Borchers, B., Uram, T., and Hendrickx, J.M.H., Tikhonov regularization of electrical conductivity depth profiles in field soils, *Soil Sci. Soc. Am. J.*, 61, 1004–1009, 1997.

Bowling, S.D., Schulte, D.D., and Woldt, W.E., A geophysical and geostatistical methodology for evaluating potential subsurface contamination from feedlot runoff retention ponds, ASAE Paper No. 972087, 1997 ASAE Winter Meetings, December 1997, Chicago, IL, ASAE, St. Joseph, MI, 1997.

Boydell, B., and McBratney, A.B., Identifying potential within-field management zones from cotton yield estimates, in *Precision Agriculture '99, Proc. Second European Conference on Precision Agriculture*, Stafford, J.V., Ed., Denmark, 11–15 July 1999, SCI, London, UK, pp. 331–341, 1999.

Brevik, E.C., and Fenton, T.E., The relative influence of soil water, clay, temperature, and carbonate minerals on soil electrical conductivity readings taken with an EM-38 along a Mollisol catena in central Iowa, *Soil Survey Horizons*, 43, 9–13, 2002.

Briggs, L.J., and McCall, A.G., An artificial root for inducing capillary movement of soil moisture, *Science*, 20, 566–569, 1904.

Brune, D.E., and Doolittle, J., Locating lagoon seepage with radar and electromagnetic survey, *Environ. Geol. Water Sci.*, 16, 195, 207, 1990.

Brune, D.E., Drapcho, C.M., Radcliff, D.E., Harter, T., and Zhang, R., Electromagnetic survey to rapidly assess water quality in agricultural watersheds, ASAE Paper No. 992176, ASAE, St. Joseph, MI, 1999.

Brus, D.J., Knotters, M., van Dooremolen, W.A., van Kernebeek, P., and van Seeters, R.J.M., The use of electromagnetic measurements of apparent soil electrical conductivity to predict the boulder clay depth, *Geoderma*, 55, 79–93, 1992.

Bullock, D.S., and Bullock, D.G., Economic optimality of input application rates in precision farming, *Prec. Agric.*, 2, 71–101, 2000.

Burger, H.R., *Exploration Geophysics of the Shallow Subsurface*, Prentice Hall PTR, Upper Saddle River, NJ, 1992.

Cameron, D.R., de Jong, E., Read, D.W.L., and Oosterveld, M., Mapping salinity using resistivity and electromagnetic inductive techniques, *Can. J. Soil Sci.*, 61, 67–78, 1981.

Cannon, M.E., McKenzie, R.C., and Lachapelle, G., Soil-salinity mapping with electromagnetic induction and satellite-based navigation methods, *Can. J. Soil Sci.*, 74, 335–343, 1994.

Carter, L.M., Rhoades, J.D., and Chesson, J.H.. Mechanization of soil salinity assessment for mapping, ASAE Paper No. 931557, 1993 ASAE Winter Meetings, 12–17 December 1993, Chicago, IL, ASAE, St. Joseph, MI, 1993.

Chanzy, A., Judge, A., Bonn, F., and Tarussov, A., Soil water content determination using a digital ground penetrating radar, *Soil Sci. Soc. Am. J.*, 60, 1318–1326, 1996.

Chernyak, G.Y., Dielectric methods for investigating moist soils, in *Works of the All-Union Research Institute of Hydrology and Engineering Geology*, Ogil'vi, N.A., Ed., New Ser. No. 5, Israel Program for Scientific Translations, Jerusalem, Israel, 1964.

Cook, P.G., and Kilty, S., A helicopter-borne electromagnetic survey to delineate groundwater recharge rates, *Water Resour. Res.*, 28, 2953–2961, 1992.

Cook, P.G., and Walker, G.R., Depth profiles of electrical conductivity from linear combinations of electromagnetic induction measurements, *Soil Sci. Soc. Am. Proc.*, 56, 1015–1022, 1992.

Cook, P.G., Walker, G.R., Buselli, G., Potts, I., and Dodds, A.R., The application of electromagnetic techniques to groundwater recharge investigations, *J. Hydrol.*, 130, 201–229, 1992.

Corwin, D.L., Solute content and concentration—Measurement of solute concentration using soil water extraction—Porous matrix sensors, in *Methods of Soil Analysis, Part 4—Physical Methods*, Dane, J.H., and Topp, G.C., Eds., Soil Sci. Soc. Am. Book Series 5. Soil Science Society of America, Madison, WI, pp. 1269–1273, 2002.

Corwin, D.L., Geospatial measurements of apparent soil electrical conductivity for characterizing soil spatial variability, in *Soil-Water-Solute Characterization: An Integrated Approach*, Álvarez-Benedí, J., and Muñoz-Carpena, R., Eds., CRC Press, Boca Raton, FL, 2005.

Corwin, D.L., and Lesch, S.M., Application of soil electrical conductivity to precision agriculture: Theory, principles, and guidelines, *Agron. J.*, 95, 455–471, 2003.

Corwin, D.L., and Lesch, S.M., Apparent soil electrical conductivity measurements in agriculture, *Comput. Electron. Agric.*, 46, 11–43, 2005a.

Corwin, D.L., and Lesch, S.M., Characterizing soil spatial variability with apparent soil electrical conductivity: I. Survey protocols, *Comp. Electron. Agric.*, 46, 103–133, 2005b.

Corwin, D.L., and Lesch, S.M., Characterizing soil spatial variability with apparent soil electrical conductivity: II. Case study. *Comp. Electron. Agric.*, 46, 135–152, 2005c.

Corwin, D.L., and Plant, R.E., Eds., Applications of apparent soil electrical conductivity in precision agriculture, *Comput. Electron. Agric.*, 46, 1–10, 2005.

Corwin, D.L., and Rhoades, J.D., An improved technique for determining soil electrical conductivity-depth relations from above-ground electromagnetic measurements, *Soil Sci. Soc. Am. J.,* 46, 517–520, 1982.

Corwin, D.L., and Rhoades, J.D., Measurement of inverted electrical conductivity profiles using electromagnetic induction, *Soil Sci. Soc. Am. J.*, 48, 288–291, 1984.

Corwin, D.L., and Rhoades, J.D., Establishing soil electrical conductivity—Depth relations from electromagnetic induction measurements, *Commun. Soil Sci. Plant Anal.*, 21, 861–901, 1990.

Corwin, D.L., Carrillo, M.L.K., Vaughan, P.J., Rhoades, J.D., and Cone, D.G., Evaluation of GIS-linked model of salt loading to groundwater. *J. Environ. Qual.,* 28, 471–480, 1999.

Corwin, D.L., Kaffka, S.R., Hopmans, J.W., Mori, Y., Lesch, S.M., and Oster, J.D., Assessment and field-scale mapping of soil quality properties of a saline-sodic soil, *Geoderma*, 114, 231–259, 2003b.

Corwin, D.L., Lesch, S.M., Oster, J.D., and Kaffka, S.R., Monitoring management-induced spatio-temporal changes in soil quality through soil sampling directed by apparent electrical conductivity, *Geoderma*, 131, 369–387, 2006.

Corwin, D.L., Lesch, S.M., Shouse, P.J., Soppe, R., and Ayars, J.E., Identifying soil properties that influence cotton yield using soil sampling directed by apparent soil electrical conductivity, *Agron. J.*, 95 (2), 352–364, 2003a.

Dalton, F.N., Development of time domain reflectometry for measuring soil water content and bulk soil electrical conductivity, in *Advances in Measurement of Soil Physical Properties: Bringing Theory into Practice*, Topp, G.C., Reynolds, W.D., and Green, R.E., Eds., SSSA Spec. Publ. 30, Soil Science Society of America, Madison, WI, pp. 143–167, 1992.

Dalton, F.N., Herkelrath, W.N., Rawlins, D.S., and Rhoades, J.D., Time-domain reflectometry: Simultaneous measurement of soil water content and electrical conductivity with a single probe, *Science*, 224, 989–990, 1984.

Davis, J.L., and Annan, A.P., Ground penetrating radar to measure soil water content, in *Methods of Soil Analysis, Part 4—Physical Methods*, Dane, J.H., and Topp, G.C., Eds., Soil Sci. Soc. Am. Book Series 5. Soil Science Society of America, Madison, WI, pp. 446–463, 2002.

Dean, T.J., Bell, J.P., and Baty, A.J.B., Soil moisture measurement by an improved capacitance technique. Part 1. Sensor design and performance, *J. Hydrol.*, 93, 67–78, 1987.

de Jong, E., Ballantyne, A.K., Caneron, D.R., and Read, D.W., Measurement of apparent electrical conductivity of soils by an electromagnetic induction probe to aid salinity surveys, *Soil Sci. Soc. Am. J.*, 43, 810–812, 1979.

Diaz, L., and Herrero, J., Salinity estimates in irrigated soils using electromagnetic induction, *Soil Sci.*, 154, 151–157, 1992.

Dobrin, M.B., *Introduction to Geophysical Prospecting*, McGraw-Hill, New York, 1960.

Doolittlc, J.A., Indorante, S.J., Potter, D.K., Hefner, S.G., and McCauley, W.M., Comparing three geophysical tools for locating sand blows in alluvial soils of southeast Missouri, *J. Soil Water Conserv.*, 57, 175–182, 2002.

Doolittle, J.A., Sudduth, K.A., Kitchen, N.R., and Indorante, S.J., Estimating depths to claypans using electromagnetic induction methods, *J. Soil Water Conserv.*, 49, 572–575, 1994.

Doran, J.W., and Parkin, T.B., Quantitative indicators of soil quality: A minimum data set, in *Methods for Assessing Soil Quality*, Doran, J.W., and Jones, A.J., Eds., SSSA Special Publication 49, SSSA, Madison, WI, pp. 25–38, 1996.

Drommerhausen, D.J., Radcliffe, D.E., Brune, D.E., and Gunter, H.D., Electromagnetic conductivity surveys of dairies for groundwater nitrate, *J. Environ. Qual.*, 24, 1083–1091, 1995.

Edlefsen, N.E., and Anderson, A.B.C., The four-electrode resistance method for measuring soil moisture content under field conditions, *Soil Sci.*, 51, 367–376, 1941.

Eigenberg, R.A., and Nienaber, J.A., Electromagnetic survey of cornfield with repeated manure applications, *J. Environ. Qual.*, 27, 1511–1515, 1998.

Eigenberg, R.A., and Nienaber, J.A., Soil conductivity map differences for monitoring temporal changes in an agronomic field, ASAE Paper No. 992176, ASAE, St. Joseph, MI, 1999.

Eigenberg, R.A., and Nienaber, J.A., Identification of nutrient distribution at abandoned livestock manure handling site using electromagnetic induction, ASAE Paper No. 012193, 2001 ASAE Annual International Meeting, 30 July–1 August 2001, Sacramento, CA, ASAE St. Joseph, MI, 2001.

Eigenberg, R.A., Doran, J.W., Nienaber, J.A., Ferguson, R.B., and Woodbury, B.L., Electrical conductivity monitoring of soil condition and available N with animal manure and a cover crop, *Agric. Ecosyst. Environ.*, 88, 183–193, 2002.

Eigenberg, R.A., Korthals, R.L., and Neinaber, J.A., Geophysical electromagnetic survey methods applied to agricultural waste sites, *J. Environ. Qual.*, 27, 215–219, 1998.

Ellsbury, M.M., Woodson, W.D., Malo, D.D., Clay D.E., Carlson, C.G., and Clay S.A., Spatial variability in corn rootworm distribution in relation to spatially variable soil factors and crop condition, in *Proc. 4th International Conference on Precision Agriculture*, Robert, P.C., Rust, R.H., and Larson, W.E., Eds., St. Paul, MN, 19–22 July 1998, ASA-CSSA-SSSA, Madison, WI, pp. 523–533, 1999.

England, C.B., Comments on "A technique using porous cups for water sampling at any depth in the unsaturated zone" by W.W. Wood, *Water Resour. Res.*, 10, 1049, 1974.

Farahani, H.J., Buchleiter, G.W., and Brodahl, M.K., Characterization of soil electrical conductivity variability in irrigated sandy and non-saline fields in Colorado, *Trans. ASAE*, 48, 155–168, 2005.

Fenton, T.E., and Lauterbach, M.A., Soil map unit composition and scale of mapping related to interpretations for precision soil and crop management in Iowa, in *Proc. 4th International Conference on Precision Agriculture*, Robert, P.C., Rust, R.H., and Larson, W.E., Eds., St. Paul, MN, 19–22 July 1998, ASA-CSSA-SSSA, Madison, WI, pp. 239–251, 1999.

Fitterman, D.V., and Stewart, M.T., Transient electromagnetic sounding for groundwater, *Geophysics*, 51, 995–1005, 1986.

Freeland, R.S., Branson, J.L., Ammons, J.T., and Leonard, L.L., Surveying perched water on anthropogenic soils using non-intrusive imagery, *Trans. ASAE*, 44, 1955–1963, 2001.

Freeland, R.S., Yoder, R.E., Ammons, J.T., and Leonard, L.L., Mobilized surveying of soil conductivity using electromagnetic induction, *Appl. Eng. Agric.*, 18, 121–126, 2002.

Gardner, C.M.K., Bell, J.P., Cooper, J.D., Dean, T.J., Hodnett, M.G., and Gardner, N., Soil water content, in *Soil Analysis—Physical Methods*, Smith, R.A. and Mullings, C.E., Eds., Marcel Dekker, New York, pp. 1–73, 1991.

Gardner, C.M.K., Dean, T.J., and Cooper, J.D., Soil water content measurement with a high-frequency capacitance sensor, *J. Agric. Eng. Res.*, 71, 395–403, 1998.

George, R.J., and Woodgate, P., Critical factors affecting the adoption of airborne geophysics for management of dryland salinity, *Exploration Geophysics*, 33, 84–89, 2002.

George, R.J., Beasley, R., Gordon, I., Heislers, D., Speed, R., Brodie, R., McConnell, C., and Woodgate, P., The national airborne geophysics project—National report. Evaluation of airborne geophysics for catchment management (see www.ndsp.gov.au), 1998.

Giese, K., and Tiemann, R., Determination of the complex permittivity from thin-sample time domain reflectometry: Improved analysis of the step response waveform, *Adv. Mol. Relax. Processes*, 7, 45–49, 1975.

Gorucu, S., Khalilian, A., Han, Y.J., Dodd, R.B., Wolak, F.J., and Keskin, M., Variable depth tillage based on geo-referenced soil compaction data in coastal plain region of South Carolina, ASAE Paper No. 011016, 2001 ASAE Annual International Meeting, 30 July–1 August 2001, Sacramento, CA, ASAE St. Joseph, MI, 2001.

Greenhouse, J.P., and Slaine, D.D., The use of reconnaissance electromagnetic methods to map contaminant migration, *Ground Water Monit. Rev.*, 3, 47–59, 1983.

Greenhouse, J.P., and Slaine, D.D., Geophysical modelling and mapping of contaminated groundwater around three waste disposal sites in southern Ontario, *Can. Geotech. J.*, 23, 372–384, 1986.

Haines, B.L., Waide, J.B., and Todd, R.L., Soil solution nutrient concentrations sampled with tension and zero-tension lysimeters: Report of discrepancies, *Soil Sci. Soc. Am. J.*, 46, 658–661, 1982.

Halvorson, A.D., and Rhoades, J.D., Field mapping soil conductivity to delineate dryland seeps with four-electrode techniques, *Soil Sci. Soc. Am. J.*, 44, 571–575, 1976.

Hanson, B.R., and Kaita, K., Response of electromagnetic conductivity meter to soil salinity and soil-water content, *J. Irrig. Drain. Eng.*, 123, 141–143, 1997.

Hart, G.L., and Lowery, B., Axial-radial influence of porous cup soil solution samplers in a sandy Soil, *Soil Sci. Soc. Am. J.* 61, 1765–1773, 1997.

Heimovaara, T.J., and Bouten, W., A computer controlled 36-channel time-domain reflectometry system for monitoring soil water contents, *Water Resour. Res.*, 26, 2311–2316, 1990.

Heimovaara, T.J., Focke, A.G., Bouten, W., and Verstraten, J.M., Assessing temporal variations in soil water composition with time domain reflectometry, *Soil Sci. Soc. Am. J.*, 59, 689–698, 1995.

Hendrickx, J.M.H., Baerends, B., Raza, Z.I., Sadig, M., and Chaudhry, M.A., Soil salinity assessment by electromagnetic induction of irrigated land, *Soil Sci. Soc. Am. J.*, 56, 1933–1941, 1992.

Hendrickx, J.M.H., Borchers, B., Corwin, D.L., Lesch, S.M., Hilgendorf, A.C., and Schlue, J., Inversion of soil conductivity profiles from electromagnetic induction measurements: Theory and experimental verification, *Soil Sci. Soc. Am. J.*, 66, 673–685, 2002.

Herkelrath, W.N., Hamburg, S.P., and Murphy, F., Automatic, real-time monitoring of soil moisture in a remote field area with time domain reflectometry, *Water Resour. Res.*, 27, 857–864, 1991.

Herrero, J., Ba, A.A., and Aragues, R., Soil salinity and its distribution determined by soil sampling and electromagnetic techniques, *Soil Use Manage.*, 19, 119–126, 2003.

Inman, D.J., Freeland, R.S., Yoder, R.E., Ammons, J.T., and Leonard, L.L., Evaluating GPR and EMI for morphological studies of loessial soils, *Soil Sci.*, 166, 622–630, 2001.

Inoue, Y., Watanabe, T., and Kitamura, K., Prototype time-domain reflectometry probes for measurement of moisture content near the soil surface for applications to "on the move" measurements, *Agric. Water Manage.*, 50, 41–52, 2001.

Jacobsen, O.H., and Schjønning, P., A laboratory calibration of time domain reflectometry for soil water measurements including effects of bulk density and texture, *J. Hydrol. (Amsterdam)*, 151, 147–157, 1993.

Jaynes, D.B., Mapping the areal distribution of soil parameters with geophysical techniques, in *Applications of GIS to the Modeling of Non-point Source Pollutants in the Vadose Zone*, Corwin, D.L., and Loague, K., Eds., SSSA Special Publication No. 48, SSSA, Madison, WI, pp. 205–216, 1996.

Jaynes, D.B., Colvin, T.S., and Ambuel, J., Soil type and crop yield determinations from ground conductivity surveys, ASAE Paper No. 933552, 1993 ASAE Winter Meetings, 14–17 December 1993, Chicago, IL, ASAE, St. Joseph, MI, 1993.

Jaynes, D.B., Colvin, T.S., and Ambuel, J., Yield mapping by electromagnetic induction, in *Proc. 2nd International Conference on Site-Specific Management for Agricultural Systems*, Robert, P.C., Rust, R.H., and Larson, W.E., Eds., ASA-CSSA-SSSA, Madison, WI, pp. 383–394, 1995b.

Jaynes, D.B., Colvin, T.S., and Kaspar, T.C., Identifying potential soybean management zones from multi-year yield data, *Comput. Electron. Agric.*, 46, 309–327, 2005.

Jaynes, D.B., Kaspar, T.C., Colvin, T.S., and James, D.E., Cluster analysis of spatiotemporal corn yield patterns in an Iowa field, *Agron. J.*, 95, 574–586, 2003.

Jaynes, D.B., Novak, J.M., Moorman, T.B., and Cambardella, C.A., Estimating herbicide partition coefficients from electromagnetic induction measurements, *J. Environ. Qual.*, 24, 36–41, 1995a.

Johnson, C.K., Doran, J.W., Duke, H.R., Weinhold, B.J., Eskridge, K.M., and Shanahan, J.F., Field-scale electrical conductivity mapping for delineating soil condition, *Soil Sci. Soc. Am. J.*, 65, 1829–1837, 2001.

Johnson, C.K., Doran, J.W., Eghball, B., Eigenberg, R.A., Wienhold, B.J., and Woodbury, B.L, Status of soil electrical conductivity studies by central state researchers, ASAE Paper No. 032339, 2003 ASAE Annual International Meeting, 27–30 July 2003, Las Vegas, NV, ASAE, St. Joseph, MI, 2003.

Johnson, C.K., Wienhold, B.J., and Doran, J.W., Linking microbial-scale findings to farm-scale outcomes in a dryland cropping system, *Precision Agric.*, 5, 311–328, 2004.

Johnston, M.A., Savage, M.J., Moolman, J.H., and du Pleiss, H.M., Evaluation of calibration methods for interpreting soil salinity from electromagnetic induction measurements, *Soil Sci. Soc. Am. J.*, 61, 1627–1633, 1997.

Kachanoski, R.G., de Jong, E., and Van-Wesenbeeck, I.J., Field scale patterns of soil water storage from non-contacting measurements of bulk electrical conductivity, *Can. J. Soil Sci.*, 70, 537–541, 1990.

Kachanoski, R.G., Gregorich, E.G., and Van-Wesenbeeck, I.J., Estimating spatial variations of soil water content using noncontacting electromagnetic inductive methods, *Can. J. Soil Sci.,* 68, 715–722, 1988.

Kaffka, S.R., Lesch, S.M., Bali, K.M., and Corwin, D.L., Relationship of electromagnetic induction measurements, soil properties, and sugar beet yield in salt-affected fields for site-specific management, *Comput. Electron. Agric.*, 46, 329–350, 2005.

Kean, W.F., Jennings Walker, M., and Layson, H.R., Monitoring moisture migration in the vadose zone with resistivity, *Ground Water*, 25, 562–571, 1987.

Khakural, B.R., Robert, P.C., and Hugins, D.R., Use of non-contacting electromagnetic inductive method for estimating soil moisture across a landscape, *Commun. Soil Sci. Plant Anal.*, 29, 2055–2065, 1998.

Kirkham, D., and Taylor, G.S., Some tests of a four-electrode probe for soil moisture measurement, *Soil Sci. Soc. Am. Proc.*, 14, 42–46, 1950.

Kitchen, N.R., Drummond, S.T., Lund, E.D., Sudduth, K.A., and Buchleiter, G.W., Soil electrical conductivity and topography related to yield for three contrasting soil-crop systems, *Agron. J.*, 95, 483–495, 2003.

Kitchen, N.R., Sudduth, K.A., and Drummond, S.T., Mapping of sand deposition from 1993 Midwest floods with electromagnetic induction measurements, *J. Soil Water Conserv.*, 51, 336–340, 1996.

Kitchen, N.R., Sudduth, K.A., Myers, D.B., Drummond, S.T., and Hong, S.Y., Delineating productivity zones on claypan soil fields using apparent soil electrical conductivity, *Comput. Electron. Agric.*, 46, 285–308, 2005.

Kohnke, H., Dreibelbis, F.A., and Davidson, J.M., A survey and discussion of lysimeters and a bibliography on their construction and performances, U.S. Dept. Agric. Misc. Publ. No. 372, U.S. Printing Office, Washington, DC, 1940.

Kravchenko, A.N., Bollero, G.A., Omonode, R.A., and Bullock, D.G., Quantitative mapping of soil drainage classes using topographical data and soil electrical conductivity, *Soil Sci. Soc. Am. J.*, 66, 235–243, 2002.

Kuràz, V., Kutilek, M., and Kaspar, I., Resonance-capacitance soil moisture meter, *Soil Sci.*, 110, 278–279, 1970.

Lesch, S.M., Sensor-directed spatial response surface sampling designs for characterizing spatial variation in soil properties, *Comp. Electron. Agric.*, 46, 153–179, 2005.

Lesch, S.M., and Corwin, D.L., Predicting EM/soil property correlation estimates via the Dual Pathway Parallel Conductance model, *Agron. J.*, 95, 365–379, 2003.

Lesch, S.M., Corwin, D.L., and Robinson, D.A., Apparent soil electrical conductivity mapping as an agricultural management tool in arid zone soils, *Comput. Electron. Agric.*, 46, 351–378, 2005.

Lesch, S.M., Herrero, J., and Rhoades, J.D., Monitoring for temporal changes in soil salinity using electromagnetic induction techniques, *Soil Sci. Soc. Am. J.*, 62, 232–242, 1998.

Lesch, S.M., Rhoades, J.D., and Corwin, D.L., ESAP-95 Version 2.10R: User manual and tutorial guide, Research Rpt. 146, USDA-ARS George E. Brown, Jr. Salinity Laboratory, Riverside, CA, 2000.

Lesch, S.M., Rhoades, J.D., Lund, L.J., and Corwin, D.L., Mapping soil salinity using calibrated electromagnetic measurements, *Soil Sci. Soc. Am. J.*, 56, 540–548, 1992.

Lesch, S.M., Strauss, D.J., and Rhoades, J.D., Spatial prediction of soil salinity using electromagnetic induction techniques: 1. Statistical prediction models: A comparison of multiple linear regression and cokriging, *Water Resour. Res.*, 31, 373–386, 1995a.

Lesch, S.M., Strauss, D.J., and Rhoades, J.D., Spatial prediction of soil salinity using electromagnetic induction techniques: 2. An efficient spatial sampling algorithm suitable for multiple linear regression model identification and estimation, *Water Resour. Res.*, 31, 387–398, 1995b.

Long, D.S., Wraith, J.M., and Kegel, G., A heavy-duty TDR soil moisture probe for use in intensive field sampling, *Soil Sci. Soc. Am. J.*, 66, 396–401, 2002.

Lund, E.D., Christy, D., and Drummond, P.E., Using yield and soil electrical conductivity (EC) maps to derive crop production performance information, in *Proc. 5th International Conference on Precision Agriculture*, Robert, P.C., Rust, R.H., and Larson, W.E., Eds., Minneapolis, MN, 16–19 July 2000, ASA-CSSA-SSSA, Madison, WI (CD-ROM), 2001.

Lund, E.D., Colin, P.E., Christy, D., and Drummond, P.E., Applying soil electrical conductivity to precision agriculture, in *Proc. 4th International Conference on Precision Agriculture*, St. Paul, MN, 19–22 July 1998. ASA-CSSA-SSSA, Madison, WI, pp. 1089–1100, 1999.

MacMillan, R.A., Pettapiece, W.W., Watson, L.D., and Goddard, T.W., A landform segmentation model for precision farming, in *Proc. 4th International Conference on Precision Agriculture*, Robert, P.C., Rust, R.H., and Larson, W.E., Eds., St. Paul, MN, 19–22 July 1998, ASA-CSSA-SSSA, Madison, WI, pp. 1335–1346, 1998.

Mallants, D., Vanclooster, M., Toride, N., Vanderborght, J., van Genuchten, M.Th., and Feyen, J., Comparison of three methods to calibrate TDR for monitoring solute movement in undisturbed soil, *Soil Sci. Soc. Am. J.*, 60, 747–754, 1996.

Mankin, K.R., and Karthikeyan, R., Field assessment of saline seep remediation using electromagnetic induction, *Trans. ASAE*, 45, 99–107, 2002.

Mankin, K.R., Ewing, K.L., Schrock, M.D., and Kluitenberg, G.J., Field measurement and mapping of soil salinity in saline seeps, ASAE Paper No. 973145, 1997 ASAE Winter Meetings, December 1997, Chicago, IL, ASAE, St. Joseph, MI, 1997.

McBratney, A.B., Bishop, T.F.A., and Teliatnikov, I.S., Two soil profile reconstruction techniques, *Geoderma*, 97, 209–221, 2000.

McBride, R.A., Gordon, A.M., and Shrive, S.C., Estimating forest soil quality from terrain measurements of apparent electrical conductivity, *Soil Sci. Soc. Am. J.*, 54, 290–293, 1990.

McNeal, B.L., Oster, J.D., and Hatcher, J.T., Calculation of electrical conductivity from solution composition data as an aid to in-situ estimation of soil salinity, *Soil Sci.*, 110, 405–414, 1970.

McNeill, J.D., Electromagnetic terrain conductivity measurement at low induction numbers, Technical Note TN-6, Geonics Limited, Mississauga, ON, Canada, 1980.

McNeill, J.D., Rapid, accurate mapping of soil salinity by electromagnetic ground conductivity meters, in *Advances in Measurements of Soil Physical Properties: Bringing Theory into Practice*, Topp, G.C., Reynolds, W.D., and Green, R.E., Eds., SSSA Special Publication No. 30, ASA-CSSA-SSSA, Madison, WI, pp. 201–229, 1992.

Morgan, C.L.S., Norman, J.M., Wolkowski, R.P., Lowery, B., Morgan, G.D., and Schuler, R., Two approaches to mapping plant available water: EM-38 measurements and inverse yield modeling, in *Proc. of the 5th International Conference on Precision Agriculture* (CD-ROM), Roberts, P.C., Rust, R.H., and Larson, W.E., Eds., Minneapolis, MN 16–19 July 2000, ASA-CSSA-SSSA, Madison, WI, p. 14, 2000.

Munday, T., Application of airborne geophysical techniques to salinity issues in the Riverland, South Australia, DWLBC Rpt. 2004/3, 2004.

Nettleton, W.D., Bushue, L., Doolittle, J.A., Wndres, T.J., and Indorante, S.J., Sodium affected soil identification in south-central Illinois by electromagnetic induction, *Soil Sci. Soc. Am. J.*, 58, 1190–1193, 1994.

Nobes, D.C., Armstrong, M.J., and Close, M.E., Delineation of a landfill leachate plume and flow channels in coastal sands near Christchurch, New Zealand, using a shallow electromagnetic survey method, *Hydrogeol. J.*, 8, 328–336, 2000.

Noborio, K., Measurement of soil water content and electrical conductivity by time domain reflectometry: A review, *Comp. Electron. Agric.*, 36, 113–132, 2001.

Nyquist, J.E., and Blair, M.S., Geophysical tracking and data logging system: Description and case history, *Geophysics* 56, 1114–1121, 1991.

Paine, J.G., Determining salinization extent, identifying salinity sources, and estimating chloride mass using surface, borehole, an airborne electromagnetic induction methods, *Water Resour. Res.*, 39, 1059, 2003.

Paltineanu, I.C., and Starr, J.L., Real-time soil water dynamics using multisensor capacitance probes: Laboratory calibration, *Soil Sci. Soc. Am. J.*, 61, 1576–1585, 1997.

Ping, J.L., and Dobermann, A., Creating spatially contiguous yield classes for site-specific management, *Agron. J.*, 95, 1121–1131, 2003.

Ranjan, R.S., Karthigesu, T., and Bulley, N.R., Evaluation of an electromagnetic method for detecting lateral seepage around manure storage lagoons, ASAE Paper No. 952440, ASAE, St. Joseph, MI, 1995.

Raulund-Rasmussen, K., Aluminum contamination and other changes of acid soil solution isolated by means of porcelain suction cups, *J. Soil Sci.*, 40, 95–102, 1989.

Reece, C.F., Simple method for determining cable length resistance in time domain reflectometry systems, *Soil Sci. Soc. Am. J.*, 62, 314–317, 1998.

Rhoades, J.D., Instrumental field methods of salinity appraisal, in *Advances in Measurement of Soil Physical Properties: Bring Theory into Practice*, Topp, G.C., Reynolds, W.D., and Green, R.E., Eds., SSSA Special Publication No. 30. Soil Science Society of America, Madison, WI, pp. 231–248, 1992.

Rhoades, J.D., Electrical conductivity methods for measuring and mapping soil salinity, in *Advances in Agronomy*, Sparks, D.L., Ed., vol. 49, Academic Press, San Diego, CA, pp. 201–251, 1993.

Rhoades, J.D., and Corwin, D.L., Determining soil electrical conductivity-depth relations using an inductive electromagnetic soil conductivity meter, *Soil Sci. Soc. Am. J.*, 45, 255–260, 1981.

Rhoades, J.D., and Corwin, D.L., Soil electrical conductivity: Effects of soil properties and application to soil salinity appraisal, *Commun. Soil Sci. Plant Anal.*, 21, 837–860, 1990.

Rhoades, J.D., and Halvorson, A.D., Electrical conductivity methods for detecting and delineating saline seeps and measuring salinity in Northern Great Plains soils, ARS W-42, USDA-ARS Western Region, Berkeley, CA, pp. 1–45, 1977.

Rhoades, J.D., and Ingvalson, R.D., Determing salinity in field soils with soil resistance measurements, *Soil Sci. Soc. Amer. Proc.*, 35, 54–60, 1971.

Rhoades, J.D., Chanduvi, F., and Lesch, S., Soil salinity assessment: Methods and interpretation of electrical conductivity measurements, FAO Irrigation and Drainage Paper #57, Food and Agriculture Organization of the United Nations, Rome, Italy, pp. 1–150, 1999a.

Rhoades, J.D., Corwin, D.L., and Lesch, S.M., Geospatial measurements of soil electrical conductivity to assess soil salinity and diffuse salt loading from irrigation, in *Assessment of Non-point Source Pollution in the Vadose Zone*, Corwin, D.L., Loague, K., and Ellsworth, T.R., Eds., Geophysical Monograph 108, American Geophysical Union, Washington, DC., pp. 197–215, 1999b.

Rhoades, J.D., Manteghi, N.A., Shouse, P.J., and Alves, W.J., Soil electrical conductivity and soil salinity: New formulations and calibrations, *Soil Sci. Soc. Am. J.*, 53, 433–439, 1989.

Rhoades, J.D., Raats, P.A.C., and Prather, R.J., Effects of liquid-phase electrical conductivity, water content and surface conductivity on bulk soil electrical conductivity, *Soil Sci. Soc. Am. J.*, 40, 651–655, 1976.

Rhoades, J.D., Shouse, P.J., Alves, W.J., Manteghi, N.M., and Lesch, S.M., Determining soil salinity from soil electrical conductivity using different models and estimates, *Soil Sci. Soc. Am. J.*, 54, 46–54, 1990.

Robinson, D.A., Gardner, C.M.K., Evans, J., Cooper, J.D., Hodnett, M.G., and Bell, J.P., The dielectric calibration of capacitance probes for soil hydrology using an oscillation frequency response mode, *Hydrol. Earth Sys. Sci.*, 2, 83–92, 1998.

Salama, R.B., Bartle, G., Farrington, P., and Wilson, V., Basin geomorphological controls on the mechanism of recharge and discharge and its effect on salt storage and mobilization: Comparative study using geophysical surveys, *J. Hydrol.*, 155, 1–26, 1994.

Scanlon, B.R., Paine, J.G., and Goldsmith, R.S., Evaluation of electromagnetic induction as a reconnaissance technique to characterize unsaturated flow in an arid setting, *Ground Water*, 37, 296–304, 1999.

Schmugge, T.J., Jackson, T.J., and McKim, H.L., Survey of methods for soil moisture determination, *Water Resour. Res.*, 16, 961–979, 1980.

Sheets, K.R., and Hendrickx, J.M.H., Non-invasive soil water content measurement using electromagnetic induction, *Water Resour. Res.*, 31, 2401–2409, 1995.

Slavich, P.G., Determining EC_a depth profiles from electromagnetic induction measurements, *Aust. J. Soil Res.*, 28, 443–452, 1990.

Slavich, P.G., and Petterson, G.H., Estimating average rootzone salinity from electromagnetic induction (EM-38) measurements, *Aust. J. Soil Res.*, 28, 453–463, 1990.

Slavich, P.G., and Yang, J., Estimation of field-scale leaching rates from chloride mass balance and electromagnetic induction measurements, *Irrig. Sci.*, 11, 7–14, 1990.

Smith, C.N., Parrish, R.S., and Brown, D.S., Conducting field studies for testing pesticide leaching models, *Int. J. Environ. Anal. Chem.*, 39, 3–21, 1990.

Smith-Rose, R.L., The electrical properties of soils for alternating currents at radio frequencies, *Proc. R. Soc. London*, 140, 359, 1933.

Spaans, E.J.A., and Baker, J.M., 1993. Simple baluns in parallel probes for time domain reflectometry, *Soil Sci. Soc. Am. J.*, 57, 668–673, 1993.

Spies, B, and Woodgate, P., Salinity mapping methods in the Australian context, Technical Rpt., Land & Water Australia, 2004.

Stogryn, A., Equations for calculating the dielectric constant of saline water, *IEEE Trans. Microwave Theory Technol. MIT* 19, 733–736, 1971.

Stroh, J.C., Archer, S.R., Doolittle, J.A., and Wilding, L.P., Detection of edaphic discontinuities with ground-penetrating radar and electromagnetic induction, *Landscape Ecol.*, 16, 377–390, 2001.

Stroh, J.C., Archer, S.R., Wilding, L.P., and Doolittle, J.A., Assessing the influence of subsoil heterogeneity on vegetation in the Rio Grande Plains of south Texas using electromagnetic induction and geographical information system, College Station, TX, *The Station*, March 1993, 39–42, 1993.

Sudduth, K.A., and Kitchen, N.R., Electromagnetic induction sensing of claypan depth, ASAE Paper No. 931531, 1993 ASAE Winter Meetings, 12–17 December 1993, Chicago, IL. ASAE, St, Joseph, MI, 1993.

Sudduth, K.A., Kitchen, N.R., Wiebold, W.J., Batchelor, W.D., Bollero, G.A., Bullock, D.G., Clay, D.E., Palm, H.L., Pierce, F.J., Schuler, R.T., and Thelen, K.D., Relating apparent electrical conductivity to soil properties across the north-central USA, *Comput. Electron. Agric.*, 46, 263–283, 2005.

Telford, W.M., Gledart, L. P., and Sheriff, R. E., *Applied Geophysics*, 2nd ed., Cambridge University Press, Cambridge, UK, 1990.

Thomas, A.M., In situ measurement of moisture in soil and similar substances by "fringe" capacitance, *J. Sci. Instr.*, 43, 21–27, 1966.

Topp, G.C., and Davis, J.L., Detecting infiltration of water through the soil cracks by time-domain reflectometry, *Geoderma*, 26, 13–23, 1981.

Topp, G.C., Davis, J.L., and Annan, A.P., Electromagnetic determination of soil water content: Measurement in coaxial transmission lines, *Water Resour. Res.*, 16, 574–582, 1980.

Topp, G.C., Davis, J.L., and Annan, A.P., Electromagnetic determination of soil water content using TDR: I. Applications to wetting fronts and steep gradients, *Soil Sci. Soc. Am. J.*, 46, 672–678, 1982a.

Topp, G.C., Davis, J.L., and Annan, A.P., Electromagnetic determination of soil water content using TDR: II. Evaluation of installation and configuration of parallel transmission lines, *Soil Sci. Soc. Am. J.*, 46, 678–684, 1982b.

Triantafilis, J., and Lesch, S.M., Mapping clay content variation using electromagnetic induction techniques, *Comput. Electron. Agric.*, 46, 203–237, 2005.

Triantafilis, J., Ahmed, M.F., and Odeh, I.O.A., Application of a mobile electromagnetic sensing system (MESS) to assess cause and management of soil salinization in an irrigated cotton-growing field, *Soil Use Manage.*, 18, 330–339, 2002.

Triantafilis, J., Huckel, A.I., and Odeh, I.O.A., Comparison of statistical prediction methods for estimating field-scale clay content using different combinations of ancillary variables, *Soil Sci.*, 166, 415–427, 2001.

U.S. Salinity Laboratory Staff, Diagnosis and improvement of saline and alkali soils, USDA Handbook 60, U.S. Government Printing Office, Washington, DC, pp. 1–160, 1954.

van der Lelij, A., Use of an electromagnetic induction instrument (type EM38) for mapping of soil salinity, Internal Report Research Branch, Water Resources Commission, NSW, Australia, 1983.

van Overmeeren, R.A., Gehrels, J.C., Sariowam, S.V., Ground penetrating radar for determining volumetric soil water content; results of comparative measurements at two test sites, *J. Hydrol.*, 197, 316–338, 1997.

Vaughan, P.J., Lesch, S.M., Corwin, D.L., and Cone, D.G., Water content on soil salinity prediction: A geostatistical study using cokriging, *Soil Sci. Soc. Am. J.*, 59, 1146–1156, 1995.

Weiler, K.W., Steenhuis, T.S., Boll, J., and Kung, K.-J.S., Comparison of ground penetrating radar and time-domain reflectometry as soil water sensors, *Soil. Sci. Soc. Am. J.*, 62, 1237–1239, 1998.

Wesseling, J., and Oster, J.D., Response of salinity sensors to rapidly changing salinity, *Soil Sci. Soc. Am. Proc.*, 37, 553–557, 1973.

Western, A.W., Blöschl, G., Grayson, R.B., Geostatistical characterization of soil moisture patterns in the Tarrawarra catchment, *J. Hydrol.*, 205, 20–37, 1998.

Williams, B.G., and Baker, G.C., An electromagnetic induction technique for reconnaissance surveys of soil salinity hazards, *Aust. J. Soil Res.*, 20, 107–118, 1982.

Williams, B.G., and Hoey, D., The use of electromagnetic induction to detect the spatial variability of the salt and clay contents of soils, *Aust. J. Soil Res.*, 25, 21–27, 1987.

Wilson, R.C., Freeland, R.S., Wilkerson, J.B., and Yoder, R.E., Imaging the lateral migration of subsurface moisture using electromagnetic induction, ASAE Paper No. 023070, 2002 ASAE Annual International Meeting, 28–31 July 2002, Chicago, IL, ASAE, St. Joseph, MI, 2002.

Wobschall, D., A frequency shift dielectric soil moisture sensor, *IEEE Trans. Geosci. Electron.*, 16, 112–118, 1978.

Wollenhaupt, N.C., Richardson, J.L., Foss, J.E., and Doll, E.C., A rapid method for estimating weighted soil salinity from apparent soil electrical conductivity measured with an aboveground electromagnetic induction meter, *Can. J. Soil Sci.* 66, 315–321, 1986.

Wraith, J.M., Solute content and concentration—Indirect measurement of solute concentration—Time domain reflectometry, in *Methods of Soil Analysis, Part 4—Physical Methods*, Dane, J.H., and Topp, G.C., Eds., Soil Science Society of America, Madison, WI, pp. 1289–1297, 2002.

Wraith, J.M., and Or, D., Temperature effects on soil bulk dielectric permittivity measured by time domain reflectometry: Experimental evidence and hypothesis development, *Water Resour. Res.*, 35, 361–369, 1999.

Wraith, J.M., Robinson, D.A., Jones, S.B., and Long, D.S., Spatially characterizing apparent electrical conductivity and water content of surface soils with time domain reflectometry, *Comput. Electron. Agric.*, 46, 239–261, 2005.

3 History of Ground-Penetrating Radar Applications in Agriculture

M. E. Collins

CONTENTS

3.1 INTRODUCTION

There has been a tremendous amount of literature written on the utilization of geophysical techniques for agricultural research in the last 20 or more years. These geophysical techniques include ground-penetrating radar (GPR), electromagnetic methods (EMs), metal detection, magnetometry, and resistivity. But much of what has been published is associated either directly or indirectly with the use of GPR for agricultural analysis.

The use of GPR as a geophysical technique to study its performance in the agricultural discipline was applied to study the variability of soils in Florida in the late 1970s. The success of GPR to investigate ranges in soil properties and facilitate the mapping of soils in Florida led to the incorporation of GPR as a routine field instrument in the Florida Cooperative Soil Survey Program. From this early achievement, the use of GPR in the agricultural arena blossomed to monitor direct and indirect applications of GPR to agricultural situations. Direct applications include those investigations that have an immediate impact on agricultural production and management. These applications may include such examples as determining limited soil depth restricted by bedrock, hardpan, or a shallow water table; surveying root biomass in a loblolly pine forest; assessing stress in citrus trees; and mapping shallow underground soil features affecting agricultural production. Indirect applications may consist of, but are not limited to, investigating hydrocarbons in soil, locating buried drainage pipes, estimating moisture contents in the vadose zone, detecting coarse layers in a sandy soil, evaluating subsurface pathways for nitrogen loss, and identifying offsite movement of agrochemicals. As you may imagine from this list, GPR has been used more to investigate soils than agricultural plants. Even though several geophysical techniques have been used for agricultural reasons, this chapter will restrict its major discussion to the use of GPR as a geophysical tool in the agricultural environment.

3.2 SOIL SURVEYING USING GPR

The application of GPR technology to study soils was begun in Florida by Benson and Glaccum (1979) and was reported by Johnson et al. (1980). The basic objective of their studies was to determine if GPR could be used to accurately identify soil features and their depths for soil survey purposes. How could this new geophysical tool be integrated into day-to-day soil mapping? They determined that the radar could accurately locate soil features such as spodic and argillic horizons as well as depth to the water table. Probably the most surprising result of the study was the brief time it took to obtain the information as compared to traditional soil mapping. This was also documented by Doolittle (1987) when he reported a decrease in cost (70 percent) and an increase in productivity (210 percent) when he compared doing transects using GPR versus conventional methods

Thus, GPR was incorporated into the Florida Cooperative Soil Survey Program in 1981, as well as in other cooperative soil survey programs, as a routine field tool to investigate subsurface features. Much has been published in this respect. Here are some of the early publications on this subject:

- Study soil microvariability (Collins and Doolittle, 1987)
- Increase quality and efficiency of soil surveys (Collins et al., 1986; Doolittle, 1982, 1987; Doolittle and Collins, 1995; Puckett et al., 1990; Schellentrager and Doolittle, 1991; Schellentrager et al., 1988)
- Determine thickness and characterize the depths of organic soil materials (Collins et al., 1986; Doolittle, 1983; Doolittle et al., 1990; Shih and Doolittle, 1984)
- Chart the depths to relatively shallow (<2 m) water tables in predominantly coarse-textured soils (Shih et al., 1985)
- Estimate depths to argillic (Asmussen et al., 1986; Collins and Doolittle, 1987; Hubbard et al., 1990; Truman et al., 1988a, 1988b) and spodic horizons (Collins and Doolittle, 1987; Doolittle, 1987)
- Improve soil-landscape modeling (Doolittle et al., 1988)

Most of the work cited above was done by the U.S. Department of Agriculture (USDA). In fact, Doolittle and Asmussen (1992) published a review of the previous ten years (1981 to 1991) on how the USDA (specifically the Soil Conservation Service, now known as the Natural Resources Conservation Service, and the Agricultural Research Service) used GPR to investigate agricultural soils. They reported the successful use of GPR to "map soils; chart the lateral extent and estimate the depth to soil horizons; and delineate hard pans, water tables, bedrock, and unsaturated flow in the vadose zone," as well as to "assess soil compaction and plow pan development; variations in soil texture, organic matter content, humification, and cementation; thickness of soil horizons, geologic layers, and peat; and movement of water and contaminants in soils." (p. 139)

Some state agricultural experiment stations were also using radar, and many were doing this in collaboration with the USDA. At the same time, other countries were getting more involved with the application of GPR to agricultural circumstances. An outcome was the establishment of the GPR international radar conferences held every even year and alternating between being hosted in the United States and other nations. At the same time, other commercial companies were developing and marketing GPR equipment which subsequently dropped the price of a standard unit. The private sector has been slower in accepting this technology.

GPR has been used by archaeological, engineering, and environmental consulting companies, but not to a great extent by agricultural businesses. One reason could be the soil conditions in which GPR performs well, as discussed next.

3.3 DIRECT GPR APPLICATIONS TO AGRICULTURAL INVESTIGATIONS

Doolittle et al. (1998) used the State Soil Geographic (STATSGO; www.ncgc.nrcs.usda.gov/products/datasets/statsgo/) database to compare soil properties within the United States and Puerto

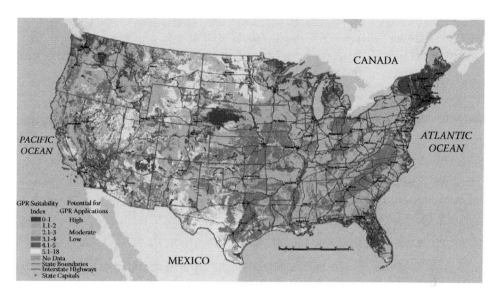

FIGURE 3.1 Ground-penetrating radar soil suitability map of the conterminous United States as presented by Doolittle et al. (2002).

Rico as to their potential for GPR applications. This was an excellent step in assessing all the soils in the United States for the radar, even though it was done on a national scale. Some of the maps they created included saline and sodic soils, and the proportion of soils with less than 18 percent clay. A more detailed map was created by Doolittle et al. (2002) and is shown in Figure 3.1. On this map, notice that the Midwest United States, the dominant agricultural region in the United States, has a low potential for GPR use. Thus, research in using GPR for agricultural purposes has not developed as quickly as it has for environmental intentions. This point also was discussed by Collins (1992) when she pointed out that Mollisols and soils with mollic epipedons (principal soils in the Midwest) have a low potential for GPR. But even so, other geophysical instruments can be used to study soils where the radar has slight potential. Doolittle and Collins (1998) compared EM induction and GPR in areas of karst in Florida and Pennsylvania. They determined that GPR was the geophysical tool to use in Florida where the soil has low pH, low cation exchange capacity (CEC), and low base saturation in comparison to the soils in Pennsylvania which were characterized by relatively high pH, high CEC, and high base saturation. The soils studied in Pennsylvania were better suited for EM applications. Agricultural applications of GPR include investigations that have a direct influence on agricultural production and management. Agricultural production and management may be affected by a limited soil depth restricted by bedrock, hardpan, or a shallow water table. In these soils, plant roots are constrained in growing volume.

Also, agricultural production and management includes using soil (e.g., peatland) as fuel. Many countries (e.g., Ireland, Sweden, Canada, Finland) mine peatlands for the organic soil that can be burned and then have a restoration process to reestablish the land for future generations. Finnish peatlands are a very valuable resource for the wealth of the country. Thus, Hanninen (1992) compared traditional drilling field methods used in Finnish peatlands to determine organic matter thickness and properties to radar data. His conclusions were that GPR could be used to locate the interface between peat and the material below provided that there is a significant difference in moisture content; and various peat layer types and aquatic sediments could be identified by radar.

The list of literature that has had success with GPR includes the following:

- Mapping bedrock depth in a glaciated landscape in Maine (Collins et al., 1989)
- Microanalyses of soil and karst on the Chiefland Limestone Plain in Florida (Collins et al., 1990)

- High-resolution mapping of soil and rock stratigraphy (Davis and Annan, 1989)
- Interpretations of a fragipan in Idaho (Doolittle et al., 2000a)
- Improvement of interpretation of water table depths and groundwater flow patterns using predictive equations (Doolittle et al., 2000b)
- Determination of forest productivity on a loamy substrata glacial drift soil in Michigan (Farrish et al., 1990)
- Assessment of Bt horizons in sandy soils (Mokma et al., 1990a) and ortstein continuity in selected Michigan soils (Mokma et al., 1990b)

All of the above used GPR to determine a soil feature that would have a direct affect on agricultural production. But there are also what may be called "indirect" applications of GPR. An example would be the determination of volumetric water contents in soil with GPR (van Overmeeren et al., 1997). This is an increasing area of radar application, and this is what I mean by indirect applications of GPR to agricultural investigations.

3.4   INDIRECT GPR APPLICATIONS TO AGRICULTURAL INVESTIGATIONS

Agricultural utilization of remote sensing to detect, identify, locate, map, predict, or estimate a buried feature or object that may affect production or management is an increasing application for GPR. This section will discuss the various ways that radar has been used for these indirect purposes. Let us begin by looking at studies involving GPR to detect and monitor groundwater.

GPR investigations to detect and monitor groundwater have several components. It has been used to estimate soil water content (Serbin and Or, 2004) during irrigation and drainage (Galagedara et al., 2005); to identify subsurface flow pathways (Collins, et al., 1994; Gish et al., 2002; Kowalsky et al., 2004; Gish et al., 2005) and nitrogen loss (Walthall et al., 2001); to estimate moisture contents in the vadose zone (Alumbaugh et al., 2002); to survey perched water on anthropogenic soils (Freeland et al., 2001); to estimate volumetric water on a field scale (Grote et al., 2003); and to map spatial variation in surface water content to compare GPR to time domain reflectometry (Huisman et al., 2002).

Serbin and Or (2004) reported using a GPR with a suspended horn antenna to obtain continuous measurements of near surface water content dynamics. These measurements were made in sand over silt loam textures. They concluded that radar enabled them to verify radar measurements at well-defined spatial scales and detailed temporal resolutions not available by other remote sensing techniques. Galagedara et al. (2005) went one step further to estimate water contents with GPR under irrigation and drainage conditions. Specifically, they were interested in determining the optimal ground wave sampling depth under irrigation and drainage situations.

Identifying subsurface flow pathways is a very important application of GPR because of what (e.g., fertilizers, pesticides) may move with the soil water. Several investigations (Collins et al., 1994; Gish et al., 2002, 2005; Kowalsky et al., 2004) documented potential subsurface water movement. Collins et al. (1994) was one of the first to do so and will be discussed as an example.

A large portion of north-central Florida is located in a "bare-karst" region. The limestone can be exposed at the surface or within a depth of a few meters. This area is known for sinkholes opening "overnight!". A series of sinkholes (mother with two daughter dolines) did open up in a matter of a few days in a field and Collins et al. (1994) used this area for their investigation.

The study site was in a pasture in which horse riding and jumping took place (Collins et al., 1994). GPR was used to identify the size of the caverns and determine subsurface flow patterns in order to locate other potential sinkholes in the immediate area. Three grids (macro, medi, and micro) were created, and GPR transects were made. Collins et al. (1994) were able to determine the size of the dolines and create three-dimensional diagrams of subsurface flow. They reported that the surface topography had preferential flow toward the dolines, and the subsurface flow patterns were more complex. There are subsurface "depressions," and the flow patterns were toward these

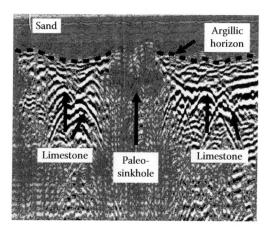

FIGURE 3.2 Ground-penetrating radar (GPR) data showing a paleo-sinkhole. The sand thickness is approximately 1 m; underlying is an argillic horizon that drapes the limestone. Notice how the radar bands point down. The limestone is shown by the hyperbolas.

low areas. These subsurface low areas are not obvious by viewing only the surface topography. A conclusion one may arrive at after looking only at the surface diagram is that soil water and nutrient movements would be toward the dolines. But by studying the subsurface topography of the clay horizon and assuming it was acting as an aquatard, one can assess that solute movement would not necessarily be in the direction of the dolines but may follow another pathway. This research would not have been possible without the use of GPR.

Using GPR in karst environments has been one of the most successful applications of this geophysical technology. Paleo-sinkholes can be identified by radar in optimum soil conditions (Figure 3.2). Optimum conditions include relatively dry sands over a moist argillic horizon that "drapes" the limestone. Breaks in the draping of the argillic horizon may indicate the existence of a void or a cavity that has been filled in with collapsed material (Figure 3.3).

Gish et al. (2002) also did a study involving GPR to evaluate its use in identifying subsurface flow pathways in a 7.5 ha agricultural production field in Maryland. They commented that understanding subsurface stratigraphy was critical to obtain accurate estimates of water fluxes in

FIGURE 3.3 An area showing the sands above the argillic horizon that drapes the limestone. The depth to limestone varies greatly in short distances. This is located in an area in Florida known as "bare" karst.

agricultural fields. They concluded that geo-referenced GPR data sets have great potential to locate soil horizons that control subsurface water pathways. Gish et al. (2005) continued with their interest in identifying subsurface flow pathways by studying such occurrences in a corn (*Zea mays* L.) field. They measured subsoil water contents that supported the GPR-identified preferential flow pathways. The impact of the subsurface water pathways was observed by the increase in corn yield during a drought season. Their conclusion was that subsurface pathways exist and influence soil moisture and corn grain yield patterns.

3.5 GPR APPLICATIONS TO STUDY PLANTS

Literature on the use of GPR applications to study plants has been extremely limited. One of the main limitations has been the wavelength and resolution associated with commercially available antennae. Until recently, the antennae were not able to distinguish very small objects such as plant roots. Truman et al. (1988b) used GPR to assess root concentrations. This may have been the earliest research in the application of GPR to evaluate plant roots.

More recently, Butnor et al. (2003) reported they were able to measure loblolly pine (*Pinus taeda* L.) root biomass to a depth of 30 cm with the aid of a digital signal processed GPR. Correlation coefficients were highly significant ($r = 0.86$, $n = 60$, $p < 0.0001$) between the GPR estimates and the measured root biomass. They concluded that GPR could decrease the number of cores needed to determine tree root biomass and biomass distribution. Hruska et al. (1999) had done a similar study in the Czech Republic. They looked at the three-dimensional distribution of oak trees (*Quercus petraea* (Mattusch.) Liebl) with DBH = 14 to 35 cm to estimate the coarse root density. The GPR unit employed in this study was able to give a resolution of approximately 3 cm in all directions. They reported satisfactory results.

Wielopolski et al. (2002) were concerned with increased CO_2 in the atmosphere. They believed that a considerable portion of CO_2 could be sequestered in plant roots. But a suitable means to measure the root morphology, distribution, and mass without destroying the roots' environment was not available. From these root characteristics, they wanted to access below-ground rates and limits of carbon accumulations. Thus, using a 1.5 GHz impulse GPR system and off-the-shelf software, they were able to image root systems (morphology and dimensions) in situ. They concluded from their study that GPR could image a 2.5 mm root twig buried in sand (under ideal conditions), but that further work is required to improve the images of the plant root system. Also, they believed that with future developments, it would be possible to routinely image roots 2 to 3 mm in diameter.

3.6 GPR GOLF COURSE STUDIES

A golf course was the study site of a very interesting investigation performed by Boniak et al. (2002). A suitable playing surface on a golf course is important for play and aesthetics. Surface watering and subsurface drainage are necessary. When the underground drainage system is not properly working, substantial damage may be done to the golf course resulting in a loss of playing time. The exact location of the underground tile drainage system is not known on many golf courses. To correct a problem, considerable destruction may be done to the course. Thus, the objective of their study was to locate and map the tile drainage system under a putting green using the radar. By doing so, it was expected that less physical destruction would take place to repair the tile because the exact locations of trouble areas could be accurately located with the GPR. Several intriguing findings were a result of this study.

Initially, the 900 MHz antenna was used because of its ability to detect shallow features with high resolution, but it was not successful because of a recent application of granular fertilizer. The fertilizer (high salt content) created noise and distortion to the radar data. As a result, the 400 MHz antenna was used for the study. A discontinuity was located where the golfers would exit the greens.

This resulted in soil compaction (higher bulk density and moisture content) in the area and affected the propagation velocity of the radar signals.

The conclusions of Boniak et al. (2002) were that GPR is very effective in locating and mapping underground drainage tile beneath golf greens. One suggestion that they offered was that a high-accuracy Global Positioning System (GPS) be used to geo-reference each location in the course of doing the radar transects. In another publication, Chong et al. (2000) also used GPR to determine the root zone of golf greens.

3.7 INCORPORATION OF GPR AND GPS INTO GIS

In the last ten years, three-dimensional modeling of radar data has increased due in part to inexpensive, commercially available, and user-friendly software. Applications of geographic information systems (GISs) have increased for the same reasons. Therefore, more users are employing the technologies. But integrating all three—GPR, GPS, and GIS—has not been routinely appreciated. One of the first to explore this opportunity was Tischler (2002).

As discussed in Tischler (2002) and Tischler et al. (2002), GIS provides a means of storing, manipulating, analyzing, and displaying spatially distributed data in a two-dimensional or three-dimensional view. Combining the efficiency and practicality of GPR with the visual appeal and interpretive power of GIS is the next reasonable step in the development of both technologies. This can be accomplished using GPS. However, few methods exist for combining geo-referenced GPR data with GIS data sets, which would reduce time and costs while increasing the interpretive quality of the information.

Four models were developed (Tischler, 2002). Model 1 consisted of categorical data that predict the presence of sand or an argillic horizon based on GPR amplitude readings. Model 2 was a numerical raster model that predicted GPR amplitude values continuously throughout the field. Models 3 and 4 were both numerical raster models that predict the depth to argillic horizon but differ in their methods of generation. The GPR processing steps taken, specific objectives, and the quantitative analysis performed on the data, differentiate all the models. Summarizing his results, Tischler commented that Model 1 met the objectives initially defined but the statistical correlation was not as strong as expected. Model 2 was much more visually appealing than Model 1, but the end value of his predicted ΔA (change in amplitude) was not a good variable to model. Models 3 and 4 were similar in generation and display. Both models predicted the same variable, depth to argillic, and are raster models. Figure 3.4 and Figure 3.5 are examples of Models 1 and 2, respectively.

3.8 CONCLUSIONS

There is no doubt that geophysical techniques have an essential function in the agricultural research world. Only GPR was discussed in detail in this chapter, but a similar chapter could be dedicated to EM use for agricultural purposes, especially in regions where the soil conditions for GPR are limited.

Over the twenty plus years that GPR has been routinely used by agriculturists, many of the studies have had direct or indirect applications to agriculture production and management. Some of the direct applications include mapping bedrock depth in a glaciated landscape; conducting microanalyses of soil and karst; high-resolution mapping soil and rock stratigraphy; interpreting a fragipan; improving interpretation of water table depths and groundwater flow patterns; determining forest productivity on a glacial drift soil; and assessing Bt horizons in sandy soils and ortstein. Indirect applications of GPR include detection, identification, location, mapping, predicting, or estimating a buried feature or object that may affect production or management in an incidental fashion. This has been a significant and increasing use for GPR and includes many investigations in detecting and monitoring groundwater and nutrients. Some of these are estimating soil water content during irrigation and drainage; identifying subsurface flow pathways; approximating moisture contents in the vadose zone;

FIGURE 3.4   Model 1. The darker gray areas are voxels with a value of 1, indicating the argillic horizon. The ligher gray areas 1 are voxels with a value of 0, indicating sand (Ap and E horizons). The white areas are meant to help visualize the layer transition. (From Tischler, M. A., 2002, Intergration of ground-penetrating radar data, global positioning systems, and geographic information systems to create three-dimensional soil models, M.S. Thesis, University of Florida. Gainesville. With permission.)

FIGURE 3.5   Model 2. Integration of ground-penetrating radar (GPR) and Global Positioning System (GPS) data into geographical information system (GIS) application as viewed in ArcScene. Darker areas are soils that are dominantly sandy in texture. Lighter areas are soils high in clay content. Each layer is a krigged interpolation of GPR values from the corresponding soil depth. (From Tischler, M. A., 2002, Intergration of ground-penetrating radar data, global positioning systems, and geographic information systems to create three-dimensional soil models, M.S. Thesis, University of Florida. Gainesville. With permission.)

surveying perched water on anthropogenic soils; determining volumetric water on a field scale; and mapping spatial variation in surface water content to compare GPR to time domain reflectometry.

The use of GPR to study plants, specifically plant roots and their biomass, has not received as much attention as the other applications. This application has slowly increased in recent years, mainly due to the development of high-frequency antennae. One of the latest applications of GPR data as well as GPS data has been to incorporate these data into GIS. This application will continue to grow as more users of GPR and GIS cooperate in their research and use the data to solve emerging agricultural issues.

REFERENCES

Alumbaugh, D., P. Y. Chang, L. Paprocki, J. R. Brainard, R. J. Glass and C. A. Rautman. 2002. Estimating moisture contents in the vadose zone using cross-borehole ground penetrating radar: A study of accuracy and repeatability. *Water Resour. Res.* 38(12).

Asmussen, L. E., H. F. Perkins, and H. D. Allison. 1986. Subsurface descriptions by ground-penetrating radar for watershed delineation. The Georgia Agricultural Experiment Stations Research Bulletin 340. University of Georgia, Athens. pp. 15.

Benson, R., and R. Glaccum. 1979. Test Report; The application of ground-penetrating radar to soil surveying for National Aeronautical and Space Administration (NASA). Technos Inc. Miami, FL.

Boniak, R., S.K. Chong, S.J. Indorante, and J.A. Doolittle. 2002. Mapping golf green drainage systems and subsurface features using ground-penetrating radar. pp. 477–481. In: Koppenjan, S. K., and L. Hua (Eds.). *Ninth International Conference on Ground-Penetrating Radar. Proceedings of SPIE,* Volume 4158. 30 April–2 May 2002. Santa Barbara, CA.

Butnor, J. R., J. A. Doolittle, K. H. Johnsen, L. Samuelson, T. Stokes, and L. Kress. 2003. Utility of ground-penetrating radar as a root biomass survey tool in forest systems. *Soil Sci. Soc. Am. J.* 67(5):1607–1615.

Chong, S. K., J. A. Doolittle, S. J. Indorante, K. Renfro, and P. Buck. 2000. Investigating without excavating. *Golf Course Manage.* 68(9):56–59.

Collins, M. E., G. W. Schellentrager, J. A. Doolittle, and S. F. Shih. 1986. Using ground-penetrating radar to study changes in soil map unit composition in selected Histosols. *Soil Sci. Soc. Am. J.* 50:408–412.

Collins, M. E., and J. A. Doolittle. 1987. Using ground-penetrating radar to study soil microvariability. *Soil Sci. Soc. Am. J.* 51:491-493.

Collins, M. E., J. A. Doolittle, and R. V. Rourke. 1989. Mapping depth to bedrock on a glaciated landscape with ground-penetrating radar. *Soil Sci. Soc. Am. J.* 53:1806-1812.

Collins, M. E., W. E. Puckett, G. W. Schellentrager, and N. A. Yust. 1990. Using GPR for micro-analyses of soils and karst features on the Chiefland Limestone Plain in Florida.

Collins, M. E. 1992. Soil Taxonomy: A useful guide for the application of ground penetrating radar. pp. 125–132. In: Hanninen, P., and S. Autio (Eds.). *Fourth International Conference on Ground Penetrating Radar.* June 8-13, 1992. Rovaniemi, Finland. Geological Survey of Finland, Special Paper 16. pp. 365.

Collins, M. E., M. Crum, and P. Hanninen. 1994. Using ground-penetrating radar to investigate a subsurface karst landscape in north-central Florida. *Geoderma.* 61:1–15. *Geoderma.* 47:159–170.

Davis, J. L., and A. P. Annan. 1989. Ground-penetrating radar for high-resolution mapping of soil and rock stratigraphy. *Geophysics.* 37:531–551.

Doolittle, J. A. 1982. Characterizing soil map units with the ground-penetrating radar. *Soil Survey Horizons.* 22(4):3–10.

Doolittle, J. A. 1983. Investing Histosols with ground-penetrating radar. *Soil Survey Horizons.* 23(3): 23–28.

Doolittle, J. A. 1987. Using ground-penetrating radar to increase the quality and efficiency of soil surveys. In *Soil Survey Techniques, Soil Sci. Soc. Am.* Special Pub. No 20. pp. 11–32.

Doolittle, J. A., R. A. Rebertus, G. B. Jordan, E. I. Swenson, and W. H. Taylor. 1988. Improving soil-landscape models by systematic sampling with ground-penetrating radar. *Soil Survey Horizons* 29(2):46–54.

Doolittle, J., P. Fletcher, and J. Turenne. 1990. Estimating the thickness of organic materials in cranberry bogs. *Soil Survey Horizons* 31(3):73–78.

Doolittle, J. A., and L. E. Asmussen. 1992. Ten years of applications of ground-penetrating radar by the United States Department of Agriculture. pp. 139–147. In: Hanninen, P., and S. Autio (Eds.). Fourth International Conference on Ground Penetrating Radar. June 8–13, 1992. Rovaniemi, Finland. Geological Survey of Finland, Special Paper 16. pp. 365.

Doolittle, J. A., and M. E. Collins. 1995. Use of soil information to determine application of ground-penetrating radar. *J. Appl. Geophys.* 33:101–108.

Doolittle, J. A., and M. E. Collins. 1998. A comparison of EM induction and GPR methods in areas of karst. *Geoderma.* 85:83–102.

Doolittle, J. A., M. E. Collins, and H. R. Mount. 1998. Assessing the appropriateness of GPR with a soil geographic database. pp. 393–400. In: Plumb, R. G. (Ed.). *Proceedings Seventh International Conference on Ground-Penetrating Radar.* May 27–30, 1998, Lawrence, KS. Radar Systems and Remote Sensing Laboratory, University of Kansas. p. 786.

Doolittle, J., G. Hoffmann, P. McDaniel, N. Peterson, B. Gardner, and E. Rowan. 2000a. Ground-penetrating radar interpretations of a fragipan in Northern Idaho. *Soil Survey Horizons.* 41(3):73–82.

Doolittle, J. A., B. J. Jenkinson, D. P. Franzmeier, and W. Lynn. 2000b. Improved radar interpretations of water table depths and groundwater flow patterns using predictive equations. pp. 488–495. In: Noon, D. (Ed.). *Proceedings Eighth International Conference on Ground-Penetrating Radar.* May 23–26, 2000, Goldcoast, Queensland, Australia. The University of Queensland. 908 p.

Doolittle, J. A., F. E. Minzenmayer, S. W. Waltman, and E. C. Benham. 2002. Ground-penetrating radar soil suitability map of the conterminous United States. pp. 7–12. In: Koppenjan, S. K., and L. Hua (Eds.). *Ninth International Conference on Ground-Penetrating Radar. Proceedings of SPIE,* Volume 4158. April 30–May 2, 2002. Santa Barbara, CA.

Farrish, K. W., J. A. Doolittle, and E. E. Gamble. 1990. Loamy substrata and forest productivity of sandy glacial drift soils in Michigan. *Can. J. Soil Sci.* 70:181–187.

Freeland, R. S., J. L. Branson, J. T. Ammons, and L. L. Leonard. 2001. Surveying perched water on anthropogenic soils using non-intrusive imagery. *Trans. ASAE* 44(6):1955–1963.

Galagedara, L. W., G. W. Parkin, J. D. Redman, P. von Bertoldi, and A. L. Endres. 2005. Field studies of the GPR ground wave method for estimating soil water content during irrigation and drainage. *J. Hydrology.* 301(1–4):182–197.

Gish, T. J., W. P. Dulaney, K. J. S. Kung, C. S. T. Daughtry, J. A. Doolittle, and P. T. Miller. 2002. Evaluating use of ground-penetrating radar for identifying subsurface flow pathways. *Soil Sci. Soc. Am. J.* 66(5):1620–1629.

Gish, T. J., C. L. Walthall, C. S. T. Daughtry, and K. J. S. Kung. 2005. Using soil moisture and spatial yield patterns to identify subsurface flow pathways. *J. Environ. Qual.* 34(1):274–286.

Grote, K., S. Hubbard, and Y. Rubin. 2003. Field-scale estimation of volumetric water content using ground-penetrating radar ground wave techniques. *Water Resour. Res.* 39(11).

Hanninen, P. 1992. Application of ground penetrating radar techniques to peatland investigations. pp. 217–221. In: Hanninen, P., and S. Autio (Eds.). *Fourth International Conference on Ground Penetrating Radar.* June 8–13, 1992. Rovaniemi, Finland. Geological Survey of Finland, Special Paper 16. pp. 365.

Hruska, J., J. Cermak, and S. Sustek. 1999. Mapping tree root systems with ground-penetrating radar. *Tree Physiol.* 19(2):125–130.

Hubbard, R. K., L. E. Asmussen, and H. F. Perkins. 1990. Use of ground-penetrating radar on upland Coastal Plain soils. *J. Soil and Water Conserv.* 45(4):399–404.

Huisman, J. A., J. Snepvangers, W. Bouten, and G. B. M. Heuvelink. 2002. Mapping spatial variation in surface soil water content: Comparison of ground-penetrating radar and time domain reflectometry. *J. Hydrology.* 269(3–4):194–207.

Johnson, R. W., R. Glaccum, and R. Wojtasinski. 1980. Application of ground penetrating radar to soil survey. *Soil and Crop Science Society of Florida Proceedings* 39:68–72.

Kowalsky, M. B., S. Finsterle, and Y. Rubin. 2004. Estimating flow parameter distributions using ground-penetrating radar and hydrological measurements during transient flow in the vadose zone. *Adv. in Water Resour.* 27(6):583–599.

Mokma, D. L., R. J. Schaetzl, E. P. Johnson, and J. A. Doolittle. 1990a. Assessing Bt horizon character in sandy soils using ground-penetrating radar: Implications for soil surveys. *Soil Survey Horizons.* 30(2):1–8.

Mokma, D. L., R. J. Schaetzl, J. A. Doolittle, and E. P. Johnson. 1990b. Ground-penetrating radar study of ortstein continuity in some Michigan Haplaquods. *Soil Sci. Soc. Am. J.* 54:936–938.

Puckett, W. E., M. E. Collins, and G. W. Schellentrager. 1990. Design of soil map units on a karst area in West Central Florida. *Soil Sci. Soc. Am. J.* 54:1068–1073.

Schellentrager, G. W., J. A. Doolittle, T. E. Calhoun, and C. A. Wettstein. 1988. Using ground-penetrating radar to update soil survey information. *Soil Sci. Soc. Am. J.* 52:746–752.

Schellentrager, G. W., and J. A. Doolittle. 1991. Systematic sampling using ground-penetrating radar to study regional variation of a soil map unit. Chapter 12. pp. 199–214. In: Mausbach, M. J., and L. P. Wilding (Eds.). *Spatial Variabilities of Soils and Landforms*. Soil Science Society of America Special Publication No. 28. pp. 270.

Serbin, G., and D. Or. 2004. Ground-penetrating radar measurement of soil water content dynamics using a suspended horn antenna. *Trans. Geoscience and Remote Sensing*. 42(8):1695–1705.

Shih, S. F., and J. A. Doolittle. 1984. Using radar to investigate organic soil thickness in the Florida Everglades. *Soil Sci. Soc. Am. J.* 48:651–656.

Shih, S. F., D. L. Myhre, G. W. Schellentrager, V. W. Carlisle, and J. A. Doolittle. 1985. Using radar to assess the soil characteristics related to citrus stress. *Soil and Crop Sci. Soc. of Fla. Proc.* 45:54–59.

Tischler, M. A. 2002. Intergration of ground-penetrating radar data, global positioning systems, and geographic information systems to create three-dimensional soil models. M.S. Thesis. University of Florida. Gainesville.

Tischler, M. A., M. E. Collins, and S. Grunwald. 2002. Intergration of ground-penetrating radar data, global positioning systems, and geographic information systems to create three-dimensional soil models. pp. 313–316. In: Koppenjan, S. K., and L. Hua (Eds.). *Ninth International Conference on Ground-Penetrating Radar. Proceedings of SPIE*, Volume 4158. April 30–May 2, 2002. Santa Barbara, CA.

Truman, C. C., H. F. Perkins, L. E. Asmussen, and H. D. Allison. 1988a. Using ground-penetrating radar to investigate variability in selected soil properties. *J. Soil and Water Conserv.* 43(4):341–345.

Truman, C. C., H. F. Perkins, L. E. Asmussen, and H. D. Allison. 1988b. Some applications of ground-penetrating radar in the southern coastal plains region of Georgia. The Georgia Agricultural Experiment Stations Research Bulletin 362. University of Georgia, Athens. pp. 27.

van Overmeeren, R. A., S. V. Sariowan, and J. C. Gehrels. 1997. Ground penetrating radar for determining volumetric soil water content; results of comparative measurements at two test sites. *J. Hydrology*. 197:316–338.

Walthall, C. L., T. J. Gish, C. S. Daughtry, W. P. Dulaney, K. J. Kung, G. McCarty, D. Timlin, J. T. Angier, P. Buss, and P. R. Houser. 2001. An innovative approach for locating and evaluating subsurface pathways for nitrogen loss. *Proceedings of the 2nd International Nitrogen Conference on Science and Policy*. **TheScientificWorld** (online journal). 2:223–229.

Wielopolski, L., G. Hendrey, M. McGuigan, and J. Daniels. 2002. Imaging tree roots systems in situ. pp. 58–62. In: Koppenjan, S. K., and L. Hua (Eds.). *Ninth International Conference on Ground-Penetrating Radar. Proceedings of SPIE*, Volume 4158. April 30–May 2, 2002. Santa Barbara, CA.

Section II

Agricultural Geophysics Measurements and Methods

4 Theoretical Insight on the Measurement of Soil Electrical Conductivity

Dennis L. Corwin, Scott M. Lesch, and Hamid J. Farahani

CONTENTS

4.1 INTRODUCTION

Due in large measure to the research that has been conducted at the U.S. Department of Agriculture–Agricultural Research Service (USDA-ARS) United States Salinity Laboratory over the past 50 years, the measurement of electrical conductivity (EC) has become a standard soil physicochemical measurement both in the laboratory and in the field to address agricultural and environmental concerns. In particular, the geospatial measurement of EC with geophysical techniques, including electrical resistivity (ER), electromagnetic induction (EMI), and time domain reflectometry (TDR), has burgeoned into one of the most useful field agricultural measurements, particularly for spatially characterizing the variability of soil properties such as salinity, water content, and texture (Corwin, 2005).

The value of spatial measurements of soil EC to agriculture is widely acknowledged due to its ability to characterize spatial variability, with applications in solute transport modeling at field and landscape scales (Corwin et al., 1999), salinity mapping and assessment (Corwin et al., 2003a), mapping soil texture (Doolittle et al., 2002) and soil type (Anderson-Cook et al., 2002; Jaynes et al.,

1993), location of claypans and depth of depositional sand (Doolittle et al., 1994; Kitchen et al., 1996; Sudduth et al., 1995), soil quality assessment (Corwin et al., 2003a; Johnson et al., 2001; McBride et al., 1990), monitoring of management-induced spatiotemporal changes in soil condition (Corwin et al., 2006), delineation of site-specific crop management units (Corwin et al., 2003b, 2008) and zones of productivity (Jaynes et al., 2003, 2005; Kitchen et al., 2005), and measuring other soil properties such as soil moisture (Kachanoski et al., 1988; Sheets and Hendrickx, 1995), clay content (Sudduth et al., 2005; Triantafilis and Lesch, 2005; Williams and Hoey, 1987), shallow subsurface available soil N (Eigenberg et al., 2002), and cation exchange capacity (Sudduth et al., 2005). The above studies mostly used the noncontact electromagnetic induction (EMI) method to measure soil EC, with a few using the four-electrode contact methods that induce electrical current into the soil through insulated metal electrodes. A complete review of EC measurements in agriculture is provided by Corwin and Lesch (2005a).

As evident from the above studies, there are significant innovations in developing useful applications of soil EC measurements in agriculture. Nearly all experimental observations similar to those listed above are based on soil EC being regarded as a surrogate measure of one or more soil properties of interest. Results are dictated by the physical and chemical properties of the soil at the time of the EC measurements. However, it is not evident that the basic principles of soil EC have been adequately examined. That is particularly the case in examining the recent interests in various applications of soil EC mapping in precision agriculture. An understanding of the spatial and temporal variability of soil EC and an appreciation for its highly complex interactions with static and dynamic soil properties, particularly at low-salt concentrations, is needed. It is the intent of this chapter to highlight the important aspects of spatial EC measurements in agriculture by providing basic principles and theory of soil EC measurement and what it actually measures, standard operating procedures for conducting a field-scale EC survey, including an outlined set of EC-directed soil sampling protocols, and examples of spatial EC surveys and their interpretation.

4.1.1 Background: Definition and Brief History of Soil Electrical Conductivity

Soil EC has its historical roots in the measurement of soil salinity. In situ measurement of soil resistivity dates back to at least the latter part of the nineteenth century when Whitney et al. (1897) attempted to infer soil water content and salinity from measurements of soil resistivity using two-probe electrodes. Gardner (1898) and Briggs (1899) reported additional measurements as part of the early group of USDA scientists investigating soil temperature, salinity, and water content effects on soil resistivity. To minimize the difficulties with the unstable two-probe method, Frank Wenner (1915) introduced the theory of utilizing four equally spaced electrodes to measure earth resistivity and wrote "A knowledge of earth resistivity (or specific resistance) may be of value in determining something of the composition of earth."

Soil salinity refers to the presence of major dissolved inorganic solutes in the aqueous phase consisting of soluble and readily dissolvable salts in soil and can be determined by measuring the total solute concentration in the soil aqueous phase, more commonly referred to as the soil solution. The determination of total solute concentration (i.e., salinity) through the measurement of EC has been well established for half a century (U.S. Salinity Laboratory Staff, 1954). Soil salinity is quantified in terms of the total concentration of soluble salts as measured by the EC of the solution in dS m^{-1}.

It is known that the EC of a pure solution (σ_w) is a function of its chemical composition and is characterized by Equation (4.1):

$$\sigma_w = k \sum_{i=1}^{n} \lambda_i M_i |v_i|$$
(4.1)

where k is the cell constant accounting for electrode geometry, λ is the molar limiting ion conductivity (S m^2 mol^{-1}), M is the molar concentration (mol m^{-3}), v is the absolute value of the ion charge, and i denotes the ion species in solution. Marion and Babcock (1976), among others, have confirmed the existence of the relationship between EC and molar concentrations of ions in the soil solution.

To determine soil EC, the soil solution is placed between two electrodes of constant geometry and distance of separation (Bohn et al., 1979). The measured conductance is a consequence of the solution's salt concentration and the electrode geometry whose effects are embodied in a cell constant. Electrical conductance was considered more suitable for salinity measurements than resistance because it increases with salt content, which simplifies the interpretation of readings. At constant potential, the electrical conductance is a reciprocal of the measured resistance as shown in Equation (4.2):

$$EC_T = k/R_T \tag{4.2}$$

where EC_T is the electrical conductivity of the solution in dS m^{-1} at temperature T (°C), k is the cell constant, and R_T is the measured resistance at temperature T. Electrolytic conductivity increases approximately 1.9 percent per degree centigrade increase in T. Customarily, EC is adjusted to a reference temperature of 25°C using Equation (4.3) from Handbook 60 (U.S. Salinity Laboratory Staff, 1954):

$$EC_{25} = f_T \cdot EC_T \tag{4.3}$$

where f_T is a temperature conversion factor that has been approximated by a polynomial form (Rhoades et al., 1999a; Stogryn, 1971; Wraith and Or, 1999) and by Equation (4.4) from Sheets and Hendrickx (1995):

$$f_T = 0.4470 + 1.4034e^{-T/26.815} \tag{4.4}$$

Soil EC is determined for an aqueous extract of a soil sample. Ideally, the EC of an extract of the soil solution (EC_w) is the most desirable, because this is the water content to which plant roots are exposed, but this is usually difficult and time consuming to obtain. The soil sample from which the extract is taken can either be disturbed or undisturbed. For disturbed samples, soil solution can be obtained in the laboratory by displacement, compaction, centrifugation, molecular adsorption, and vacuum- or pressure-extraction methods. Because of the difficulty in extracting soil solution from soil samples at typical field water contents, soil solution extracts are most commonly from higher than normal water contents. The most common extract obtained is that from a saturated soil paste (EC_e), but other commonly used extract ratios include 1:1 ($EC_{1:1}$), 1:2 ($EC_{1:2}$), and 1:5 ($EC_{1:5}$) soil-to-water mixtures. Unfortunately, the partitioning of solutes over the three soil phases (i.e., gas, liquid, and solid) is influenced by the soil-to-water ratio at which the extract is made, which confounds comparisons between ratios and interpretations; consequently, standardization is needed for comparison of EC measurements. For undisturbed soil samples, EC_w can be determined either in the laboratory on a soil solution sample collected with a soil-solution extractor installed in the field or directly in the field using in situ, imbibing-type, porous-matrix, salinity sensors. All of these approaches for measuring soil EC are time and labor intensive; as a result, they are not practical for the characterization of the spatial variability of soil salinity at field extents and larger.

Because of the time, labor, and cost of obtaining soil solution extracts, developments in soil salinity measurement at field and landscape scales over the past 30 years have shifted to EC measurement of the bulk soil, referred to as the apparent soil electrical conductivity (EC_a). The measurement of EC_a is an indirect method for the determination of soil salinity because EC_a measures conductance not only through the soil solution, but also through solid soil particles and via exchangeable cations that exist at the solid–liquid interface of clay minerals. The shift away from extracts to the measurement of EC_a occurred because the time and cost of obtaining soil solution extracts prohibited their practical use at field scales and the high local-scale variability of soil rendered salinity sensors and small-volume soil core samples of limited quantitative value. Historically, the utility of EC_a has been in identifying geological features in geophysical sciences and explorations (McNeill, 1980; Zalasiewicz, et al., 1985) and in agricultural soil salinity surveys for diagnostics, leaching, and salt loading (Corwin et al., 1996; Rhoades and Ingvalson, 1971; Rhoades et al., 1990). During the past decade, there has been an increased interest in using EC_a maps to infer the spatial variability of soil properties important to crop production. In particular, there is an emerging interest in utilizing the spatial variability in EC_a for the purposes of guiding soil sampling (as opposed to systematic grid sampling) and developing management zones to vary agricultural inputs.

The measurement of soil EC_a is primarily through the use of the geophysical techniques of ER, EMI, and TDR. Among the many advanced sensors recently introduced in precision agriculture, EMI and ER EC_a measuring devices provide the simplest and least expensive soil variability measurement. Electrical resistivity introduces an electrical current into the soil through current electrodes at the soil surface, and the difference in current flow potential is measured at potential electrodes that are placed in the vicinity of the current flow. Generally, there are four electrodes inserted in the soil in a straight line at the soil surface, with the two outer electrodes serving as the current electrodes and the two inner electrodes serving as the potential electrodes (Figure 4.1). A resistance meter is used to measure the potential gradient. For a homogeneous soil, the volume of measurement with ER is roughly πa^3, where a is the interelectrode spacing when the electrodes are equally spaced. The most commonly used ER equipment is the Veris Soil EC Mapping System (Veris Technologies, Salina, KS). The Veris 3100 unit has six coulter electrodes mounted on a platform that can be pulled by a pickup truck. It uses a modified Wenner configuration to measure EC_a by inducing current in the soil through two coulter electrodes and measuring the voltage drop across the two pairs of coulters that are spaced to measure EC_a for the top 0.3 m (shallow) and 0.9 m (deep) of soil (Lund et al., 2000). The shallow and deep EC_a readings at each measurement point in the field are useful in examining soil profile changes. Although soil compaction affects EC_a due to the reduced porosity and increased soil particle-to-particle contact, compaction is not easily identified from a Veris EC_a map, as the compacted layer represents only a small percentage of the domain of EC_a measurements.

The Veris unit interfaces with a differential Global Positioning System (GPS) and provides simultaneous and geo-referenced readings of EC_a. The Veris unit is designed to operate in tilled or untilled conditions, where the coulters penetrate the soil 20 to 50 mm (more penetration for drier

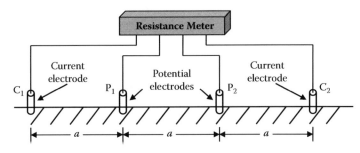

FIGURE 4.1 Electrical resistivity with a Wenner array electrode configuration where the interelectrode spacing is equal between current and potential electrodes: C_1 and C_2 represent the current electrodes, P_1 and P_2 represent the potential electrodes, and a represents the interelectrode spacing. (From Rhoades, J.D., and Halvorson, A.D., Electrical conductivity methods for detecting and delineating saline seeps and measuring salinity in Northern Great Plains soils, ARS W-42, USDA-ARS Western Region, Berkeley, CA, pp. 1–45, 1977. With permission.)

and looser soil surface conditions). It is good practice to map fields when they are not very dry. Soil EC_a mapping with the Veris unit should not be attempted when the soil is frozen, or in the presence of any frost layers. Frozen soil has significantly different conductive properties, and the EC_a data collected will not be valid. The Veris unit is a rugged and reliable system with no known difficulties in mapping fields in the spring prior to tillage and planting operations or in the fall after harvest with heavy standing and flat-lying surface residue conditions (Farahani and Buchleiter, 2004). For ease of maneuvering, fields are normally traversed in the direction of crop rows, but the resulting map is not affected by the direction of travel. On average, travel speeds through the field range between 7 and 16 km h^{-1} with measurements taken every second, corresponding to 2 to 4 m spacing between measurements in the direction of travel, respectively. A parallel swather (such as AgGPS Parallel Swathing Option, Trimble Navigation Ltd., Sunnyvale, CA) mounted inside the vehicle pulling the Veris unit may be used to guide parallel passes through the field at desired (i.e., 12 to 18 m) swath widths.

The direct contact method used by EC_a equipment like Veris has a distinct advantage over the EMI method in that there is no possibility of ambient electrical (for instance from power lines), metallic (operator's belt buckle), or engine noise interferences. Other important advantages of ER-type methods over EMI are that there is no calibration or nulling procedures required prior to mapping, and there is no known report of any observed drift in the measured soil EC_a by ER. Regular "drift runs" that involve traversing the same location in a field are needed for EMI in order to determine the drift resulting from air temperature effects on the instrument throughout the day. Because the electrodes in the ER method directly inject the signal into the soil, changes in air temperature have virtually no effect on the readings. It is noted that EC_a data collected with either method are affected by soil temperature. The most obvious disadvantage of direct ER methods is their intrusiveness as compared to the nonintrusive EMI methods. The invasive ER method requires solid contact between the coulters and soil; consequently, dry conditions or irregular microtopography can prevent contact. Although the distinction between the two differing EC_a measuring methods of ER and EMI is important, side-by-side measurements of soil EC_a by contact electrodes and EMI methods has given highly correlated values (Sudduth et al., 2003) and has provided similar maps (Doolittle et al., 2002).

In the case of EMI, EC is measured remotely using a frequency signal in the range of 0.4 to 40 kHz and primarily measures signal loss to determine EC_a. The EMI measurement is made with the instrument at or above the soil surface. An EMI transmitter coil located at one end of the instrument induces circular eddy-current loops in the soil (Figure 4.2). The magnitude of these loops is directly proportional to the EC_a of the soil in the vicinity of that loop. Each current loop generates a secondary electromagnetic field that is proportional to the current flowing within the loop. A portion of the secondary-induced electromagnetic field from each loop is intercepted by the receiver coil, and the sum of these signals is related to a depth-weighted EC_a.

For TDR, an applied electromagnetic pulse is guided along a transmission line embedded in the soil. The time delay between the reflections of the pulse from the beginning and the end of the

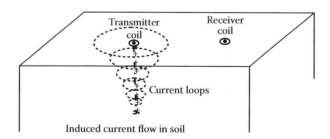

FIGURE 4.2 The principle of operation electromagnetic induction. (From Corwin, D.L., and Lesch, S.M., *Agron. J.*, 95, 455–471, 2003. With permission.)

transmission line is used to determine the velocity of propagation through soil, which is controlled by the relative dielectric permittivity or dielectric constant. By measuring the resistive load impedance across the probe, EC_a can be determined. Although TDR has been demonstrated to compare closely with other accepted methods of EC_a measurement (Heimovaara et al., 1995; Mallants et al., 1996; Reece, 1998; Spaans and Baker, 1993), it has not been widely used for geospatial measurements of EC_a at field and larger spatial extents. Only ER and EMI are commonly used at these spatial extents (Rhoades et al., 1999a, 1999b).

Field measurements of EC_a to determine soil salinity began in the early 1970s with the use of ER (Halvorson and Rhoades, 1976; Rhoades and Halvorson, 1977; Rhoades and Ingvalson, 1971). However, geospatial measurements of EC_a in the field did not occur in earnest until the 1980s, primarily with the use of EMI, which had definite advantages over ER because it was noninvasive. Observational research through the 1980s and early 1990s largely correlated EC_a measurements to soil properties in an effort to sort out what soil properties were measured by EC_a (Table 4.1). From the late 1990s to the present, the complex spatial relationship between EC_a, edaphic properties, and within-field variations in crop yield for site-specific crop management has increasingly become the focus of EC_a research. However, over the past three decades and even today, measurements of EC_a are often misunderstood and misinterpreted. The misconceptions regarding EC_a are the consequence of incomplete knowledge of the basic principles and theory of the EC_a measurement.

4.1.2 Misconceptions Surrounding the Apparent Soil Electrical Conductivity (EC_a) Measurement

When scientists began to take EC_a measurements in the field and correlate them to soil properties, there were preconceived notions about what was being measured. Those scientists in the arid southwestern United States felt that salinity was being measured, and those in the Midwest felt water content and texture were being measured. In reality, both were correct, but each failed to acknowledge that EC_a is a complex physicochemical measurement influenced by any soil property that influences electrical conductance pathways in soil. Additional research produced correlations between EC_a and soil properties that were not directly measured but indirectly related to properties that were measured by EC_a (Table 4.1).

With a few exceptions, the EC_a related observations suggest salinity, soil water content, clay content, exchange cations, temperature, and organic matter content are the dominating soil properties affecting EC_a, but the strength of the reported correlations varies widely with coefficients ranging from below 0.4 to above 0.8. Contrasting findings are evident in literature; for example, Dalgaard et al. (2001) reported a higher correlation between EC_a and clay at higher water contents, and Banton et al. (1997), in a detailed study on an experimental farm near Quebec City, found texture parameters to have a stronger correlation with EC_a for dry than wet soil conditions. Johnson et al. (2001), on the other hand, found no strong correlations between EC_a and a host of soil properties in a 250 ha dryland no-till field in eastern Colorado, concluding that EC_a delineations were useful in identifying overall soil variability but not in producing specific maps of any individual soil property. In nonsaline fields in Missouri, depth to claypans (a sublayer with 50 to 60 percent in clay and varying in depth from 0.1 to 1 m) was found highly correlated to EC_a (Doolittle et al., 1994). As EC_a increased, depth to claypan decreased. A more extensive study on the Missouri claypan, however, produced EC_a maps that exhibited little resemblance to the maps of measured depth to claypans (Sudduth et al., 1995), concluding that the EC_a data were strongly influenced by the crop and farming system, soil water, and crop biomass at the time of EC_a measurements. Additional difficulties with interpreting literature are that most of the identified soil properties that dominate EC_a variability exhibit significant codependency and thus provide overlapping (or redundant), but confusing, information about EC_a. Generally speaking, the degree of EC_a association with a given soil property

TABLE 4.1

Compilation of Literature Measuring ECa with Geophysical Techniques (ER or EMI) Categorized According to the Soil-Related Properties Directly or Indirectly Measured by ECa

Soil Property	References
Directly measured soil properties:salinity (and nutrients, e.g., NO3-)	Bennett and George (1995); Cameron et al. (1981); Cannon et al. (1994); Corwin and Rhoades (1982, 1984, 1990); de Jong et al. (1979); Diaz and Herrero (1992); Drommerhausen et al. (1995); Eigenberg and Nienaber (1998, 1999, 2001); Eigenberg et al. (1998, 2002); Greenhouse and Slaine (1983); Halvorson and Rhoades (1976); Hanson and Kaita (1997); Hendrickx et al. (1992); Herrero et al. (2003); Johnston et al. (1997); Kaffka et al. (2005); Lesch et al. (1992, 1995a, 1995b, 1998, 2005); Mankin and Karthikeyan (2002); Mankin et al. (1997); Nettleton et al. (1994); Paine (2003); Ranjan et al. (1995); Rhoades (1992, 1993); Rhoades and Corwin (1981, 1990); Rhoades and Halvorson (1977); Rhoades et al. (1976, 1989, 1990, 1999a, 1999b); Slavich and Petterson (1990); Sudduth et al. (2005); van der Lelij (1983); Williams and Baker (1982); Williams and Hoey (1987); Wollenhaupt et al. (1986)
Water content	Brevik and Fenton (2002); Farahani et al. (2005); Fitterman and Stewart (1986); Freeland et al. (2001); Hanson and Kaita (1997); Kachanoski et al. (1988, 1990); Kaffka et al. (2005); Kean et al. (1987); Khakural et al. (1998); Morgan et al. (2000); Sheets and Hendrickx (1995); Sudduth et al. (2005); Vaughan et al. (1995); Wilson et al. (2002)
Texture related (e.g., sand, clay, depth to claypans or sand layers)	Anderson-Cook et al. (2002); Banton et al. (1997); Boettinger et al. (1997); Brevik and Fenton (2002); Brus et al. (1992); Doolittle et al. (1994, 2002); Inman et al. (2001); Jaynes et al. (1993); Kitchen et al. (1996); Lesch et al. (2005); Rhoades et al. (1999a); Scanlon et al. (1999); Stroh et al. (1993); Sudduth and Kitchen (1993); Sudduth et al. (2005); Triantafilis and Lesch (2005); Triantafilis et al. (2001); Williams and Hoey (1987);
Bulk density related (e.g., compaction)	Gorucu et al. (2001); Rhoades et al. (1999a)
Indirectly Measured Soil Properties:Organic matter related (including soil, organic carbon, and organic chemical plumes)	Benson et al. (1997); Bowling et al. (1997); Brune and Doolittle (1990); Brune et al. (1999); Farahani et al. (2005); Greenhouse and Slaine (1983, 1986); Jaynes (1996); Nobes et al. (2000); Nyquist and Blair (1991); Sudduth et al. (2005)
Cation exchange capacity	Farahani et al. (2005); McBride et al. (1990); Sudduth et al. (2005); Triantafilis et al. (2002)
Leaching	Corwin et al. (1999); Lesch et al. (2005); Rhoades et al. (1999a); Slavich and Yang (1990)
Groundwater recharge	Cook and Kilty (1992); Cook et al. (1992); Salama et al. (1994)
Herbicide partition coefficients	Jaynes et al. (1995)
Soil map unit boundaries	Fenton and Lauterbach (1999); Stroh et al. (2001)
Corn rootworm distributions	Ellsbury et al. (1999)
Soil drainage classes	Kravchenko et al. (2002)

Source: From Corwin, D.L., and Lesch, S.M., *Comput. Electron. Agric.*, 46, 11–43, 2005. With permission.

is time-of-measurement dependent. That dependency is mainly due to the dynamic nature of some soil properties (such as soil water content, solution concentration, and temperature).

Although there was nothing technically unsound concerning the research relating EC_a to directly or indirectly measured soil properties, an impression was created that EC_a was a vague, ethereal measurement that was less than robust due to the spatially heterogeneous and complex nature of soil. There is some truth to this notion, because electrical conductance through soil is complex due to the complex nature of soil. As a result, without a scientific understanding and explanation for what was being measured, geospatial EC_a readings and their correlations with soil properties are easily misinterpreted and misused to spatially characterize properties that are only loosely related or completely unrelated to actual properties being measured. This fact became evident in the early application of spatial EC_a measurements to site-specific crop management when correlations between EC_a and crop yields were found that were positive, negative, and unrelated. If, in fact, EC_a was measuring just water content and texture, or just salinity, then why were correlations with crop yield so erratic? The answer lies in the principles and theory behind the EC_a measurement.

4.2 BASIC PRINCIPLES AND THEORY OF EC_a

Electrical conduction through soil is due to the presence of free salts in the soil solution and exchangeable ions at the surfaces of solid particles. The soil equivalent resistance model (Sauer et al., 1955) provides the basis for the formulation of a mechanistic soil EC_a model applicable to the entire range of soil solution concentrations (Rhoades et al., 1989; Shainberg et al., 1980). Three pathways of current flow contribute to the EC_a of a soil: (1) a conductance pathway traveling through alternating layers of soil particles and soil solution, (2) a conductance pathway traveling through the continuous soil solution, and (3) a conductance pathway traveling through or along the surface of soil particles in direct and continuous contact (Rhoades et al., 1989, 1999a). These three pathways of current flow are illustrated in Figure 4.3. Conceptually, the first pathway can be thought of as a solid–liquid, series-coupled element, and the second and third pathways represent continuous liquid and solid elements, respectively.

Rhoades et al. (1989) introduced a model for EC_a describing the three separate current-flow pathways acting in parallel:

$$EC_a = \left[\frac{(\theta_{ss} + \theta_{ws})^2 \cdot EC_{ws} \cdot EC_{ss}}{\theta_{ss} \cdot EC_{ws} + \theta_{ws} \cdot EC_{ss}} \right] + (\theta_{sc} \cdot EC_{sc}) + (\theta_{wc} \cdot EC_{wc}) \qquad (4.5)$$

where θ_{ws} and θ_{wc} are the volumetric soil water contents in the series-coupled soil–water pathway (cm^3 cm^{-3}) and in the continuous liquid pathway (cm^3 cm^{-3}), respectively; θ_{ss} and θ_{sc} are the volumetric contents of the surface-conductance (cm^3 cm^{-3}) and indurated solid phases of the soil (cm^3 cm^{-3}), respectively; EC_{ws} and EC_{wc} are the specific electrical conductivities of the series-coupled soil–water pathway (dS m^{-1}) and continuous-liquid pathway (dS m^{-1}); and EC_{ss} and EC_{sc} are the electrical conductivities of the surface-conductance (dS m^{-1}) and indurated solid phases (dS m^{-1}), respectively. Equation (4.5) was reformulated by Rhoades et al. (1989) into Equation (4.6):

$$EC_a = \left[\frac{(\theta_{ss} + \theta_{ws})^2 \cdot EC_w \cdot EC_{ss}}{(\theta_{ss} \cdot EC_{ws}) + (\theta_{ws} \cdot EC_{ss})} \right] + (\theta_w - \theta_{ws}) \cdot EC_w \qquad (4.6)$$

where $\theta_w = \theta_{ws} + \theta_{wc}$ = total volumetric water content (cm^3 cm^{-3}), $\theta_{sc} \cdot EC_{sc}$ was assumed to be negligible, and solution conductivity equilibrium was assumed (i.e., $EC_w = EC_{ws} = EC_{wc}$, where EC_w is average electrical conductivity of the soil water assuming equilibrium). According to Equation (4.6), EC_a is determined by the following five parameters: θ_w, θ_{ws}, θ_{ss}, EC_{ss}, and EC_w. Using the following empirical approximations, Rhoades et al. (1989) showed that these five parameters are related to four measurable soil properties, which include soil salinity (EC_e; dS m^{-1}), saturation percentage (SP; SP is the gravimetric soil water content at saturation), bulk density (ρ_b; Mg m^{-3}), and gravimetric water content (θ_g; kg kg^{-1}):

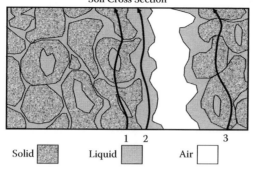

Pathways of Electrical Conductance
Soil Cross Section

FIGURE 4.3 The three conductance pathways for the EC_a measurement. (Modified from Rhoades, J.D., Manteghi, N.A., Shouse, P.J., and Alves, W.J., *Soil Sci. Soc. Am. J.*, 53, 433–439, 1989. With permission.)

$$\theta_w = \theta_g \cdot \rho_b \tag{4.7}$$

$$\theta_{ws} = 0.639\theta_w + 0.011 \tag{4.8}$$

$$\theta_{ss} = \frac{\rho_b}{2.65} \tag{4.9}$$

$$EC_{ss} = 0.019(SP) - 0.434 \tag{4.10}$$

$$EC_w = \left[\frac{EC_e \cdot \rho_b \cdot SP}{100 \cdot \theta_w}\right] = EC_e \cdot \left(\frac{SP}{\theta_g}\right) \tag{4.11}$$

The reliability of Equation (4.6) through Equation (4.11) has been evaluated by Corwin and Lesch (2003). These equations are reliable except under extremely dry soil conditions. However, Lesch and Corwin (2003) developed a means of extending equations for extremely dry soil conditions by dynamically adjusting the assumed water content function. Using the above theory, Farahani et al. (2005) deduced the importance of various soil properties to explain EC_a variability in nonsaline soils of three Colorado fields. Their examination of the above theoretically based EC_a model was useful in highlighting the general complexity of EC_a, its major pathways in the soil, and the concept as a whole, even though crude assumptions were made in approximating the parameters in Equation (4.6) through Equation (4.11).

Equation (4.3) and Equation (4.6) through Equation (4.11) indicate that EC_a is influenced by EC_e, SP, θ_g, ρ_b, and temperature. The SP and ρ_b are both directly influenced by clay content and organic matter (OM). Furthermore, the exchange surfaces on clays and OM provide a solid–liquid phase pathway primarily via exchangeable cations; consequently, clay content and mineralogy, cation exchange capacity (CEC), and OM are recognized as additional factors influencing EC_a measurements. Soil EC_a is expected to increase with increasing clay and OM contents, because EC_{ss} and θ_w (or more correctly θ_{ws}) increase with clay and OM. With EC_{ss} regarded as a function of clay, CEC, and OM (Rhoades et al., 1976; Shainberg et al., 1980), it is not surprising that most field observations of EC_a versus soil properties have identified clay, CEC, and OM as main factors dominating EC_a variability in nonsaline soils. It is noted from Equation (4.6) that conductance through alternating layers of soil particles and soil solution pathway (the first term) is complicated, and its contribution to EC_a is not obvious as the dynamic soil properties of EC_w and θ_w change. As soil water content changes, EC_w is expected to change. As soil water evaporates, EC_w is expected to increase due to increased concentration of free ions in solution. These are counteracting mechanisms contributing to the complexity of EC_a and soil property relationships. In other words, empirical EC_a versus soil property functions are expected to be temporally variable unless EC_w and θ_w remain relatively unchanged. As discussed previously (Equation (4.3), the other important soil variable causing change in EC_a is temperature, with EC_a increasing by approximately 1.9 percent per degree centigrade. This could be significant for shallow depths that may exhibit the greatest temperature variation. Apparent soil electrical conductivity is a complex physicochemical property that must be interpreted with these influencing factors in mind.

Field measurements of EC_a are the product of both static and dynamic factors, which include soil salinity, clay content and mineralogy, water content, bulk density, and temperature. Although the effect of soil static and dynamic factors on spatial variability of EC_a is of significant importance, understanding their influence on the temporal variability of EC_a is equally important. That is particularly true if delineated EC_a zones are to be used to manage agricultural inputs across the field for multiple years. Johnson et al. (2003) described the observed dynamics of the general interaction of these factors. In general, the magnitude and spatial heterogeneity of EC_a in a field are dominated by one or two of these factors, which will vary from one field to the next, making the interpretation of EC_a measurements highly site specific. In instances where dynamic soil properties (e.g., salinity) dominate the EC_a measurement, temporal changes in spatial patterns exhibit more fluidity than systems that are dominated by static factors (e.g., texture). In texture-driven systems, spatial patterns remain consistent because variations in dynamic soil properties (i.e., water content) affect only the magnitude of measured EC_a (Johnson et al., 2003). This was clearly demonstrated by Farahani and Buchleiter (2004), who used multiyear measurements of EC_a from three irrigated and nonsaline sandy fields in eastern Colorado and quantified their degree of temporal change. For each field, soil EC_a values were highly correlated between measurement days (for periods of a few days to 4 years between measurements), but significant deviations from the 1:1 line (indicative of temporal variability) were exhibited. In spite of the temporal variability of the absolute magnitudes of EC_a, delineating spatial patterns of EC_a into low, medium, and high zones across each field was highly stable over time, mainly because they reflect the static soil properties. Johnson et al. (2003) warn that EC_a maps of static-driven systems convey very different information from those of less-stable dynamic-driven systems. For this reason, it is imperative that the soil properties dominating EC_a measurements within a field are established to be able to correctly interpret spatial EC_a survey data.

Numerous EC_a studies have been conducted that revealed the site specificity and complexity of spatial EC_a measurements with respect to the particular property or properties influencing the EC_a measurement at that study site. Table 4.1 is a compilation of various laboratory and field studies and the associated dominant soil property or properties measured.

Many of the misinterpretations and misunderstandings regarding past field EC_a surveys have been due to a disregard for the complex and dynamic interrelationship and influence of various physical and chemical properties on EC_a, which are quantified in Equation (4.6) through Equation (4.11). Because EC_a does not measure an individual soil property such as salinity or water content, but rather is a product of the influence of several properties, geospatial EC_a surveys are best used to direct soil sampling in order to characterize the spatial variability of those properties that correlate with EC_a at a given study site (Corwin, 2005). Characterizing spatial variability with EC_a-directed soil sampling is based on the hypothesis that when EC_a correlates with a soil property or properties, then spatial EC_a information can be used to identify sites that reflect the range and variability of the property or properties. In instances where EC_a correlates with a particular soil property, an EC_a-directed soil sampling approach will establish the spatial distribution of that property with an optimum number of site locations to characterize the variability and keep labor costs minimal (Corwin et al., 2003a; Lesch, 2005). Also, if EC_a is correlated with crop yield, then an EC_a-directed soil sampling approach can be used to identify those soil properties that are causing the variability in crop yield (Corwin et al., 2003b).

4.3   GUIDELINES FOR CONDUCTING A FIELD-SCALE EC$_a$-DIRECTED SOIL SAMPLING SURVEY FOR AGRICULTURE

The basic steps of a field-scale EC$_a$ survey for characterizing spatial variability include (1) EC$_a$ survey design, (2) geo-referenced EC$_a$ data collection, (3) soil sample design based on geo-referenced EC$_a$ data, (4) soil sample collection, (5) physicochemical analysis of pertinent soil properties, (6) stochastic or deterministic calibration of EC$_a$ to soil properties, (7) determination of the soil properties influencing the EC$_a$ measurements at the study site, and (8) GIS development. Details on EC$_a$-directed soil sampling protocols are presented by Corwin and Lesch (2005b, 2005c). Outlined protocols are provided in Table 4.2. Of the eight basic steps, EC$_a$-directed soil sample design, stochastic or deterministic calibration of EC$_a$, and determination of the soil properties influencing the geospatial EC$_a$ measurements are the least understood and yet are crucial for correctly understanding and interpreting spatial EC$_a$ data. Ideally, efforts must be directed toward mapping EC$_a$ when the soil property of interest is expected to have its greatest influence on EC$_a$ values. This maximizes the likelihood of inferring the spatial patterns of the soil property of interest from the EC$_a$ map. For instance, the effect of texture (or clay content) on EC$_a$ is more pronounced at higher water contents (Dalgaard et al., 2001), suggesting EC$_a$ field mapping when the soil is wet rather than dry.

4.3.1   EC$_a$-DIRECTED SOIL SAMPLE DESIGN

An EC$_a$ survey of a field is most often conducted with either mobile ER or EMI equipment that has been coupled to a GPS. Depending on the level of detail desired, from 100 to several thousand spatial measurements of EC$_a$ are taken generally in regularly spaced traverses across the field of interest. The use of mobile EMI equipment has one slight advantage over the use of mobile ER equipment due to the fact that EMI is noninvasive, which is the ability to take measurements on dry and stony soils.

Once a geo-referenced EC$_a$ survey is conducted, the data are used to establish the locations of the soil core sample sites for (1) calibration of EC$_a$ to a correlated soil sample property (e.g., salinity, water content, and clay content) and (2) delineation of the spatial distribution of soil properties correlated to EC$_a$ within the field surveyed. Currently, two different sampling schemes are used to establish the locations where soil cores are to be taken: design-based and model-based sampling schemes. Design-based sampling schemes have historically been the most commonly used and, hence, are more familiar to most research scientists. Design-based methods include simple random sampling, stratified random sampling, multistage sampling, cluster sampling, and network sampling schemes. The use of unsupervised classification by Fraisse et al. (2001) and Johnson et al. (2001) is an example of design-based sampling. Model-based sampling schemes are less common. Specific model-based sampling approaches that have direct application to agricultural and environmental survey work are described by McBratney and Webster (1983), Russo (1984), and Lesch et al. (1995a, 1995b, 2005).

The sampling approach introduced by Lesch et al. (1995a, 1995b, 2005) is specifically designed for use with ground-based soil EC$_a$ data. This sampling approach attempts to optimize the estimation of a regression model (i.e., minimize the mean square prediction error produced by the calibration function), while simultaneously insuring that the independent regression model residual error assumption remains approximately valid. This, in turn, allows an ordinary regression model to be used to predict soil property levels at all remaining (i.e., nonsampled) conductivity survey sites.

There are two main advantages to the response-surface approach. First, a substantial reduction in the number of samples required for effectively estimating a calibration function can be achieved, in comparison to more traditional design-based sampling schemes. Second, this approach lends itself naturally to the analysis of remotely sensed EC$_a$ data. Many types of ground-, airborne-, and satellite-based remotely sensed data are often collected specifically because one expects this data

TABLE 4.2

Outline of Steps for an EC$_a$-Directed Soil Sampling Survey

1. Site description and EC$_a$ survey design
   - (a) Record site metadata
   - (b) Establish site boundaries
   - (c) Select Global Positioning System (GPS) coordinate system
   - (d) Establish EC$_a$ measurement intensity
2. EC$_a$ data collection with mobile GPS-based equipment
   - (a) Geo-reference site boundaries and significant physical geographic features with GPS
   - (b) Measure geo-referenced EC$_a$ data at the predetermined spatial intensity and record associated metadata
3. Soil sample design based on geo-referenced EC$_a$ data
   - (a) Statistically analyze EC$_a$ data using an appropriate statistical sampling design to establish the soil sample site locations
   - (b) Establish site locations, depth of sampling, sample depth increments, and number of cores per site
4. Soil core sampling at specified sites designated by the sample design
   - (a) Obtain measurements of soil temperature through the profile at selected sites
   - (b) At randomly selected locations obtain duplicate soil cores within a 1 m distance of one another to establish local-scale variation of soil properties
   - (c) Record soil core observations (e.g., mottling, horizonation, textural discontinuities, etc.)
5. Laboratory analysis of appropriate soil physical and chemical properties defined by project objectives
6. If needed, stochastic or deterministic calibration of EC$_a$ to EC$_c$ or to other soil properties (e.g., water content and texture)
7. Spatial statistical analysis to determine the soil properties influencing EC$_a$ and crop yield
   - (a) Soil quality assessment:
     - (1) Perform a basic statistical analysis of physical and chemical data by depth increment and by composite depth over the depth of measurement of EC$_a$
     - (2) Determine the correlation between EC$_a$ and physical and chemical soil properties by composite depth over the depth of measurement of EC$_a$
   - (b) Site-specific crop management (if EC$_a$ correlates with crop yield, then)
     - (1) Perform a basic statistical analysis of physical and chemical data by depth increment and by composite depths
     - (2) Determine the correlation between EC$_a$ and physical and chemical soil properties by depth increment and by composite depths
     - (3) Determine the correlation between crop yield and physical and chemical soil properties by depth and by composite depths to determine depth of concern (i.e., depth with consistently highest correlation, whether positive or negative, of soil properties to yield) and the significant soil properties influencing crop yield (or crop quality)
     - (4) Conduct an exploratory graphical analysis to determine the relationship between the significant physical and chemical properties and crop yield (or crop quality)
     - (5) Formulate a spatial linear regression (SLR) model that relates soil properties (independent variables) to crop yield or crop quality (dependent variable)
     - (6) Adjust this model for spatial autocorrelation, if necessary, using Restricted Maximum Likelihood or some other technique
     - (7) Conduct a sensitivity analysis to establish dominant soil property influencing yield or quality
8. Geographic information system (GIS) database development and graphic display of spatial distribution of soil properties

Source: Corwin, D.L., and Lesch, S.M., *Comp. Electron. Agric.,* 46, 103–133, 2005. With permission

to correlate strongly with some parameter of interest (e.g., crop stress, soil type, soil salinity, etc.), but the exact parameter estimates (associated with the calibration model) may still need to be determined via some type of site-specific sampling design. This approach explicitly optimizes this site selection process.

A user-friendly software package (ESAP) developed by Lesch et al. (2000), which uses a response-surface sampling design, has proven to be particularly effective in delineating spatial distributions of soil properties from EC_a survey data (Corwin and Lesch, 2003; Corwin et al., 2003a, 2003b, 2006). The ESAP software package identifies the optimal locations for soil sample sites from the EC_a survey data. These sites are selected based on spatial statistics to reflect the observed spatial variability in EC_a survey measurements. Generally, six to twenty sites are selected depending on the level of variability of the EC_a measurements for a site. The optimal locations of a minimal subset of EC_a survey sites are identified to obtain soil samples.

In a detailed study of the utility of EC_a mapping, Farahani and Buchleiter (2004) used two different soil sampling designs for comparison. Measured EC_a data from three center-pivot fields, two near the town of Wiggins and one near Yuma in eastern Colorado, were used to identify sample locations using ESAP and a combination of cluster analysis and random soil sampling within clusters. For the latter sampling design, three to five sample locations were randomly selected from each of five delineated EC_a zones (according to cluster analysis). For the sandy and nonsaline fields in eastern Colorado, both sampling methods effectively captured the spatial variability. The random sampling from within EC_a clusters was found to be simple and subsequently used in a number of dryland and irrigated fields to characterize EC_a delineated zones for the purposes of site-specific management.

4.3.2 CALIBRATION OF EC_a TO SOIL PROPERTIES

Apparent soil electrical conductivity can be calibrated to any soil property that significantly influences the EC_a measurement, such as salinity, θ_g, clay content, SP, ρ_b, and OM. As indicated in Table 4.1, there are numerous studies that document the relationships between soil electrical conductivity and various soil physical and chemical properties. All of the data analysis and interpretation presented in these papers can be classified into two data modeling categories: deterministic and stochastic.

In general, stochastic models are based on some form of objective sampling methodology used in conjunction with various statistical calibration techniques. The most common types of calibration equations are geostatistical models (generalized universal kriging models and cokriging models) and spatially referenced regression models.

Traditionally, universal kriging models have been viewed as an extension of the ordinary kriging technique and used primarily to account for large-scale (nonstationary) trends in spatial data. However, this modeling technique can be easily generalized to model ancillary survey data (such EMI or ER data) when this data correlates well with some spatially varying soil property of interest (e.g., soil salinity). This generalization is commonly referred to as a "spatial linear model" or "spatial random field model" in the statistical literature. This modeling approach requires the estimation of a regression equation with a spatially correlated error structure. This type of model probably represents the most versatile and accurate statistical calibration approach, provided enough calibration sample sites are collected ($n > 50$) to ensure a good estimate of the correlated error structure.

Regardless of their versatility, spatial linear models are typically used in regional situations. Such an approach is rarely used for field-scale survey work, due to the large number of required calibration soil samples, which makes this approach economically impractical. Instead, most calibration equations of soil properties are spatially referenced regression models. A spatially referenced regression model is just an ordinary regression equation that includes the soil property being calibrated with EC_a and trend surface parameters. The model assumes an independent error structure that can usually be achieved through carefully designed sampling plans, such as the response-surface sampling design. In practice, these are the only models that can be reasonably estimated with a limited number of soil samples ($n < 15$).

Deterministic conductivity data modeling and interpretation have been carried out either from a geophysical or a soil science approach. In the geophysical approach, mathematically sophisticated inversion algorithms are generally employed. These approaches, which rely heavily on geophysical theory, have met with limited success for the interpretation of near-surface EC_a data. Part of the reason for the lack of success is that most geophysical inversion approaches assume that (1) there are multiple conductivity signal readings available for each survey point and (2) that distinct, physical strata differences exist within the near-surface soil horizon. Neither of these conditions is typically satisfied in most EC_a surveys.

A more common technique, which is used in soil science, is to employ some form of deterministic conversion model (i.e., an equation that converts EC_a to a soil property based on knowledge of other soil properties). One model of this type that has been shown to be useful is the model developed by Rhoades et al. (1989, 1990; see Equation (4.6) through Equation (4.11)) and extended by Lesch and Corwin (2003). The model demonstrates that soil EC can be reduced to a nonlinear function of five soil properties: EC_e, SP, θ_g, ρ_b, and soil temperature. In Rhoades et al. (1990), the model was used to estimate field soil salinity levels based on EC_a survey data and measured or inferred information about the remaining soil physical properties. Corwin and Lesch (2003) and Lesch et al. (2000) showed that this model can also be used to assess the degree of influence that each of these soil properties has on the acquired EC_a-survey data.

4.3.3 DETERMINATION OF THE SOIL PROPERTIES INFLUENCING EC_a

In the past, the fact that EC_a is a function of several soil properties (i.e., soil salinity, texture, water content, etc.) has sometimes been overlooked in the application of EC_a measurements to agriculture. For instance, precision agriculture studies relating EC_a to crop yield have met with inconsistent results due to the fact that a combination of factors influence EC_a measurements to varying degrees across units of management, thereby confounding interpretation. In areas of saline soils, salinity dominates the EC_a measurements, and interpretations are often straightforward. However, in areas other than arid zone soils, texture and water content or even OM may be the dominant properties measured by EC_a. To use spatial measurements of EC_a in a soil quality or site-specific crop management context, it is necessary to understand what factors are most significantly influencing the EC_a measurements within the field of study. There are two commonly used approaches for determining the predominant factors influencing EC_a measurement: (1) wavelet analysis and (2) simple statistical correlation.

An explanation of the use of wavelet analysis for determining the soil properties influencing EC_a measurements is provided by Lark et al. (2003). Even though wavelet analysis is a powerful tool for determining the dominant complex interrelated factors influencing EC_a measurement, it requires soil sample data collected on a regular grid or equal-spaced transect. Grid or equal-spaced transect sampling schemes are not as practical for determining spatial distributions of soil salinity or other correlated soil properties from EC_a measurements as the statistical and graphical approach developed by Lesch et al. (1995a, 1995b, 2000).

The most practical means of interpreting and understanding the tremendous volume of spatial data from an EC_a survey is through statistical analysis and graphic display. For a soil quality assessment, a basic statistical analysis of all physical and chemical data by depth increment provides an understanding of the vertical profile distribution. A basic statistical analysis consists of the determination of the mean, minimum, maximum, range, standard deviation, standard error, coefficient of variation, and skewness for each depth increment (e.g., 0 to 0.3, 0.3 to 0.6, 0.6 to 0.9, and 0.9 to 1.2 m) and by composite depth (e.g., 0 to 1.2 m) over the depth of measurement of EC_a. In the case of EC_a measured with ER, the composite depth over the depth of measurement of EC_a is based on the spacing between the electrodes, while in the case of EMI measurements of EC_a, the composite depth over the depth of measurement of EC_a is based on the spacing between the coils and the orientation of the coils (i.e., vertical or horizontal). The calculation of the correlation coefficient between EC_a and mean value of each soil property by depth increment and composite depth over multiple sample sites determines those soil properties that correlate best with EC_a and those soil properties that are spatially represented by the EC_a-directed sampling design. Those properties not correlated with EC_a are not spatially characterized with the EC_a-directed sampling design, indicating

TABLE 4.3

Correlation Coefficients (r) between Shallow (0 to 0.3 m) and Deep (0 to 0.9 m) EC_a (mS m^{-1}) and Their Corresponding (Same Depth) Soil Properties at Three Colorado Fields

Soil Property	Wiggins 1		Wiggins 2		Yuma	
	Shallow EC_a	Deep EC_a	Shallow EC_a	Deep EC_a	Shallow EC_a	Deep EC_a
Sand (%)	−0.96	−0.95	−0.84	−0.90	−0.76	−0.90
Silt (%)	0.82	0.91	0.67	0.80	0.68	0.84
Clay (%)	0.96	0.94	0.89	0.94	0.82	0.93
Bulk density ((b, Mg m^{-3})	−0.13	−0.53	−0.34	−0.52	—	—
Organic matter (g g^{-1}%)	0.92	0.92	0.75	0.79	0.80	0.89
Ca^{+2} (meq L^{-1})	0.85	0.92	0.94	0.88	0.82	0.82
Mg^{+2} (meq L^{-1})	0.91	0.94	0.93	0.87	0.58	0.75
K^{+1} (meq L^{-1})	0.80	0.76	0.73	0.75	0.67	0.86
Na^{+1} (meq L^{-1})	0.65	0.87	0.62	0.79	0.58	0.26
CEC (meq 100 g^{-1})	0.86	0.93	0.94	0.88	0.87	0.87
pH	−0.81	−0.76	−0.48	−0.48	0.50	0.23
Soluble salts ($EC_{1:1}$, mS m^{-1})	0.86	0.95	0.24	0.66	0.86	0.78

a design-based sampling scheme such as stratified random sampling is probably needed to better spatially characterize these soil properties.

An example of a simple statistical approach to infer EC_a versus soil properties relations is given by the correlation coefficients between EC_a and soil properties by depth (the terms "shallow" and "deep" refer to soil depths of 0 to 0.3 m and 0 to 0.9 m, respectively) given in Table 4.3 (Farahani et al., 2005). As given, the EC_a measurements from these sandy soils are very useful in inferring texture variability with correlation coefficient values between EC_a and clay (or sand) well over 0.8.

Crop yield monitoring data in conjunction with EC_a survey data can be used from a site-specific crop management perspective (1) to identify those soil properties influencing yield and (2) to delineate site-specific management units (SSMU). For site-specific crop management, an understanding of the influence of spatial variation in soil properties on within-field crop-yield (or crop quality) variation is crucial. To accomplish this using EC_a, crop yield (or crop quality) must correlate with EC_a within a field. If crop yield (or crop quality) and EC_a are correlated, then basic statistical analyses by depth increment (e.g., 0 to 0.3, 0.3 to 0.6, 0.6 to 0.9, and 0.9 to 1.2 m) and by composite depths (e.g., 0 to 0.3, 0 to 0.6, 0 to 0.9, and 0 to 1.2 m) are performed. As before, the correlation between EC_a and mean values of each physical and chemical soil property for each depth increment and each composite depth establishes those soil properties that are spatially characterized with the EC_a-directed sampling design. The correlations between crop yield (or crop quality) and soil properties will also establish the depth of concern (i.e., the root zone of the crop), which will be the composite depth that consistently has the highest correlation of each soil property (i.e., each soil property determined to be significant to influencing yield) with crop yield (or crop quality). Exploratory graphical analyses (i.e., scatter plots of crop yield or crop quality and each soil property) are then conducted for the depth of concern to determine the linear or curvilinear relationship between the significant physical and chemical properties and crop yield (or crop quality). A spatial linear regression is formulated that relates the significant soil properties as the independent variables to crop yield (or crop quality) as the dependent variable. The functional form of the model is developed from the exploratory graphic analysis. If necessary, the model is adjusted for spatial autocorrelation using restricted maximum likelihood or some other technique. This entire spatial statistical analysis process is clearly demonstrated by Corwin et al. (2003b) and Corwin and Lesch (2005c).

To use spatial measurements of EC_a in a site-specific crop management context, it is not only necessary to understand those soil-related factors that influence within-field variation in crop yield (or crop quality), but also to pinpoint the dominant soil-related factors influencing within-field crop variation. Corwin et al. (2003b) used sensitivity analysis simulations to arrive at the dominant edaphic and anthropogenic factors influencing within-field cotton yield variations. Sensitivity analysis involves increasing a single independent variable (i.e., edaphic factors) and observing the resultant effect on the dependent variable (i.e., crop yield or crop quality). This is done for each independent variable. The relative effect of each independent variable on the dependent variable determines the independent variable that most significantly influences the dependent variable.

4.3.4   CASE STUDIES

Table 4.4 is a compilation of correlation data for six field study sites where EC_a surveys using EMI were performed for the purpose of salinity appraisal. Table 4.4 shows the variation in the influence of various soil properties upon EC_a for different field locations. In all cases, the surveys were performed as outlined in Table 4.2. An intensive EC_a survey was performed, followed by soil core sample site selection where from six to twenty sites were selected for sampling. An analysis was performed on the soil cores for various physical and chemical properties (e.g., saturation percentage, salinity, and water content). The correlations in Table 4.2 were determined using the EC_a survey and soil sample data. The correlations in Table 4.2 indicate the soil properties influencing the EC_a reading most. Following is a discussion about the six EC_a surveys presented in Table 4.2.

4.3.4.1   Coachella Valley Wheat (*Triticum aestvum* L.) Field

This is an example of a survey where the salinity represented by ln (EC_e), the soil texture reflected by the saturation percentage (SP), and the volumetric water content (θ_w) correlate with the EMI data, which are represented as $\ln(EMI_{ave})$, where EMI_{ave} is the geometric mean of the vertical and horizontal EMI readings (i.e., $\sqrt{EMI_h \cdot EMI_v}$).

4.3.4.2   Coachella Valley Sorghum Field

This field is an example of where only salinity correlates well with the EMI data. Note from Table 4.5 that neither the soil texture nor volumetric water content correlate with salinity with $r = -0.10$ and $r = 0.28$, respectively. Because of this lack of correlation with salinity and because the texture and water content exhibit minimal sample variation (i.e., sample range for SP is 51.0 to 61.1 percent; sample range for θ_w is 0.33 to 0.41 $cm^3\ cm^{-3}$), they correlate poorly with the EMI data with $r = -0.20$ and $r = 0.25$, respectively (Table 4.2).

4.3.4.3   Broadview Water District (Quarter Sections 16-2 and 16-3)

These combined quarter sections display large variability in soil texture, as indicated by SPs ranging from 33.2 to 85 percent, and in water content (θ_w ranges from 0.21 to 0.39 $cm^3\ cm^{-3}$) with relatively minimal salinity variation (80 percent of the samples fell below the mean value of 3.65 dS m^{-1}). Salinity, SP, and water content correlate with the EMI data. Saturation percentage and water content are highly correlated with $r = 0.84$ and $r = 0.86$, respectively (Table 4.2).

4.3.4.4   Fresno Cotton (*Gossypium hirsutum* L.) Field

The Fresno cotton field is of particular interest because of the high positive correlation of EMI data with salinity ($r = 0.87$) and moderate positive correlation with SP ($r = 0.71$) but a negative correlation with water content ($r = -0.65$). The negative correlation between SP and water content (see Table 4.5; $r = -0.78$) suggests an unexpected inverse relationship between texture and water content.

TABLE 4.4

Means and Ranges of Soil Factors (EC$_e$, SP, and θ$_w$) and the Correlations of ln(EMI$_{ave}$) with ln(EC$_e$), SP, and θ$_w$ for Six Field-Scale Surveys

| Field | Soil Factors (EC$_e$, SP, and θ$_w$) | | |
	Mean	Range	Correlation[a]
Coachella Valley Wheat Field			
ln(EMI$_{ave}$) and ln(EC$_e$)	2.33	0.85–6.64	0.87
ln(EMI$_{ave}$) and SP	40.4	36.5–45.8	0.78
ln(EMI$_{ave}$) and θ$_w$	0.24	0.18–0.32	0.77
Coachella Valley Sorghum Field			
ln(EMI$_{ave}$) and ln(EC$_e$)	10.1	5.37–16.8	0.88
ln(EMI$_{ave}$) and SP	57.0	51.0–61.1	−0.20
ln(EMI$_{ave}$) and θ$_w$	0.38	0.33–0.41	0.25
Broadview Water District (Quarter Sections 16-2 and 16-3)			
ln(EMI$_{ave}$) and ln(EC$_e$)	3.65	1.61–8.19	0.62
ln(EMI$_{ave}$) and SP	50.3	33.2–85.0	0.84
ln(EMI$_{ave}$) and θ$_w$	0.31	0.21–0.39	0.86
Fresno Cotton Field			
ln(EMI$_{ave}$) and ln(EC$_e$)	5.42	1.28–9.57	0.87
ln(EMI$_{ave}$) and SP	79.4	59.3–103.0	0.71
ln(EMI$_{ave}$) and θ$_w$	0.28	0.22–0.33	−0.65
Coachella Valley—Kohl Ranch Field			
ln(EMI$_{ave}$) and ln(EC$_e$)	11.8	3.73–22.9	0.94
ln(EMI$_{ave}$) and SP	63.3	59.7–66.7	−0.33
ln(EMI$_{ave}$) and θ$_w$	0.33	0.30–0.36	0.76
ln(EMI$_{ave}$) and ln(SAR)	23.2	5.55–40.2	0.89
ln(EMI$_{ave}$) and ln(B)	1.44	0.52–2.57	0.91
Broadview Water District (Quarter Section 10-2)			
ln(EMI$_{ave}$) and ln(EC$_e$)	2.66	0.90–5.69	0.80
ln(EMI$_{ave}$) and SP	55.4	40.6–67.4	0.49
ln(EMI$_{ave}$) and θ$_w$	0.38	0.29–0.42	0.59
ln(EMI$_{ave}$) and ρ$_b$	1.35	1.26–1.44	−0.35
ln(EMI$_{ave}$) and sand	25.5	8.35–49.9	−0.38
ln(EMI$_{ave}$) and silt	39.3	26.3–51.6	0.42
ln(EMI$_{ave}$) and clay	35.3	23.8–44.5	0.29

Note: EM$_{ave}$ is the geometric mean (i.e., sqrt[EMI$_h$ · EMI$_v$]) of the EC$_a$ taken in the horizontal (EMI$_h$) and vertical (EMI$_v$) coil configurations using EMI, EC$_e$ is the electrical conductivity of the saturation paste (dS m^{-1}), SP is the saturation percentage, θ$_w$ is the volumetric water content (cm^3 cm^{-3}), SAR is the sodium adsorption ratio, B is boron (mg kg^{-1}), ρ$_b$ is the bulk density (g cm^{-3}), sand is the percent sand, silt is the percent silt, and clay is the percent clay.

[a] The correlation column corresponds to the correlation between the measured EC$_a$ and the specified soil property.

Source: Modified from Corwin, D.L., and Lesch, S.M., *Agron. J.*, 95, 455–471, 2003. With permission.

TABLE 4.5
Correlation Matrix of Soil Properties for the Six Field-Scale EC$_a$ Surveys

Field	ln(EC$_e$)	SP	θ_w
Coachella Valley Wheat Field			
ln(EC$_e$)	1.00	0.69	0.66
SP		1.00	0.91
θ_w			1.00
Coachella Valley Sorghum Field			
ln(EC$_e$)	1.00	−0.10	0.28
SP		1.00	−0.47
θ_w			1.00
Broadview Water District (Quarter Sections 16-2 and 16-3)			
ln(EC$_e$)	1.00	0.23	0.33
SP		1.00	0.82
θ_w			1.00
Fresno Cotton Field			
ln(EC$_e$)	1.00	0.38	−0.37
SP		1.00	−0.78
θ_w			1.00
Coachella Valley—Kohl Ranch Field			
ln(EC$_e$)	1.00	−0.39	0.72
SP		1.00	−0.04
θ_w			1.00
Broadview Water District (Quarter Section 10-2)			
ln(EC$_e$)	1.00	0.08	0.14
SP		1.00	0.91
θ_w		1.00	

Source: Modified from Corwin, D.L., and Lesch, S.M., Agron. J., 95, 455–471, 2003. With permission.

Note: ECe is the electrical conductivity of the saturation extract (dS m^{-1}), SP is the saturation percentage. θ_w is the volumetric water content (cm^3 cm^{-3}).

4.3.4.5   Coachella Valley—Kohl Ranch Field

This field displays a range of correlations between EMI and soil properties (Table 4.2). Salinity correlates very well, water content fairly well, and soil texture exhibits weak negative correlation indicating that the dominant soil properties influencing the EMI reading are salinity and water content. In addition, two secondary properties, SAR and boron, were measured. The fact that these correlated quite well with the EMI data suggests the close association of these properties with salinity in this particular field because the EMI reading does not directly measure SAR or boron but is rather an artifact of solute flow.

4.3.4.6   Broadview Water District (Quarter Section 10-2)

The dominant soil property influencing the EMI reading is salinity, with a correlation between ln(EMI$_{ave}$) and ln(EC$_e$) of 0.80. No strong correlation was found between EMI data and a variety of soil properties, including SP, water content, bulk density, and separates of sand, silt, and clay.

From Table 4.4 and Table 4.5, what is known about the interrelationship of soil properties influencing the EC_a measurement for agricultural soils in the arid southwest? First, it is clear that the inner-correlation structure of the various primary soil properties (EC_e, SP, θ_w) determines how well each property ultimately correlates with the EC_a signal data. However, the variability of each soil property also influences the final correlation estimates, because increased variability in any given soil property directly translates into increased variation in the EC_a data. Obviously, one may encounter many diverse types of inner-correlation structures and different degrees of specific soil property variation as shown in Table 4.4 and Table 4.5. Thus, the ultimate correlation between the EC_a signal data and any specific soil property may be quite different from field to field. For example, this effect is clearly evident in the $\ln(EMI_{ave})$ and SP correlation estimates shown in Table 4.4, where the observed estimates range from -0.33 to 0.84. Second, with respect to EC_e data, the best scenario for the prediction of salinity from EC_a signal data occurs when the EC_e, SP, and θ_w cross-correlation estimates are all positive and high (i.e., near 1), and the SP and θ_w variation is minimal.

4.4 CLOSING REMARKS

The need for a means of measuring within-field variation in soil salinity within the root zone in a quick, reliable, and cost-effective manner resulted in the development of GPS-based mobile ER and EMI techniques to measure and map EC_a. However, the measurement of EC_a is complicated by the influence of several soil properties aside from soil salinity, including soil texture, temperature, and water content. This has enabled geospatial measurements of EC_a to become a tool for directing soil sampling to characterize spatial variability of soil properties correlated with EC_a within a given field. When maps of EC_a are properly understood, they can be used to (1) provide a graphic inventory of the scope of the soil salinity problem, (2) provide useful spatial information concerning soil texture and water content, (3) identify potential areas in need of improved irrigation and drainage management, (4) identify areas in need of reclamation, (5) provide a means of monitoring management-induced spatiotemporal changes in soil properties that potentially influence crop production, or (6) provide a means to identify edaphic factors influencing within-field crop variation.

REFERENCES

Anderson-Cook, C.M., Alley, M.M., Roygard, J.K.F., Khosla, R., Noble, R.B., and Doolittle, J.A., Differentiating soil types using electromagnetic conductivity and crop yield maps, *Soil Sci. Soc. Am. J.*, 66, 1562–1570, 2002.

Banton, O., Seguin, M.K., and Cimon, M.A., Mapping field-scale physical properties of soil with electrical resistivity, *Soil Sci. Soc. Am. J.*, 61, 1010–1017, 1997.

Bennett, D.L., and George, R.J., Using the EM38 to measure the effect of soil salinity on *Eucalyptus globulus* in south-western Australia, *Agric. Water Manage.*, 27, 69–86, 1995.

Benson, A.K., Payne, K.L., and Stubben, M.A., Mapping groundwater contamination using DC resistivity and VLF geophysical methods—A case study, *Geophysics,* 62, 80–86, 1997.

Boettinger, J.L., Doolittle, J.A., West, N.E., Bork, E.W., and Schupp, E.W., Nondestructive assessment of rangeland soil depth to petrocalcic horizon using electromagnetic induction, *Arid Soil Res. Rehabil.*, 11, 372–390, 1997.

Bohn, H.L., McNeal, B.L., and O'Connor, G.A., *Soil Chemistry*, John Wiley & Sons, New York, 1979.

Bowling, S.D., Schulte, D.D., and Woldt, W.E., A geophysical and geostatistical methodology for evaluating potential subsurface contamination from feedlot runoff retention ponds, ASAE Paper No. 972087, 1997 ASAE Winter Meetings, December 1997, Chicago, IL, ASAE, St. Joseph, MI, 1997.

Brevik, E.C., and Fenton, T.E., The relative influence of soil water, clay, temperature, and carbonate minerals on soil electrical conductivity readings taken with an EM-38 along a Mollisol catena in central Iowa, *Soil Survey Horizons*, 43, 9–13, 2002.

Briggs, L.J., Electrical instruments for determining the moisture, temperature, and solute salt content of soils, *U.S. Dept. Agric. Bulletin 15*, 1899.

Brune, D.E., and Doolittle, J., Locating lagoon seepage with radar and electromagnetic survey, *Environ. Geol. Water Sci.,* 16, 195 and 207, 1990.

Brune, D.E., Drapcho, C.M., Radcliff, D.E., Harter, T., and Zhang, R., Electromagnetic survey to rapidly assess water quality in agricultural watersheds, ASAE Paper No. 992176, ASAE, St. Joseph, MI, 1999.

Brus, D.J., Knotters, M., van Dooremolen, W.A., van Kernebeek, P., and van Seeters, R.J.M., The use of electromagnetic measurements of apparent soil electrical conductivity to predict the boulder clay depth, *Geoderma*, 55, 79–93, 1992.

Cameron, D.R., de Jong, E., Read, D.W.L., and Oosterveld, M., Mapping salinity using resistivity and electromagnetic inductive techniques, *Can. J. Soil Sci.*, 61, 67–78, 1981.

Cannon, M.E., McKenzie, R.C., and Lachapelle, G., Soil-salinity mapping with electromagnetic induction and satellite-based navigation methods, *Can. J. Soil Sci.*, 74, 335–343, 1994.

Cook, P.G., and Kilty, S., A helicopter-borne electromagnetic survey to delineate groundwater recharge rates, *Water Resour. Res.*, 28, 2953–2961, 1992.

Cook, P.G., Walker, G.R., Buselli, G., Potts, I., and Dodds, A.R., The application of electromagnetic techniques to groundwater recharge investigations, *J. Hydrol.*, 130, 201–229, 1992.

Corwin, D.L., Geospatial measurements of apparent soil electrical conductivity for characterizing soil spatial variability, in *Soil-Water-Solute Characterization: An Integrated Approach*, Álvarez-Benedí, J., and Muñoz-Carpena, R., Eds., CRC Press, Boca Raton, FL, pp. 639–672, 2005.

Corwin, D.L., and Lesch, S.M., Application of soil electrical conductivity to precision agriculture: Theory, principles, and guidelines, *Agron. J.*, 95, 455–471, 2003.

Corwin, D.L., and Lesch, S.M., Apparent soil electrical conductivity measurements in agriculture, *Comput. Electron. Agric.*, 46, 11–43, 2005a.

Corwin, D.L., and Lesch, S.M., Characterizing soil spatial variability with apparent soil electrical conductivity: I. Survey protocols, *Comp. Electron. Agric.*, 46, 103–133, 2005b.

Corwin, D.L., and Lesch, S.M., Characterizing soil spatial variability with apparent soil electrical conductivity: II. Case study. *Comp. Electron. Agric.*, 46, 135–152, 2005c.

Corwin, D.L., and Rhoades, J.D., An improved technique for determining soil electrical conductivity-depth relations from above-ground electromagnetic measurements, *Soil Sci. Soc. Am. J.,* 46, 517–520, 1982.

Corwin, D.L., and Rhoades, J.D., Measurement of inverted electrical conductivity profiles using electromagnetic induction, *Soil Sci. Soc. Am. J.*, 48, 288–291, 1984.

Corwin, D.L., and Rhoades, J.D., Establishing soil electrical conductivity-depth relations from electromagnetic induction measurements, *Commun. Soil Sci. Plant Anal.*, 21, 861–901, 1990.

Corwin, D.L., Carrillo, M.L.K., Vaughan, P.J., Rhoades, J.D., and Cone, D.G., Evaluation of GIS-linked model of salt loading to groundwater. *J. Environ. Qual.,* 28, 471–480, 1999.

Corwin, D.L., Kaffka, S.R., Hopmans, J.W., Mori, Y., Lesch, S.M., and Oster, J.D., Assessment and field-scale mapping of soil quality properties of a saline-sodic soil, *Geoderma*, 114, 231–259, 2003a.

Corwin, D.L., Lesch, S.M., Oster, J.D., and Kaffka, S.R., Monitoring management-induced spatio-temporal changes in soil quality through soil sampling directed by apparent electrical conductivity, *Geoderma*, 131, 369–387, 2006.

Corwin, D.L., Lesch, S.M., Shouse, P.J., Soppe, R., and Ayars, J.E., Delineating site-specific management units using geospatial EC$_a$ measurements, in *Handbook of Agricultural Geophysics*, Allred, B., Daniels, J., and Ehsani, R., Eds., CRC Press, Boca Raton, FL, 2008.

Corwin, D.L., Lesch, S.M., Shouse, P.J., Soppe, R., and Ayars, J.E., Identifying soil properties that influence cotton yield using soil sampling directed by apparent soil electrical conductivity, *Agron. J.*, 95, 352–364, 2003b.

Corwin, D.L., Rhoades, J.D., and Vaughan, P.J., GIS applications to the basin-scale assessment of soil salinity and salt loading to groundwater, in *Applications of GIS to the Modeling of Non-Point Source Pollutants in the Vadose Zone*, Corwin, D.L., and Loague, K., Eds., SSSA Spec. Publ. 48, SSSA, Madison, WI, pp. 295–313, 1996.

Dalgaard, M., Have, H., and Nehmdahl, H., Soil clay mapping by measurement of electromagnetic conductivity, in *Third European Conference on Precision Agriculture,* Vol. 1, Agro Montpellier, Montpellier, France, pp. 349–354, 2001.

de Jong, E., Ballantyne, A.K., Cameron, D.R., and Read, D.W., Measurement of apparent electrical conductivity of soils by an electromagnetic induction probe to aid salinity surveys, *Soil Sci. Soc. Am. J.,* 43, 810–812, 1979.

Diaz, L., and Herrero, J., Salinity estimates in irrigated soils using electromagnetic induction, *Soil Sci.*, 154, 151–157, 1992.

Doolittle, J.A., Indorante, S.J., Potter, D.K., Hefner, S.G., and McCauley, W.M., Comparing three geophysical tools for locating sand blows in alluvial soils of southeast Missouri, *J. Soil Water Conserv.*, 57, 175–182, 2002.

Doolittle, J.A., Sudduth, K.A., Kitchen, N.R., and Indorante, S.J., Estimating depths to claypans using electromagnetic induction methods, *J. Soil Water Conserv.*, 49, 572–575, 1994.

Drommerhausen, D.J., Radcliffe, D.E., Brune, D.E., and Gunter, H.D., Electromagnetic conductivity surveys of dairies for groundwater nitrate, *J. Environ. Qual.*, 24, 1083–1091, 1995.

Eigenberg, R.A., Doran, J.W., Nienaber, J.A., Ferguson, R.B., and Woodbury, B.L., Electrical conductivity monitoring of soil condition and available N with animal manure and a cover crop, *Agric. Ecosyst. Environ.*, 88, 183–193, 2002.

Eigenberg, R.A., Korthals, R.L., and Neinaber, J.A., Geophysical electromagnetic survey methods applied to agricultural waste sites, *J. Environ. Qual.*, 27, 215–219, 1998.

Eigenberg, R.A., and Nienaber, J.A., Electromagnetic survey of cornfield with repeated manure applications, *J. Environ. Qual.*, 27, 1511–1515, 1998.

Eigenberg, R.A., and Nienaber, J.A., Soil conductivity map differences for monitoring temporal changes in an agronomic field, ASAE Paper No. 992176, ASAE, St. Joseph, MI, 1999.

Eigenberg, R.A., and Nienaber, J.A., Identification of nutrient distribution at abandoned livestock manure handling site using electromagnetic induction, ASAE Paper No. 012193, 2001 ASAE Annual International Meeting, 30 July–1 August 2001, Sacramento, CA, ASAE St. Joseph, MI, 2001.

Ellsbury, M.M., Woodson, W.D., Malo, D.D., Clay, D.E., Carlson, C.G., and Clay, S.A., Spatial variability in corn rootworm distribution in relation to spatially variable soil factors and crop condition, in *Proc. 4th International Conference on Precision Agriculture*, Robert, P.C., Rust, R.H., and Larson, W.E., Eds., St. Paul, MN, 19–22 July 1998, ASA-CSSA-SSSA, Madison, WI, pp. 523–533, 1999.

Farahani, H.J., and Buchleiter, G.W., Temporal stability of soil electrical conductivity in irrigated sandy fields in Colorado, *Trans. ASAE,* 47, 79–90, 2004.

Farahani, H.J., Buchleiter, G.W., and Brodahl, M.K., Characterization of soil electrical conductivity variability in irrigated sandy and non-saline fields in Colorado, *Trans. ASAE*, 48, 155–168, 2005.

Fenton, T.E., and Lauterbach, M.A., Soil map unit composition and scale of mapping related to interpretations for precision soil and crop management in Iowa, in *Proc. 4th International Conference on Precision Agriculture*, Robert, P.C., Rust, R.H., and Larson, W.E., Eds., St. Paul, MN, 19–22 July 1998, ASA-CSSA-SSSA, Madison, WI, pp. 239–251, 1999.

Fitterman, D.V., and Stewart, M.T., Transient electromagnetic sounding for groundwater, *Geophysics,* 51, 995–1005, 1986.

Fraisse, C.W., Sudduth, K.A., and Kitchen, N.R., Delineation of site-specific management zones by unsupervised classification of topographic attributes and soil electrical conductivity, *Trans. ASAE*, 44, 155–166, 2001.

Freeland, R.S., Branson, J.L., Ammons, J.T., and Leonard, L.L., Surveying perched water on anthropogenic soils using non-intrusive imagery, *Trans. ASAE*, 44, 1955–1963, 2001.

Gardner, F.D., The electrical method of moisture determination in soils: Results and modifications in 1897, U.S. Dept. Agric. Bulletin 12, 1898.

Gorucu, S., Khalilian, A., Han, Y.J., Dodd, R.B., Wolak, F.J., and Keskin, M., Variable depth tillage based on geo-referenced soil compaction data in coastal plain region of South Carolina, ASAE Paper No. 011016, 2001 ASAE Annual International Meeting, 30 July–1 Aug. 2001, Sacramento, CA, ASAE St. Joseph, MI, 2001.

Greenhouse, J.P., and Slaine, D.D., The use of reconnaissance electromagnetic methods to map contaminant migration, *Ground Water Monit. Rev.*, 3, 47–59, 1983.

Greenhouse, J.P., and Slaine, D.D., Geophysical modelling and mapping of contaminated groundwater around three waste disposal sites in southern Ontario, *Can. Geotech. J.*, 23, 372–384, 1986.

Halvorson, A.D., and Rhoades, J.D., Field mapping soil conductivity to delineate dryland seeps with four-electrode techniques, *Soil Sci. Soc. Am. J.*, 44, 571–575, 1976.

Hanson, B.R., and Kaita, K., Response of electromagnetic conductivity meter to soil salinity and soil-water content, *J. Irrig. Drain. Eng.*, 123, 141–143, 1997.

Heimovaara, T.J., Focke, A.G., Bouten, W., and Verstraten, J.M., Assessing temporal variations in soil water composition with time domain reflectometry, *Soil Sci. Soc. Am. J.*, 59, 689–698, 1995.

Hendrickx, J.M.H., Baerends, B., Raza, Z.I., Sadig, M., and Chaudhry, M.A., Soil salinity assessment by electromagnetic induction of irrigated land, *Soil Sci. Soc. Am. J.*, 56, 1933–1941, 1992.

Herrero, J., Ba, A.A., and Aragues, R., Soil salinity and its distribution determined by soil sampling and electromagnetic techniques, *Soil Use Manage.*, 19, 119–126, 2003.

Inman, D.J., Freeland, R.S., Yoder, R.E., Ammons, J.T., and Leonard, L.L., Evaluating GPR and EMI for morphological studies of loessial soils, *Soil Sci.*, 166, 622–630, 2001.

Jaynes, D.B., Mapping the areal distribution of soil parameters with geophysical techniques, in *Applications of GIS to the Modeling of Non-Point Source Pollutants in the Vadose Zone*, Corwin, D.L., and Loague, K., Eds., SSSA Special Publication No. 48, SSSA, Madison, WI, pp.205–216, 1996.

Jaynes, D.B., Colvin, T.S., and Ambuel, J., Soil type and crop yield: Determinations from ground conductivity surveys, ASAE Paper No. 933552, 1993 ASAE Winter Meetings, 14–17 December 1993, Chicago, IL, ASAE, St. Joseph, MI, 1993.

Jaynes, D.B., Colvin, T.S., and Kaspar, T.C., Identifying potential soybean management zones from multi-year yield data, *Comput. Electron. Agric.*, 46, 309–327, 2005.

Jaynes, D.B., Kaspar, T.C., Colvin, T.S., and James, D.E., Cluster analysis of spatiotemporal corn yield patterns in an Iowa field, *Agron. J.*, 95, 574–586, 2003.

Jaynes, D.B., Novak, J.M., Moorman, T.B., and Cambardella, C.A., Estimating herbicide partition coefficients from electromagnetic induction measurements, *J. Environ. Qual.*, 24, 36–41, 1995.

Johnson, C.K., Doran, J.W., Duke, H.R., Weinhold, B.J., Eskridge, K.M., and Shanahan, J.F., Field-scale electrical conductivity mapping for delineating soil condition, *Soil Sci. Soc. Am. J.*, 65, 1829–1837, 2001.

Johnson, C.K., Doran, J.W., Eghball, B., Eigenberg, R.A., Wienhold, B.J., and Woodbury, B.L, Status of soil electrical conductivity studies by central state researchers, ASAE Paper No. 032339, 2003 ASAE Annual International Meeting, 27–30 July 2003, Las Vegas, NV, ASAE, St. Joseph, MI, 2003.

Johnston, M.A., Savage, M.J., Moolman, J.H., and du Pleiss, H.M., Evaluation of calibration methods for interpreting soil salinity from electromagnetic induction measurements, *Soil Sci. Soc. Am. J.*, 61, 1627–1633, 1997.

Kachanoski, R.G., de Jong, E., and Van-Wesenbeeck, I.J., Field scale patterns of soil water storage from non-contacting measurements of bulk electrical conductivity, *Can. J. Soil Sci.*, 70, 537–541, 1990.

Kachanoski, R.G., Gregorich, E.G., and Van-Wesenbeeck, I.J., Estimating spatial variations of soil water content using noncontacting electromagnetic inductive methods, *Can. J. Soil Sci.*, 68, 715–722, 1988.

Kaffka, S.R., Lesch, S.M., Bali, K.M., and Corwin, D.L., Relationship of electromagnetic induction measurements, soil properties, and sugar beet yield in salt-affected fields for site-specific management, *Comput. Electron. Agric.*, 46, 329–350, 2005.

Kean, W.F., Jennings Walker, M., and Layson, H.R., Monitoring moisture migration in the vadose zone with resistivity, *Ground Water*, 25, 562–571, 1987.

Khakural, B.R., Robert, P.C., and Hugins, D.R., Use of non-contacting electromagnetic inductive method for estimating soil moisture across a landscape, *Commun. Soil Sci. Plant Anal.*, 29, 2055–2065, 1998.

Kitchen, N.R., Sudduth, K.A., and Drummond, S.T., Mapping of sand deposition from 1993 Midwest floods with electromagnetic induction measurements, *J. Soil Water Conserv.*, 51, 336–340, 1996.

Kitchen, N.R., Sudduth, K.A., Myers, D.B., Drummond, S.T., and Hong, S.Y., Delineating productivity zones on claypan soil fields using apparent soil electrical conductivity, *Comput. Electron. Agric.*, 46, 285–308, 2005.

Kravchenko, A.N., Bollero, G.A., Omonode, R.A., and Bullock, D.G., Quantitative mapping of soil drainage classes using topographical data and soil electrical conductivity, *Soil Sci. Soc. Am. J.*, 66, 235–243, 2002.

Lark, R.M., Kaffka, S.R., and Corwin, D.L., Multiresolution analysis of data of electrical conductivity of soil using wavelets, *J. Hydrol.*, 272, 276–290, 2003.

Lesch, S.M., Sensor-directed spatial response surface sampling designs for characterizing spatial variation in soil properties, *Comp. Electron. Agric.*, 46, 153–179, 2005.

Lesch, S.M., and Corwin, D.L., Using the dual-pathway parallel conductance model to determine how different soil properties influence conductivity survey data, *Agron. J.*, 95, 365–379, 2003.

Lesch, S.M., Corwin, D.L., and Robinson, D.A., Apparent soil electrical conductivity mapping as an agricultural management tool in arid zone soils, *Comput. Electron. Agric.*, 46, 351–378, 2005.

Lesch, S.M., Herrero, J., and Rhoades, J.D., Monitoring for temporal changes in soil salinity using electromagnetic induction techniques, *Soil Sci. Soc. Am. J.*, 62, 232–242, 1998.

Lesch, S.M., Rhoades, J.D., Lund, L.J., and Corwin, D.L., Mapping soil salinity using calibrated electromagnetic measurements, *Soil Sci. Soc. Am. J.*, 56, 540–548, 1992.

Lesch, S.M., Rhoades, J.D., and Corwin, D.L., ESAP-95 Version 2.10R: User manual and tutorial guide, *Research Rpt. 146*, USDA-ARS George E. Brown, Jr. Salinity Laboratory, Riverside, CA, 2000.

Lesch, S.M., Strauss, D.J., and Rhoades, J.D., Spatial prediction of soil salinity using electromagnetic induction techniques: 1. Statistical prediction models: A comparison of multiple linear regression and cokriging, *Water Resour. Res.*, 31, 373–386, 1995a.

Lesch, S.M., Strauss, D.J., and Rhoades, J.D., Spatial prediction of soil salinity using electromagnetic induction techniques: 2. An efficient spatial sampling algorithm suitable for multiple linear regression model identification and estimation, *Water Resour. Res.*, 31, 387–398, 1995b.

Lund, E.D., Christy, C.D., and Drummond, P.E., Using yield and soil electrical conductivity (ECa) maps to derive crop production performance information, in *Proceedings of the 5th International Conference on Precision Agriculture* (CD-ROM), Roberts, P.C., Rust, R.H., and Larson, W.E., Eds., Minneapolis, MN 16–19 July 2000, ASA-CSSA-SSSA, Madison, WI, 2000.

Mallants, D., Vanclooster, M., Toride, N., Vanderborght, J., van Genuchten, M.Th., and Feyen, J., Comparison of three methods to calibrate TDR for monitoring solute movement in undisturbed soil, *Soil Sci. Soc. Am. J.*, 60, 747–754, 1996.

Mankin, K.R., Ewing, K.L., Schrock, M.D., and Kluitenberg, G.J., Field measurement and mapping of soil salinity in saline seeps, ASAE Paper No. 973145, 1997 ASAE Winter Meetings, December 1997, Chicago, IL, ASAE, St. Joseph, MI, 1997.

Mankin, K.R., and Karthikeyan, R., Field assessment of saline seep remediation using electromagnetic induction, *Trans. ASAE*, 45, 99–107, 2002.

Marion, G.M., and Babcock, K.L., Predicting specific conductance and salt concentration in dilute aqueous solutions, *Soil Sci.*, 122, 181–187, 1976.

McBratney, A.B., and Webster, R., Optimal interpolation and isarithmic mapping of soil properties V. Co-regionalization and multiple sampling strategy, *J. Soil Sci.*, 34, 137–162, 1983.

McBride, R.A., Gordon, A.M., and Shrive, S.C., Estimating forest soil quality from terrain measurements of apparent electrical conductivity, *Soil Sci. Soc. Am. J.*, 54, 290–293, 1990.

McNeill, J.D., Electrical conductivity of soils and rocks, Technical Note TN-5, Geonics Ltd., Mississauga, Ontario, Canada, 1980.

Morgan, C.L.S., Norman, J.M., Wolkowski, R.P., Lowery, B., Morgan, G.D., and Schuler, R., Two approaches to mapping plant available water: EM-38 measurements and inverse yield modeling, in *Proceedings of the 5th International Conference on Precision Agriculture* (CD-ROM), Roberts, P.C., Rust, R.H., and Larson, W.E., Eds., Minneapolis, MN 16–19 July 2000, ASA-CSSA-SSSA, Madison, WI, p. 14, 2000.

Nettleton, W.D., Bushue, L., Doolittle, J.A., Wndres, T.J., Indorante, S.J., Sodium affected soil identification in south-central Illinois by electromagnetic induction, *Soil Sci. Soc. Am. J.*, 58, 1190–1193, 1994.

Nobes, D.C., Armstrong, M.J., and Close, M.E., Delineation of a landfill leachate plume and flow channels in coastal sands near Christchurch, New Zealand, using a shallow electromagnetic survey method, *Hydrogeol. J.*, 8, 328–336, 2000.

Nyquist, J.E., and Blair, M.S., Geophysical tracking and data logging system: Description and case history, *Geophysics* 56, 1114–1121, 1991.

Paine, J.G., Determining salinization extent, identifying salinity sources, and estimating chloride mass using surface, borehole, an airborne electromagnetic induction methods, *Water Resour. Res.*, 39, 1059, 2003.

Ranjan, R.S., Karthigesu, T., and Bulley, N.R., Evaluation of an electromagnetic method for detecting lateral seepage around manure storage lagoons, ASAE Paper No. 952440, ASAE, St. Joseph, MI, 1995.

Reece, C.F., Simple method for determining cable length resistance in time domain reflectometry systems, *Soil Sci. Soc. Am. J.*, 62, 314–317, 1998.

Rhoades, J.D., Instrumental field methods of salinity appraisal, in *Advances in Measurement of Soil Physical Properties: Bring Theory into Practice*, Topp, G.C., Reynolds, W.D., and Green, R.E., Eds., SSSA Special Publication No. 30. Soil Science Society of America, Madison, WI, pp. 231–248, 1992.

Rhoades, J.D., Electrical conductivity methods for measuring and mapping soil salinity, in *Advances in Agronomy*, Sparks, D.L., Ed., vol. 49, Academic Press, San Diego, CA, pp. 201–251, 1993.

Rhoades, J.D., and Corwin, D.L., Determining soil electrical conductivity–depth relations using an inductive electromagnetic soil conductivity meter, *Soil Sci. Soc. Am. J.*, 45, 255–260, 1981.

Rhoades, J.D., and Corwin, D.L., Soil electrical conductivity: Effects of soil properties and application to soil salinity appraisal, *Commun. Soil Sci. Plant Anal.*, 21, 837–860, 1990.

Rhoades, J.D., and Halvorson, A.D., Electrical conductivity methods for detecting and delineating saline seeps and measuring salinity in Northern Great Plains soils, ARS W-42, USDA-ARS Western Region, Berkeley, CA, pp. 1–45, 1977.

Rhoades, J.D., and Ingvalson, R.D., Determining salinity in field soils with soil resistance measurements, *Soil Sci. Soc. Amer. Proc.*, 35, 54–60, 1971.

Rhoades, J.D., Chanduvi, F., and Lesch, S., Soil salinity assessment: Methods and interpretation of electrical conductivity measurements, *FAO Irrigation and Drainage Paper #57*, Food and Agriculture Organization of the United Nations, Rome, Italy, pp. 1–150, 1999a.

Rhoades, J.D., Corwin, D.L., and Lesch, S.M., Geospatial measurements of soil electrical conductivity to assess soil salinity and diffuse salt loading from irrigation, in *Assessment of Non-Point Source Pollution in the Vadose Zone*, Corwin, D.L., Loague, K., and Ellsworth, T.R., Eds., Geophysical Monograph 108, American Geophysical Union, Washington, D.C., pp. 197–215, 1999b.

Rhoades, J.D., Manteghi, N.A., Shouse, P.J., and Alves, W.J., Soil electrical conductivity and soil salinity: New formulations and calibrations, *Soil Sci. Soc. Am. J.*, 53, 433–439, 1989.

Rhoades, J.D., Raats, P.A.C., and Prather, R.J., Effects of liquid-phase electrical conductivity, water content and surface conductivity on bulk soil electrical conductivity, *Soil Sci. Soc. Am. J.*, 40, 651–655, 1976.

Rhoades, J.D., Shouse, P.J., Alves, W.J., Manteghi, N.M., and Lesch, S.M., Determining soil salinity from soil electrical conductivity using different models and estimates, *Soil Sci. Soc. Am. J.*, 54, 46–54, 1990.

Russo, D., Design of an optimal sampling network for estimating the variogram, *Soil Sci. Soc. Am. J.*, 48, 708–716, 1984.

Salama, R.B., Bartle, G., Farrington, P., and Wilson, V., Basin geomorphological controls on the mechanism of recharge and discharge and its effect on salt storage and mobilization: Comparative study using geophysical surveys, *J. Hydrol.*, 155 (1 / 2), 1–26, 1994.

Sauer, Jr., M.C., Southwick, P.F., Spiegler, K.S., and Wyllie, M.R.J., Electrical conductance of porous plugs: Ion exchange resin-solution systems, *Ind. Eng. Chem*, 47, 2187–2193, 1955.

Scanlon, B.R., Paine, J.G., and Goldsmith, R.S., Evaluation of electromagnetic induction as a reconnaissance technique to characterize unsaturated flow in an arid setting, *Ground Water*, 37, 296–304, 1999.

Shainberg, J., Rhoades, J.D., and Prather, R.J., Effect of exchangeable sodium percentage, cation exchange capacity, and soil solution concentration on soil electrical conductivity, *Soil Sci. Soc. Am. J.*, 44, 469–473, 1980.

Sheets, K.R., and Hendrickx, J.M.H., Non-invasive soil water content measurement using electromagnetic induction, *Water Resour. Res.*, 31, 2401–2409, 1995.

Slavich, P.G., and Petterson, G.H., Estimating average rootzone salinity from electromagnetic induction (EM-38) measurements, *Aust. J. Soil Res.*, 28, 453–463, 1990.

Slavich, P.G., and Yang, J., Estimation of field-scale leaching rates from chloride mass balance and electromagnetic induction measurements, *Irrig. Sci.*, 11, 7–14, 1990.

Spaans, E.J.A., and Baker, J.M., 1993. Simple baluns in parallel probes for time domain reflectometry, *Soil Sci. Soc. Am. J.*, 57, 668–673, 1993.

Stogryn, A., Equations for calculating the dielectric constant of saline water, *IEEE Trans. Microwave Theory Technol. MIT* 19, 733–736, 1971.

Stroh, J.C., Archer, S.R., Doolittle, J.A., and Wilding, L.P., Detection of edaphic discontinuities with ground-penetrating radar and electromagnetic induction, *Landscape Ecology*, 16, 377–390, 2001.

Stroh, J.C., Archer, S.R., Wilding, L.P., and Doolittle, J.A., Assessing the influence of subsoil heterogeneity on vegetation in the Rio Grande Plains of south Texas using electromagnetic induction and geographical information system, College Station, Texas, *The Station*, March 1993, 39–42, 1993.

Sudduth, K.A., and Kitchen, N.R., Electromagnetic induction sensing of claypan depth, ASAE Paper No. 931531, 1993 ASAE Winter Meetings, 12–17 Dec. 1993, Chicago, IL. ASAE, St, Joseph, MI, 1993.

Sudduth, K.A., Kitchen, N.R., Bollero, G.A., Bullock, D.G., and Wiebold, W.J., Comparison of electromagnetic induction and direct sensing of soil electrical conductivity, *Agron. J.*, 95, 472–482, 2003.

Sudduth, K.A., Kitchen, N.R., Hughes, D.F., and Drummond, S.T., Electromagnetic induction sensing as an indicator of productivity on claypan soils, in *Proc. Second International Conference on Site-Specific Management of Agricultural Systems*, Robert, P.C., Rust, R.H., and Larson, W.E., Eds., Minneapolis, MN, 27–30 Mar. 1994, ASA-CSSA-SSSA, Madison, WI, pp. 671–681, 1995.

Sudduth, K.A., Kitchen, N.R., Wiebold, W.J., Batchelor, W.D., Bollero, G.A., Bullock, D.G., Clay, D.E., Palm, H.L., Pierce, F.J., Schuler, R.T., and Thelen, K.D., Relating apparent electrical conductivity to soil properties across the north-central USA, *Comput. Electron. Agric.*, 46, 263–283, 2005.

Triantafilis, J., and Lesch, S.M., Mapping clay content variation using electromagnetic induction techniques, *Comput. Electron. Agric.*, 46, 203–237, 2005.

Triantafilis, J., Ahmed, M.F., and Odeh, I.O.A., Application of a mobile electromagnetic sensing system (MESS) to assess cause and management of soil salinization in an irrigated cotton-growing field, *Soil Use Manage.*, 18, 330–339, 2002.

Triantafilis, J., Huckel, A.I., and Odeh, I.O.A., Comparison of statistical prediction methods for estimating field-scale clay content using different combinations of ancillary variables, *Soil Sci.*, 166, 415–427, 2001.

U.S. Salinity Laboratory Staff, Diagnosis and improvement of saline and alkali soils, *USDA Handbook 60*, U.S. Government Printing Office, Washington, DC, pp. 1–160, 1954.

van der Lelij, A., Use of an electromagnetic induction instrument (type EM38) for mapping of soil salinity, Internal Report Research Branch, Water Resources Commission, NSW, Australia, 1983.

Vaughan, P.J., Lesch, S.M., Corwin, D.L., and Cone, D.G., Water content on soil salinity prediction: A geostatistical study using cokriging, *Soil Sci. Soc. Am. J.*, 59, 1146–1156, 1995.

Wenner, F., A method of measuring earth resistivity, in scientific papers of the U.S. Bureau of Standards, No. 258, pp. 469–78, 1915.

Whitney, M., Gardner, F.D., and Briggs, L.J., An electrical method of determining the moisture content of arable soils, *U.S. Dept. Agric., Division of Soils, Bulletin 6*, 5–26, 1897.

Williams, B.G., and Baker, G.C., An electromagnetic induction technique for reconnaissance surveys of soil salinity hazards, *Aust. J. Soil Res.*, 20, 107–118, 1982.

Williams, B.G., and Hoey, D., The use of electromagnetic induction to detect the spatial variability of the salt and clay contents of soils, *Aust. J. Soil Res.*, 25, 21–27, 1987.

Wilson, R.C., Freeland, R.S., Wilkerson, J.B., and Yoder, R.E., Imaging the lateral migration of subsurface moisture using electromagnetic induction, ASAE Paper No. 023070, 2002 ASAE Annual International Meeting, 28–31 July 2002, Chicago, IL, ASAE, St. Joseph, MI, 2002.

Wollenhaupt, N.C., Richardson, J.L., Foss, J.E., and Doll, E.C., A rapid method for estimating weighted soil salinity from apparent soil electrical conductivity measured with an aboveground electromagnetic induction meter, *Can. J. Soil Sci.* 66, 315–321, 1986.

Wraith, J.M., and Or, D., Temperature effects on soil bulk dielectric permittivity measured by time domain reflectometry: Experimental evidence and hypothesis development, *Water Resour. Res.*, 35, 361–369, 1999.

Zalasiewicz, J.A., Mathers, S.J., and Cornwell, J.D., The application of ground conductivity measurements to geological mapping, *Quaternary J.f Eng. Geol. London*, 18, 139–148, 1985.

5 Resistivity Methods

Barry J. Allred, Douglas Groom, M. Reza Ehsani,
and Jeffrey J. Daniels

CONTENTS

5.1   INTRODUCTION

Resistivity methods were among the first geophysical techniques developed. The basic concept originated with Conrad Schlumberger, who conducted the initial resistivity field tests in Normandy, France, during 1912 (Sharma, 1997). As an historical note, Conrad Schlumberger and his brother, Marcel Schlumberger, later founded the global oilfield services company, Schlumberger Limited, which has been a leader in the development of borehole geophysical systems. Resistivity methods were originally employed in the petroleum and mining industries, and afterward found use in archeological, hydrological, environmental, and geotechnical investigations. Development of continuous resistivity measurement techniques in the late 1980s and early 1990s has transformed the basic resistivity method into an effective and efficient tool to assess soil conditions in large agricultural fields.

The resistivity method, employed in its earliest and most conventional form, uses an external power source to supply electrical current between two "current" electrodes inserted at the ground surface. The propagation of current in the subsurface is three-dimensional, and so, too, is the associated electric field. Information on the electric field is obtained by measuring the voltage between a second pair of "potential" electrodes also inserted at the ground surface. The two current and two potential electrodes together compose a four-electrode array. The magnitude of the current applied and the measured voltage are then used in conjunction with data on electrode spacing and arrangement to determine a subsurface resistivity value. The resistivity measurement obtained is a bulk value representing a continuous volume of the subsurface beneath the four-electrode array.

5.2 OHM'S LAW AND RESISTIVITY (OR ELECTRICAL CONDUCTIVITY)

Consider a cylinder, as shown in Figure 5.1, composed of uniform material, having a length, L, and at each of its ends, a cross-sectional area, A. An electric current, I, defined as the flow rate of electric charge, is applied at one end of the cylinder and exits the other. The cylinder, to a greater or lesser extent, opposes this through-flow of electric current, thereby causing a drop in electric potential, ΔV, which occurs along the column's length from the end where I enters to the end where I leaves. Electric potential can be described as the potential energy for a unit charge resulting from its position within an electric field. As indicated by Equation (5.1), ΔV is proportional to I, and the proportionality constant is the resistance, R, which is a characteristic of the cylinder's overall ability to oppose current flow:

$$\Delta V = -RI \tag{5.1}$$

The minus sign in Equation (5.1) simply indicates that current flow is in a direction opposite to that of increasing electric potential. Equation (5.1) is referred to as Ohm's law, and the resistance, R, of the cylinder can itself be expressed:

$$R = \frac{\rho L}{A} \tag{5.2}$$

where again, L and A, respectively, are the length and cross-sectional area of the cylinder, and ρ is the resistivity.

Resistivity is a property only of the material composing the cylinder (Figure 5.1) and represents the capability of that particular material to oppose the flow of electric current. The ρ values for soil and rock materials are typically reported in units of ohm-meters (Ωm). Electrical conductivity, σ, is the reciprocal of ρ (= $1/\rho$), and is a property indicative of a material's ability to convey electrical current, not oppose it. Instead of ρ, the value of σ, in units of millisiemens/meter (mS/m), is most commonly reported in agriculture. For reference, a ρ value of 1 Ωm corresponds to a σ value of 1000 mS/m. Furthermore,

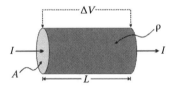

FIGURE 5.1 Flow of electric current, I, through a cylinder composed of uniform material with resistivity , ρ, which produces a difference in an electric potential, ΔV, from one end of the cylinder to the other.

it should be pointed out that agricultural literature often employs EC as the symbol for electrical conductivity rather than, σ, which is the symbol for electrical conductivity routinely found in near-surface geophysics texts. The resistivity methods utilized for agricultural purposes are focused largely on the determination of σ (EC) values for soil materials.

5.3 FACTORS INFLUENCING RESISTIVITY (OR ELECTRICAL CONDUCTIVITY) IN SOIL MATERIALS

This section provides a brief overview of the factors influencing soil resistivity (or electrical conductivity). Additional discussion regarding this subject is given in Chapter 2 and Chapter 4. The ability of a soil material to transfer electric current, as indicated by the resistivity (or electrical conductivity) of the soil, is determined by the components that make up the soil. Soil typically consists of solid, gas, and liquid phases (Figure 5.2). The solid phase of the soil includes both mineral and organic matter and, excluding the larger fragments (generally rock materials), can be divided by

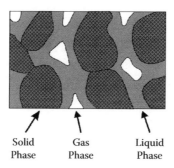

Solid   Gas   Liquid
Phase   Phase   Phase

FIGURE 5.2 Magnified thin section of a soil. Solid-phase quartz particles are good electric current insula-tors, and clay mineral and organic matter particles transmit current if surfaces are wetted. The gas phase is essentially made up of air, which does not transmit electric current. The soil solution liquid phase contains dissolved ions and therefore has the capability to deliver electric current.

particle size into sand (2.0 to 0.05 mm), silt (0.05 to 0.002 mm), and clay (less than 0.002 mm) frac-tions. Quartz, considered an excellent electric insulator, usually dominates a soil's sand and silt size fractions. The clay size fraction is made up primarily of clay minerals and organic matter. Given sufficiently wet conditions, clay minerals and organic matter contribute significantly to electric cur-rent flow in soil, and more information on this topic will be provided later in the section. The soil gas phase is mostly air, which is a good insulator, and like quartz, will oppose the flow of electric current. The soil liquid phase is an electrolytic aqueous solution, referred to as the "soil solution." An electrolyte is a chemical substance that will dissociate into ions within a solution. There are usu-ally a variety of dissolved anions and cations in the soil solution, and some of the most common are SO_4^{2-}, Cl^-, HCO_3^-, NO_3^-, PO_4^{3-}, Ca^{2+}, Mg^{2+}, K^+, Na^+, and NH_4^+.

Unlike a copper wire, where the electric current charge carriers are electrons, dissolved ions within the soil solution serve as the electric current charge carriers in a soil. Therefore, the electric current in soil is largely electrolytic, meaning that the flow of electric current is governed sub-stantially by the movement of dissolved ions in the soil solution. Insight regarding the resistivity behavior of sandy and silty soils is provided by Archie's law, which is based on electrolytic current flow through a porous media. Archie's law is empirical, and it quantifies the relationship between the overall porous media resistivity, the resistivity of the electrolytic aqueous solution present in the porous media, and the amount of electrolytic solution present per unit volume of porous media. The law was developed particularly for clay-free rocks and sediment, and therefore, can also be used for soils that contain essentially no clay minerals or organic matter. The form of Archie's law most applicable to both saturated and unsaturated conditions for a sandy or silty soil is given as follows:

$$\rho = z_1 \frac{\rho_W}{\varphi^{z_2} S^{z_3}} \tag{5.3}$$

where ρ is the overall soil resistivity; ρ_W is the resistivity of the soil solution; φ is the porosity (vol-ume fraction of soil not part of solid phase); S is the saturation (volume fraction of the porosity filled with soil solution); z_1 is a constant with a value between 0.5 and 2.5, often initially approximated by 1.0; z_2 is a constant ranging from 1.3 to 2.5, but closer to 1.3 for loose sediments; and z_3 is a constant with a value usually close to 2 when S is greater than 0.3 (Keller and Frischknecht, 1966; Parasnis, 1986; Reynolds, 1997). Inspection of Equation (5.3) shows that ρ for a sandy or silty soil depends on ρ_W and the volumetric water content, θ, which can be defined as the volume of soil solution per unit volume of soil, equaling the product of φ and S ($\theta = \varphi S$).

It warrants pointing out that the values of ρ_W and θ are often, but not always, interrelated. One clear example is where higher-temperature conditions lead to evapotranspiration-associated losses of soil solution (decreased θ), resulting in the soil solution dissolved ions becoming more concentrated within the soil solution that remains. This increased ion concentration decreases ρ_W because there are now a greater number of electric current charge carriers per unit amount of soil solution. Interestingly, the likely overall result for this soil drying scenario (decreased θ and ρ_W) is a rise in ρ, the reason for which will be discussed in more depth later in the section. Although there is an understanding that ρ_W and θ are often interrelated, in order to focus the discussion within the next three paragraphs strictly on ρ_W, especially regarding potential effects on overall soil resistivity, ρ, it is assumed, for the sake of argument, θ remains constant.

The ρ_W in Equation (5.3) is itself a function not only of the dissolved ion concentrations as previously implied, but also the dissolved ion mobilities. Dissolved ion mobility is, in turn, governed by the ion type and temperature conditions. Equation (5.3) indicates that ρ_W and ρ are directly related when there is no change in θ. Consider a case where there is an increase in the total dissolved ion concentration (decreased ρ_W), while θ remains constant. This case obviously results in a greater number of charge carriers per unit volume of soil, thereby enhancing the capacity of the sandy or silty soil to transmit electrolytic current, in turn leading to a decrease in ρ. One possible agricultural situation involving a scenario in which ρ_W and ρ are reduced, while the beginning versus ending θ conditions are the same, is a fertigation event where a soil initially at field capacity and having a dilute soil solution is intensely flushed with a more concentrated solution containing nutrients (NO_3^-, PO_4^{3-}, and K^+), followed by the soil then being allowed to drain back to field capacity. (Field capacity for a particular soil corresponds to the remaining θ value that occurs when all the possible gravity drainable water has been leached from the initially saturated to near-saturated soil.)

The capacity of a sandy or silty soil to deliver electrolytic current depends not only on the total amount of ions present but also on the mobility of the various ions within the soil solution. Accordingly, the distribution of the types of dissolved ions within the soil solution potentially has a strong impact on ρ_W, and likewise ρ, given constant θ. The reason for this impact is that the different dissolved ions typically found in the soil solution have different mobilities. As an example, at 25°C, given a constant electric potential difference driving electrolytic current in an aqueous solution, the mobility of SO_4^{2-} is approximately twice the mobility of HCO_3^- (Keller and Frischknecht, 1966). Therefore, assuming all other aspects are equal, a soil solution with SO_4^{2-} as the dominant anion will have a lower ρ_W value than soil solution with HCO_3^- as the dominant anion.

Temperature conditions additionally influence ion mobility and, therefore, ρ_W. Ion mobility in an aqueous solution is inversely dependent on the viscosity of the solution, which is inversely dependent on solution temperature. For instance, as temperature decreases, soil solution viscosity increases, ion mobility is reduced, and ρ_W rises (soil solution electrical conductivity is lowered). An equation commonly used to adjust ρ_W due to changes in temperature is

$$\rho_{W-T} = \frac{\rho_{W-25°C}}{1 + z_4\left(T - 25°\right)} \tag{5.4}$$

where ρ_{W-T} is the soil solution resistivity at a temperature of T (°C), $\rho_{W-25°C}$ is the soil solution resistivity at a reference temperature of 25°C, and z_4 is a temperature coefficient with a value of 0.022/°C (McNeill, 1980). Equation (5.4) reveals that a decrease in temperature from 35°C to 15°C would increase ρ_W by 56 percent. Based on the direct relationship between ρ_W and ρ in Equation (5.3), this temperature change would also cause the overall sandy or silty soil resistivity to increase by 56 percent (given negligible evapotranspiration losses and θ remains constant). The important implication of Equation (5.4) is that seasonal and even daily temperature fluctuations can have a significant effect on near-surface soil resistivity values. (Daily temperature fluctuations in near-surface soil of 10°C are not uncommon.) As a special case, extremely cold conditions, in which the soil solution freezes, will cause the overall sandy or silty soil resistivity to become extremely high (conductivity falls to zero), due to the difficulty in transmitting electric current through ice.

Equation (5.3) indicates an inverse relationship between ρ for a sandy or silty soil and the volumetric water (soil solution) content, θ (= φS). Although this inverse relationship is usually found for most soils, instances do occur when this relationship does not hold, implying that other factors need to be taken into account. There are two fairly intuitive effects that θ can have on ρ. First, given set concentrations for the dissolved ions within the soil solution (ρ_w is now assumed to be constant), a change in θ causes a change in the total number of charge carriers (dissolved ions) per unit volume of soil, in turn altering the soil's capacity to distribute electric current as defined by ρ (or σ, EC). Waterlogged soil conditions produced by an irrigation event conducted to flush salts from the soil profile, followed next by complete gravity drainage, is one agricultural scenario corresponding to ρ_w remaining constant, while θ changes from near-saturation to field capacity. Under this scenario, ρ gets larger as the soil drains.

Regarding the second effect of θ on ρ, as θ varies significantly, so too does the continuity of soil solution through which electrolytic current is transferred. These changes in the soil solution continuity in turn alter the soil's ability to transfer electric current as quantified by ρ. To further emphasize this second θ effect, a substantial decrease in soil wetness (lower θ) reduces the thickness of the soil solution films covering solid soil particles, thereby lengthening the electric current flow travel paths within the soil solution (increased tortuosity), and as indicted by Equation (5.2), diminishing the overall capability of the soil to convey current (higher ρ). Wet soils near saturation or at field capacity will exhibit good soil solution continuity for distributing current, but for extremely dry soils, soil solution films may not completely cover all the solid surfaces, thus severely reducing the connectivity of travel paths for electrolytic current flow. In an example discussed previously, soil drying reduces the amount of soil solution present (lower θ), narrowing and lengthening the soil solution conduits for electrolytic current flow, which in turn almost always results in an increased ρ value, even though there is a drop in ρ_w due to evapotranspiration concentrating dissolved ions in the remaining soil solution.

Up to this point, the discussion regarding soil resistivity has focused on sandy or silty materials containing no clay minerals and organic matter. Most soils, however, have significant amounts of clay minerals (layered aluminosilicates) and organic matter. Soil organic matter can be divided into two components. The first component includes only a few percent of the total soil organic matter and includes living organisms (worms, bacteria, fungi, etc.) and nondecomposed substances such as dead plant roots. The second component, representing the large majority of soil organic matter, is the stable, decomposed residue called "humus" (Bohn et al., 1985). It is the stable, decomposed residue portion of the soil organic matter which affects the overall soil resistivity, ρ. Clay minerals and organic matter often coat the larger sand- and silt-sized particles, resulting in the sides of soil pores being dominantly composed of these clay mineral and organic matter materials, consequently giving these materials a much larger impact on soil processes than would be inferred based on their weight percent alone. The general manner in which the previously discussed factors influence ρ_w and θ, and likewise ρ, is the same regardless of whether a soil does or does not contain clay minerals and organic matter. The major difference regarding soils containing clay minerals and organic matter is that given sufficiently wet conditions, there is an additional effect caused by the presence of clay minerals and organic matter, which tends to enhance electric current flow.

Substitution of ions different from the ones normally composing the clay mineral crystal lattice and the functional groups that are a part of the humus chemical structure typically yield large numbers of discrete negatively charged sites on the surfaces of clay mineral and organic matter particles. Cations are electrostatically attracted and become attached at the negatively charged surface sites. The quantity of the negatively charged surface sites per unit amount of dry soil is referred to as the cation exchange capacity (CEC). Given a sufficient amount of soil solution covering the clay mineral and organic matter surfaces, these attached cations are exchangeable and often displaced by other cations temporarily present in the soil solution. The displaced cations move freely within the soil solution adjacent to clay mineral and organic matter surfaces and can then displace cations at different negatively charged exchange sites. The displacement and movement of cations from exchange site to exchange site produces a mechanism by which electric current is essentially transmitted electrolytically through a soil along clay mineral and organic matter surfaces. As previously indicated, assuming conditions are wet enough, the presence of clay minerals and organic matter in significant amounts will increase the capability of the soil to deliver electric current, corresponding to a ρ value lower than what would have been obtained if the clay minerals and organic matter were absent.

The mechanism of electrolytic current transfer facilitated by clay minerals and organic matter depends on the number of negatively charged surface sites. Accordingly, if this mechanism of electric current delivery dominates, which is certainly not always the case, then a strong inverse correlation between the cation exchange capacity (CEC) and ρ is to be expected. However, different clay minerals exhibit different CEC values. Furthermore, pH governs part of the clay mineral CEC and all of the soil organic matter CEC. The resulting implication regarding CEC dependence on clay mineral type and pH is that an inverse relationship between ρ and the total amount of clay minerals and organic matter present, although anticipated, probably will not be as strong as the inverse relationship between CEC and ρ, again assuming this clay mineral and organic matter facilitated, surface-based, electrolytic current transport mechanism dominates.

Water molecules orient themselves within the electric field adjacent to clay mineral and organic matter surfaces. This phenomenon prevents freezing of the soil solution portion near the clay mineral and organic matter surfaces. Furthermore, as temperatures drop below 0°C, dissolved ions tend to migrate out of the soil solution portion that begins to freeze, causing the unfrozen soil solution next to clay mineral and organic matter surfaces to become much more concentrated with dissolved ions. The main consequence for soils containing significant amounts of clay minerals and organic matter is that these soils maintain their ability to deliver electric current, even when temperatures drop below 0°C for prolonged periods. Therefore, when temperatures stay at 0°C or below for an extended time, soils containing significant amounts of clay minerals and organic matter will typically have lower ρ values than soils having little clay mineral and organic matter material.

The following equation for soil resistivity, valid for temperatures above 0°C, was developed by Rhoades et al. (1976), and takes into account clay mineral and organic matter facilitated, surface-based, electric current:

$$\frac{1}{\rho} = \frac{\theta\left(z_5\theta + z_6\right)}{\rho_W} + \frac{1}{\rho_S} \tag{5.5}$$

The ρ_S value in Equation (5.5) is the clay mineral and organic matter resistivity component, and z_5 and z_6 are empirically derived constants. The validity of Equation (5.5) is supported, in part, because of similarities to Equation (5.6) (Schlumberger Wireline and Testing, 1991), which is commonly used in the petroleum industry to determine the resistivity value of a shaly (clayey) sand.

$$\frac{1}{\rho} = \frac{\phi^{z_2} S^{z_3}\left(1 - V_{SH}\right)}{z_1\rho_W} + \frac{W\left(V_{SH}\right)}{R_{SH}} \tag{5.6}$$

For Equation (5.6), the variables ρ, ρ_W, ϕ, and S along with constants, z_1, z_2, and z_3 were previously defined, and V_{SH} is a shale (clay) volumetric characteristic, the value of W is a function of S, and R_{SH} is a quantity related to the shale resistivity. The first terms to the right of the equal signs in Equation (5.5) and Equation (5.6) certainly exhibit a degree of equivalence to one another, because each contains θ in the numerator and ρ_W in the denominator. (With respect to Equation (5.6), remember, $\theta = \phi S$.) The quantities that make up the second term on the right side of Equation (5.6) indicate that this term is a resistivity component related to the presence of clay minerals, much the same as the second term on the right side of Equation (5.5).

Equation (5.5) and Equation (5.6) imply a soil model having two separate but continuous electric current pathways, effectively corresponding to two parallel resistors within an electric circuit, where the first of the pathways involves electrolytic current delivery strictly through the bulk soil solution, and the second pathway involves electrolytic current delivery along clay mineral and organic matter surfaces (Knight and Endres, 2005). Here, the term "bulk soil solution" is used to indicate all of the soil solution excluding the thin layers of soil solution adjacent to clay mineral and organic matter surfaces, which contain the "cloud" of cations attracted to the negatively charged surface sites. Rhoades et al. (1989) developed a soil resistivity equation based on another somewhat different model having two separate electric current pathways, where the first pathway is constrained to the larger soil pores and involves electrolytic current transfer strictly through the "mobile" portion of the bulk soil solution. The second pathway is constrained to the smaller pores and electrolytic current transfer through the "immobile" portion of the bulk soil solution alternates with electrolytic current transfer along clay mineral and organic matter surfaces. As an electric circuit, this second model again corresponds overall to two resistors in parallel, but the difference with respect to the first model is that one of these parallel resistors is in effect representative of two resistors in series. The second model considers current flow to be negligible via a "continuous" pathway involving just the electrolytic current transport along clay mineral and organic matter surfaces. More information on the soil resistivity equation derived from this second model is provided in Chapter 2 and Chapter 4. Regardless of the model, there is general agreement that the flow of electric current in soil occurs by two mechanisms; electrolytic current transmission through the bulk soil solution and electrolytic current transmission along particle surfaces composed of clay minerals and organic matter.

Discussions involving Equation (5.3), Equation (5.4), and Equation (5.5) imply that the soil resistivity, ρ, is influenced by complex interactions among a number of different factors. These complex interactions make it entirely possible for the correlation between one specific factor and ρ to be much weaker or even the inverse of what might be expected (Allred et al., 2005; Banton et al., 1997; Johnson et al., 2001). The occurrence of this type of result simply indicates that there are other factors, either individually or as a group, that have a greater impact on ρ than the specific factor being considered. Some factors affecting ρ, such as soil temperature and soil volumetric water content, are very transient, often exhibiting substantial changes over periods of a few hours or days. Other factors affecting ρ, if they vary temporally at all, do so at a slower rate, and in this category are soil properties including pH, organic matter content, clay content, CEC, and specific surface. Factors like nutrient level and salinity sometimes exhibit little variation over long periods, but will then change rapidly with an irrigation or fertilizer application event. Changes in soil temperature and shallow hydrologic conditions can cause the average ρ value within an agricultural field to increase or decrease (Allred et al., 2005). Soil resistivity spatial patterns within an agricultural field, however, have been found to remain relatively consistent over time, regardless of temperature and shallow hydrologic conditions, indicating that ρ spatial patterns are governed predominantly by the spatial variations in soil properties (Allred et al., 2005, 2006; Lund et al., 1999).

5.4 THEORETICAL CURRENT FLOW IN A HOMOGENEOUS EARTH AND APPARENT RESISTIVITY

Resistivity methods produce three-dimensional patterns of electric current flow and electric potential within the subsurface. Figure 5.3 two-dimensionally illustrates the distributions of current flow (solid black arrowhead lines) and potential (circular dashed black lines) in a vertical plane that intersects the positions of four electrodes, C_1, C_2, P_1, and P_2 located at the ground surface. For simplicity, the lines of equal potential shown in Figure 5.3a represent only the electric field due to current applied at C_1, and the lines of equal potential shown in Figure 5.3b represent only the electric field due to current collected at C_2. Figure 5.3a shows that electric current, $+I$, applied at the surface

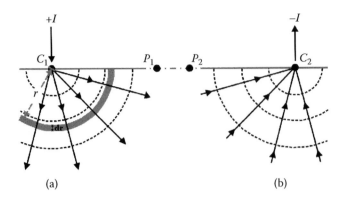

FIGURE 5.3 Electric current flow (solid black arrowhead lines) and electric potential (circular dashed black lines) due to (a) an isolated electrode, C_1, through which current, $+I$, is applied into the ground, and (b) an isolated electrode, C_2, through which current, $-I$, is collected from the ground. The electrode locations for measurement of potential are P_1 and P_2.

through a single isolated electrode, C_1, will spread out in a radial pattern away from C_1 within the ground beneath it (again, solid black arrowhead lines), given that the soil or rock material through which the current travels is homogenous and isotropic. Figure 5.3b shows that for an isolated electrode, C_2, that collects current, $-I$, from the ground, there is again a radial pattern of subsurface current flow, but in a direction toward C_2. Regardless of whether electric current is being applied or collected by an isolated electrode, surfaces of equal electric potential due to a particular current electrode will have a hemispherical shape centered about the position of that current electrode, represented two-dimensionally in Figure 5.3 by dashed circular lines.

Consider the gray hemispherical shell with a wall thickness of dr that is depicted in Figure 5.3a. For illustration, the gray hemispherical shell appears thick, but for the sake of discussion, assume it is infinitesimally thin, so dr is actually extremely small. The current application electrode, C_1, is located at the shell's radial center. Let r be the radial distance from C_1 to the inside surface of the gray hemispherical shell. The area of the shell's inside surface is $2\pi r^2$. The resistivity of the homogeneous and isotropic material making up the subsurface is designated as ρ. The total electric current flowing across this hemispherical shell equals $+I$, which is the electric current applied at C_1. The electric potential difference from the inside to the outside surfaces of the thin hemispherical shell is dV.

A form of Ohm's law, based on the scenario just described, can be given as follows:

$$dV = -\frac{\rho I dr}{2\pi r^2} \qquad (5.7)$$

Upon integration, the electric potential at a distance r from C_1 is expressed as follows:

$$V_r = \frac{\rho I}{2\pi}\frac{1}{r} + z_7 \qquad (5.8)$$

where $z_7 = 0$, because $V_r = 0$ at infinite distance from C_1. If r is set to equal the distance between C_1 and P_1 ($r = \overline{C_1P_1}$), then the potential at P_1 due to the current applied at C_1 is:

$$V_{C_1P_1} = \frac{\rho I}{2\pi}\left(\frac{1}{\overline{C_1P_1}}\right) \qquad (5.9)$$

The equation for the electric potential at P_1 due to C_2 has a similar arrangement:

$$V_{C_2P_1} = \frac{-\rho I}{2\pi}\left(\frac{1}{\overline{C_2P_1}}\right) \tag{5.10}$$

where the minus sign is introduced to reflect the fact that the current flow at C_2 (negative current electrode) is opposite in direction compared to the current flow for C_1 (positive current electrode). The equations for the electric potential at P_2 due to C_1 and C_2 are comparable in form to those of Equation (5.9) and Equation (5.10), and as a result, the total difference in electric potential, ΔV, between P_1 and P_2 can be expressed as follows:

$$\Delta V = \frac{\rho I}{2\pi}\left(\frac{1}{\overline{C_1P_1}} - \frac{1}{\overline{C_2P_1}} - \frac{1}{\overline{C_1P_2}} + \frac{1}{\overline{C_2P_2}}\right) \tag{5.11}$$

Equation (5.11) can be rearranged with respect to ρ as follows:

$$\rho = 2\pi\frac{\Delta V}{IG} \tag{5.12}$$

where G is an abbreviation for the term in Equation (5.11) contained within parentheses. Upon inspection, it is apparent that G is fundamentally a geometric factor, the value of which is governed by the arrangement and spacing of the current electrodes (C_1 and C_2) and potential electrodes (P_1 and P_2).

The development of Equation (5.12) is important because it indicates, given a specific four-electrode array setup on the ground surface, that the known electric current applied between C_1 and C_2 along with the measured electric potential difference (voltage) between P_1 and P_2 can be used to calculate the resistivity, ρ, of the subsurface beneath the electrode array. For homogeneous and isotropic soil or rock material, the calculated ρ represents the true subsurface resistivity. However, subsurface soil and rock material are more commonly heterogeneous and anisotropic. The resistivity calculated with Equation (5.12) is therefore considered a bulk measurement for the subsurface beneath the electrode array, the value of which can be influenced significantly by the nature of the soil or rock heterogeneity and anisotropy that is present. It is for this reason that the resistivity calculated by Equation (5.12) is typically referred to as the "apparent resistivity," designated by ρ_a. The reciprocal of the apparent resistivity (= $1/\rho_a$) is called the apparent electrical conductivity, which is given the symbol, σ_a, or for agriculture, EC_a.

5.5 COMMON ELECTRODE ARRAYS AND CORRESPONDING APPARENT RESISTIVITY EQUATIONS

The conventional, basic resistivity methods use a variety of electrode arrays that have different arrangement and spacing characteristics for the two current electrodes (C_1 and C_2) and two potential electrodes (P_1 and P_2). The three resistivity electrode arrays most commonly utilized are illustrated in Figure 5.4. All the electrodes for these three arrays are inserted at the ground surface and oriented along a single line. Regardless of whether the subsurface is homogeneous or nonhomogeneous, interchanging the current electrodes with the potential electrodes does not alter the geometric factor, G, or even the apparent resistivity value, although in actuality, it is best to keep the distance between potential electrodes as small as possible to avoid the effects of electric potential

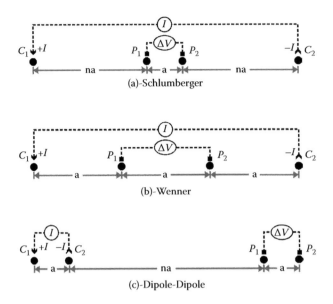

FIGURE 5.4 Arrangement and spacing of current electrodes (C_1 and C_2) and potential electrodes (P_1 and P_2) for the three resistivity electrode arrays most commonly employed.

"noise" caused by naturally occurring electric currents (Dobrin, 1976; Milsom, 2003; Parasnis, 1986). When using any of the three electrode arrays shown in Figure 5.4 for the purpose of areal resistivity mapping, the measured apparent resistivity value, ρ_a, is referenced to the midpoint location between the two outer electrodes, although it is clearly understood that ρ_a in fact represents a continuous volume of the subsurface beneath the entire electrode array.

The traditional Schlumberger array is symmetric, and as shown in Figure 5.4a, typically has the current electrodes on the outside of the array and the potential electrodes placed within the array's interior. The spacing between the current electrodes is by a large factor greater than the spacing between the potential electrodes. Being able in practice to move the outer current electrodes further apart, while potential electrode positions are kept constant, makes the Schlumberger array one of the best available for determining variations of resistivity with depth. Referring to the Figure 5.4a setup for the Schlumberger array, the apparent resistivity is calculated as follows:

$$\rho_a = \frac{\Delta V}{I} \pi (n)(n+1) a \qquad (5.13)$$

where a is a distance value, and n is a factor by which that distance is multiplied.

The Wenner array (Figure 5.4b) is also symmetric, again typically with the current electrodes on the outside and the potential electrodes on the inside of the array. The spacing between all adjacent electrode pairs is the same within the Wenner array $(\overline{C_1P_1} = \overline{P_1P_2} = \overline{P_2C_2} = a)$. This electrode array is often used for mapping lateral changes in ρ_a, partly because when the array is moved from one location to the next, there is less chance of setup error if the spacing between all adjacent electrode pairs is the same for the entire geophysical survey. Based on the Wenner array setup shown in Figure 5.4b, the equation for apparent resistivity is

$$\rho_a = \frac{\Delta V}{I} 2 \pi a \qquad (5.14)$$

The dipole-dipole array (Figure 5.4c) is normally configured to have a relatively large separation between the pair of current electrodes on one side of the array and the pair of potential electrodes on the other side of the array. The dipole-dipole array is employed both for mapping lateral changes in ρ_a and for assessing the variation of resistivity with depth. When using this array for areal mapping of ρ_a, the spacing between electrodes remains constant as the array is moved from one location to another, but for measuring resistivity changes with depth, the array midpoint stays at the same location and the current electrode pair and the potential electrode pair are moved further apart. As deduced from inspection of Figure 5.4, if the overall electrode array length is large, the dipole-dipole array has a decided advantage over other arrays, due to the lesser amount of electric cable that is needed for transferring current or measuring voltage. The amount of electric cable required by the Schlumberger or Wenner arrays, in particular, can become quite unmanageable for long electrode arrays. An additional advantage of the dipole-dipole array is that the cables for the current electrodes are more easily kept separate from the cables for the potential electrodes, which reduces the electric potential noise due to electromagnetic coupling (Sharma, 1997). The apparent resistivity for the dipole-dipole array (Figure 5.4c) is expressed as follows:

$$\rho_a = \frac{\Delta V}{I} \pi (n)(n+1)(n+2) a \tag{5.15}$$

5.6 CONVENTIONAL EQUIPMENT

Field resistivity surveys carried out with one of the more common electrode arrays (Shlumberger, Wenner, or dipole-dipole) utilizes equipment that is fairly basic. All that is actually needed is an electric current power source, a transmitter to regulate and measure the electric current, a receiver to determine the electric potential difference (essentially a high-impedance voltmeter that draws very little current), four insulated single-core copper wire cables, and four electrodes. The transmitter and receiver are often but not always combined into a single unit. Iron, steel, or copper stakes are commonly used for the electrodes that are inserted into the soil at the ground surface.

One problem does occur with the use of metal stakes for electrodes, and it involves the formation of unwanted electric potentials due to electrode polarization caused by contact between the metal in the stake and the electrolytic aqueous solution present within the soil. This problem is alleviated by using low-frequency alternating electric current (AC) to cancel out these spurious electric potentials. Typically, AC with a frequency less than 10 Hz is used for conventional resistivity surveys, so that these simple metal stake electrodes can be employed. Lower-frequency AC is used instead of higher-frequency AC, because higher-frequency AC reduces the electric current density with depth and, in turn, the depth of investigation (Sharma, 1997). Direct electric current (DC) is employed for very deep investigations and requires the use of nonpolarizing electrodes. A nonpolarizing electrode is typically composed of a copper rod inserted through the lid of a container that is porous at its base and filled with an aqueous copper sulfate solution (Milsom, 2003).

5.7 COMPUTER-CONTROLLED MULTIELECTRODE SYSTEMS

Conventional resistivity methods can be enhanced by the group installation of twenty-five or more electrodes at a time, which are inserted at the ground surface along a line or in a grid pattern. These electrodes are then connected via switching units and multicore cable to a computer-controlled transmitter and receiver unit (Sharma, 1997). This computer-controlled multielectrode system allows an electric current to be applied between any two electrodes along a line or within a grid, while at the same time the electric potential difference is measured with respect to another electrode pair along the line or within the grid. The system is normally programmed to collect a sequence of

ρ_a measurements from successive sets of four electrodes (two current and two potential) that are a part of the multielectrode line or grid. The main advantage of a computer-controlled multielectrode system over conventional resistivity methods is in the reduced amount of time it takes to carry out a resistivity survey.

5.8 CONTINUOUS MEASUREMENT SYSTEMS

Conventional resistivity surveys, whether for areal ρ_a mapping or assessing soil resistivity changes with depth, proceed at a relatively slow pace, because electrodes need to be pulled from the ground, moved, and then reinserted before the next ρ_a measurement is obtained. Computer-controlled multielectrode systems are certainly an improvement over conventional equipment with respect to resistivity survey efficiency; however, there is still a considerable amount of initial setup time needed, especially in cases where 100 or more electrodes need to be installed. Development of continuous resistivity measurement systems began in the middle 1980s to the early 1990s, with researchers first in France and then Denmark, and this technological innovation dramatically expanded the use of resistivity methods for agricultural purposes. These continuous resistivity measurement systems allow ρ_a to be measured at a fast rate along a transect without the need to repeatedly insert and remove electrodes. Continuous resistivity techniques can be divided into two types based on the manner in which electric current is applied to the subsurface; one uses a galvanic contact approach, and the other employs a capacitive-coupling approach.

5.8.1 GALVANIC CONTACT RESISTIVITY

The term "galvanic contact," when used to describe resistivity methods, implies that the current electrodes are in contact with the soil surface, and there is a direct transfer of electric current between the current electrodes and the soil. Conventional resistivity surveys and the computer-controlled multielectrode systems previously described both employ galvanic contact to introduce electric current into the subsurface. Perhaps the greatest advance with galvanic contact continuous measurement systems was in the design of electrodes capable of maintaining good electrical contact with soil while continuous ρ_a measurements are collected as the electrode array is pulled from one location to another.

It is important to note that contact resistance can be a problem with all galvanic contact methods, and it occurs when resistivity is very high at or just below the ground surface where the electrodes are placed or inserted. This situation can arise when the ground surface is extremely dry, frozen, or covered with snow or ice, thereby making it difficult to transfer electric current from the current electrode into the adjacent soil (Reynolds, 1997). When conventional resistivity equipment or computer-controlled multielectrode systems are being used, this problem can often be solved simply by using a saline solution to wet the soil surrounding the inserted current electrode. This tactic for addressing the contact resistance problem is not feasible when conducting resistivity surveys with galvanic contact continuous measurement systems, and as a result, these types of surveys should be avoided under snow- or ice-covered or frozen ground conditions, and maybe even when the soil surface is extremely dry.

Sorensen (1996) was one of the first to develop a galvanic contact continuous resistivity measurement system having electrode arrays capable of being continuously pulled along the ground surface. The system engineered by Sorensen (1996) included a line of cylindrical steel tube current and potential electrodes connected from one to the other via a multicore cable (Figure 5.5). Because the electrodes were composed of steel tubes, they could be easily pulled in a line over the ground surface. The system's transmitter and receiver unit supplied a maximum 30 mA alternating electric current at a frequency between 15 and 25 Hz. This electric current passed through the ground between the two current electrodes and voltage was measured at two different pairs of potential electrodes (Figure 5.5). The spacing and arrangement of the one pair of current electrodes with

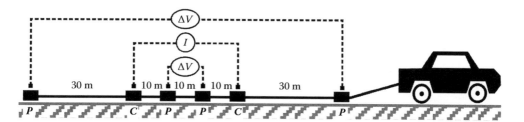

FIGURE 5.5 Electrode spacing and arrangement (C = current electrode, P = potential electrode) for the galvanic contact continuous resistivity measurement system developed by Sorensen (Sorensen, K., *First Break*. v. 14, no. 3, pp. 85–90, 1996.)

respect to the two pairs of potential electrodes enabled the system to measure ρ_a with two Wenner arrays, one 30 m in length and the other 90 m in length, thereby providing two different depths of investigation. Based on information provided by Loke (2004), a 30 m Wenner array has a median investigation depth of approximately 5 m, and the median investigation depth for a 90 m Wenner array is about 16 m.

Galvanic contact continuous resistivity measurement systems with investigation depths more appropriate for use in agriculture have been developed in both the United States and France. One system popular within the United States is the Veris 3100 Soil EC Mapping System (Veris Technologies) shown in Figure 5.6a. A data-logger placed inside the vehicle collects one ρ_a measurement every second. Apparent resistivity measurement locations are determined using an integrated Global Positioning System (GPS) receiver. The electrodes are mounted on a steel frame (Figure 5.6b) and are composed of 43 cm diameter steel coulters (disks) that cut through the soil to depths of approximately 2.5 to 7.5 cm as they are pulled along behind the vehicle at field speeds of up to 25 km/h. The Veris 3100 Soil EC Mapping System has six coulters with nonadjustable spacing (two for electric current application and two pairs for voltage measurement), essentially providing two Wenner

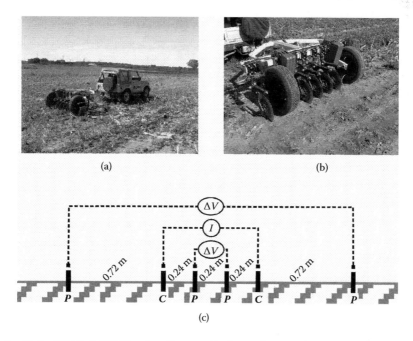

FIGURE 5.6 Veris 3100 Soil EC Mapping System: (a) photo of system in operation, (b) close-up of steel coulters used for current and potential electrodes, and (c) schematic of electrode spacing and arrangement (C = current electrode, P = potential electrode).

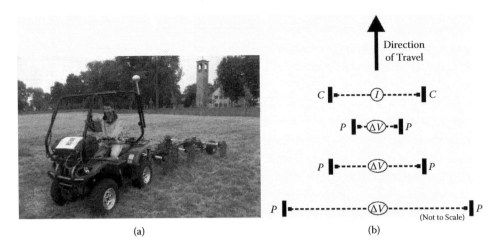

(a)   (b)

FIGURE 5.7 Geocarta ARP-03: (a) photo of system in operation, and (b) plan view of electrode arrangement (C = current electrode, P = potential electrode). (Courtesy of Geocarta SA, Paris, France.)

electrode array configurations (Figure 5.6c). The shorter Wenner array (0.7 m) maps the top 0.3 m of the soil profile, and a longer one (2.1 m) maps the top 0.9 m of the soil profile. As indicated by Figure 5.6a and Figure 5.6b, these electrode arrays are oriented perpendicular to the direction of travel while ρ_a is being measured.

The first galvanic contact continuous resistivity measurement systems were probably developed in France (Dabas et al., 1994). The ARP-03 (Figure 5.7a) is one of these systems and is marketed by Geocarta SA. This system is integrated with a GPS receiver for accurate determination of ρ_a measurement locations. A radar triggered ρ_a measurement is obtained for every 0.2 m of travel. Steel coulters serve as electrodes for the ARP-03. To improve ground contact, metal spike extensions are attached along the outer circumference of the coulter electrodes. The plan view arrangement of the ARP-03 electrodes is depicted in Figure 5.7b. As shown, there is one pair of current electrodes and three pairs of potential electrodes. The arrangement shown in Figure 5.7b provides for three electrode arrays (one current electrode pair and one potential electrode pair), with each array having a trapezoidal configuration. Investigation depths for the three ARP-03 electrode arrays are 0.5, 1, and 2 m.

5.8.2 CAPACITIVELY COUPLED RESISTIVITY

Capacitively coupled systems are also capable of collecting continuous resistivity measurements. As the name implies, these systems use a capacitive-coupling approach to introduce electric current into the ground and to measure potential differences at the soil surface. This capacitive-coupling is accomplished using coaxial cables. Essentially, a large capacitor is formed by the coaxial cable and the soil surface. The metal shield of the coaxial cable is one of the capacitor plates, and the soil surface is the other capacitor plate, with the outer insulation of the coaxial cable acting as the dielectric material separating the two plates. The system transmitter applies an alternating current (AC) to the coaxial cable side of the capacitor, in turn generating AC in the soil on the other side of the capacitor. With regard to the receiver, a similar phenomenon occurs, except in reverse. The AC in the soil charges up the capacitance of the coaxial cable, which is measured to determine the potential difference (voltage) generated by the flow of electric current within the soil.

Two coaxial cables are attached to the transmitter, one on each side, to form a current dipole. Two coaxial cables are also attached to the receiver, one on each side, to form a potential dipole. This equipment setup, along with some initial data processing, allows a capacitively coupled system to mimic a conventional galvanic contact dipole-dipole electrode array having one pair of current electrodes (current dipole) and one pair of potential electrodes (potential dipole). Similar to a gal-vanic contact system, a capacitively coupled system determines ρ_a (or σ_a, EC_a) using measured electric current, measured voltage, and array characteristics, which in this case, are the coaxial cable dipole lengths and the distance between the two dipoles. The spacing between the two dipoles governs the soil investigation depth, given that the dipole lengths remain constant.

As previously mentioned, the transmitter in a capacitively coupled resistivity measurement system generates alternating electric current. The higher the AC frequency, the better the capacitive coupling, and the more electric current is coupled to the ground. However, higher frequencies also generate some unwanted electromagnetic effects that adversely influence resistivity measurement. A transmission frequency of between 12 kHz and 20 kHz is a good compromise between the need for a low enough frequency to avoid unfavorable electromagnetic effects on resistivity measurement and the need for a high enough frequency to get good current coupling to the ground.

There are some advantages and disadvantages for capacitively coupled resistivity measurement systems when compared to galvanic contact systems. Capacitively coupled systems can be used in areas covered with pavement, whereas galvanic contact systems, with electrodes that need to be inserted into the ground, usually cannot be employed in these settings. In soil environments with high resistivity, it is often difficult when using galvanic contact continuous measurement systems to adequately transfer electric current from a current electrode into the ground, a problem previously referred to as contact resistance. This problem does not occur with capacitively coupled systems; therefore, these systems are a good choice for use in high-resistivity soil environments. Conversely, capacitively coupled systems do not work well in low-resistivity soil environments because the potential dipole voltage becomes too small to be measured reliably. Increasing the amount of electric current transferred into the ground would solve this problem, but this tactic is not easily accom-plished with capacitively coupled systems. Furthermore, due to limitations on the amount of electric current that can be transferred into the ground, the maximum investigation depth achievable with a capacitively coupled system is around 20 m, which is much less than what can be obtained with a galvanic contact system.

The OhmMapper TR1 (Geometrics, Inc.) displayed in Figure 5.8 is an example of a capaci-tively coupled continuous resistivity measurement system. The current and potential dipoles can be 5 or 10 m in length and are composed of two coaxial cables integrated with either transmitter or receiver electronics. The transmitter generates a 16 kHz alternating current. The OhmMapper TR1 also has a battery power supply, a data logger console, and rope that separates the two dipoles from one another (Figure 5.8). The depth of investigation can be adjusted by lengthening or shortening the rope separating the current and potential dipoles. Continuous ρ_a measurements are collected at time intervals as short as 1 sec. The OhmMapper TR1 can be integrated with a GPS receiver for accurate determination of the resistivity measurement locations.

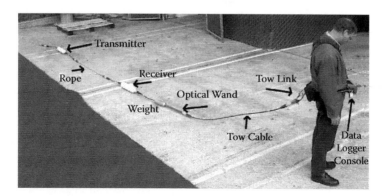

FIGURE 5.8 OhmMapper TR1 capacitively coupled continuous resistivity measurement system. (Courtesy of Geometrics, Inc., San Jose, California.)

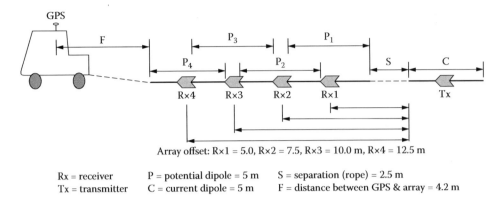

FIGURE 5.9 Schematic of OhmMapper TR4 with one current dipole and four potential dipoles. (Courtesy of Geometrics, Inc., San Jose, California.)

The OhmMapper TR1 mimics a single galvanic contact dipole-dipole electrode array and therefore has only one receiver (potential) dipole. Different OhmMapper systems can have up to five potential dipoles. Figure 5.9 is a schematic that illustrates the setup for a system, the OhmMapper TR4, having four potential dipoles. These systems will accordingly mimic up to five dipole-dipole electrode arrays at a time, with each array having a different length and therefore different resistivity measurement investigation depth. The value of these multiple potential dipole OhmMapper systems is in their ability to continuously measure both lateral and vertical soil resistivity variations simultaneously.

5.9 FIELD DATA COLLECTION MODES

As previously implied in Section 5.5 where the most common electrode arrays were discussed, there are two main modes for resistivity data collection that are used in the field: one is "constant separation traversing," and the other is "vertical electric sounding" (Reynolds, 1997). With constant separation traversing, the spacing between electrodes remains constant as the whole electrode array is moved along a transect (Figure 5.10a). Again, for the Schlumberger, Wenner, and dipole-dipole arrays, apparent resistivity, ρ_a, measurement locations are referenced to the midpoint position between the two outer electrodes. Constant separation traversing resistivity measurements are usually collected along a series of transects forming a grid that covers a study area, thereby making this data collection mode ideal for highlighting horizontal variations in ρ_a. The development of continuous resistivity measurement equipment has made it possible to efficiently conduct constant separation traversing surveys on large agricultural fields.

Information regarding resistivity changes with depth can be obtained through vertical electric sounding. The procedure for vertical electric sounding involves keeping the midpoint of the resistivity array stationary, while the overall array length is successively increased (Figure 5.10b). As the electrode array is lengthened, electric current penetrates further into the subsurface, providing a greater depth of investigation. Consequently, changes in ρ_a that occur with successive increases in the electrode array length indicate variations of resistivity with depth. Vertical electric sounding is done with either conventional equipment or computer-controlled multielectrode systems.

Horizontal and vertical resistivity data can be collected together. One procedure based on conventional equipment or a computer-controlled multielectrode system is to conduct a complete vertical electric sounding at each regularly spaced location along lines within a grid. Collection of horizontal and vertical resistivity data together can also be accomplished by conducting several resistivity survey passes over a study area. For each successive pass, the electrode array is lengthened, and investigation depth is increased. Continuous measurement resistivity systems are especially applicable to this type of approach. Additionally, with the continuous measurement resistivity systems that have

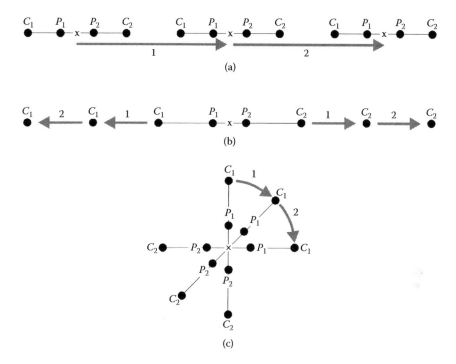

FIGURE 5.10   Field data collection modes: (a) constant separation traversing, (b) vertical electric sounding, and (c) azimuthal rotation (C = current electrode, P = potential electrode, × = electrode array midpoint).

more than one pair of potential electrodes (or more than one potential dipole), a single survey pass is sometimes all that is needed, because in effect, apparent resistivity values are being obtained simultaneously for two or more electrode arrays of different length and investigation depth.

Although not used to the extent of constant separation traversing and vertical electric sounding, there is a third mode of resistivity data collection, perhaps best referred to as "azimuthal rotation." Azimuthal rotation data collection, as shown in Figure 5.10c, is carried out with either the electrode array midpoint or one of the array's end points kept stationary, while successive ρ_a measurements are made as the usually linear electrode array is pivoted about this stationary point in increments of 10° to 20° through a complete sweep of 180° or 360° (Watson and Barker, 1999). Wenner and Schlumberger electrode configurations are most commonly used (Steinich and Marín, 1997; Taylor and Fleming, 1988; Watson and Barker, 1999), but square electrode array patterns have also occasionally been employed (Lane et al., 1995). The spacing between electrodes remains constant as the array is pivoted about a point. A sweep of 180° is all that is required, but 360° sweeps are useful because two measurements are obtained for each array orientation. (A line rotated 180° has the same trend as it did before rotation.) Once the complete 180° or 360° sweep of ρ_a measurements is completed, the entire electrode array is moved to a new stationary pivot point where another set of azimuthal measurements are collected. Azimuthal rotation data collection is employed to quantify the directional components of resistivity (resistivity anisotropy), which may itself provide useful information on the presence and trends of aligned features in the subsurface. These azimuthal measurements are usually obtained with conventional four-electrode array equipment.

5.10   DATA ANALYSIS PRODUCTS

Figure 5.11 through Figure 5.13 illustrate some of the fundamental data analysis products produced from constant separation traversing, vertical electric sounding, and azimuthal rotation resistivity surveys. The apparent resistivity, ρ_a, measurements obtained along a transect during constant

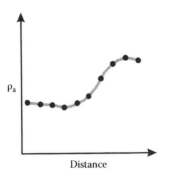

FIGURE 5.11 A horizontal profile of apparent resistivity along a transect, which can be produced from data collected by a constant separation traversing survey.

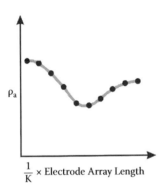

FIGURE 5.12 A plotted vertical electric sounding curve.

separation traversing can be graphed to produce a profile of the lateral changes in ρ_a found along the transect (Figure 5.11). In an agricultural setting, assuming the resistivity survey had a shallow depth of investigation (~2 m), the fairly sharp step increase in ρ_a shown in Figure 5.11 most likely represents some spatially abrupt soil property change (e.g., a decrease in soil clay content over a short distance).

The measurements of ρ_a acquired during a vertical electric sounding are usually plotted versus some fraction, $1/K$, of the electrode array length (Figure 5.12). For the Schlumberger array, the K value is typically 2; therefore, ρ_a in the case of a Schlumberger array is normally graphed with respect to half the array length. The plotted vertical sounding curve depicted in Figure 5.12, if obtained for agricultural purposes, might indicate a soil profile trend from the surface downward, in which soil resistivity first decreases with depth and then increases with depth. The presence of a clay-pan within the soil profile is one scenario that would account for this type of vertical resistivity trend. During the past, changes in resistivity with depth were determined using manual graphical procedures, where type curves usually representing simple two- or three-layer resistivity depth models were overlaid and fit to the measured vertical electric sounding plot. These older graphical procedures are rarely used today.

Forward computer modeling techniques superseded these graphical procedures and provided substantially improved capabilities for quantifying the vertical distribution of resistivity. The one-dimensional forward modeling approach involves the following four steps:

1. The person responsible for data interpretation constructs a one-dimensional resistivity depth model based on the available information regarding subsurface conditions.
2. Computer software is used to generate a synthetic vertical electric sounding curve corresponding to the initial one-dimensional model of soil resistivity variation with depth.
3. The computer-generated synthetic vertical sounding curve is then compared to the vertical sounding curve measured in the field.
4. The interpreter then adjusts the initial one-dimensional resistivity model, followed by the generation of a new synthetic vertical sounding curve.

The last forward modeling step is repeated until there is a good fit between the measured vertical electric sounding curve and the synthetic vertical electric sounding curve. Once a good enough fit is achieved, the final one-dimensional resistivity depth model is considered to be a reasonable estimate for the true vertical resistivity distribution.

In recent years, inverse modeling techniques have gained widespread acceptance for use in determining resistivity variations with depth. The inverse computer modeling approach is completely automated, using iterative procedures coupled with optimization protocols to produce a resistivity depth model from which synthetically generated apparent resistivity data are in good agreement with actual field measurements. An important issue needing special emphasis with respect to both forward and inverse computer modeling is that within the limits of measurement error for ρ_a, there are often several resistivity depth models, differing from one another to a certain extent, that are able to provide a good fit between the measured apparent resistivity data and synthetically generated apparent resistivity data. Consequently, the resistivity depth model obtained through these computer analysis procedures may not always be "unique," and it is worth keeping in mind the possibility that other models exist that can

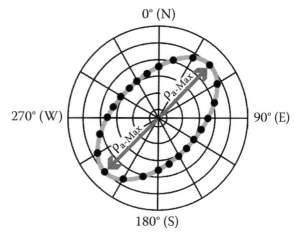

FIGURE 5.13 Azimuthal rotation apparent resistivity data plotted on a polar coordinate graph.

adequately account for the observed apparent resistivity data.

The apparent resistivity measurements from an azimuthal rotation survey carried out at a particular location are plotted on a polar coordinate graph (Figure 5.13). Each plotted ρ_a measurement shown in Figure 5.13 is associated with an angle and a radial length. Given a specific ρ_a measurement, the orientation of the electrode array, clockwise from true north, is specified by the angle. As an example, a ρ_a value plotted with an angle of 45° represents a measurement obtained with a northeast–southwest-oriented electrode array. Figure 5.13 is representative of an azimuthal rotation survey where ρ_a measurements were made as the electrode array was pivoted about a stationary point in increments of 15° through a complete sweep of 360°. The radial length from the center of the graph to a data point represents the magnitude of the ρ_a measurement.

For any location where an azimuthal rotation survey is conducted, a circular pattern for the ρ_a measurements plotted on a polar graph implies that resistivity is the same in all directions (resistivity is isotropic). If the plotted ρ_a measurements instead have the pattern of an ellipse, then resistivity varies with direction (resistivity is anisotropic). The principle (longest) axis of the ellipse indicates the electrode array orientation for which the maximum ρ_a value or values, ρ_{a-Max}, were acquired during the azimuthal rotation survey. The orientation of this principle axis often corresponds with the trend of aligned features in the subsurface. The fact that ρ_{a-Max} is found along an orientation coinciding with linear subsurface trends is somewhat counterintuitive but is explained by the "anisotropy paradox" (Keller and Frischknecht, 1966; Parasnis, 1986). Taylor and Fleming (1988) determined azimuthal rotation resistivity surveys to be useful for characterizing fracture systems in glacial till. For vertical fractures with an average length greater than the length of the electrode array, the overall fracture system trend was found to coincide with the principle axis of the polar graph ρ_a ellipse (the orientation corresponding to ρ_{a-Max}). The magnitude of ρ_{a-Max} proved to be a good indicator of fracture density. The presence of two major fracture systems results in the polar graph of ρ_a having two superimposed ellipses and, therefore, two principle axes (Taylor and Fleming, 1988).

Two of the more detailed and comprehensive data analysis products from resistivity surveying include horizontal (areal) apparent resistivity (or apparent electrical conductivity) maps and resistivity (or electrical conductivity) depth sections. Measurements from a constant separation traversing survey are commonly collected along a set of equally spaced transects covering a study area. With a complete data set (all ρ_a or σ_a, EC_a values for all transects), interpolation and contouring procedures can be applied to produce a map showing horizontal changes in ρ_a or σ_a, EC_a.

Figure 5.14 shows two horizontal EC_a maps of the same agricultural test plot located in Columbus, Ohio. Measurement transects were spaced 3.1 m apart. The data used to produce Figure 5.14a

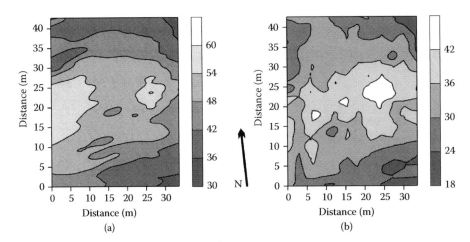

FIGURE 5.14 Apparent soil electrical conductivity (EC$_a$) maps of the same agricultural test plot from data collected with (a) the Veris 3100 Soil EC Mapping System and (b) the OhmMapper TR1. The EC$_a$ scales for each map are different, and the EC$_a$ values are in mS/m.

were collected by the Veris 3100 Soil EC Mapping System (continuous galvanic contact method) with its 2.1 m electrode array. The data used to produce Figure 5.14b were collected by an Ohm-Mapper TR1 (continuous capacitively coupled method) with a 5 m current dipole, a 5 m potential dipole, and a 1.25 m separation between the dipoles. The reported investigation depth for the Veris 3100 with its 2.1 m electrode array is 0.9 m, and the OhmMapper TR1 in the configuration described had a median investigation depth of 0.8 m. There was an interval of two days between the Veris 3100 and OhmMapper TR1 surveys, but soil temperature and moisture conditions did not change drastically during this time interval.

When compared to one another, the spatial EC$_a$ patterns exhibited by both Figure 5.14 maps appear consistent, and this observation is confirmed by a correlation coefficient (r) between the two maps that equals 0.79. The average test plot EC$_a$ from the Veris 3100 survey was 46.5 mS/m, and the average test plot EC$_a$ from the OhmMapper TR1 survey was 33 mS/m. The difference in the average test plot EC$_a$ may reflect dissimilarities between the Veris 3100 and OhmMapper TR1 systems in regard to the soil volume influencing the instrument response, the magnitude of effects due to small-scale features, the sensitivity to small changes in field conditions, the relative impact of unwanted electric signal (noise), procedures for calculating EC$_a$, among others. Consequently, given a particular survey area, the horizontal ρ_a or σ_a, EC$_a$ maps produced with different measurement systems having similar depths of investigation will usually display similar spatial patterns, but average ρ_a or σ_a, EC$_a$ values may be significantly different. Importantly though, the ρ_a or σ_a, EC$_a$ horizontal spatial patterns prove useful in assessing lateral changes in soil properties.

Two-dimensional resistivity (or electrical conductivity) depth sections characterize the distribution of resistivity (or electrical conductivity) with depth beneath a measurement transect along the surface. The resistivity (or electrical conductivity) values shown in a depth section are considered to represent true values, not apparent values. The data needed to create a depth section can be obtained several ways, such as through a set of vertical electric soundings conducted at a number of regularly spaced locations along a transect, several resistivity survey passes over a transect with a one-electrode array whose length is changed each pass, or one resistivity survey pass over the transect using several different length electrode arrays at once. The data acquired are then used as input for the forward or inverse computer modeling programs that generate the resistivity (or electrical conductivity) depth sections.

Two examples of EC depth sections are displayed in Figure 5.15. Each EC depth section is from a separate agricultural test plot. The two agricultural test plots are situated adjacent to one another

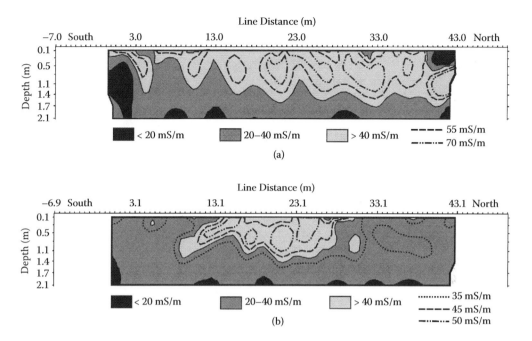

FIGURE 5.15   Two examples of electrical conductivity depth sections.

in Columbus, Ohio. The transect for the Figure 5.15a EC depth section was separated from the transect for the Figure 5.15b EC depth section by a distance of 41.5 m. Both EC depth sections show complex EC patterns within the soil profile from the surface down to 2 m. Furthermore, significant differences are exhibited among the two Figure 5.15 depth sections even though the distance between their transects is relatively short. For reference, Figure 5.15b shows EC variations at depth beneath a south-to-north transect that passes through the center of the test plot where the Figure 5.14 EC_a map data were obtained.

The data used to produce Figure 5.15a and Figure 5.15b were acquired with the OhmMapper TR1. The OhmMapper TR1 was configured with a 5 m current dipole and a 5 m potential dipole. There were four OhmMapper survey passes over each transect corresponding to a depth section on Figure 5.15. The separation between the current and potential dipoles was increased from 0.625 m to 1.25 and then to 2.5 m, and again to 5 m for the four successive passes over a transect. Successive increases in the OhmMapper TR1 dipole-dipole array length provided for greater depths of investigation. The EC depth sections depicted in Figure 5.15 were generated using the OhmMapper TR1 data as input to a two-dimensional, least-squares optimization inverse computer modeling program, RES2DINV, developed by Loke (2007). Although not used extensively in agriculture at present, resistivity (or electrical conductivity) depth sections can potentially provide some very useful agricultural information, such as the vertical position of salinity buildup within the soil profile or the depth to a clay-pan, fragipan, or caliche layer.

If a number of azimuthal rotation resistivity surveys are carried out within an area of interest, the results of these surveys could be incorporated into some form of a map product. One possibility is to plot line segments on a map corresponding to the locations of each azimuthal rotation resistivity survey. The orientation of a particular line segment on the map would coincide with the principle axis of the polar graph ρ_a ellipse, assuming resistivity anisotropy existed at that survey location. Line segment lengths would in turn reflect ρ_{a-Max} values. A map such as this could provide valuable information on the pervasiveness, orientation, and intensity of aligned features present within the soil profile, for which one example might be the extent, trend, and density of a fracture system.

5.11   RESISTIVITY METHOD APPLICATIONS IN AGRICULTURE

Research investigations have provided some valuable insight on potential applications and limitations with regard to using resistivity methods for agricultural purposes. There are clear indications that resistivity methods can be a valuable tool for assessing salinity conditions in farm fields (Rhoades et al., 1976, 1990). Kravchenko et al. (2002) combined topographic information and EC_a data (obtained with a Veris 3100 Soil EC Mapping System) to map soil drainage classes. Studies have also focused on determining the relationship between various soil properties and the apparent soil electrical conductivity (σ_a, EC_a) as measured with galvanic contact resistivity methods. Using conventional resistivity equipment at a location near Quebec City, Canada, Banton et al. (1997) found that EC_a was moderately correlated with soil texture (% sand, % silt, and % clay) and organic matter, but not with porosity, bulk density, or hydraulic conductivity. Johnson et al. (2001) found a positive correlation at their Colorado test site between EC_a (measured with a Veris 3100 Soil EC Mapping System) and bulk density, percentage clay, laboratory-measured soil electrical conductivity, and pH; but a negative correlation between EC_a and total and particulate organic matter, total carbon, total nitrogen, extractable phosphorous, microbial biomass carbon, microbial biomass nitrogen, potentially mineralizable nitrogen, and surface residue mass. These two studies (Banton et al., 1997; Johnson et al., 2001) imply that soil properties can interact in a complex manner to affect the EC_a measured with resistivity methods.

The apparent resistivity (or apparent electrical conductivity) of a soil can also be measured with electromagnetic induction methods. Studies carried out using electromagnetic induction methods have shown the feasibility of using EC_a for estimating herbicide partition coefficients (Jaynes et al., 1995), determining clay-pan depth (Doolittle et al., 1994), and monitoring soil nutrient buildup from manure applications (Eigenberg and Nienaber 1998). Consequently, with the proper equipment, it is probable that resistivity methods could also be employed to estimate herbicide partition coefficients, determine clay-pan depth, and monitor soil nutrient buildup.

A rather interesting example regarding an agricultural application of resistivity methods involved using a constant separation traversing resistivity survey to provide insight at a field research facility on the soil salinity impact due to different drainage water management practices. Figure 5.16 shows the field research site setup composed of four test plots. No buried drainage pipe network was installed at the C1 test plot. Test plot C2 contained a buried drainage pipe network, but this drainage pipe network was used only to remove excess water from the soil. Test plots S1 and S2 were subirrigated during the growing season. Subirrigation can be described as the addition of water to a buried drainage pipe network for the purpose of irrigating crops through the root zone.

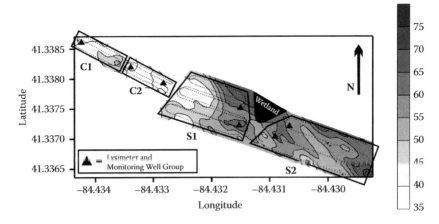

FIGURE 5.16   Apparent soil electrical conductivity (EC_a) map of field research facility used for assessing different drainage water management strategies. Values for EC_a are in mS/m.

Suction lysimeters were utilized during a five-year period (2001 to 2006) to obtain monthly soil solution samples at depths of 0.6 and 1.2 m beneath the surface. The C1 test plot had six suction lysimeters (three of 0.6 m depth and three of 1.2 m depth) installed at its one lysimeter and monitoring well group location marked by a black triangle in Figure 5.16. In the C2, S1, and S2 test plots, there were four suction lysimeters (two at 0.6 m depth and two at 1.2 m depth) installed at each lysimeter and monitoring well group location. The average soil solution electrical conductivity values measured in the laboratory for test plots C1, C2, S1, and S2, respectively, were 1.17, 1.29, 3.28, and 3.24 dS/m. These average soil solution electrical conductivity values initially created concern that the practice of subirrigation might be causing a salinity buildup.

To gain further information on the possibility of a subirrigation-induced salinity increase, a constant separation traversing resistivity survey was carried out at the site using a Veris 3100 Soil EC Mapping System. Measurements of EC_a acquired by resistivity methods are expected to be strongly correlated with soil solution electrical conductivity. The EC_a spatial pattern at the site is shown in Figure 5.16, with the small black dots representing EC_a measurement locations. The mapped EC_a pattern indicates that the larger average soil solution electrical conductivity values found in test plots S1 and S2 are not due to subirrigation practices, but are instead the result of some east-to-west transition in natural soil conditions causing the S1 and S2 soil solution electrical conductivities to be greater than those for C1 and C2. This finding becomes more evident by focusing strictly on the S1 test plot. If subirrigation produces a salinity increase, then the entire S1 test plot should exhibit high EC_a values, which is definitely not the case. Although the east side of the S1 test plot, where the suction lysimeters are located, does have high EC_a values, the west side of the S1 test plot has EC_a values that are much lower and similar to those measured in test plots C1 and C2. Viewing the site as a whole, the change from high to low EC_a values occurs somewhat abruptly over a short east–west distance interval passing through the center of the S1 test plot. Undoubtably, if the S1 suction lysimeters had instead been installed on the west side of the test plot, the average soil solution electrical conductivity value would have been substantially lower. Again, the EC_a mapping results clearly indicate that subirrigation is not producing a salinity buildup, and the high S1 and S2 soil solution electrical conductivity values are due to the S1 and S2 suction lysimeters being located in the part of the field research facility where natural soil conditions produce high soil solution electrical conductivities.

Finally, resistivity method applications in environmental and hydrological disciplines may also prove useful in agriculture. Resistivity methods can be a valuable tool in regard to characterization and leak detection for landfills and chemical disposal pits (Reynolds, 1997; Sharma, 1997). These resistivity methods should likewise be equally useful for characterizing and detecting leaks for animal waste storage ponds and treatment lagoons. The use of resistivity methods to determine the trends and density of fracture systems in glacial till was previously discussed. Azimuthal rotation surveys could also be employed to gather the same type of information on fracture systems within the soil profile. This soil profile fracture system information might have some worth in designing subsurface drainage systems, because orienting drain lines to intersect soil fractures will probably improve the soil water removal efficiency of the overall drainage pipe network.

REFERENCES

Allred, B. J., M. R. Ehsani, and D. Saraswat. 2005. The impact of temperature and shallow hydrologic conditions on the magnitude and spatial pattern consistency of electromagnetic induction measured soil electrical conductivity. *Trans. ASAE.* v. 48, pp. 2123–2135.

Allred, B. J., M. R. Ehsani, and D. Saraswat. 2006. Comparison of electromagnetic induction, capacitively coupled resistivity, and galvanic contact resistivity methods for soil electrical conductivity measurement. *Applied Eng. Agric.* v. 22, pp. 215–230.

Banton, O., M. K. Seguin, and M. A. Cimon. 1997. Mapping field-scale physical properties of soil with electrical resistivity. *Soil Sci. Soc. Am. J.* v. 61, pp. 1010–1017.

Bohn, H. L., B. L. McNeal, and G. A. O'Connor. 1985. *Soil Chemistry,* 2nd Edition. John Wiley & Sons, New York.

Dabas, M., J. P. Decriaud, G. Ducomet, A. Hesse, A. Mounir, and T. Tabbagh. 1994. Continuous recording of resistivity with towed arrays for systematic mapping of buried structures at shallow depth. *Revue d'Archéométrie.* v. 18, pp. 13–19.

Dobrin, M. B. 1976. *Introduction to Geophysical Prospecting,* 3rd Edition. McGraw-Hill, New York.

Doolittle, J. A., K. A. Sudduth, N. R. Kitchen, and S. J. Indorante. 1994. Estimating depths to claypans using electromagnetic induction methods. *J. Soil and Water Cons.* v. 49, pp. 572–575.

Eigenberg, R. A., and J. A. Nienaber. 1998. Electromagnetic survey of cornfield with repeated manure applications. *J. Eniviron. Qual.* v. 27, pp. 1511–1515.

Jaynes, D. B., J. M. Novak, T. B. Moorman, and C. A. Cambardella. 1995. Estimating herbicide partition coefficients from electromagnetic induction measurements. *J. Environ. Qual.* v. 24, pp. 36–41.

Johnson, C. K., J. W. Doran, H. R. Duke, B. J. Wienhold, K. M. Eskridge, and J. F. Shanahan. 2001. Field-scale electrical conductivity mapping for delineating soil condition. *Soil Sci. Soc. Am. J.* v. 65, pp. 1829–1837.

Keller, G. V., and F. C. Frischknecht. 1966. *Electrical Methods in Geophysical Prospecting.* Pergamon Press, Oxford, UK.

Knight, R. J., and A. L. Endres. 2005. An introduction to rock physics principles for near-surface geophysics. In *Near-Surface Geophysics.* 31–70. D. K. Butler, editor. Investigations in Geophysics No. 13. Soc. Expl. Geophy. Tulsa, OK.

Kravchenko, A. N., G. A. Bollero, R. A. Omonode, and D. G. Bullock. 2002. Quantitative mapping of soil drainage classes using topographical data and soil electrical conductivity. *Soil Sci. Soc. Am. J.* v. 66, pp. 235–243.

Lane, J. W., F. P. Haeni, and W. M. Watson. 1995. Use of a square-array direct-current resistivity method to detect fractures in crystalline bedrock in New Hampshire. *Ground Water.* v. 33, pp. 476–485.

Loke, M. H., 2004. *Tutorial: 2-D and 3-D Electrical Imaging Surveys.* Geotomo Software. www.geoelectrical. com. Gelugor, Penang, Malaysia.

Loke, M. H. 2007. RES2DINV—Computer inversion software, http://www.geoelectrical.com.

Lund, E. D., P. E. Colin, D. Christy, and P. E. Drummond. 1999. Applying soil electrical conductivity technology to precision agriculture. In *Proc. 4th Int. Conf. Precision Agric.* 1089–1100. P. C. Robert, R. H. Rust, and W. E. Larson, eds. July 19–22. St. Paul, MN. ASA, CSSA, and SSSA. Madison, WI.

McNeill, J. D. 1980. *Electrical Conductivity of Soils and Rocks.* Technical Note TN-5. Geonics Ltd. Mississauga, Ontario, Canada.

Milsom, J. 2003. *Field Geophysics,* 3rd Edition. John Wiley & Sons, Chichester, UK.

Parasnis, D. S. 1986. *Principles of Applied Geophysics.* Chapman & Hall, London, UK.

Reynolds, J. M. 1997. *An Introduction to Applied and Environmental Geophysics.* John Wiley & Sons, Chichester, UK.

Rhoades, J. D., P. A. C. Raats, and R. J. Prather. 1976. Effects of liquid-phase electrical conductivity, water content, and surface conductivity on bulk soil electrical conductivity. *Soil Sci. Soc. Am. J.* v. 40, pp. 651–655.

Rhoades, J. D., N. A. Manteghi, P. J. Shouse, and W. J. Alves. 1989. Soil electrical conductivity and soil salinity: New formulations and calibrations. *Soil Sci. Soc. Am. J.* v. 53, pp. 433–439.

Rhoades, J. D., P. J. Shouse, W. J. Alves, N. A. Manteeghi, and S. M. Lesch. 1990. Determining soil salinity from soil electrical conductivity using different models and estimates. *Soil Sci. Soc. Am. J.* v. 54, pp. 46–54.

Schlumberger Wireline and Testing. 1991. *Log Interpretation Principles/Applications.* Schlumberger Wireline and Testing. Sugar Land, TX.

Sharma, P. V. 1997. *Environmental and Engineering Geophysics.* Cambridge University Press, Cambridge, UK.

Sorensen, K. 1996. Pulled array continuous electrical profiling. *First Break.* v. 14, no. 3, pp. 85–90.

Steinich, B., and L. E. Marín. 1997. Determination of flow characteristics in the aquifer of the Northwestern Peninsula of Yucatan, Mexico. *J. Hydrology.* v. 191, pp. 315–331.

Taylor, R. W., and A. H. Fleming. 1988. Characterizing jointed systems by azimuthal resistivity surveys. *Ground Water.* v. 26, pp. 464–474.

Watson, K. A., and R. D. Barker. 1999. Differentiating anisotropy and lateral effects using azimuthal resistivity offset Wenner soundings. *Geophysics.* v. 64, pp. 739–745.

6 Electromagnetic Induction Methods

Jeffrey J. Daniels, Mark Vendl, M. Reza Ehsani, and Barry J. Allred

CONTENTS

6.1   INTRODUCTION

Electromagnetic (EM) techniques are the methods widely utilized for locating conductive and metallic objects in the subsurface. The foundation concept of EM induction is that any time-varying EM field will cause current to flow in any conductive (or semiconductive) object that it encounters. The EM field propagates freely through space and most semiconducting materials, as described by the wave equation. When the low-frequency EM field encounters an object with a different conductivity from the host material, then the magnetic field of the EM field induces eddy currents (vortex currents) in an object that has a conductivity contrast with the surrounding medium, along with galvanic currents caused by the electric field components. The direction of the current flow in the conductive object will tend to oppose the direction of the original field (sometimes called the primary, or inducing field), and will cause its own EM field (sometimes called the secondary field). The presence or absence of the galvanic currents is a function of the size, shape, and electrical properties of the buried object and host material. The eddy and galvanic currents emit a secondary EM field, and this process is often referred to as induction. Sensors (coils, or long wires) are used to measure this secondary EM field, and the task of the geophysicist is to determine the cause of the anomalies that are present in the data.

6.2 BASIC PRINCIPLES

The fundamental principle of EM induction is illustrated in Figure 6.1. In environmental, agricultural, and engineering applications, EM methods are used primarily for locating either metallic conductors or conductive fluids, and determining soil moisture and mineralogy variations. The underlying concept of induction is widely known, and it is used for a variety of common tasks that include metal detectors at airport security entrances, the measurement of electrical current flowing in wires, and hunting for buried treasure with metal detectors.

The EM theory that underlies EM techniques has been developed since the later part of the nineteenth century, with the work Lenz, Biot, Ampere, Faraday, and Maxwell. There are numerous excellent physics and engineering texts on the subject at all levels of mathematical complexity, including works by Stratton (1941), Balanis (1989), Morse and Feshbach (1953), and many others. Excellent summaries are present in the geophysical literature (e.g., Keller and Frischknecht, 1966, and Ward and Hohmann, 1987). The following discussion is intended to cover some of the fundamentals that are essential to a basic understanding of the concepts underlying EM methods.

Ampere was the first investigator to recognize that a current passing through a wire has an associated magnetic field. A simple experiment, shown in Figure 6.1a, is often used to demonstrate the principle of Ampere's law. Iron filings laying in a plane perpendicular to a current-carrying wire (e.g., a wire sticking up through a table containing iron filings) will form circular patterns around the wire when current flows through the wire. The path of the line integral for the case of a straight current-carrying wire becomes a circle with its center on the wire. The integral sum of the magnetic field along this path is $2\pi r$, which is the circumference of a circle of radius r, and the relationship between the EM field caused by current flow can be stated as follows:

$$B = \frac{I}{2\pi r} \tag{6.1}$$

where B is the magnetic induction (magnetic field strength), I is the current flowing in the wire, and r is the distance from the wire.

The magnitude of the magnetic field increases proportionally to the electric current, and there are no magnetic fields in any direction other than the circular path around the wire. Furthermore, the direction of the magnetic field is in a path determined by the "right-hand-rule," which states that

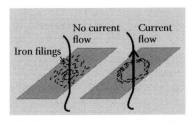

(a) Effect of electric current on iron filings

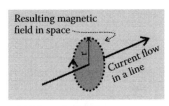

(b) Magnetic field from current along a straight wire

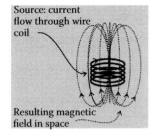

(c) Magnetic field from current through a coil

FIGURE 6.1 Ampere's law: (a) current flowing through metal filings resting on a table will cause magnetic filings to arrange in a circumferential pattern around the current-carrying wire; (b) current flowing through a wire causes a magnetic field that is circumferential to the wire carrying the current; and (c) current flowing through a coil of wire causes a perpendicular magnetic field that is in the shape of a toroid around the coil. Note the current flow and magnetic field directions in all cases of the primary and secondary fields follow the right-hand rule.

if we think of placing our right hand around the current-carrying wire, with our thumb in the direction of current flow, then the direction of the magnetic field is in the same direction as our fingers point. We often call the directional lines that represent the magnetic field, *lines of force*. Maxwell's contribution was to modify Ampere's law to include time-varying EM fields (a changing electric field) along with conduction currents.

Ampere looked at induction from the point of view of the magnetic field caused by the flow of electric current and only perceived the phenomenon in terms of his concept of action-at-a-distance. Conversely, Michael Faraday turned the problem around and considered it from the perspective of the magnetic field. He developed the concept of lines of magnetic induction. Faraday's concept is the basis for our way of visualizing the magnetic field in terms of lines-of-force and flux-density. Faraday's law states that a moving magnetic field can change current flow in a conductor in a manner that is the converse of Ampere's law. The simplest experiment to demonstrate Faraday's law is to move a magnet through a loop of wire, as shown in Figure 6.2a. If a current-flow meter (Amp meter) is attached to the loop, then the current flow that is measured is proportional to how fast the magnet is moved through the loop. The current in the loop is called the induced current, and the moving magnet that is inducing the current is called an induced electromotive force. An analogous experiment utilizes two loops, as shown in Figure 6.2b, with current flowing through the loop on the

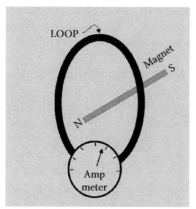

(a) Moving bar magnet inducing an electric current in a loop

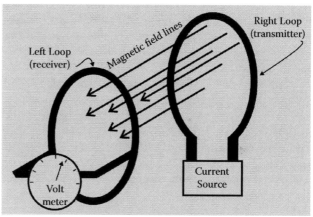

(b) Current flow in right loop causes a magnetic field which induces current flow in the left loop

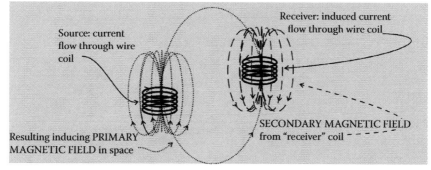

(c) Current source causes magnetic field inducing a current flow in receiver coil

FIGURE 6.2 Principles of induction: (a) a magnet moving through a loop causes current flow in the coil; (b) current generated in one wire loop causes a magnetic field that induces current flow in a second coil; and (c) the combination of Ampere's and Faraday's laws as used in geophysical electromagnetic equipment. (Parts (a) and (b) modified from Halliday, D., and Resnick, R., 1960, *Physics* (parts I and II), John Wiley & Sons, New York. With permission.)

right-hand side of the figure caused by a battery. The right loop emits a magnetic field according to Ampere's observations, with the vector direction of the magnetic lines of flux governed by the right-hand-rule. The induced magnetic field propagates through the air and induces current to flow in the passive loop that is shown on the left-hand side of the figure. Figure 6.2c shows how these laws can be combined in a "source" and "receiver" arrangement, with the receiver "detecting" the EM field, as follows: (1) current flowing in the source coil creates a "primary" magnetic field, (2) flux lines from the primary field flow through the second coil on the right creating a flow of current in the second coil (called an induced current), and (3) the flow of current in the second coil, in turn, creates a "secondary" magnetic field. It logically follows that this secondary magnetic field can, in turn, create another secondary magnetic field that will induce another current flow in the source coil. This is called mutual coupling. It should also be noted that if the secondary coil is oriented perpendicular to the primary coil, then no current will be induced in the secondary coil. The vector nature of the EM field is often used to determine the maximum and minimum direction of the EM field by rotating the receiver coil.

So far in this discussion, we have only been concerned with static, or steady, fields (i.e., constant, or direct, current; magnetic moving at a constant velocity, or no acceleration, through the loop). If we consider that the field is time varying (e.g., and alternating, or AC current), then Faraday's law expressed as a time-varying magnetic field can be combined with Ampere's law to form the wave equation. We will not show the differential equations for these expressions, but they show that any time-varying EM energy (AC current, or sinusoidal magnetic field) will move (propagate) through time and space. The wave equation governs the propagation of all EM waves, including radio waves. There is also a mechanical analog of the EM wave equation for acoustic and seismic wave propagation.

Ampere's and Faraday's laws are the two fundamental laws for all of EM theory, and the governing principles for the EM induction geophysical method. The practical challenge for the geophysical interpreter is to visualize how this applies in the subsurface. We need to combine the principles of Faraday's and Ampere's laws, visualizing the EM field interacting with an object in the subsurface, as illustrated in Figure 6.3. The source, or EM transmitter, energized by a time-varying current, emits an EM field that propagates into the subsurface. If the EM field encounters a change in electrical conductivity, then a change in the EM field is induced in the object. The object-induced

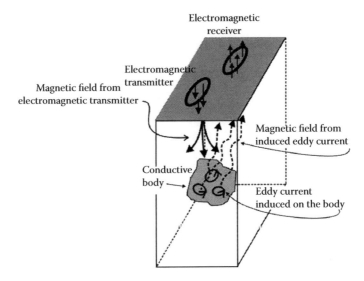

FIGURE 6.3 Basic principle of geophysical induction. A *primary electromagnetic field* emitted from a transmitter propagates through space until it encounters a conductor. Eddy currents induced in the conductor radiate a *secondary field* that is measured at the receiver location.

secondary EM field propagates into the surrounding media. Some of the secondary EM energy (flux) is detected by an EM receiver on the surface. The receiver records both the primary field from the source and the secondary field from the object in the subsurface. These two signals must be separated in the electronics of an EM system.

6.3 EQUIPMENT

There are many implementations of EM methods that have been developed for geophysical applications over the years. In addition to the type and orientation of transmitter and receiver, EM methods may be classified in a fundamental way as those that use an artificial source of EM energy, and those that use the earth's natural EM field (telluric, magnetotelluric [MT], audio magnetotelluric [AMT] methods). Artificial source methods are often classified by the type of source and detector (ground electric field line, or magnetic field coils or magnetometers), and the nature of the transmitter and receiver (e.g., long line or dipole electric field, small loop or large loop). Artificial source methods may be further subdivided by the type of wave that is transmitted and received: a system that transmits and receives a wave of a single frequency is called a frequency domain system, and a system that transmits and receives a multiple-frequency pulse of energy is called a time-domain system. Artificial and natural systems are further classified by the range of frequencies (e.g., low frequency, very low frequency [VLF], extremely low frequency [ELF], audio frequency, etc.), and the characteristics of the EM field vector that is measured (e.g., ellipticity, tilt, amplitude, or phase). To further complicate things, methods have come to be known by their commercial names, and EM techniques have been developed for use on the ground, in the air, and in boreholes.

The EM field at any point in space is a vector, with a magnitude and a direction. The magnitude of the field at a given point in space is a function of the orientation of the transmitted field, the modification of the direction of the field by materials between the transmitter and receiver (the object of the measurements), and the direction of the field measured by the receiver (receiver orientation). EM measurement instruments and field surveys are designed to exploit the vector nature of the EM field, and to measure attributes of the EM field that indicate the size, depth, and orientation of the objects in the subsurface. The attributes measured relate to either the spatial or time relationships of the measured secondary EM field. The field attributes include the amplitude, time delay (phase), and orientation of the received field with respect to the primary transmitted field. Specifically, the following parameters can be measured: (1) the phase of the spatial components with respect to the source, (2) the orientation of the field (tilt, and the axis of the field ellipsoid), (3) the relative amplitudes of measurements at different frequencies, (4) the amplitude ratio and phase differences between different spatial components, or (5) the phase and amplitude of individual spatial components with respect to the source. Methods are designed to "normalize" the effect of the primary field. The separation of most transmitter–receiver pairs is fixed to eliminate geometric effects from the primary field caused by varying the relative positions of the transmitter and receiver.

If the transmitter and receiver are located above the surface of the ground, then the field measured at the receiver is a combination of the field that propagates directly through the air (the primary field), the EM field that is influenced by the background material in the subsurface (sometimes referred to as the terrain), and the induced secondary field from the object of a contrasting conductivity located in the subsurface. The field that propagates directly from the transmitter is called the primary field, and the field caused by the eddy currents induced on the subsurface terrain and any buried object in the subsurface is called the secondary field, as shown in Figure 6.4a. The direction (sign) of the secondary field induced by the conductive object in the subsurface is opposite to the primary field, in accordance with Lenz's law.

The secondary field is also "shifted" in time, and this shift is called a phase shift. The terminology describing the phase of a wave is not discussed very clearly in most of the literature. However, popular usage of the term "phase" in describing induction EM methods usually refers to the fact that the secondary field has a certain phase relationship to the primary field. The phase shift is

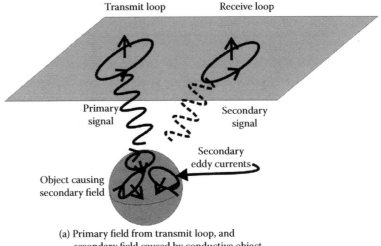

(a) Primary field from transmit loop, and
secondary field caused by conductive object

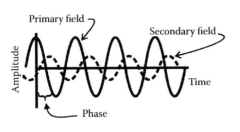

(a) Primary and secondary fields
arriving at the receive antenna

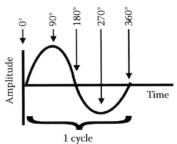

(c) Amplitude, time, phase relationships
of 1 cycle of oscillation of the primary signal

FIGURE 6.4 The primary and secondary electomagnetic field: (a) the received signal referenced to the primary field, and (b) the amplitude and phase of the received signal related to several descriptions of the field. A single cycle representing a continuously oscillating frequency domain wave for a single point on the spreading primary field and the induced secondary field is shown.

meaningful only when it is related to some standard time reference, which is usually the primary field. For example, a secondary field that is 90° out of phase with the primary field has its maximum amplitude (or zero crossings) shifted one-quarter cycle in time with respect to the primary field. A phase shift of 180° means the secondary field is the mirror image of the primary field. The relative phase measurement points on the primary wave are shown in Figure 6.4c.

The field measured at the receiver is the vector sum of the primary and the secondary fields, as shown in Figure 6.5. The electric and magnetic fields are three-dimensional vectors with a magnitude and direction that vary with each cycle of the combined primary and secondary fields measured at the receiver. At any location away from the source, the tips of the electric and magnetic field vectors can be viewed as tracing out ellipsoids in space over a period of time, as shown in Figure 6.5a. The tracing of the ellipsoid is repeated for each new cycle of each frequency of the transmitted wave. The size of the ellipsoid at any point on its surface is proportional to the magnitude of the electric (or magnetic) field vector at that particular point. This can be called a polarization ellipsoid. There is a minimum and maximum value of the ellipsoid that can be measured at any point. Furthermore, the ellipsoid can be tilted in space, with the angle of the major axis oriented at an angle that is a function of the position and electrical properties of the object in the subsurface.

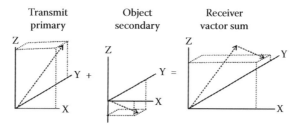

(b) Received field as a vector sum of the primary and secondary fields

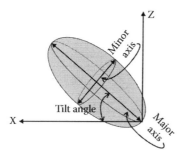

(a) Parameters associated with ellipsoidal representation of the
amplitude and vertical orientation of the field

FIGURE 6.5 The vector nature of the electromagnetic field: (a) the primary magnetic field vector, the secondary field vector, and their sum; and (b) the ellipsoid that represents the direction, amplitude, and tilt of the measured field.

If the amplitude and orientation of the field are measured, then the field can be viewed as an ellipsoid of revolution, as shown in Figure 6.5b. The parameters measured at the receiver are (a) the maximum amplitude corresponding to the major axis of the ellipsoid; (b) the minimum amplitude, which corresponds to the minor axis of the ellipsoid; and (c) the tilt angle of the major axis with respect to the horizontal ground surface. In some cases, the azimuthal (x–y position) orientation of the major axis may also be measured.

Field instruments are designed to separate (or normalize) the primary and secondary EM fields, because the secondary field is most important for detecting an object in the subsurface. In addition to the primary field and secondary field, some systems utilize the in-phase and out-of-phase components, or the real and imaginary components. The out-of-phase component is also sometimes referred to as the quadrature component. All of these definitions are related to the complex nature (oscillating, time-varying) of the EM field when it is referenced to the primary field.

One distinguishing characteristic of various EM methods is their operating domain and the frequency, or time period, of the signal transmitted and received. The operating domain refers to time or frequency domain designation, which in theory should provide equivalent results because the two domains are mathematically related by the Fourier Transform. The principles of the time and frequency domains are illustrated and contrasted in Figure 6.6. A time domain input signal is generally a square wave with a positive and negative polarity. The input signal is typically a few hundred milliseconds long, and may be as long as a second or more; the time period depends upon the application, with longer times used to investigate deeper into the earth. As shown in Figure 6.6a, the received signal is no longer a square wave. After the input signal is turned off, there is a decay and delay of the signal over time. This shape of the decay curve is a function of changes in the signal as it travels into the earth and encounters objects with different electrical conductivity values. The received signal (output decay) is recorded on a separate coil and the signal amplitude measured at different delay times. These amplitude decay signals are interpreted as a function of the subsurface distribution of electrical conductivity.

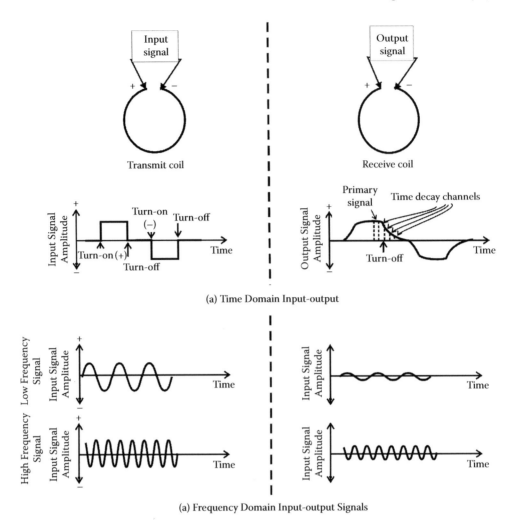

FIGURE 6.6 The time and frequency domains transmit and receive signals: (a) time domain transmits (input) and receives (output) signals, and (b) frequency domain transmits and receives signals.

Frequency domain systems utilize an input signal that is a sinusoidal wave with a fixed frequency (a square wave time domain signal is composed of many frequencies), as shown in Figure 6.6b. The signal recorded at the receive coil is delayed (called a phase shift) in time with a decreased amplitude. The amplitude and phase of the received signals are used to interpret the propagation path and induction of the signals that have traveled from the transmit coil, through the earth, to the receive coil. Some frequency domain systems transmit and receive several different frequencies in sequence, as illustrated by the low-frequency and high-frequency signals shown in the figure. Lower-frequency signals penetrate more deeply into the earth than higher-frequency signals; hence, the depth of investigation can be somewhat predicted by the chosen frequency. Typical frequencies range from a few hundred Hz for intermediate-depth investigations (approximately 10 to 100 m) to several hundred kilohertz for very shallow soil conductivity determinations.

So far, we only illustrated the use of coils for transmitting and receiving EM fields. However, in many geophysical applications, EM signals are transmitted and received by straight wires (commonly called lines) or by loops (or coils) of wire. Most lines have electrodes in the ground at each end of the line, as shown in Figure 6.7. If the grounded ends are very close together with respect

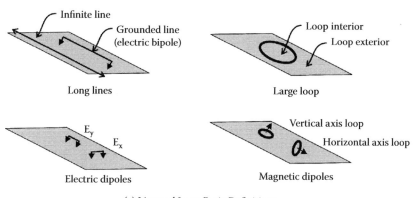

(a) Line and Loop Basic Definitions

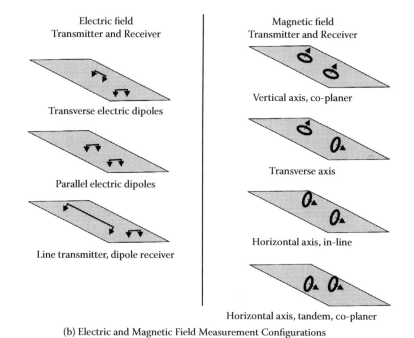

(b) Electric and Magnetic Field Measurement Configurations

FIGURE 6.7 Electromagnetic electric field line and magnetic field loop (a) definitions and (b) configurations.

to the distance from the transceiver, as shown in Figure 6.7b, then the electric field transceiver is called an electric dipole. The designation of a transceiver as a dipole, or conversely a long line, is determined by the proximity of the observer to the transceiver. A magnetic loop is called a dipole if the receiver is placed a long distance from the loop relative to the diameter of the loop. A good rule of thumb is that if the receiver is placed a distance that is greater than ten times the diameter of the transmitter loop, then the transmitter can be considered to be a dipole. Theoretically, geophysicists often refer to this as the far-field. In practice, the EM fields are more uniformly distributed in the far-field, and this makes interpretation of measurements in the far-field much easier than those in the near-field. The same far-field rule holds for an electric field dipole. Conversely, if the diameter of a loop is very large, and the observer is very close to the loop, then the "loop" approximates a

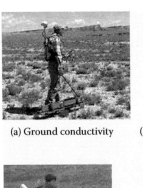

(a) Ground conductivity   (b) Shallow-intermediate   (c) Shallow-intermediate depth
                                  depth   (person-pulled, or ATV towed)

(d) Shallow-intermediate   (e) Intermediate depth   (f) Towed array with GPS
    depth continuous mode   fixed station
    person carried

FIGURE 6.8 Examples of commercially available time- and frequency-domain electromagnetic (EM) systems: (a) ground conductivity system (www.geonics.com/html/products.html), (b) frequency-domain intermediate-depth EM system (www.geophex.com/GEM-2/GEM-220home.htm), (c) shallow–intermediate-depth time-domain system (www.geonics.com/html/metaldetectors.html), (d) shallow–intermediate-depth person-portable system (www.geophex.com/GEM203/GEM-320home.htm), (e) intermediate-depth fixed station (www.geonics.com/html/conductivitymeters.html), and (f) towed array of five frequency-domain sensors (www.geophex.com/GEM-2/GEM-220home.htm).

line from the observer's viewpoint, and the EM fields measured by the observer are approximately the same as those that would be measured from a long line. In most cases, the manufacturer supplies an instrument with the transmitter and receiver located at a fixed size/spacing ratio, and the geophysicist does not have to be concerned with interpretation problems associated with not being in the far-field.

For purposes of agricultural applications, the most practical systems are handheld, or cart-mounted loop-loop systems. Examples of a few of the systems currently available are shown in Figure 6.8. Note that some of the systems operate in the time domain, and other systems utilize multiple frequencies and are called frequency domain systems.

6.4   FIELD OPERATION CONSIDERATIONS

Field procedures for EM measurements follow the basic design that must be employed for most geophysical methods. The process of making a series of geophysical measurements in the field is commonly called a survey. The field procedure for a specific objective is determined by answering a number of questions, including the following:

- What are the objectives of the survey?
- What is the nature of the subsurface environment?
- What are the electrical properties of the materials at the site?
- How large is the survey area?
- What is the nature of the site access?
- How will cultural features affect the measurements?

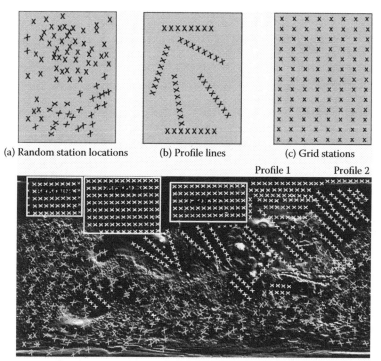

(a) Random station locations   (b) Profile lines   (c) Grid stations

Profile 1   Profile 2

(d) Station distribution in grids, profile lines and randomly in an area with mixed topography
(image modified from htt://ssed.gsfc.nasa.gov/tharsis/shademap.html)

FIGURE 6.9   Station arrangements: (a) random distribution of stations, (b) stations along profile lines, (c) a grid of stations, and (d) mixed station arrangements in a complex topographic setting.

The fundamental questions that must be asked prior to any survey are as follows: (1) What is the maximum depth of penetration?, (2) What are the line and trace spacing (horizontal resolution)?, and (3) What is the resolution needed to achieve the goals of a survey?

6.4.1   MODE OF SENSOR TRANSPORT

EM system sensors and recording equipment may be hand carried, backpacked, pulled along in a handcart, or mounted on a vehicle (aircraft, all-terrain-vehicle, or some other type of motorized vehicle). In any case, registering the measurement locations for future reference is one of the most difficult issues related to any type of field sensor measurements. This may be accomplished by establishing measurement locations (stations) over a grid or along profile lines, as shown in Figure 6.9, or by recording a Global Positioning System (GPS) measurement at each station. If the topography is rugged, or if the landscape is cluttered with trees, roads, fences, or other cultural features, then it may be necessary to locate stations in pattern that appears to be more-or-less randomly dispersed, as shown in Figure 6.9a. GPS measurements are usually employed to record the location of dispersed stations. Figure 6.9d illustrates an environment where grids of stations and profile lines are used in the flat terrain, and randomly dispersed station locations are employed in the rugged terrain. Measurements along profile lines are often utilized for locating linear features, such as buried pipes, in the subsurface.

6.4.2   STATION SPACING

A good rule of thumb for establishing an adequate trace measurement spacing is that the spacing should be less than one quarter of the size of the smallest object that is to be detected by the survey.

This value is twice the Nyquist sampling frequency and should be adequate for most situations. Technically, "oversampling" is never a real problem, but measuring more stations than necessary costs extra money and should be avoided for economic reasons. Economics and other practical considerations (e.g., spatial survey accuracy) are the only limitations on using very small distances between stations.

6.4.3 DEPTH OF PENETRATION

Surveys should be designed so that the instrument depth of penetration is twice that required to see the deepest object anticipated in the survey area. The depth of penetration varies with the separation of transmit and receive coils (Geonics Limited [Date Unknown]—Technical Note TN-31). However, from a practical point of view, the only way to know the depth of penetration for a particular survey site is to make an initial guess based on estimations from the skin depth calculations (Equation (6.2)). The skin depth can be calculated using the following relationship:

$$\text{one skin depth} = \frac{1}{\alpha} = \frac{1}{\omega \left\{ \frac{\mu \varepsilon}{2} \left[\left(1 + \frac{\sigma^2}{\varepsilon^2 \omega^2} \right)^{1/2} - 2 \right] \right\}^{1/2}} \text{ m} \qquad (6.2)$$

where α is the attenuation constant in the wave equation (Balanis, 1989), σ is the subsurface electrical conductivity, μ is the subsurface magnetic permeability, ω is angular frequency, and ε is the subsurface electrical permittivity. A plot of the skin depth versus frequency is provided in Figure 6.10.

6.4.4 INTERFERENCE EFFECTS

The physical surroundings of a survey and the proximity of interfering features should be a consideration for any survey. Some of these features are shown in Figure 6.11. Overhead or underground power lines can give off a signal that interferes with commercial systems, although most systems have 60 Hz filters that eliminate most of the power line related noise. There are no foolproof rules of thumb for eliminating these noise sources. However, they should be avoided whenever possible.

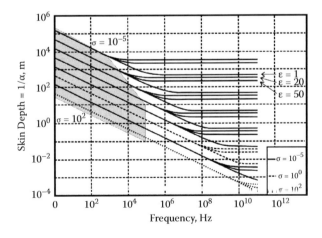

FIGURE 6.10 Skin depth plotted as a function of frequency. One skin depth is the depth at which the signal is attenuated by the amount $1/e$. The area in gray is the electromagnetic induction region, where attenuation is primarily influenced by the conductivity (σ) and is minimally affected by the permittivity (ε).

FIGURE 6.11   Field features that can interfere with electromagnetic measurements.

Furthermore, structures containing metal are always more problematic than nonmetallic objects; a building constructed from some types of stone may not create any noticeable interference, but a building constructed with a lot of steel-reinforcing bar or from rocks containing a lot of magnetite may cause serious interference for any EM system. A practical guideline is to avoid making measurements any closer to an object than five times the height of the object. A simple field test consisting of running a profile line perpendicular to the potentially interfering feature can determine the limit of influence of overhead structures.

6.5   PROCESSING, DISPLAY, AND INTERPRETATION

With the exception of metal detectors used for locating buried utilities, EM data must be displayed and interpreted. The level of sophistication varies from visual identification of anomalies on the data to more complex modeling and inversion to extract specific features from a buried object of interest. The following outlines these approaches to data preparation and interpretation.

6.5.1   DATA PREPARATION AND PROCESSING

Assuming the interpreter has chosen the type of data to be displayed (e.g., phase, amplitude, ellipticity, etc.), EM data can be displayed effectively as profile lines or two-dimensional maps. A critical stage in the analysis and processing of geophysical data is their presentation in a manner that permits the analyst to obtain a comprehensive, integrated view of the data set. The use of contour maps is based on experience that indicates they are very useful in presenting the spatial and amplitude character of geophysical fields and subsurface features and their relationship to one another. Two-dimensional displays permit rapid visual identification of patterns, trends, gradients, and other characteristics of a three-dimensional field. In a similar manner, data observed in one dimension, along a linear or curvilinear transect, are displayed as profiles with the geophysical variable as the vertical component and the relative spatial position of the observation site as the horizontal axis. Machine processing and presentation of digital data, which is the norm, have opened up numerous

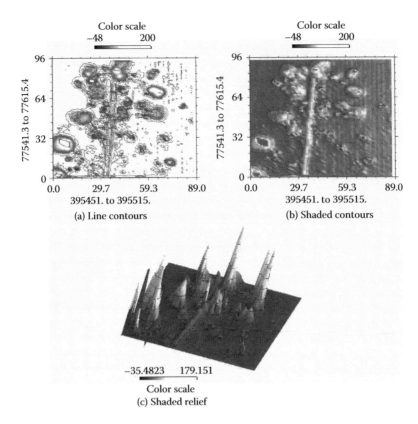

FIGURE 6.12   Basic presentations of gridded data: (a) line contours, (b) shaded contours, and (c) perspective view.

alternative schemes of presentation (e.g., gray scale, perspective views, shaded relief, color contouring) that are widely used to supplement or complement the traditional black-and-white contour maps (or profiles).

Gridded data can be presented in numerous useful forms that complement the contour map and assist the analyst in evaluating and interpreting the data set. The most basic form is the contour map, where lines are drawn through data with the same values. The simplest presentation of data is the line contour map (Figure 6.12a), but the shaded contours (Figure 6.12b) can provide a view that is easier to interpret. Perspective presentations (Figure 6.12c) of gridded data provide a three-dimensional view of the data set that generally is more pictorial than a contour map. The three-dimensional view is pictured from a specified azimuth and angle above the zero level (horizon) with variable amplitudes plotted as heights as in a three-dimensional view of topography. The information shown is highly dependent on the position of the viewing site and the vertical scale exaggeration. Thus, finding an optimum presentation often is an iterative process.

Amplitude filtering is probably the most commonly used processing and interpretation tool. In fact, it is so common that we do not even think of it as filtering. Amplitude filtering is best illustrated by example. Figure 6.13 shows a magnetic contour map that has been plotted at two different amplitude levels. Clearly, a few anomalies are enhanced when we filter out the low-intensity (low-amplitude) anomalies. However, the trade-off is always that we lose detail and may eliminate anomalies of interest when we select the cutoff values. It is usually best to display and interpret several maps with different cutoff values.

There are many other forms of automated spatial filtering of gridded data that we will not detail in this chapter. Examples and explanations of filtering based on linear systems analysis and statistics can be found in Odgers (2007) and Cressie (1993). Profiles of EM measurements are useful to

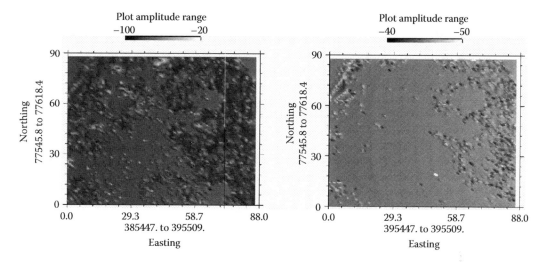

FIGURE 6.13   Contour shaded magnetic maps using different amplitude cutoff values.

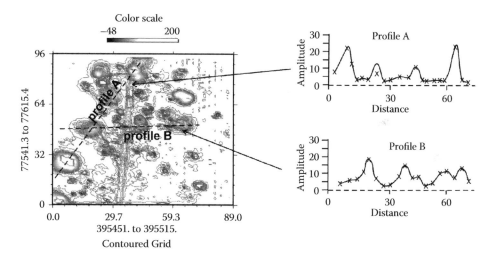

FIGURE 6.14   Extraction of profile lines from gridded data.

locate and define anomalies. If the orientation of an object of search is known, and the object is elongated in one direction, then profile lines may be the preferred means to find the location of the object. Profile lines may also be used to extract data from gridded data for modeling. Profiles can be extracted from the gridded data and can be stacked in their relative spatial position based on a level of the variable representing the position of the profile. The resulting stacked profile map gives a three-dimensional view of the data set focusing on the shorter wavelength components that may be lost in the normal contouring process. A simulated profile extraction from a set of gridded data is shown in Figure 6.14.

6.5.2   INTERPRETATION

Interpretation of EM data is conducted as different levels of complexity, depending upon the objective. A simple metal detector beeps when it supposedly passes over a very shallow buried object, and EM data measured with more sophisticated geophysical instruments need to be plotted and analyzed. The simplest analysis consists of visual inspection of the data, and the most complex interpretations involve inverse modeling.

6.5.2.1 Visual Interpretation

A visual interpretation consists of plotting the data and mapping the location of the anomalies from profiles or a contour map of the data. Clearly the data in Figure 6.14 indicate the location of two types of anomalies: the long linear feature that trends from the top to bottom of the contoured grid is a pipeline, and the other features are buried pits filled with metallic debris. If anomalies stand out on the data, like they do in Figure 6.14, and there is no need to know the depth and size of the objects, then a visual interpretation may be adequate. However, if more detail is required from the data, then it is necessary to model the data, using a numerical simulation of the EM response of a buried object to an external EM field.

6.5.2.2 Forward Modeling

In geophysics, we have the disadvantage of not being able to directly measure all of the forces and responses for a particular phenomenon, because most of the lines of force are buried and inaccessible to direct measurement. Therefore, we must resort to making a few measurements on the surface of the earth, a borehole, or air, and deduce the remaining points from the observed measurements. In order to make these deductions, we must determine the distribution of objects in the subsurface that created the distribution of EM fields that were measured on the surface, in the air, or in the borehole. The procedure that we use to simulate the response from an idealized distribution of objects in the subsurface is called the process of mathematical, or numerical, modeling.

In a more general sense, a model is either a physical or a mathematical analog of the distribution of physical properties in the subsurface that gave rise to the observed measurements. The physical analog may consist of a test pit or water tank containing the objects that have been hypothesized to cause the observed measurements. The objects and the geophysical measurement techniques are scaled-down versions of the objects buried in the earth and the equipment that was used to make the observed measurements. A mathematical model is often used rather than a physical model. The mathematical model consists of a solution to a mathematical description of the diffusion of energy in the case of EM induction and thermal methods.

Mathematical models are computed using the differential equations that describe wave propagation as discussed earlier in this chapter. The models generated from solving these equations for buried objects in the presence of a particular geophysical field are called theoretical, or forward models. These models consist of spatial and physical property parameters. The spatial parameters of a model are the size and location of the objects, and the physical property parameters depend upon the type of geophysical measurement being modeled.

Models are usually composed of bodies (sometimes referred to as objects or targets) of an idealized shape. These idealized objects are represented by boundaries between physical properties. The difference between the value of the physical property within the object and the value of the physical property surrounding the object is called the contrast in the physical property. The boundary is the surface that separates the object from the surrounding (or host) material. Models are classified as zero, one, two, or three dimensional, depending upon how many dimensions are used to define the object. Examples of one-, two-, and three-dimensional objects are shown in Figure 6.15. The simplest model is a whole-space, where there are no boundaries (only a host material). There are no true whole-spaces, but outer space is a close approximation for most physical properties.

The surface of the earth can be modeled as the boundary between two half-spaces: an upper half-space (the air), and a lower half-space (the solid earth). The earth is an imperfect example of a lower half-space. We know that the earth's surface is not flat. It is a spheroid. However, within the variations of shallow crustal and near-surface measurements, the surface of the earth is a good

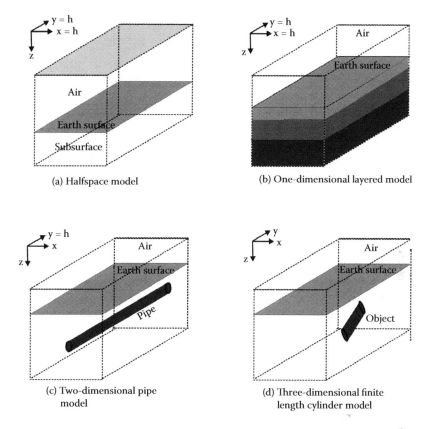

FIGURE 6.15 Dimensionality of models: (a) a homogenous half-space that extends to infinity in the x and y directions, (b) a layered-earth model that extends to infinity in the x and y directions, (c) a two-dimensional pipe model that extends to infinity in the y direction, and (d) a three-dimensional object that has finite dimensions in all three directions.

approximation to the boundary between two half-spaces. This brings us to a very important point about models: *models are idealized approximations to the distribution of objects in the subsurface.* Models make it possible to simulate geophysical measurements mathematically, and they are never an exact replication of conditions in the subsurface.

The layered earth (Figure 6.15b) is an example of a one-dimensional model. It is one dimensional because the physical properties change in only one dimension. A pipe that is infinitely long (Figure 6.15c) is an example of a two-dimensional object, and a spheriod (Figure 6.15d) is an example of a three-dimensional object.

Forward models like that shown in Figure 6.15 are used by geophysicists in a variety of ways, including the following:

- Presurvey prediction of the response expected from targets in the surface. Calculating models prior to a survey prevents using a particular technique when the theory predicts that the resulting field measurements will not detect a target.
- Design of field surveys to optimize line and station spacing.
- Inversion of field data.

Nearly all geophysical measurements that follow the basic laws of Newtonian and Maxwellian physics can be approximated by mathematical models. These mathematical models describe the spatial and temporal (time-dependent) "state" of the EM field. In other words, the models describe the physical concepts of changes the direction the EM field propagates as a function of time into the mathematical language by the use of derivatives and vectors. All of the fundamental equations in Newtonian and Maxwellian physics describe the physical phenomenon occurring at a particular point in space, and at a particular instant in time. The fundamental question that needs to be answered is: "What is the reaction to a propagating EM wave that is acting on a particular point at a particular instant in time?" These changes in the EM fields are described by a partial differential equation, which mathematically can be expressed in a general way as $f(x,y,z,t)$, where f is the functional expression, x,y,z represent the spatial changes in Cartesian coordinates, and t is time. The differential equation is needed because fields (force and electric) change with time, and the derivatives are the mathematical way to express changes in time and space. It should be noted that geophysical forward models for all geophysical measurements that follow the laws Newton and Maxwell are related to, in the form of, or can be derived from the wave equation.

6.5.2.3 Inverse Modeling

Geophysical data are generally very simple measurements made at a single point below, on, or above the surface of the earth. These relatively simple measurements are generally used to infer some pretty complicated events, or physical property distributions in the subsurface. Inverse modeling involves the process of manipulating the parameters of a theoretical model until the values computed from the mathematical model match the field measurements. The process of inverse modeling for some geophysical methods (e.g., gravity, magnetic, resistivity, thermal) is often called curve matching, because the process involves matching, or fitting, the curve computed from mathematical modeling with the curve of values from field measurements. Figure 6.16 illustrates field data along a profile and the computed response from a hypothetical model. The generalized procedure that this used in interpretation (cut-and-try, analog curve match, or inversion) is shown in Figure 6.16c. The parameters of the model are adjusted in an iterative manner until the computed model response matches the field measurements.

Curve matching (or in the case of wave propagation, trace matching) can be achieved using several different approaches, including cut-and-try, analog overlays and templates, and automated inversion. The cut-and-try technique is a simple process that involves generating individual forward models, comparing the model to the field data visually, and iteratively changing the model parameters until the values for the theoretical model are close to the field measurement values. The analog overlay technique involves looking through a catalog of theoretical curves until the interpreter finds a curve that is very close to the field measurement curve. The analog overlay method was the only method available to interpreters prior to the widespread use of digital computers in the early 1960s. These are approximation techniques because of the simplifying assumptions made in arriving at a solution. Automated, or computer, inversion has become so standard in geophysics that it is nearly a specialization in itself. Most inverse modeling has been developed over the years for potential field and resistivity methods, and the following discussion reflects this fact. More recently, inverse models have been developed for EM methods.

Least-squares inversion is a numerical way to find the physical properties that generate a forward model response that most closely approximates the field measurements. Least-squares inversion is an automated way to implement the curve-matching flowchart shown in Figure 6.16c. Least-squares inversion can be applied to any data and model where changing the parameters in the mathematical equation changes the result of the equation in a linear fashion. Therefore, least-squares inversion can be applied to resistivity, gravity, and EM data by "linearizing" the model. An excellent explanation and summary of inverse modeling is provided at the UBC Geophysical Inversion Facility Web site (www.eos.ubc.ca/research/ubcgif/).

Inversion by the process of least-squares is a numerical problem of adjusting the physical properties of the model until the model curve matches (or comes close to matching) the curve traced

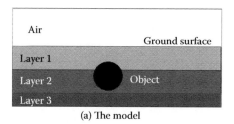

(a) The model

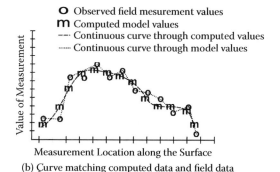

(b) Curve matching computed data and field data

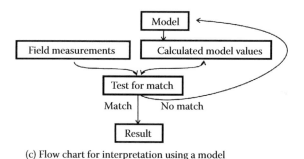

(c) Flow chart for interpretation using a model

FIGURE 6.16   Generalized process of inversion: (a) computed and field measurement values plotted as a function of distance along the surface of the ground, (b) the model used to compute the values for the *computed curve*, and (c) a flowchart of the process of inversion.

through the observed field data points. There are some basic rules for deciding on the type of model to be used for the inversion:

- The model must be appropriate for the situation at hand. You would not use the numerical model that represents a sphere to determine the parameters of a pipe.
- The model must be relatively easy to compute. It is generally inappropriate to use a model when the forward model takes an enormous amount of computer memory and time to compute. Several iterations to enable convergence should only take a short period of time and not require a supercomputer, or should not require a system with massive parallel processors. Of course, there is the rare situation where it might be appropriate to utilize a very complicated model.

6.6   SUMMARY

EM methods can be an effective means to locate conductive (metallic) and semiconductive (e.g., water and clays) materials in the subsurface. EM methods are rooted in well-known theory and are easily implemented in the field. The EM technique is highly adaptable to a variety of mobile plat-forms (handheld, vehicle mounted, or airborne) and does not require direct contact with the ground. Interpretation can range from simple visual inspection of the data to more complex geophysical inversion to determine the depth and size of the object causing the anomaly.

REFERENCES

Balanis, C.A., 1989, *Advanced Engineering Electromagnetics*, John Wiley & Sons, New York, 981 pp.

Cressie, N., 1993, *Statistics for Spatial Data, revised edition*, Wiley, New York, 900 pp.

Geonics Limited, (date unknown), *Applications of "Dipole-Dipole" Electromagnetic Systems for Geological Depth Sounding*, Geonics Ltd. Technical Note TN-31, www.geonics.com/pdfs/technicalnotes/tn31.pdf.

Halliday, D., and Resnick, R., 1960, *Physics* (parts I and II): John Wiley & Sons, New York, 1214 pp.

Keller, G.V., and Frischknecht, F.C., 1966, *Electrical Methods in Geophysical Prospecting*: International Series in Electromagnetic Waves, V. 10, Pergamon Press, New York, 517 pp.

Morse, P.M., and Feshbach, H., 1953, *Methods of Theoretical Physics* (Vol. I & II): McGraw-Hill, New York, 1978 pp.

Odgers, T., 2007, Gridding and Contouring Tutorial, www.geoafrica.co.za/contouring/contour_tutorial_gridding_methods.htm.

Stratton, J., 1941, *Electromagnetic Theory*, McGraw-Hill, New York, 615 pp.

Ward, S.H., and Hohmann, G.W., 1987, *Electromagnetic Theory for Geophysical Applications: Electromagnetic Methods in Applied Geophysics, Vol. I, Investigations in Geophysics*, Society of Exploration Geophysicists, Tulsa, OK, pp. 131–311.

7 Ground-Penetrating Radar Methods (GPR)

Jeffrey Daniels, M. Reza Ehsani, and Barry J. Allred

CONTENTS

7.1   INTRODUCTION

Ground-penetrating radar (commonly called GPR) is a high-resolution electromagnetic technique designed primarily to investigate the shallow subsurface of the earth, building materials, and roads and bridges. GPR has been developed over the past thirty years for shallow, high-resolution investigations of the subsurface. GPR is a time-dependent geophysical technique that can provide a three-dimensional pseudo-image of the subsurface, including the fourth dimension of color, and can also provide accurate depth estimates for many common subsurface objects. Under favorable conditions, GPR can provide precise information concerning the nature of buried objects. It has also proven to be a tool that can be operated in boreholes to extend the range of investigations away from the boundary of the hole.

7.2 BASIC PRINCIPLES

GPR uses the principle of scattering electromagnetic energy in the form of an electromagnetic wave to locate buried objects. The basic principles and theory of operation for GPR have evolved through the disciplines of electrical engineering and seismic exploration, and GPR specialists tend to have backgrounds either in geophysical exploration or electrical engineering. The fundamental principle of operation is the same as that used to detect aircraft overhead, but with GPR, antennas are moved over the surface, similar to a sonic fish-finder, rather than rotating about a fixed point. This has led to the application of field operational principles that are analogous to the seismic reflection method.

It is not necessary to understand electromagnetic theory to use and interpret GPR data, but it is important to know a few basic principles and have an empirical understanding of how electromagnetic energy travels (propagates) in the subsurface. We begin with the most basic description of a wave: a propagating wave is described by a frequency, a velocity, and a wavelength, as shown in Figure 7.1a. If we have two waves and add them together, then we also have to consider the phase, or time offset, of the wave, as shown in Figure 7.1b. We also know from basic physics that we can add waves with different frequencies and different phases together, then we can form any wave shape that we want; and, this wave will still propagate. A wave composed of multiple frequencies, with a resulting finite time duration (e.g., a pulse of electromagnetic energy), is called a time-domain wave. A time-domain wave, similar to the waveform commonly applied to GPR systems, is shown in Figure 7.1c. The most fundamental underpinning principle of GPR is as follows: (a) a pulse of time-domain electromagnetic energy can be formed by a transmit antenna, propagate into the earth with a particular velocity and amplitude, and be recorded by a receive antenna, and (b) the energy of pulse recorded over time provides a time-history of the pulse traveling through the subsurface (Figure 7.1c).

The theory of GPR is based on Maxwell's equations and the vector form of the wave equation, which is the same fundamental theory as the seismic method, with a major difference being that the seismic method is based on the scalar wave equation, and GPR is based on the vector wave equation. In a practical sense, this means that the propagating GPR wave has both a magnitude and an orientation.

7.2.1 PROPAGATION AND SCATTERING

The practical result of the radiation of electromagnetic waves into the subsurface for GPR measurements is shown by the basic operating principle illustrated in Figure 7.1. The electromagnetic wave is radiated from a transmitting antenna, spreads out over time in the form of a spherical wavefront (Figure 7.1a), and travels through the material at a velocity determined, primarily, by the permittivity (sometimes called the dielectric constant or electric permittivity) of the material.

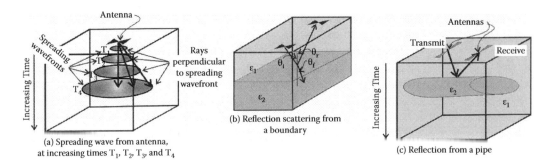

FIGURE 7.1 Simple wave reflection scattering in the subsurface: (a) simplified view of the process of a wave spreading from an antenna, (b) rays demonstrating by the ray method of a transmitted electromagnetic wave scattered from a buried layer with a contrasting permittivity, and (c) reflection scattering from a buried pipe. Permittivity of the host media is ε_1, and the permittivity of the reflecting target is ε_2.

When a wave encounters a material with a different permittivity, then the electromagnetic energy will change direction and character. This transformation at a boundary is called scattering. When a wave impinges on interface, it scatters the energy according to the shape and roughness of the interface and the contrast of electrical properties between the host material and the object. Part of the energy is scattered back into the host material, and the other portion of the energy may travel into, and through, the object. Scattering at the interface between an object and the host material is of four main types: (1) specular reflection scattering, (2) diffraction scattering, (3) resonant scattering, and (4) refraction scattering.

A vector line drawn from the transmit antenna to the reflection point on a layer (Figure 7.1b) or object (Figure 7.1c) is called a ray. Because the reflected energy follows the *Law of Reflection*, as illustrated by the rays in Figure 7.2, where the angle of reflection is equal to the angle of incidence, then the reflection point for any given transmit–receive antenna pair over a flat layer (Figure 7.1b, Figure 7.2) is at a point in the subsurface that is halfway between the transmit and receive antennas. The wave energy that propagates into the object, or layer, enters at an angle determined by contrast in electrical properties and is called the refracted energy. The angle that the wave enters into the second (lower) object or layer is called the angle of refraction and is determined by Snell's law, as shown in Figure 7.2. Refraction scattering is not generally an important consideration for GPR measurements over the surface of the earth, because GPR waves attenuate very rapidly in most near-surface earth materials, and velocities decrease with increasing depth. However, refraction may be important for determining layer thicknesses in building materials and roadbeds, where the velocity of an overlying layer is sometimes less than the velocity of a lower layer.

Diffraction scattering occurs when a wave is partially blocked by a sharp boundary. The wave scatters off of a point, and the wave spreads out in different directions, as first noted by Fresnel (1788–1827). The nature of the diffracted energy depends upon the sharpness of the boundary and the shape of the object relative to the wavelength of the incident wave. Diffractions commonly can be seen on GPR and seismic data as semicoherent energy patterns that splay out in several directions from a point or along a line. Geologically, they often are measured in the vicinity of a vertical fault, or a discontinuity in a geologic layer (abrupt horizontal change in the geology).

Resonant scattering occurs when a wave impinges on a closed object (e.g., a cylinder), and the wave bounces back and forth between different points of the boundary of the object. Every time the wave hits a boundary, part of the energy is refracted back into the host material, and part of the energy is reflected back into the object. This causes the electromagnetic energy to resonate (sometimes called ringing) within the object. The resonant energy trapped inside of the object quickly

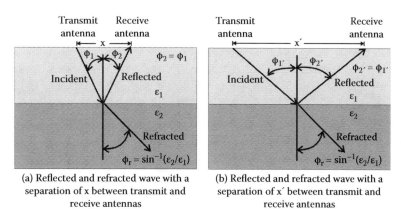

(a) Reflected and refracted wave with a separation of x between transmit and receive antennas

(b) Reflected and refracted wave with a separation of x′ between transmit and receive antennas

FIGURE 7.2 Transmitted electromagnetic wavefront reflected and refracted from a buried layer with contrasting electrical permittivities. Electrical permittivity of the host media is ε_1, and the permittivity of the lower layer is ε_2. Following the law of reflection, the angle of reflection is equal to the angle of incidence $\phi1 = \phi2$, and $\phi1' = \phi2'$. The angle of refraction follows Snell's law and is constant irrespective of the incident angle.

dissipates as part of it is reradiated to the outside of the object. Closed objects are said to have a resonant frequency based on the size of the object, and the electrical properties of the object and the surrounding material. However, the ability of an object to resonate depends on the wavelength (velocity of the object, divided by the frequency of the wave) with respect to dimensions of the object. The length of time that an object resonates is determined by the permittivity contrast between the object and the surrounding material.

Diffraction and resonant scattering are complicated phenomena that depend on the properties of the incident wave (including polarization, amplitude, and frequency content) and the size, shape, and electrical properties of the scattering object. Diffraction scattering can be seen on some GPR records where the boundaries between media are sharp (e.g., engineering investigations of foundations), but resonant scattering is very difficult to discern in all but the most ideal conditions.

7.2.2 GPR RECORDING

Considering the wave scattered from the object in Figure 7.3, if a receive antenna is switched on at precisely the instant that the pulse is transmitted, then the pulses will be recorded by the receive antenna as a function of time. The first pulse will be the wave that travels directly through the air (because the velocity of air is greater than any other material), and the second pulse that is recorded will be the pulse that travels through the material and is scattered back to the surface, traveling at a velocity determined by the permittivity (ε) of the material. The resulting record measured at the receive antenna is similar to one of the time–amplitude plots shown in Figure 7.3b, with the "input" wave consisting of the direct wave that travels through air, and the "output" pulse consisting of the wave reflected from the buried scattering body. The recording of both pulses over a period of time with receive antenna system is called a "trace," which can be thought of as a time-history of the

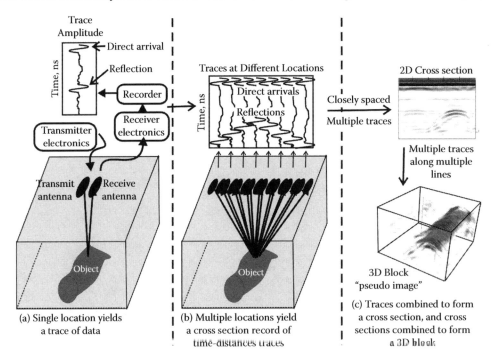

FIGURE 7.3 Steps in the GPR process: (a) formation of a single time trace of data, showing the direct arrival and reflection from the object; and (b) multiple traces for a wiggle-trace display, with each trace representing a different position on the earth's surface. Multiple traces can be color coded, with different colors (gray scales) assigned to amplitudes, and the result displayed as a cross section, or as a three-dimensional block view as shown on the right side of the figure.

travel of a single pulse from the transmit antenna to the receive antenna, and includes all of its different travel paths. The trace is the basic measurement for all time-domain GPR surveys. A scan is a trace where a color scale, or a gray scale, has been applied to the amplitude values.

7.2.3 POLARIZATION

The electromagnetic field at a given point in space, at a specific time, has both a magnitude and a direction, and can be described by a vector. As the electromagnetic wave propagates, the orientation and magnitude of these vectors change as a function of time. Polarization describes the magnitude and direction of the electromagnetic field as a function of time and space. Commercial antenna systems are designed to take advantage of the polarization of waves. However, one must be aware of polarization when interpreting data, because a polarized wave has a preferred orientation to detecting objects, particularly linear objects like pipes.

The importance of polarization can be seen by considering the fact that because a linearly polarized (electrical current oscillates back and forth along a straight line as the wave propagates perpendicular to the direction of oscillation) signal has an electric field component in only one direction, then any linear metallic object oriented perpendicular to the direction of polarization will not scatter the signal, and the object will not be detected. This effect is illustrated in Figure 7.4. The practical solution to this potential problem is to run multiple surveys to cover several potential

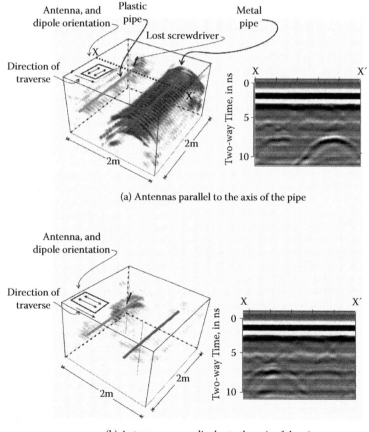

(a) Antennas parallel to the axis of the pipe

(b) Antennas perpendicular to the axis of the pipe

FIGURE 7.4   Three-dimensional block and cross sections over plastic (dielectric) and copper (conductor) cylinders when the antennas are (a) parallel to the long axis and (b) perpendicular to the long axis of the cylinder. Note the buried screwdriver that was located above the plastic cylinder.

orientations of subsurface objects, or to measure more than one component of the transmitted linearly polarized signal. Examples of the effect of polarization are presented in the following section on interpretation of GPR data.

7.2.4 VELOCITY

Electromagnetic waves travel at a specific velocity determined primarily by the permittivity of the material. The relationship between the velocity of the wave and material properties is the fundamental basis for using GPR to investigate the subsurface. To state this fundamental physical principle in a different way: the velocity is different between materials with different electrical properties, and a signal passed through two materials with different electrical properties over the same distance will arrive at different times. The interval of time that it takes for the wave to travel from the transmit antenna to the receive antenna is simply called the travel time. The basic unit of electromagnetic wave travel time is the nanosecond (ns), where $1 \text{ ns} = 10^{-9} \text{ s}$.

The time it takes an electromagnetic wave to travel from one point to another is called the travel time, which is measured as an inverse function of velocity. Because the velocity of an electromagnetic wave in air is 3×10^8 m/s (0.3 m/ns), then the travel time for an electromagnetic wave in air is approximately 3.3333 ns per m traveled. The velocity is proportional to the inverse square root of the permittivity of the material, and because the permittivity of earth materials is always greater than the permittivity of the air, the travel time of a wave in a material other than air is always greater than 3.3333 ns/m.

The permittivity is also commonly referred to as the dielectric permittivity, but for simplicity it will be referred to as the permittivity in this book. The permittivity has been explained as a measure of polarizability of a material, which causes displacement currents to flow, which in turn affects the propagation of an electromagnetic wave. The effect of the permittivity on attenuation was previously discussed.

The permittivity is also directly related to the velocity of propagation of an electromagnetic wave, which is a very important property for analyzing and processing GPR data. The permittivity is related to the velocity by the following relationship:

$$v_m = \frac{v_0}{\sqrt{\dfrac{\varepsilon_m}{\varepsilon_0}}} \tag{7.1}$$

where v_m is the velocity of the wave through any material, v_0 is the speed of light in air (3×10^8 m/s), ε_m is the permittivity of the material, and ε_0 is the permittivity of free space (air in a vacuum), with a value of 8.85×10^{-12} Farads/m.

The ratio of the permittivity of the material to the permittivity of air ($\varepsilon_r = \varepsilon_m/\varepsilon_0$) is called the relative permittivity. The range of values of relative permittivity is from 1 for air to ~81 for water. The high permittivity of water is caused by the rotational polarization of the water molecule. Not surprisingly, the quantity of water tends to dominate the relative permittivity of porous rocks and minerals.

7.3 EQUIPMENT

7.3.1 SYSTEM OVERVIEW

GPR equipment consists of antennas, electronics, and a recording device, as shown in Figure 7.5. The transmitter and receiver electronics are always separate, but in a fixed-mode configuration, they are often contained in different boxes, and in some systems designed for moving-mode operation, all

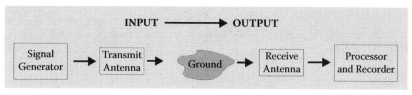

(a) Input/output of a GPR system

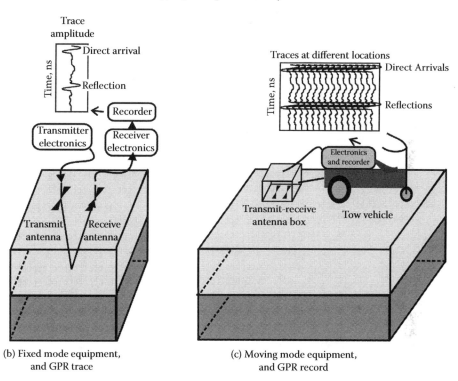

(b) Fixed mode equipment,
and GPR trace

(c) Moving mode equipment,
and GPR record

FIGURE 7.5   Operational components of a GPR system: (a) input/output components; (b) operation in fixed-offset mode, showing a single trace output; and (c) operation in moving mode, whereby a series of traces are recorded at different positions along the surface, producing a time–distance cross section over the earth.

of the electronics are contained in one box. In some cases, the electronics may be mounted on top of the antennas, making for a compact system but decreasing the operational flexibility of the system.

GPR systems are digitally controlled, and data are usually recorded digitally for postsurvey processing and display. The digital control and display part of a GPR system generally consist of a microprocessor, memory, and a mass storage medium to store the field measurements. A small microcomputer and standard operating system are often utilized to control the measurement process, store the data, and serve as a user interface. Data may be filtered in the field to remove noise, or the raw data may be recorded and the data processed for noise removal at a later time. Field filtering for noise removal may consist of electronic filtering and digital filtering prior to recording the data on the mass data storage medium. Field filtering should normally be minimized except in those cases where the data are to be interpreted immediately after recording.

7.3.2   ANTENNAS

The most important components of a GPR system are transmit and receive antennas. The purpose of a transmit antenna is to launch an electromagnetic wave into the ground; the purpose of the receive

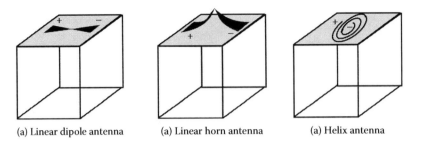

(a) Linear dipole antenna   (a) Linear horn antenna   (a) Helix antenna

FIGURE 7.6 Basic antenna designs: (a) linear bowtie dipole, (b) horn, and (c) helix.

antenna is to detect the electromagnetic wave after it has traveled through the material in the subsurface. The details of antenna design and implementation are very complex, and the performance of a GPR antenna may be somewhat different from the theoretical design envisioned by the design engineer.

Antennas can be considered to be transducers that convert electric currents on the metallic antenna elements to transmit electromagnetic waves that propagate into a material. Antennas radiate electromagnetic energy when there is a change in the acceleration of the current on the antenna. The acceleration that causes radiation may be either linear (e.g., a time-varying electromagnetic wave traveling on the antenna) or angular acceleration. Radiation occurs along a curved path, and radiation occurs anytime the current changes direction (e.g., at the end of the antenna element).

The primary design objective for a GPR antenna is to launch an electromagnetic wave into the ground that has a known shape and amplitude. There are many types of antennas, and they are all designed to optimize the characteristics of the wave that is launched into the ground. An antenna is designed to propagate a wavefront with a simple pattern, with particular input pulse characteristics (in the case of a time-domain system), at a high level of power of the pulse. A few of the more common designs include the linear dipole, the horn, and the helix, as shown in Figure 7.6. The linear dipole and horn antennas launch a linearly polarized wave, and the helix launches a circularly polarized wave.

Ideally, all of the energy from the input pulse will be expended during one trip of the electric currents on the antenna from the input feed to the tip of the antenna, with the wavelength of the propagating wave equal to twice the length of one half of the total length of the antenna. However, some of the energy in the form of electric current remains on the antenna after the pulse has traveled the full length of the antenna. The electric currents can reverberate between the tip of the antenna and the input feed, transmitting another pulse of propagating energy each time the currents travel from the tip to the input feed. This phenomenon is called antenna ringing, and part of the task of an antenna design engineer is to minimize this ringing effect. There are a lot of design "tricks" to reducing the antenna ringing, including placing resistors on the tips of the antenna elements to absorb some of the energy (this is called "loading," and the energy in the electric current is converted to heat by the resistors on loaded dipole antennas), and changing the shape of the dipoles so that most of the energy on the antenna element reflects at an angle and the energy of reflections from different parts of the antenna cancels by destructive interference. This is where the art of antenna design comes in, and a pair of tin-snips is the antenna designer's secret tool.

Impedance mismatch is another cause of antenna ringing. GPR antennas are designed to operate on the surface of the ground, and the ground becomes part of the antenna. Antennas are designed to minimize the contrast in electrical properties between the antenna and the ground by "matching" the electrical impedance between the antenna and the ground surface. A perfect impedance match maximizes the amount of energy that goes into the ground, and an imperfect match means that a lot of the energy is reflected back (backscattered) off of the surface, which also causes antenna ringing. In practice, a perfect impedance match that covers all conditions is not possible, and antenna designers must compromise on the issue of obtaining a perfect impedance match by designing the antennas with an intermediate impedance value that will match many types of rock, soil, and material electrical properties. In cases where the electrical conductivity is high (e.g., clay soil), it is impossible to match the impedance, and all antennas over these types of "lossy" materials will ring like bells.

GPR antennas are designed to maximize the amount of energy that propagates in the ground, but antennas naturally radiate energy in all directions, including above the surface of the ground. The effect of this fact is that "reflections" will appear on data from anything above the surface of the ground that can reflect the electromagnetic energy (e.g., buildings, people, trees, ceilings, walls, etc.). In order to avoid this effect, the antennas must be shielded from transmitting and receiving a signal above the antenna. *Shielding* is relatively easy to accomplish for signals with a short wavelength (high frequency), but it is difficult to achieve good shielding of antennas that transmit and receive long wavelengths. In practice, it is very difficult to shield an antenna with a center-band frequency below about 300 Mhz. The interpreter of GPR data must be aware of the fact that reflections from above the ground will appear on the data, and these artifacts will need to be identified so they are not interpreted as scattering from subsurface objects.

7.4 FIELD OPERATIONS

In practice, GPR measurements can be made by towing the antennas continuously over the ground, or at discreet points along the surface. These two modes of operation are illustrated in Figure 7.5. The fixed-mode antenna arrangement consists of moving antennas independently to different points and making discrete measurements. The moving-mode arrangement keeps the transmit and receive antennas at a fixed distance with the antenna pair moved along the surface by pulling them by hand or with a vehicle. Transmit and receive antennas are moved independently in the fixed mode of operation. This allows more flexibility of field operation than when transmit and receive antennas are contained in a single box. For example, different polarization components can be recorded easily when transmit and receive antennas are separate. In the fixed mode of operation, a trace is recorded at each discrete position of transmit and receive antennas through the following sequence of events in the GPR system: (1) a wave is transmitted, (2) the receiver is turned on to receive and record the received signals, and (3) after a certain period of time the receiver is turned off. The resulting measurements recorded during the period of time that the receiver is turned on are called a trace, and the spacing between measurement points is called the trace spacing. The chosen trace spacing should be a function of the target size and the objectives of the survey. Traces displayed side by side form a GPR time-distance record, or GPR cross section, which shows how the reflections vary in the subsurface. If the contrasts in electrical properties (e.g., changes in permittivity) are relatively simple, then the GPR time–distance record can be viewed as a two-dimensional pseudo-image of the earth, with the horizontal axis the distance along the surface, and the vertical axis the two-way travel time of the radar wave. The two-way travel time on the vertical axis can be converted to depth, if the permittivity (which can be converted to velocity) is known. The GPR time–distance record is the simplest display of GPR data that can be interpreted in terms of subsurface features. A GPR time–distance record can also be produced by making a series of fixed-mode measurements at a constant interval between traces on the surface.

7.4.1 FIELD PROCEDURES

Field procedures for GPR measurements follow the basic design that must be followed for most other geophysical methods. The field procedures for a specific objective are determined by answering a number of questions, including the following:

- What are the objectives of the survey?
- What is the nature of the subsurface environment?
- What are the electrical properties of the materials at the site?

- How large is the survey area?
- What is the nature of the site access?
- How will cultural features affect the measurements?

The fundamental questions that must be asked prior to any survey are as follows: (1) What is the maximum depth of penetration?, (2) What are the line and trace spacing (horizontal resolution)?, and (3) What is the vertical resolution needed to achieve the goals of a survey?

7.4.1.1 Station and Trace Spacing

The survey objectives dictate the depth of investigation, the lateral resolution that must be achieved, and the orientation of the antennas. However, the depth of penetration and resolution are also determined by the electrical properties of the material that contains the objects that are the targets of the survey. These combined factors help to establish the operating frequencies of the antenna that must be used, and the optimum spacing between measurements on the surface. A good rule of thumb for establishing adaquate trace measurement spacing is that the spacing should be less than one quarter of the size of the smallest object that is to be detected by the survey. This value is twice the Nyquist sampling frequency and should be adequate for most situations. Contrary to popular statements, it is impossible to "oversample": oversampling is a myth propagated by lazy people, which is the spatial analog to the popular myths of "overstudying" for exams, or being "overeducated." Also, the rule that "less is best" when it comes to determining the trace measurement spacing is erroneous. Economics and other practical considerations (e.g., spatial survey accuracy) are the only limitations to using a very small measurement spacing between traces and lines.

7.4.1.2 Depth of Penetration

Surveys should be designed so they record a two-way travel time that is twice the amount of time required to see the deepest object anticipated in the survey area. There is nothing more frustrating than to run a survey and discover that the objects of interest were below (beyond, in time) the data that were recorded. The depth of penetration of a radar wave depends upon the electrical properties and the center-band frequency of the antenna. The theoretical depth of penetration for a given antenna frequency can be computed from the equation for the skin depth, if the values of conductivity and permittivity are known. However, from a practical point of view, the only way to know the depth of penetration for a particular survey site is to make an initial guess based on past experience and to determine the actual depth of penetration by testing at the site. Dry homogenous rocks and soils (e.g., beach sand, nonvuggy limestone, granite, etc.), permafrost regions, lignite, and peat bogs, can yield penetration depths up to 20 m, and sometimes even greater depths. In contrast, the normal penetration depth for most soils is on the order of 1 to 3 m, with a low range of a few centimeters in a soil that is predominantly montmorillonite clay, up to several tens of meters for a clean sand.

7.4.1.3 Antenna Frequency

There are no fixed rules for determining the optimum antenna frequency range that should be used for a given survey, but the choice of antenna frequency should be based on the survey objectives and the electrical properties at the site. Because the depth of penetration decreases as the center-band frequency increases, the choice of antenna frequency is often determined by the depth of penetration that is needed. Lower frequencies generally improve the depth of penetration for most soil and rock types. However, a lower frequency has an adverse effect on the resolution of the radar wave for shallow investigations. A high-frequency antenna (500 MHz) would provide a detailed image of very shallow features, but it would not penetrate very deeply. A low-frequency antenna (50 MHz) would significantly improve the depth of penetration of the radar wave, but the resolution of very shallow features would be less distinct. The center-band frequency of the antenna should be computed so that the wavelength is smaller than one half the size of the smallest target. However, this does not mean that you will necessarily miss objects that are smaller than twice the size of the wavelength, because scattering depends upon a number of factors in addition to target size.

The size of the site, the access to the survey area, and the nature of surface features at the site influence the location and spacing of measurements. A large site area may necessitate the use of a vehicle to tow the antennas and preclude the possibility of making fixed-mode measurements. Conversely, site access problems (e.g., confined space) may make it necessary to make fixed-mode measurements. Surface metallic objects (e.g., fences), buildings, overhead utilities, and underground objects may influence both the frequency of the antenna that is selected and the location and spacing of measurements. For example, a small confined area containing a lot of cultural features above the surface may require the use of a high-frequency antenna that can be shielded from scattered energy above the surface.

Finally, the desired detail of the output display should strongly influence the survey design. If a three-dimensional pseudo-image is the desired output, then it is necessary to make measurements at a very close spacing. However, if the objective is to simply detect a large object, then measurements at a wide spacing may be adequate.

7.4.2 Filtering and Amplification

Filtering GPR raw field data is a necessary step to obtaining good data, and all field recording systems involve some type of analog and digital filtering. This is clearly illustrated by the three traces shown in Figure 7.7. The first trace is the raw field data, the second trace is the trace with a low-cut filter, the third trace has had a band-pass filter applied, and the fourth trace is that with band-pass filtering and amplification. The raw field data (Figure 7.7) do not even look like they contain any useful information. The low-cut frequency filter, which removes the low-frequency components, improves the appearance of the trace so that it looks like a GPR trace with positive and negative polarities situated properly above and below the zero line. The process of applying a low-cut filter in the early stages of processing is call "de-wowing" by GPR processing practitioners. The final trace filtering step removes the externally generated high-frequency noise that may be present on the data, and removes more of the low-frequency components beyond the basic de-wow filtering. Energy that occurs on a trace at later times can be enhanced by applying amplification to the trace that increases with time. This amplitude increase as a function of time is called gain. The output trace after applying these two filtering stages and gain is data that can be used for display and interpretation.

7.4.3 Static Corrections

Static corrections consist of applying a time correction to data measured at a different distance from the object, when the difference in distance is caused by changes in elevation on the surface. The static correction simply consists of subtracting the travel time for each trace, using the elevation and velocity of the surface material, as follows: $t_s = 2d/v$, where t_s is the two-way static time correction and d is the elevation between the baseline and the elevation where the trace is measured, as defined in Figure 7.8. Figure 7.8 also illustrates applying static corrections to field data where the elevation is changing.

7.4.4 Velocity Analysis and Two-Dimensional Filtering

One of the primary features of GPR is the fact that data can be displayed, processed, and interpreted in two and three spatial dimensions. Data are commonly measured along a line to create a cross section that is analogous to a seismic cross section. The vertical axis is the two-way travel time of the pulse that was transmitted and received at each position along the line. If the velocity of the host

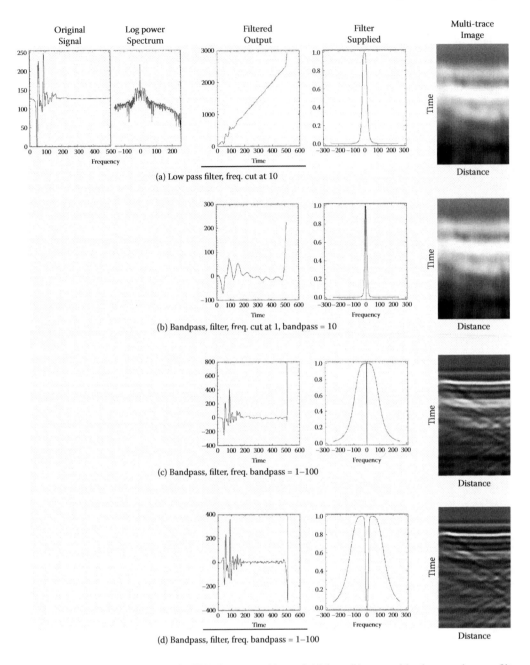

FIGURE 7.7 Frequency filtering of a GPR data trace: (a) raw field data, (b) trace with a low-cut de-wow filter applied, and (c) trace with low-cut de-wow filter and band-pass filter applied.

material is known, then the time–distance GPR cross section can be converted to a depth–distance cross section, which we can call a pseudo-section. Anomalies on the pseudo-section can be interpreted in terms of depth and horizontal location. The ideal situation is for clutter and noise to be removed from the depth–distance cross section (or three-dimensional display) so that the data are a true representation of objects in the subsurface. The three steps that can help to achieve this goal are velocity analysis, migration, and multidimensional frequency-wavenumber filtering. The discussion in this section is confined to two-dimensional data analysis, but these concepts can be extended to three-dimensional data sets.

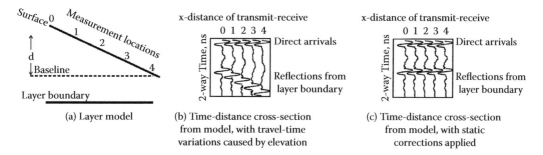

FIGURE 7.8 Static corrections: (a) definitions and model, (b) field measurements, and (c) field measurements with static corrections applied. "d" is the distance from the baseline to the measurement locations on the surface.

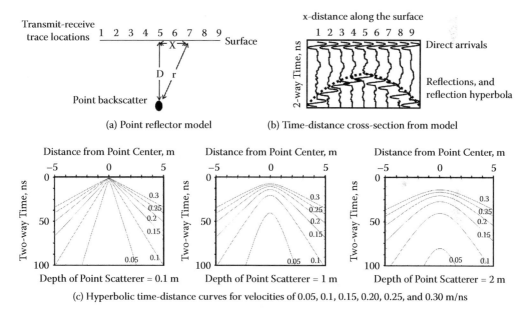

(c) Hyperbolic time-distance curves for velocities of 0.05, 0.1, 0.15, 0.20, 0.25, and 0.30 m/ns

FIGURE 7.9 Backscattered reflection from a point in the subsurface, as a function of surface position: (a) point model and hypothetical time–distance cross section, (b) change in hyperbola shape as a function of velocity and depth, and (c) time–distance plots of different velocities of the host material and depths of a point scatterer.

Spatial effects on GPR data are directly related to the velocity of the material. These effects are summarized in Figure 7.9. A signal reflected from a point in the subsurface (Figure 7.9a) appears as a hyperbola on the GPR record, because a backscattered reflection occurs when transmit–receive antennas approach the center of the buried object and when the antennas are moving away from the object. The shape of the hyperbola is directly related to the velocity, with a flattening of the hyperbola corresponding to an increase in velocity. The effect of changing the velocity on the shape of the hyperbola is shown in Figure 7.9b, and the simple equation that determines the shape of the hyperbola on the time–distance cross section is $t(x) = 2r/v$, where t is the two-way travel time of the backscattered GPR pulse at a distance x away from the center of the point, for a velocity v of the material.

Figure 7.9c suggests that the velocity of a material can be determined directly from GPR data, when the data contain an anomaly from a point scatterer (e.g., as approximated by a buried pipe, or the sharp edge of an object that is crossed by a GPR line of measurement traces). The velocity can be calculated from time–distance relationships by taking the two-way travel time values at the maximum point over the reflection hyperbola (corresponding to $x = 0$, with the corresponding two-way time t_0) and another point (e.g., any distance from the center of the hyperbola, call it x_1, and the corresponding two-way time t_1) along the reflection hyperbola, and applying the following relationship:

$$v = 2\sqrt{\frac{\left(x_1 - x_0\right)^2}{\left(t_1^2 - t_0^2\right)}} \tag{7.2}$$

An analogous field computation of velocity can be obtained over a layer.

7.5 DISPLAY AND INTERPRETATION

7.5.1 DISPLAY PROCESSING

Processing and display are an integral part of being able to effectively interpret GPR data. GPR data can rarely be interpreted without some type of processing to improve the resolution of coherent signals that represent the targets that are the objective of the survey, and a display that enables the interpreter to easily identify the anomalies that identify the time–space location of the targets. Processing and display can be conducted in the field, but in many cases, it is more convenient to process and display data at a later time in the office or laboratory.

Data display is a critical step toward providing an effective interpretation of GPR data. A poor display masks anomalies, and a good display enhances the target anomalies above the noise and coherent clutter. The spatial distribution of the field data determines the lateral resolution of the data. Three-dimensional displays can only be produced if field data are measured on a two-dimensional grid. The objective of GPR data presentation is to provide a display of the processed data which closely approximates an image of the subsurface, with the anomalies that are associated with the objects of interest located in their proper spatial positions. Data display is central to data interpretation, and an integral part of interpretation.

There are three types of displays of surface data, including a one-dimensional trace, a two-dimensional cross section, and a three-dimensional display. A one-dimensional trace is not of very much value until several traces are placed side by side to produce a two-dimensional cross section, or placed in a three-dimensional block view.

7.5.1.1 Two-Dimensional Displays

The wiggle trace (or scan) is the building block of all displays. A single trace can be used to detect objects (and determine their depth) below a spot on the surface. By towing the antenna over the surface and recording traces at a fixed spacing, a record section of traces is obtained. The horizontal axis of the record section is surface position, and the vertical axis is round-trip travel time of the electromagnetic wave. A GPR record section is similar to the display for an acoustic sonogram, or the display for a fish finder. Two types of wiggle-trace cross sections of GPR traces are shown in Figure 7.10. Wiggle-trace displays are a natural connection to other common displays used in engineering (e.g., oscilloscope display), but it is often impractical to display the numerous traces measured along a GPR transect in wiggle-trace form. Therefore, scan displays have become the normal mode of two-dimensional data presentation for GPR data. A scan display is obtained by simply assigning a color (or a variation of color intensity) to amplitude ranges on the trace, as shown in Figure 7.11. Scan displays are generally used for GPR data because of the high data volume (large number of traces/m).

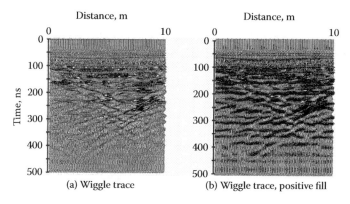

FIGURE 7.10 Wiggle-trace displays without and with positive amplitude fill.

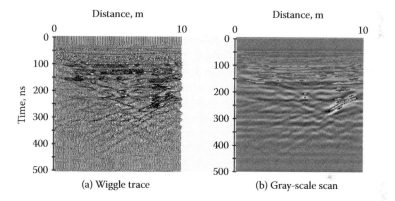

FIGURE 7.11 Scan displays: (a) the conversion of a wiggle-trace display to a color scan display and (b) a gray scale scan display.

7.5.1.2 Three-Dimensional Displays

Three-dimensional displays are fundamentally block views of GPR traces recorded at different positions on the surface. Data are usually recorded along profile lines, in the case of a continuous recording system, or at discrete points on the surface in fixed-mode recording (Figure 7.12). In either case, the antennas must be oriented in the same polarization orientation direction for each recording position if linearly polarized dipole antennas are used.

Obtaining a good three-dimensional display is a critical part of interpreting GPR data. Targets of interest are generally easier to identify and isolate on three-dimensional data sets than on conventional two-dimensional profile lines. Simplifying the image by eliminating the noise and clutter is the most important factor for optimizing the interpretation. Image simplification may be achieved by carefully assigning the amplitude-color ranges, displaying only one polarity of the GPR signal, using a limited number of colors, decreasing the size of the data set displayed as the complexity of the target increases, displaying a limited time range (finite-thickness time slice), and carefully selecting the viewing angle. Further image simplification in cases of very complex (or multiple) targets may also be achieved by displaying only the peak values (maximum and minimum values) for each trace. Finite-thickness time slices and cross sections have many advantages over the infinitesimal thin slices routinely used for interpreting GPR data. These principles are illustrated in a stepwise fashion in Figure 7.13.

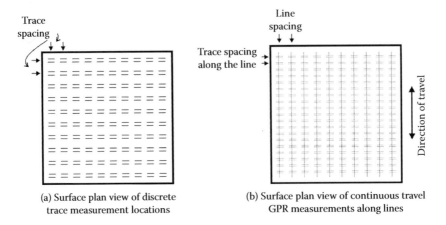

FIGURE 7.12   Grids and lines for fixed mode and moving mode for three-dimensional data measurements. Note that the polarization orientation of the antennas is the same for each measurement point on the grid, or along the profile lines.

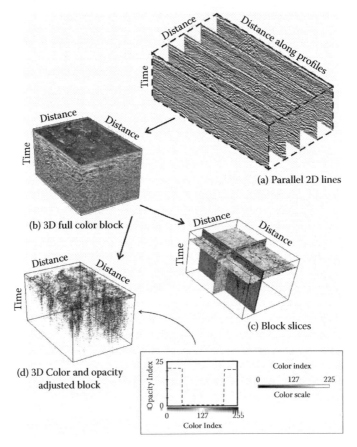

FIGURE 7.13   The process of constructing a three-dimensional display from a series of two-dimensional lines: (a) series of two-dimensional lines and (b) three-dimensional full-color (or gray scale) block view, (c) horizontal slices and vertical cross sections, and (d) adjustment of color scale (gray scale) and opacity to emphasize the important features in a three-dimensional block.

7.5.2   INTERPRETATION

GPR can be used to locate any object that has an electrical properties contrast with the surrounding ground, is within the detection range of the radar waves, and is not masked by clutter or noise. Scattering reflections are caused by an abrupt change in the electrical properties (primarily electric permittivity) in the subsurface. Some common features that have a high contrast include empty cavities, voids, or tunnels; changes in rock porosity; the water table; metal objects (e.g., barrels, tanks, pipes, etc.); plastic containers; concrete foundations; oil, petroleum, dense nonaqueous phase liquid (DNAPL) spills; or changes in geology.

Interpretation is the intellectual (human or intelligent computer) process of identifying anomalies on the GPR data and determining the nature (size, shape, and physical properties) of the object in the subsurface that is causing each anomaly. A good interpretation is the result of the skill of the interpreter (or sophistication of the pattern recognition algorithms), the quality of the data recorded in the field, and the clarity of the processed display used for interpretation. The interpretation begins with a good display that makes it easy to identify anomalies, with interpretation and processing inevitably overlapping each other. Data processed to the point where ready for interpretation should contain a minimum amount of noise (either random noise or coherent noise). Coherent noise can consist of features that are a part of the system (e.g., antenna ringing) or objects that are not a target of the survey (e.g., geologic features, overhead cultural features, etc.). The objects that are not a target of the survey are often called clutter. GPR data interpretation should progress along the following stages, with some overlap and feedback between stages:

1. Optimize the two-dimensional display to isolate the distinctive anomalies in the data.
2. Identify and classify the anomalies on the two-dimensional displays. Isolate the anomalies of interest from the clutter in the data.
3. Formulate the three-dimensional display and optimize the display to isolate the trends in the data.
4. Plot out time slices and cross sections of the three-dimensional data display to provide a final interpretative view of the anomalies.
5. Classify and identify features on the two-dimensional and three-dimensional data displays.
6. Map the map location and depth of identified objects.

Reflection anomalies on GPR records caused by linear objects (e.g., fences, overhead powerlines, corners of buildings, etc.) located above the surface of the ground are easy to identify by calculating the velocity from the reflection hyperbolas. The velocity of a radar wave in air is approximately 1 ft/ns in English length units (0.305 m/ns in metric units), and the velocity of electromagnetic waves through all other materials is much slower than the velocity through air. Linear surfaces (e.g., parallel fences, walls of buildings, overhead pipes, etc.) are more difficult to identify directly from the data and may require the use of good field notes in order to identify them.

Noise from external radio wave sources can be identified by the fact that it is usually semicontinuous and tends to contaminate a series of traces. Radio wave, or microwave, frequency noise on GPR records can sometimes be minimized by digital signal processing, if the frequency of the noise is outside the operating frequency range of the antennas.

8 Magnetometry, Self-Potential, and Seismic

Additional Geophysical Methods Having Potentially Significant Future Use in Agriculture

Barry J. Allred, Michael Rogers, M. Reza Ehsani, and Jeffrey J. Daniels

CONTENTS

8.1 MAGNETOMETRY METHODS

8.1.1 MAGNETOMETRY METHOD INTRODUCTION

Magnetometry is a passive remote sensing method that records the magnitude of Earth's local magnetic field at a sensor location. The Earth's overall magnetic field is a dipole field with the North Magnetic Pole and the South Magnetic Pole acting much like the ends of a bar magnet. There are secondary regional and local variations to the primary dipole field caused by soils and objects with different magnetic properties located above, on, or beneath the ground surface. Historically, the oersted and gamma have been the common units for measuring variations in Earth's magnetic field, and some of the older magnetometry data are displayed in these units. However, more recently, the common geophysical unit of measure for a magnetic field is called a tesla, where one tesla is equal to 10^4 oersteds, or 10^9 gammas. The Earth's magnetic field intensity varies between 0.25 and 0.65 oersteds (25,000 to 65,000 nanoteslas [nT], or 25,000 to 65,000 gammas).

Magnetometers are passive remote sensing devices that measure the magnitude of Earth's local magnetic field at the location of the sensor. Magnetometers may be placed on the ground surface, in the air (airborne magnetometer), in satellites (satellite magnetometry), or beneath the surface of Earth (borehole magnetometry). In practice, for agricultural purposes, handheld magnetometers are used with the sensor positioned within a couple meters of the ground surface. Commonly used magnetometers measure The Earth's local magnetic field with a precision between 0.1 nT and 1.0 nT. To put this sensitivity into context, a magnetometer operator cannot wear glasses, zippers, or other objects containing ferromagnetic metals (Clark, 1996; Gaffney and Gater, 2003), because these objects are likely to interfere with measurements.

8.1.2 TYPES OF MAGNETOMETERS

Fluxgate and optically pumped magnetometers are the two most commonly used instruments. Proton precession magnetometers, a third type, are rarely used due to long data acquisition times. Fluxgate magnetometers use a nickel–iron alloy core surrounded by a primary and secondary wire wrapped around the core. The core is magnetized by the component of Earth's magnetic field parallel to the core and by an alternating current in the primary winding. The alternating current in the primary winding creates an alternating magnetic field, which induces a current in the secondary winding that is proportional to Earth's local magnetic field. Fluxgate magnetometers have a 0.1 nT precision and fast acquisition times but require careful alignment of the sensor to avoid introducing anomalies (Scollar et al., 1990).

Optically pumped magnetometers commonly use cesium vapor and the Zeeman-effect to measure the magnitude of the Earth's local magnetic field. The Zeeman-effect arises when atoms containing a magnetic moment are placed in an external magnetic field. An oversimplified model of such an atom is a bar magnet with a north–south dipole. In the absence of an external magnetic field, the bar magnet has the same energy regardless of its alignment. In the presence of an external field, a bar magnet will become reoriented to align with the external magnetic field. The aligned state of the magnet has a lower energy than the nonaligned state. The difference in energy between these two states is related to the magnitude of the external magnetic field. Optically pumped magnetometers have a 0.1 nT precision, acquisition times of ten readings per second, and do not have the fluxgate alignment issues. Because both instruments use different methods of measuring Earth's magnetic field, they have the potential to record different data when used at the same site (Scollar et al., 1990).

Both fluxgate and optically pumped magnetometers can be used in single mode or gradient mode. In single mode, a single sensor is used to record the magnetic field. Because the background magnetic field varies with time, some means of compensating for these background changes (called temporal variations) must be incorporated into the design of each magnetometry field survey. Compensation is achieved by removing the temporal magnetic field variations from the data collected during a magnetometry field survey, and the temporal magnetic field variations are determined by maintaining a separate continuously recording stationary magnetometer at a base station or by reoccupying a fixed location periodically (e.g., every few minutes) throughout the time of the survey. An alternative is to use two magnetometers that are horizontally or vertically separated (Figure 8.1) in a setup called a gradiometer. The difference found between the magnetic fields recorded by each sensor is divided by the sensor separation resulting in the gradient of the magnetic field. Because the gradient is a spatial measurement that can be considered to be independent of time, there is no need to compensate for temporal variations in the Earth's magnetic field. Additionally, the gradient

FIGURE 8.1   Magnetometry field survey employing the gradient mode for data collection.

is a measure of how rapidly the magnetic field is changing, resulting in an enhanced view of the shallow subsurface.

8.1.3 MAGNETOMETRY FIELD PRINCIPLES

Data are collected discreetly or continuously while moving along transects. In discreet mode, the magnetometer is held stationary over a position along the transect while the magnitude of Earth's local magnetic field is recorded. In continuous mode, the magnetometer moves at a normal walking pace along the transect taking approximately ten readings every second. At a normal walking speed, this corresponds to a reading approximately every 10 cm. Transects are commonly separated by 25 to 50 cm when looking for human alteration of the landscape (Figure 8.1), and larger transect spacing when looking at geologic features. Data can be acquired in a unidirectional or bidirectional format. Unidirectional is where one travels along the first line from south to north, walks back to the start of the second line while not gathering data, and then gathers data along the second line while walking south to north. During a bidirectional survey, one collects data south to north along the first transect, turns around, and moves over to the second transect and gathers data while walking north to south.

8.1.4 MAGNETOMETRY DATA PROCESSING AND ANALYSIS

Magnetometer surveys produce x, y, and z data, where x and y are horizontal locations, and z is the magnitude of the vertical component (fluxgate) or the total field (optical pumping) of Earth's local magnetic field. The first step in postsurvey analysis of magnetic data is to examine the data using mapping software such as Golden Software's Surfer, Geoscan's Geoplot, or Geosoft's Oasis Montaj. These programs allow one to present the data as contour, image, shaded-relief, and surface plots. Large changes in the magnetic field caused by highly magnetic materials should be identified and "despiked." Large magnetic changes create scaling problems that obscure smaller changes of interest, and despiking removes any magnetic features above or below a selected value (Figure 8.2). Another filter can aid in removing large geologic trends.

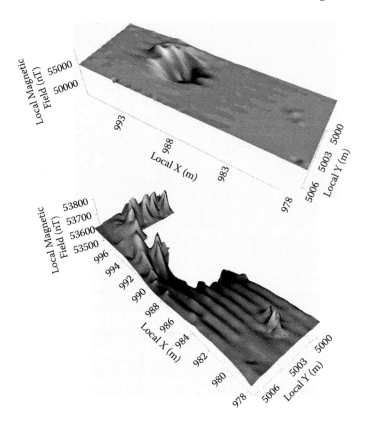

FIGURE 8.2 Example of the "despiking" process to remove a large magnetic field anomaly shown in top plot, thereby enhancing some of the more subtle magnetic field features that are present as shown in bottom plot.

Additional defects needing attention are drift, edge matching, striping, positional errors, and periodicity due to walking gait. Drift is caused by thermal changes within the sensor and appears as a slow temporal increase or decrease of the baseline magnetic field. Edge-matching is the process of matching the magnetic baseline at the end of one survey unit to the beginning of the next. Because there is some time delay between surveying each unit, the magnetic baseline changes due to natural fluctuations caused by the Sun (Figure 8.3). Both of these defects can be removed in the field by surveying in gradient mode or eliminated postsurvey using software. Striping appears during bidirectional surveys with every other transect having a slightly higher or lower magnetic baseline when compared to its neighbor (Figure 8.3). This defect becomes large if the magnetometer operator accidentally carries ferrous metals on their person. As the operator walks in one direction, the magnetometer is located in front of the operator and sits in one end of the operator's magnetic dipole. (The operator will have a magnetic field similar to a bar magnet.) When the operator turns around, his or her dipole does not rotate with him or her due to the operator's magnetic field being induced by and aligned with the direction of Earth's magnetic field. As the operator walks in the opposite direction, the magnetometer sits in the other side of his or her magnetic dipole. Because the Earth's magnetic field is not parallel with the ground surface, the magnetometer will record a different magnetic baseline when in the operator's north pole compared to the south pole. Positional errors arise due to a range of sources such as changing walking speed, reaction time, and computer timing. These errors give the data a zig-zag look. Periodicity is caused by the natural up and down motion when walking and appears as highs and lows in the magnetic baseline on a regular interval related to the operator's stride (Gaffney and Gater, 2003; Scollar et al., 1990).

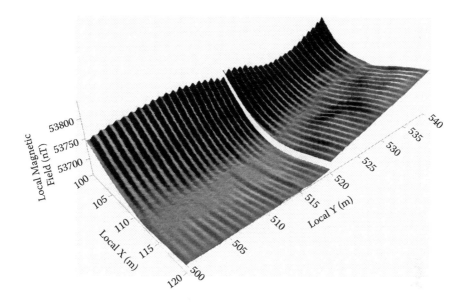

FIGURE 8.3   A magnetometry data collection effort in the field is often broken up into separate survey units that together cover a study area. The data from the individual survey units are later merged together to produce a magnetic field map for the entire study area. Edge-matching is the process of matching the magnetic baseline at the end of one survey unit to the beginning of the next. Bidirectional surveys can produce striping, with every other transect having a slightly higher or lower magnetic baseline when compared to its neighbor.

8.1.5   MAGNETOMETRY AGRICULTURAL APPLICATION EXAMPLE

The Oregon State University Research Dairy in Corvallis, Oregon, sits on 73 ha of working land 2.4 km west of the main campus. The dairy contains 40 registered Jersey cows and 130 registered Holstein cows and houses 20 to 140 various aged heifers. Near-surface geophysical surveys were conducted on a 1 ha portion of a pasture that is clay-tile drained with the most recent installation approximately forty years ago. This field was selected for study because excessive levels of *Escherichia coli* were found in adjacent Oak Creek that seemed related to the method of spraying liquid effluent as part of the nutrient and manure management strategy. Successful mapping of agricultural drainage systems was essential in understanding the relationship between effluent spraying and creek contamination.

Drainage pipe locations could not be imaged using a 500 MHz ground-penetrating radar antenna due to extreme attenuation of the signal caused by soils with high clay content (Rogers, 2003). Drainage pipe locations were successfully mapped using a Geometrics G-868 optically pumped gradiometer. A 100-meter-square study area was divided into twenty-five 20-meter-square subunits using nonmagnetic polyvinyl chloride (PVC) stakes to mark the subunit corners. Nonmagnetic (fiberglass) survey tapes marked the subunit perimeter, and nineteen blaze-orange, 0.95-gauge plastic weed trimmer line were used to mark north–south running transect lines spaced every meter. The magnetic survey was bidirectional with the sensors separated by 0.5 m in a vertical gradient mode. Under ideal conditions, establishing the survey grid and conducting the aforementioned magnetic survey could be accomplished by an experienced team of three in 4 to 5 days. Golden Software's Surfer 7 (Golden Software Inc., 1999) was used to create a shaded-relief of the plot with no postacquisition processing (Figure 8.4). An iron water pipe, the clay tile drainage system, and other magnetic signals spatially associated with subsurface objects are evident in the shaded-relief plot (Rogers et al., 2005).

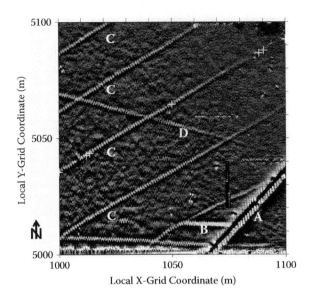

FIGURE 8.4   Shaded-relief map showing drain line locations detected with magnetometry methods. A is an iron water pipe, B is an unknown feature, C are associated with a known clay tile drainage system, and D is associated with an unexpected drainage system. The crosses mark known clay tile drainage pipe.

Although the magnetic signal spatially associated with the drainage system is clearly seen in the western portion of Figure 8.4, the signal fades away when traveling to the east of the survey region. Examination of the soils identified a change in soil type between the western and eastern portions of the site. A detailed examination of soil iron concentration extracted from trenches identified different levels of disturbed iron-rich soils related to drainage pipe installation. More iron-rich soils were disturbed in the western portion of the site and may explain why the signal is stronger in this region (Rogers et al., 2006). The magnetic surveys at the Oregon State University Research Dairy demonstrate that agricultural drainage systems can be magnetically imaged, but success in doing so appears strongly related to the amount of iron-rich soil disturbed during installation.

8.2   SELF-POTENTIAL METHODS

8.2.1   SELF-POTENTIAL METHOD INTRODUCTION

Self-potential is a passive geophysical method—that is, it does not rely on the application of energy from an artificial source. Self-potential, from an operational standpoint, is probably the simplest geophysical method, essentially requiring only the measurement of a naturally occurring electric potential difference between two locations on the ground surface. The electric potential difference measured is associated with nonartificial electric current transmitted through the ground. These naturally generated electric potential differences range in magnitude from less than a millivolt (mV) to over one volt (Reynolds, 1997). The magnitude and sign (positive or negative) of a delineated self-potential anomaly can provide indications as to the character of the subsurface feature producing the anomaly. Naturally occurring electric potential gradients tend to be consistent and unidirectional or fluctuate with time. Although somewhat of an oversimplification, the focus of the self-potential method is usually to measure the consistent, unidirectional electric potential differences, and fluctuating potential differences are typically considered to be "noise" or unwanted signal. Caution is warranted; however, because some self-potential noise is relatively consistent and unidirectional. Self-potential noise is often associated with alternating electric current (AC) generated from thunderstorm activity, variations in the Earth's magnetic field, heavy rainfall effects, and so forth (Reynolds, 1997).

8.2.2 Sources of Consistent and Unidirectional Natural Electric Potential Gradients

Consistent and unidirectional natural potential differences generally fall into three main categories—electrokinetic, electrochemical, and mineral:

1. *Electrokinetic*—This type of potential is also called electrofiltration potential or streaming potential and is associated with electric current generated by the movement of a dissolved ion containing water through a porous media such as soil or rock. The underlying mechanism creating current and potential differences is believed to be electrokinetic coupling between pore side walls and dissolved ions (Sharma, 1997). Electrokinetic potentials increase in the direction of water flow, and electric charge moves in the opposite direction (Reynolds, 1997). Therefore, profiled or mapped self-potential anomalies caused by electrokinetic processes will be positive for locations where there is convergence of subsurface water flow and negative where there is divergence of subsurface water flow.

2. *Electrochemical*—Differences in pore water dissolved ion concentrations between two subsurface locations results in spontaneous transport of electric charge carriers (dissolved ions) and development of spatial gradients of electric potential. Diffusion potential gradients exist with this spatial charge imbalance scenario when pore water cation and anion mobilities differ. Nernst potential gradients exist under these circumstances when there are no differences in dissolved cation and anion mobilities. For diffusion and Nernst electric potential gradients to persist, subsurface spatial differences in pore water dissolved ion concentrations must be maintained. The mechanism for maintaining spatial differences in pore water dissolved ion concentrations has not yet been identified, but the most likely candidate is some form of below-ground chemical interaction involving atmospheric oxygen (Parasnis, 1986). Electrochemical potential gradients are also formed due to the subsurface presence of quartz veins, pegmatite intrusives, or clay deposits, and all three cases are probably a response to cation adsorption processes.

3. *Mineral*—Spatial self-potential anomalies are often measured at the ground surface over massive metal ore bodies. The mechanism generating these anomalies is not completely understood. However, the explanation having the greatest acceptance involves natural electric currents originating from chemical oxidation and reduction reactions between a metal ore body straddling the water table and the pore water dissolved ions present in the surrounding rock or soil.

It should be noted that electric potential differences measured at the surface in agricultural settings are in all likelihood due to either electrokinetic or electrochemical sources.

8.2.3 Self-Potential Equipment, Field Procedures, and Data Analysis Products

The basic equipment setup needed for self-potential surveying is very simple and includes only three components: electrodes, a voltmeter, and electric cable. The voltmeter measures the electric potential difference between two electrodes placed at the ground surface that are connected to the voltmeter via electric cable. The metal stakes commonly used as electrodes for resistivity geophysical surveys are not acceptable for obtaining self-potential measurements. The problem with using metal stakes for electrodes is the formation of unwanted electric potentials due to electrode polarization caused by contact between the bare metal in the stake and the electrolytic aqueous solution present within the soil. For resistivity surveys, the unwanted potentials produced by metal

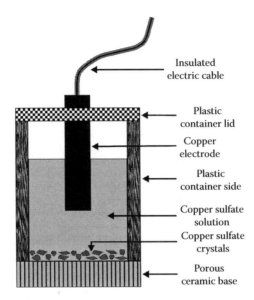

FIGURE 8.5   Typical components making up a nonpolarizing electrode.

stake electrodes are canceled out by injecting low-frequency alternating current into the subsurface, which is an option not relevant for the self-potential method. Therefore, nonpolarizing electrodes are required for self-potential data collection. A nonpolarizing electrode, as shown in Figure 8.5, is typically made of a copper rod inserted through the lid of a container that is porous at its base and filled with a saturated, aqueous, copper sulfate solution (Milsom, 2003). The voltmeter employed to measure the electric potential difference between the two nonpolarizing electrodes needs to have a high input impedence of at least 1×10^8 ohms and a measurement resolution of 1 mV (Reynolds, 1997; Sharma, 1997). The electric cable that connects the voltmeter between the two nonpolarizing electrodes should be insulated (polyethylene or Teflon coating) copper, copper/steel, or cadmium/bronze wire with an American Wire Gauge (AWG) thickness between 18 and 26. The thicker insulated wire with the lower gauge of 18 should be employed for rough field conditions that require an electric cable with more durability.

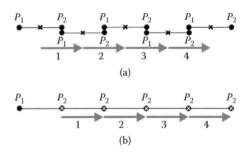

FIGURE 8.6   The two modes of self-potential data collection: (a) both electrodes are moved along a transect, with the separation distance between electrodes kept constant; and (b) one electrode remains stationary at a base station, while the second electrode is moved along a transect or a series of transects. (P_1 and P_2 are potential electrodes.)

There are two procedures for collecting self-potential data in the field. The first mode of data collection is illustrated in Figure 8.6a and involves consecutively moving both electrodes (P_1 and P_2) together along a transect, with the separation distance between the electrodes kept constant. For each move of the electrode pair, the new position of the trailing electrode (P_1) corresponds with the previous position of the leading electrode (P_2). The reference location for each self-potential voltage measurement is assumed to be the midpoint between the two electrodes (x-position in Figure 8.6a). Each measurement obtained with this first data collection mode is reported either as an electric potential difference (voltage) or as an electric potential gradient (voltage divided by the separation distance between electrodes). The second mode of self-potential data collection is shown in Figure 8.6b. With this second data collection mode, one electrode (P_1) remains stationary at a base station, while the second electrode (P_2) is moved along a transect or a series of transects. The location referenced for each measurement obtained is assumed to coincide with the position of the moving electrode (P_2 in Figure 8.6b). Potential differences (voltage) are the values normally reported for this second self-potential field survey procedure.

Self-potential measurement transects, where possible, should be oriented perpendicular to the known directional trends of subsurface features that are being investigated. Self-potential data collected along a single transect can be plotted as a profile showing the potential difference or potential gradient changes with respect to transect distance. If there are a series of self-potential measurement transects, contour maps can be generated detailing horizontal spatial variations in electric potential difference or electric potential gradient.

8.2.4 POSSIBLE AGRICULTURAL APPLICATIONS FOR SELF-POTENTIAL METHODS

Self-potential anomalies associated with electrokinetic processes have been used to locate leaks in dams, reservoirs, and landfills (Reynolds, 1997; Sharma, 1997). Self-potential methods may likewise be useful for locating leaks in animal waste storage ponds and treatment lagoons. Again, profiled or mapped self-potential anomalies caused by electrokinetic processes will be positive for locations where there is convergence of subsurface water flow and negative where there is divergence of subsurface water flow. Consequently, self-potential surveys could prove valuable for determining drainage pipe functioning condition with respect to water removal (positive anomalies should be generated) or subirrigation (negative anomalies should be generated). In addition, based on the same convergent flow electrokinetic potential considerations, self-potential methods might be employed to provide insight on the magnitude and extent of the water table depression surrounding an irrigation well. Finally, horizontal spatial patterns for soil salinity and soil clay content could be evaluated with self-potential methods, if salinity and clay content spatial variations produce corresponding spatial changes in electrochemically generated potential differences or gradients.

8.3 SEISMIC METHODS

8.3.1 SEISMIC METHOD INTRODUCTION

Seismic waves are essentially elastic vibrations that propagate through soil and rock materials. Seismic waves can be introduced into the subsurface naturally, with earthquakes being a prime example. Earthquakes are seismic waves, oftentimes extremely destructive, that typically result from the energy released due to movement along large fractures (faults) in the Earth's crust. Data obtained from earthquakes have allowed seismologists to resolve the overall structure of the Earth (solid iron/nickel inner core, liquid iron/nickel outer core, mantle, and lithosphere).

Explosive, impact, vibratory, and acoustic artificial energy sources can also be used to introduce seismic waves into the ground for the purpose of investigating subsurface conditions or features. The use of active (artificial energy source) seismic geophysical methods is widespread in the petroleum and mining industries. Active seismic methods have additionally been employed for hydrological, environmental, geotechnical engineering, and archeological investigations. Although presently used very little for agricultural purposes, seismic methods are likely to find significant agricultural applications in the near future.

For seismic geophysical methods where artificial energy is supplied, the seismic waves generated are timed as they travel through the subsurface from the energy source to the sensors, which are called geophones. Incoming seismic wave amplitudes, and hence energy, are also measured at the geophones. The energy source is ordinarily positioned on the surface or at a shallow depth, and the geophones are normally inserted at the ground surface. Data on the timed arrivals and amplitudes of the seismic waves measured by the geophones are then used to gain insight on belowground conditions or to characterize and locate subsurface features. Numerous texts provide detailed descriptions of seismic methods (Coffeen, 1978; Dobrin and Savit, 1988; Lines and Newrick, 2004; Pelton, 2005a, 2005b; Telford et al., 1976), and readers are referred to these for more information. However, some basic attributes of the seismic waves and their propagation through Earth materials along with general data collection and analysis considerations and potential seismic method agricultural applications are discussed as follows.

8.3.2 Seismic Wave Types

There are two categories of seismic waves: "surface waves" and "body waves." Surface waves, as the name implies, are seismic waves that travel only along Earth's surface. Seismic body waves, although capable of traveling along the surface directly from source to sensor, can also travel with a vertical component through soil and rock well below ground. P-waves (also called primary waves, compressional waves, and longitudinal waves) are a type of seismic body wave having an elastic back-and-forth particle motion orientation that coincides with the direction of wave propagation. P-waves can be transmitted through solid, liquid, and gas materials. P-waves are the fastest seismic waves, and their velocity, V_P, within a soil or rock material is given by the following equation:

$$V_P = \sqrt{\frac{k + \frac{4}{3}\mu}{\rho}} \qquad (8.1)$$

where k is the bulk modulus, μ is the rigidity modulus (or shear modulus), and ρ is density. As indicated by Equation (8.1), the P-wave velocity in soil or rock depends only on elastic moduli and density of the soil or rock.

S-waves (also called secondary waves, shear waves, and traverse waves) are the second seismic body wave type and have an elastic particle motion that is perpendicular to the direction of wave propagation. There are two kinds of S-waves: the SV-wave and the SH-wave. The particle motion for an SV-wave has a vertical component. SH-waves, on the other hand, have a particle motion that is completely horizontal. S-waves are only capable of traveling through solid material, not liquids or gases. S-waves are slower than P-waves and have a velocity, V_S, given by

$$V_S = \sqrt{\frac{\mu}{\rho}} \qquad (8.2)$$

where all quantities have been previously defined. The S-wave velocity, as indicated by Equation (8.2), is governed strictly by shear stress elastic behavior and density of the soil or rock through which the S-wave travels.

There are two types of surface waves: Rayleigh waves and Love waves. Rayleigh wave particle motion is elliptical retrograde in a vertical plane oriented coincident with the direction of wave propagation. The Rayleigh wave amplitude decreases exponentially with depth. For a given soil or rock material, the Rayleigh wave velocity is approximately nine-tenths of the S-wave velocity for the same material ($= 0.9V_S$). Love waves occur only where there is a low S-wave velocity layer at the surface that is underlain by a layer with a much higher S-wave velocity. Love waves are essentially SH-waves transmitted via multiple reflections between the top and bottom of the low seismic velocity surface layer. Accordingly, the Love wave particle motion is horizontal and perpendicular to the direction of wave propagation. The overall Love wave velocity for a particular soil or rock material is less than the S-wave velocity for the same material. Neither surface wave is capable of being transmitted through liquids, such as water.

   Both surface wave types are dispersive. Dispersion occurs when different frequency components of the surface wave travel at different velocities, which causes the surface wave to become more spread out in length the farther the wave propagates. Dispersion is a result of different surface wave frequency components having different penetration depths coupled with vertical changes in soil and rock elastic moduli and density, which are the properties governing seismic wave velocity. Consequently, surface wave dispersion can provide insight as to the vertical seismic velocity structure in the shallow subsurface. (Note: The dispersion of body waves is considered to be negligible.)

8.3.3 ASPECTS OF SEISMIC WAVE PROPAGATION

The seismic energy source is usually assumed for analysis purposes to occur at a point location, especially in situations where artificial energy is applied for seismic investigation of the subsurface. Again, this point source is positioned at or near the ground surface for most seismic geophysical surveys. Seismic surface waves generated at the source propagate away from the source as a continually expanding circular wavefront along Earth's surface. Seismic body waves propagate away from the source into the subsurface as a continually expanding hemispherical wavefront (Figure 8.7). Geometrical spreading of the seismic surface or body wavefront results in a decrease with time of the seismic wave amplitude, and hence energy intensity, at any point along the wavefront as it continues to expand. Basically, as the seismic waves propagate outwards, the total seismic energy generated at the source is being distributed over a greater and greater diameter circle for surface waves and over a larger and larger hemisphere for body waves. Seismic wave amplitude and energy are also reduced due to frictional dissipation of elastic energy into heat. The amount of seismic wave frictional dissipation is dependent on the nature of the soil or rock material through which the wave passes. The combined effect of geometrical spreading and frictional dissipation is called attenuation, A, and for body waves can be expressed with the following relationship:

$$A = \frac{A_0 e^{-\alpha r}}{r} \tag{8.3}$$

where A_0 is the initial body wave amplitude, α is the frictional dissipation adsorption coefficient, and r is the radial distance of the wavefront from the source location (Sharma, 1997). Attenuation is frequency dependent, with the higher-frequency components of a seismic wave having been found to attenuate more rapidly with distance traveled than the lower-frequency seismic wave components.

   Transformations of seismic body waves incident on a subsurface interface between two soil and rock layers are depicted in Figure 8.8. Solid and dashed arrowhead lines represent seismic wave travel paths, and the Figure 8.8 schematics are based on the assumption that Layer 2 P-wave

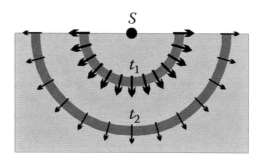

FIGURE 8.7   In a uniform soil and rock medium, seismic body waves propagate away from the source (S) as a continually expanding hemispherical wavefront. Arrows represent the direction of travel at points along the wavefront. The size of the arrows represents attenuation effects causing the wavefront amplitude and energy intensity (energy per unit wavefront area) to be less at time t_2 than at earlier time t_1.

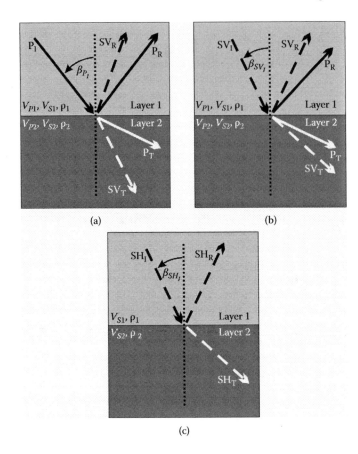

FIGURE 8.8 Body wave transformations at a subsurface interface separating two layers having different seismic velocities (V_P and V_S) and density (ρ) properties: (a) incident P-wave, (b) incident SV-wave, and (c) incident SH-wave.

and S-wave velocities (V_{P2} and V_{S2}, respectively) are greater than the Layer 1 P-wave and S-wave velocities (V_{P1} and V_{S1}, respectively). The Layer 1 density is ρ_1, and the density of Layer 2 is ρ_2. As shown in Figure 8.8a, for $\beta_{P_I} > 0$, a P-wave initially traveling through Layer 1 and incident at the interface (P_I) will generate a P-wave reflected back through Layer 1 (P_R), an SV-wave reflected back through Layer 1 (SV_R), a P-wave transmitted into Layer 2 (P_T), and an SV-wave transmitted into Layer 2 (SV_T). An incident SV-wave (SV_I), assuming $\beta_{SV_I} > 0$, likewise produces P_R, SV_R, P_T, and SV_T (Figure 8.8b). However, as depicted in Figure 8.8c, an incident SH-wave (SH_I) generates only reflected SH-waves (SH_R) and transmitted SH-waves (SH_T).

The travel directions of the reflected and transmitted seismic waves in Figure 8.8 are given as an angle referenced to an imaginary line extending through and oriented perpendicular to the interface (dotted vertical lines in Figure 8.8) and can be determined by Snell's law based only on the incident seismic wave angle and the seismic velocities of the two layers adjacent to the interface. For an incident P-wave, Snell's law gives.

$$\frac{\sin\left(\beta_{P_I}\right)}{V_{P1}} = \frac{\sin\left(\beta_{P_R}\right)}{V_{P1}} = \frac{\sin\left(\beta_{SV_R}\right)}{V_{S1}} = \frac{\sin\left(\beta_{P_T}\right)}{V_{P2}} = \frac{\sin\left(\beta_{SV_T}\right)}{V_{S2}} \qquad (8.4)$$

where the reflected P-wave travel direction angle, β_{P_R}, equals the incident P-wave travel direction angel, β_{P_I}; the reflected SV-wave travel direction angle is β_{SV_R}; the transmitted P-wave travel direction angle is β_{P_T}; the transmitted SV-wave travel direction angle is β_{SV_T}; and all other quantities have been previously defined. With respect to an incident SV-wave, Snell's law gives

$$\frac{\sin\left(\beta_{SV_I}\right)}{V_{S1}} = \frac{\sin\left(\beta_{P_R}\right)}{V_{P1}} = \frac{\sin\left(\beta_{SV_R}\right)}{V_{S1}} = \frac{\sin\left(\beta_{P_T}\right)}{V_{P2}} = \frac{\sin\left(\beta_{SV_T}\right)}{V_{S2}} \tag{8.5}$$

where the reflected SV-wave angle, β_{SV_R}, equals the incident SV-wave angle, β_{SV_I}. Finally, in regard to an incident SH-wave, Snell's law gives:

$$\frac{\sin\left(\beta_{SH_I}\right)}{V_{S1}} = \frac{\sin\left(\beta_{SH_R}\right)}{V_{S1}} = \frac{\sin\left(\beta_{SH_T}\right)}{V_{S2}} \tag{8.6}$$

where the reflected SH-wave angle, β_{SH_R}, equals the incident SH-wave angle, β_{SH_I}, and the transmitted SH-wave travel direction angle is β_{SH_T}.

When the transmitted angle of a seismic wave equals 90°, a process called refraction occurs in which the seismic wave travels along the interface at the V_2 velocity while continually causing oscillations at the interface that transmit derivative seismic waves back into Layer 1. A refracted wave is often called the head wave. The incident seismic wave angle that produces a refracted wave (or head wave) is referred to as the critical angle. For a seismic wave initially traveling downward, this refraction phenomenon can occur only at an interface where the seismic wave velocity in the layer below the interface (V_2) is greater than the seismic wave velocity in the layer above the interface (V_1).

Incident P-waves that are normal (perpendicular) to the interface ($\beta_{P_I} = 0°$) produce only P_R and P_T waves (no S-waves) with travel paths that are also perpendicular to the interface ($\beta_{P_R} = \beta_{P_T} = 0°$). Incident SV-waves that are normal to the interface ($\beta_{SV_I} = 0°$) produce only SV_R and SV_T waves (no P-waves) with travel paths normal to the interface ($\beta_{SV_R} = \beta_{SV_T} = 0°$). Incident SH-waves that are normal to the interface ($\beta_{SH_I} = 0°$) still produce only SH_R and SH_T waves (Figure 8.8c), with travel paths normal to the interface ($\beta_{SH_R} = \beta_{SH_T} = 0°$). The ratio of the reflected wave amplitude to the incident wave amplitude is called the reflection coefficient. For incident P-waves normal to the interface, the reflection coefficient, R_{C-P}, is expressed as follows:

$$R_{C-P} = \frac{\rho_2 V_{P2} - \rho_1 V_{P1}}{\rho_1 V_{P1} + \rho_2 V_{P2}} \tag{8.7}$$

The reflection coefficient given an incident S-wave (SV or SH) normal to the interface is

$$R_{C-S} = \frac{\rho_1 V_{S1} - \rho_2 V_{S2}}{\rho_1 V_{S1} + \rho_2 V_{S2}} \tag{8.8}$$

(Note the difference in form of the numerators for Equation (8.7) and Equation (8.8). Negative values of R_{C-P} or R_{C-S} simply imply that the polarity of the reflected seismic wave has been reversed with respect to the incident seismic wave. The transmission coefficient, T_C, is the ratio of the transmitted wave amplitude to the incident wave amplitude, and given an incident seismic wave normal to the interface (P, SV, and SH), is quantified using the following relationship:

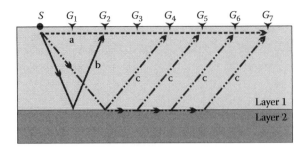

FIGURE 8.9   Three main seismic wave travel paths from source (*S*) to geophone sensors (*G₁, G₂, G₃, G₄, G₅,* G_6, and G_7): (a) direct path, (b) reflected path, and (c) refracted path.

$$T_C = \frac{2\rho_1 V_1}{\rho_1 V_1 + \rho_2 V_2} \tag{8.9}$$

where V_1 is the P-wave or S-wave velocity of Layer 1, V_2 is the P-wave or S-wave velocity of Layer 2, and all other quantities have been previously defined. The ρV product term in Equation (8.7) through Equation (8.9) is referred to as the "acoustic impedance" for a particular soil and rock material.

The three main seismic wave travel paths from source to geophones are depicted in Figure 8.9. The direct travel path (a) along the surface from source to geophones can be taken by all seismic wave types including Rayleigh and Love surface waves along with P and S body waves. Only seismic body waves take the reflected (b) and refracted (c) travel paths. Furthermore, as illustrated in Figure 8.8, conversions between P-waves and S-waves can occur along these reflected and refracted seismic wave travel paths. Figure 8.9 represents source-to-geophone travel paths based on a subsurface having two layers. As the number of distinct subsurface layers increases, so too do the number of possible source-to-geophone reflected and refracted seismic body wave travel paths.

8.3.4   Seismic Method Data Collection, Data Analysis, and Potential Agricultural Applications

Seismic survey data are typically collected as the energy source and an array of geophone sensors are moved along a transect or series of transects. Explosive, impact, vibratory, and acoustic energy are commonly employed to generate seismic waves. Geophones convert seismic ground vibrations into an electrical potential difference (voltage) that is then recorded (Sheriff, 2002). As previously mentioned, the energy source is ordinarily positioned on the surface or at a shallow depth, and the array of geophones is normally inserted at the ground surface. Seismic reflection surveys measure the travel times and amplitudes associated with seismic waves taking reflected travel paths (b in Figure 8.9). Relative to the depth of investigation, the source and the geophone array are located fairly close together when conducting a seismic reflection survey. Seismic refraction surveys are focused predominantly on measuring travel times for seismic waves that take refracted travel paths (c in Figure 8.9). For a seismic refraction survey, the source and the geophone array are located a considerable distance apart relative to the depth of investigation. Some seismic surveys concentrate only on seismic waves traveling directly from source to geophone (a in Figure 8.9), particularly in regard to measurement of body wave travel times and amplitudes or surface wave travel times and dispersion characteristics (especially for Rayleigh waves).

There is continuous development of new and improved seismic equipment for field data acquisition. Safe, portable seismic energy sources are now available for shallow subsurface investigations. Land-based geophone array streamers have been recently produced that can be pulled along the ground surface without the inconvenience of having to detach and then reinsert individual geophones.

Consequently, capabilities presently exist for obtaining continuous seismic measurements along a transect, thereby dramatically increasing the speed at which a seismic survey can be conducted.

Seismic data analysis products include an assortment of different depth sections, maps, and three-dimensional displays that depict various seismic wave attributes. Complex computer-processing procedures are often required to generate these different seismic data analysis products. A detailed description of seismic theory, data collection, and data analysis are beyond the scope of this book, and readers requiring more information on these topics are again referred to a number of excellent texts that cover seismic methods (Coffeen, 1978; Dobrin and Savit, 1988; Lines and Newrick, 2004; Pelton, 2005a, 2005b; Telford et al., 1976).

Equation (8.1) and Equation (8.2) show P-wave and S-wave velocities to be dependent on the elastic moduli and density of the soil and rock material through which the waves travel. These elastic moduli and density may be significantly correlated with soil properties and conditions. Consequently, shallow investigation (<2 m) seismic methods that obtain information on spatial variations of P-wave and S-wave velocities could provide important insight on spatial patterns of soil properties or conditions, such as horizontal and vertical changes in soil texture or compaction. In a similar manner, the seismic wave adsorption coefficient in Equation (8.3) is a property that may also correlate with soil properties and conditions; therefore, seismic methods capable of measuring spatial changes in the adsorption coefficient could additionally supply information on soil property and condition spatial patterns.

Up to this point, there has been no discussion regarding seismic wave frequency. Earthquake seismic waves have frequencies ranging from 0.03 to 20 Hz (Cleveland Museum of Natural History, 2006). Traditional seismic methods, especially those used for exploration purposes in the petroleum industry, typically employ seismic waves with frequencies less than 100 Hz (Coffeen, 1978). Sharma (1997) has a much more inclusive definition of seismic waves, which incorporates elastic waves with frequencies up to 10 kHz. Based on this broader definition, a large percentage of acoustic waves would be considered seismic waves. Acoustic waves are simply P-waves that are audible to the average human and have a frequency range from 20 Hz to 20 kHz. High-frequency seismic waves, as indicated previously, attenuate more rapidly with distance traveled than do lower-frequency seismic waves. Consequently, where investigation depths are large, as with petroleum exploration, geophysical methods employing low-frequency seismic waves are appropriate; however, for agricultural applications having extremely shallow investigation depths, geophysical methods using higher-frequency seismic waves may provide useful information. Traditional seismic methods have rarely been utilized for agricultural purposes, but laboratory studies employing 2 to 7 kHz acoustic waves have provided evidence that acoustic wave velocities correlate significantly with soil compaction, soil porosity, and soil water content, and acoustic wave adorption coefficients exhibit significant correlation with soil bulk density and soil water content (Lu et al., 2004; Oelze et al., 2002). Seismic tomographic imaging techniques have been proven effective for differentiating healthy and decayed woods in tree trunks (al Hagrey, 2007). Therefore, seismic methods have promise in regard to agricultural applications and will likely find greater use within the near future.

REFERENCES

al Hagrey, S.A. 2007. Geophysical imaging of root-zone, trunk, and moisture heterogeneity. *Journal of Experimental Botany.* v. 58, pp. 839–854.

Clark, A. 1996. *Seeing Beneath the Soil: Prospecting Methods in Archaeology.* B.T. Batsford. London, UK.

Cleveland Museum of Natural History. 2006. Earthshaking terminology. www/cmnh.org/site/researchand-collections_SeismicObservatory_EarthquakeTerminology.aspx.

Coffeen, J. A. 1978. *Seismic Exploration Fundamentals.* PennWell. Tulsa, OK.

Dobrin, M. B., and C. H. Savit. 1988. *Introduction of Geophysical Prospecting,* 4th edition. McGraw-Hill. New York.

Gaffney, C., and J. Gater. 2003. *Revealing the Buried Past: Geophysics for Archaeologists.* Tempus. Glouces-tershire, UK.

Golden Software, Inc. 1999. *Surfer 7 User's Guide: Contouring and 3D Surface Mapping for Scientists and Engineers.* Golden Software, Inc. Golden, CO.

Lines, L. R., and R. T. Newrick. 2004. *Fundamentals of Geophysical Interpretation.* Geophysical Monograph Series No. 13. Soc. Expl. Geophy. Tulsa, OK.

Lu, Z., C. J. Hickey, and J. M. Sabatier. 2004. Effects of compaction on the acoustic velocity in soil. *Soil Sci. Soc. Am. J.* v. 68, pp. 7–16.

Milsom, J. 2003. *Field Geophysics*, 3rd edition. John Wiley & Sons. Chichester, UK.

Oelze, M. L., W. D. O'Brien Jr., and R. G. Darmody. 2002. Measurement of attenuation and speed of sound in soils. *Soil Sci. Soc. Am. J.* v. 66, pp. 788–796.

Parasnis, D. S. 1986. *Principles of Applied Geophysics.* Chapman & Hall. London, UK.

Pelton, J. R. 2005a. Near-surface seismology: Wave propagation. In *Near-Surface Geophysics.* 177–217. D. K. Butler, editor. Investigations in Geophysics No. 13. Soc. Expl. Geophy. Tulsa, OK.

Pelton, J. R. 2005b. Near-surface seismology: Surface-based methods. In *Near-Surface Geophysics.* 219–263. D. K. Butler, editor. Investigations in Geophysics No. 13. Soc. Expl. Geophy. Tulsa, OK.

Reynolds, J. M. 1997. *An Introduction to Applied and Environmental Geophysics.* John Wiley & Sons. Chichester, UK.

Rogers, M. 2003. Effect of Iron Redistribution in Soils on Cesium Magnetometer Surveys at the Oregon State University Research Dairy. Ph.D. Dissertation. Oregon State University. Corvallis, OR.

Rogers, M. B., J. R. Cassidy, and M. I. Dragila. 2005. Ground-based magnetic surveys as a new technique to locate subsurface drainage pipes: A case study. *Appl. Eng. in Agric.* v. 21, pp. 421–426.

Rogers, M. B., J. E. Baham, and M. I. Dragila. 2006. Soil iron content effects on the ability of magnetometer surveying to locate buried agricultural drainage pipes. *Appl. Eng. in Agric.* v. 22, pp. 701–704.

Scollar, I., A. Tabbagh, A. Hesse, I. Herzog, and R. E. Arvidson. 1990. *Archaeological Prospecting and Remote Sensing, Topics in Remote Sensing; 2.* Cambridge University Press. New York.

Sharma, P. V. 1997. *Environmental and Engineering Geophysics.* Cambridge University Press. Cambridge, UK.

Sheriff, R. E. 2002. *Encyclopedic Dictionary of Applied Geophysics*, 4th edition. Society of Exploration Geophysics. Tulsa, OK.

Telford, W. M., L. P. Geldart, R. E. Sheriff, and D. A. Keys. 1976. *Applied Geophysics.* Cambridge University Press. Cambridge, UK.

Section III

The Global Positioning System and Geographic Information Systems

9 Integration of the Global Positioning System (GPS) into Agricultural Geophysics

Dorota A. Grejner-Brzezinska

CONTENTS

9.1 INTRODUCTION

Much of the geophysics equipment used today for agricultural applications has the capability to be integrated with Global Positioning System (GPS) receivers so that positional data can be obtained at the same time geophysical measurements are collected. Integration of GPS with resistivity and electromagnetic induction equipment already allows soil electrical conductivity to be mapped over large farm fields in just a few hours time. The GPS is a satellite-based, all-weather, continuous, global radionavigation and time transfer system, designed, financed, deployed, and operated by the U.S. Department of Defense (DOD). The system was originally intended for military applications, but in the 1980s, the U.S. government made it available to civilian users, free of any subscription fees or setup charges. The first GPS satellite was launched on February 22, 1978, and in 1993, the system was declared fully operational. During the following years, the breadth and the scope of the GPS applications exploded in the civilian market, taking advantage of the systems' full and sustained operability. As a result, radionavigation-based positioning and tracking is currently ever-present in a number of science, engineering, mapping, and everyday life applications.

GPS is an example of a ubiquitous technology responsible for the paradigm shift in contemporary navigation, positioning, surveying, and mapping techniques. As a result of progressive innovation and a significant drop in the price of the equipment in the last decade, GPS technology currently supports a variety of applications. These applications range from precise positioning, cadastral mapping and engineering, to remote sensing, environmental and GIS (geographic information system) surveys, and law enforcement. GPS radionavigation signals are used to navigate spaceships, aircrafts, and land-based vehicles, including transportation fleets and emergency cars, and to guide and track individual pedestrians. GPS is also used to monitor wildlife, track the race cars and paragliders, and monitor deformation of large structures. The system has also been effectively used in traffic monitoring, location-based services (LBS) and, in recent years, in monitoring the space weather and tropospheric conditions; and the number of new applications is still growing.

From the standpoint of agricultural geophysics, GPS is essential in providing a means for geo-registration (geolocation) of the ground-based sensors used to monitor soil conditions and locate subsurface features. This chapter will introduce the primary definitions, concepts, and mathematical models related to GPS positioning and sensor orientation applications. The overview of GPS design, implementation, and modernization is provided, followed by the primary positioning modes and associated mathematical models, including the real-time kinematic (RTK) and differential GPS (DGPS) concepts. The primary error sources, positioning accuracy, and basic aspects of GPS instrumentation are also addressed.

9.2 HOW GPS WORKS

The primary concept related to navigation with satellites is triangulation in space. The GPS satellites that serve as a space-based reference for the user's positioning solution transmit a continuous signal toward Earth during their 12-hour revolution around the globe. GPS receivers use signals from multiple satellites to determine the distances (or ranges) to the satellites that are subsequently used to triangulate the user's position coordinates (see Figure 9.1). The range observation is recovered by measuring the travel time of the signal between the satellite and the user's receiver. To perform a positioning or a navigation task, a GPS receiver must be locked onto the signal of at least three satellites to calculate a two-dimensional (2D) position (latitude and longitude); with four or more satellites in view, the receiver can determine three-dimensional (3D) position coordinates (latitude, longitude, and height) of the user. If continuous lock to multiple satellites is maintained, the receiver can provide an uninterrupted position solution, as well as additional information, such as speed, bearing, distance traveled, and distance to destination. The receiver can even provide location and directions to the nearest post office or gas station if it is equipped with suitable GIS databases and digital maps.

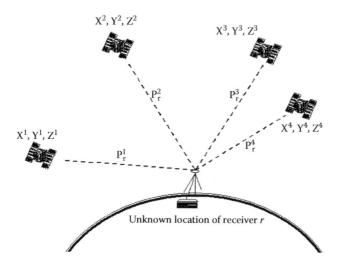

FIGURE 9.1   Determination of the position of user r by triangulation, using range measurements P to multiple satellites.

The concept of GPS technology was formulated with the following primary objectives:

- Suitability for all classes of platforms, such as spaceborne, airborne, marine, land-based, and individual pedestrian, under a wide variety of dynamics
- Availability any time, any weather, anywhere on Earth and its vicinity
- Real-time positioning, velocity, and time determination capability
- Providing the service to an unlimited number of users worldwide
- Positioning on a single global geodetic datum (World Geodetic System, WGS84)
- Redundancy provisions to ensure the survivability of the system
- Restricting the highest accuracy to a certain class of users (military)
- Low cost and low power users' unit

9.3   THE MAIN COMPONENTS OF GPS

The main components of the GPS system are *Space, Control,* and *User* segments. The GPS operational constellation is made up of twenty-four satellites that orbit Earth at the altitude of ~20,000 km (Figure 9.2). The satellites are placed in six nearly circular orbital planes, inclined at 55 degrees with respect to the equatorial plane, with nominally four satellites in each plane. This configuration assures the simultaneous visibility of five to eight satellites at any point on Earth. The constellation establishes the *Space Segment.*

GPS satellites are powered by solar energy. During the solar eclipses, they use the backup batteries carried onboard. Because the satellites tend to drift from their assigned orbital positions, primarily due to orbit perturbations caused by Earth, the Moon, and planets' gravitational pull, solar radiation pressure, and so forth, they have to be constantly monitored by the *Control Segment* to determine their exact location in space. In order to keep the satellites as close as possible to the predesigned orbits, each satellite is equipped with small rocket boosters that can be fired when the orbit correction is needed. The *Control Segment* consists of 11 monitoring stations, each checking the exact altitude, position, speed, and overall health of the orbiting satellites 24 hours a day. In 2005 the National Geospatial-Intelligence Agency (NGA) added six more stations to the initial network of five. Based on these observations, the position coordinates and clock bias, drift, and drift rate can be predicted for each satellite, and then broadcast to each satellite for retransmission back to the users. The satellite position is parameterized in terms of predicted ephemeris, expressed in a

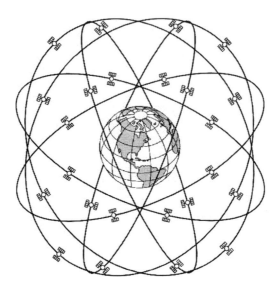

FIGURE 9.2   The Global Positioning System (GPS) constellation. (From Rizos, C., in *Manual of Geospatial Science and Technology*, pp. 75–94, Taylor & Francis, London, 2002. With permission.)

Earth-Centered-Earth-Fixed (ECEF) reference frame, known as the World Geodetic System 1984 (WGS84) (www.wgs84.com). The clock parameters are provided in the form of polynomial coefficients and, together with the predicted ephemeris, are broadcast to the users in the GPS navigation message. The accuracy of the predicted orbit is typically several meters.

Since the launch of the first GPS Block I satellite in 1978, the system evolved through several spacecraft designs, focused primarily on increased design life, extended autonomous operation time, better frequency standards (clocks), and the provision for accuracy manipulation, and was controlled by the DOD. The current constellation, as of February 2008, consists of 31 Block II/IIA/IIR/IIR-M satellites. Information about the current status of the constellation can be found, for example, at the U.S. Naval Observatory (USNO) Web site (http://tycho.usno.navy.mil/gpscurr.html). The *User Segment*, including all GPS equipment used in a variety of civilian and military applications, establishes the third component of the GPS system (Figure 9.3).

9.4   GPS SIGNAL CHARACTERISTICS

GPS satellites transmit low-power radio signals on two carrier frequencies designated as L1 and L2. The L1 carrier is 1575.42 MHz, and the L2 carrier is 1227.60 MHz in the ultrahigh frequency (UHF) band. GPS is a line-of-sight (LOS) system, and thus the signals can penetrate clouds, glass, and plastic but will not go through most solid objects, such as buildings and dense foliage. A GPS signal contains three different types of information: a Pseudo-Random Noise (PRN) code, ephemeris data (navigation message), and almanac data. The coarse-acquisition (C/A) code is available on the L1 frequency and the precise (P) code is available on both L1 and L2 signal. Due to the spread spectrum characteristic of the signals, the system provides a large margin of resistance to interference. All signals transmitted by GPS satellites are coherently derived from a fundamental frequency of 10.23 MHz as shown in Table 9.1.

Each satellite transmits a C/A-code and a unique segment of a P-code, which is the satellite's designated identification (ID), ranging from 1 to 32. This carrier modulation enables the measurement of the signal travel time between the satellite and the receiver (user), which is converted to *pseudorange* (i.e., raw distance between the satellite and the receiver, see Sections 9.6 and 9.7). The travel time, and thus the pseudorange, measured using the P code is up to ten times more accurate

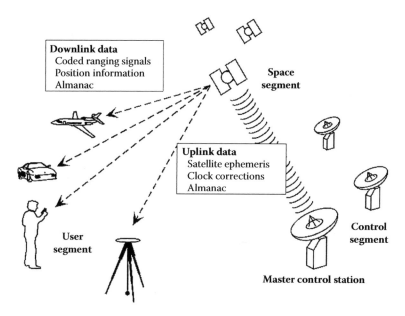

FIGURE 9.3 Primary segments of the Global Positioning System (GPS). (From Rizos, C., in *Manual of Geospatial Science and Technology*, pp. 75–94, Taylor & Francis, London, 2002. With permission.)

TABLE 9.1
Basic Components of the GPS Satellite Signal

Component	Frequency (MHz)	Ratio of Fundamental Frequency (f_o)	Wavelength (cm)
Fundamental frequency f_o	10.23	1	2,932.6
L1 carrier	1,575.42	154	19.04
L2 carrier	1,227.60	120	24.45
L5 carrier[a]	1176.45	115	25.50
P-code	10.23	1	2,932.6
C/A code	1.023	1/10	29,326
W-code	0.5115	1/20	58,651
Navigation message	50×10^{-6}	1/204,600	N/A

[a] New frequency planned under GPS Modernization Plan; not available as of June 2005 (see Section 9.11).

---

than the C/A-code–based measurement. Ephemeris data, which are continuously transmitted by each satellite, provide information regarding the satellite position in space, the status of the satellite (healthy or unhealthy), and current date and time. In addition to the broadcast ephemeris, each satellite transmits the almanac data showing the orbital information for that satellite and for every other satellite in the system. The almanac is used by the GPS receiver to search the sky for the satellites to acquire the signals. After the codes and data removal by the receiver's processing electronics, the signal's carrier phase can be measured; it provides GPS measurement of the highest accuracy. In addition, a Doppler shift can be observed, providing a measure of the range rate to the satellite between the consecutive epochs of observation. In general, dual-frequency receivers can provide carrier phase, pseudorange, and Doppler on L1 and L2. Single-frequency receivers measure pseudorange, carrier phase, and Doppler on L1 only.

C/A is the civilian code available to all users of the Standard Positioning Service (SPS). Under the *Anti-Spoofing** (AS) policy imposed by the DOD, the P-code is encrypted by an additional "W-code," resulting in the "Y-code," available exclusively to military users, and designated as Precise Positioning Service (PPS). PPS guarantees positioning accuracy of at least 22 m (95 percent of the time) horizontally and 27.7 m vertically, while guaranteed positioning accuracy of SPS is 100 m (95 percent of the time) horizontally and 156 m (95 percent of the time) vertically. However, the practical accuracy of SPS is usually higher (see Section 9.5). Under AS policy, the civilian receivers must use special signal tracking techniques to recover observables on L2, because no C/A code is available on L2 (Hofman-Wellenhof et al., 2001).

9.5 THE PRIMARY ERROR SOURCES

In general, biases and errors affect GPS measurements. Biases are defined as measurement errors that cause the observation to differ from the "true" distance by a systematic amount, and errors are considered synonymous to instrumental noise. Biases may occur due to imperfect knowledge of constants, such as "fixed" parameters, for example satellite or station coordinates, or may have a physical basis such as atmospheric effects on signal propagation, and can be normally accounted for by using special mathematical models or data-processing techniques. The primary error sources affecting GPS signals can be grouped into four basic categories based on the source of error: (1) atmospheric and environmental (signal propagation), (2) satellite related, (3) receiver related, and (4) reference station related. In addition, geometric factors and number of satellites tracked have a significant impact on the final positioning accuracy. Radio interference and jamming, as well as intentional signal degradation, have a considerable (yet random in magnitude), impact on GPS signals, which significantly affects the positioning accuracy.

9.5.1 ATMOSPHERIC AND ENVIRONMENTAL EFFECTS

9.5.1.1 Ionosphere and Troposphere Delays

The satellite signal is subject to atmospheric delays (Figure 9.4) as it passes through the atmospheric layers of the ionosphere and troposphere. The amount of delay depends on the state of the ionosphere and troposphere and can be estimated from models in the case of the troposphere or removed by using dual-frequency GPS measurements and relative (differential) processing (Hofman-Wellenhof et al., 2001). An approximated ionospheric model is also included in the satellite navigation message and can be used to partially remove ionospheric delay from GPS range measurements. The accuracy of this model is about 50 percent of the total ionospheric delay (Klobuchar, 1987).

The presence of free electrons in the ionosphere causes a nonlinear dispersion of electromagnetic waves traveling through the ionized medium, affecting their speed, direction, and frequency. The largest effect is on the signal speed, and as a result, GPS pseudorange is delayed and is thus measured too long, while the carrier phase advances and is thus measured too short. The total effect can reach up to 150 m and is a function of the Total Electron Content (TEC) along the signal's path, the frequency of the signal, the geographic location, the time of observation, the time of the year, and the period within the 11-year sun spot cycle (the last peak in ionospheric activity was 2001; see Figure 9.5). Because the ionospheric delay is different for L1 and L2, its effect can be removed from the GPS observables by combining dual-frequency data into iono-free linear combination and by relative processing, as will be explained in Section 9.7 (Hofman-Wellenhof et al., 2001).

Because the troposphere is a nondispersive medium for all frequencies below 15 GHz, the L1 and L2 carrier phase and pseudorange measurements are delayed by the same amount. Consequently, the elimination of the tropospheric effect by dual-frequency measurements is not possible; instead,

---

* Anti-spoofing policy guards against fake transmissions of satellite data by encrypting the P-code to form the Y-code.

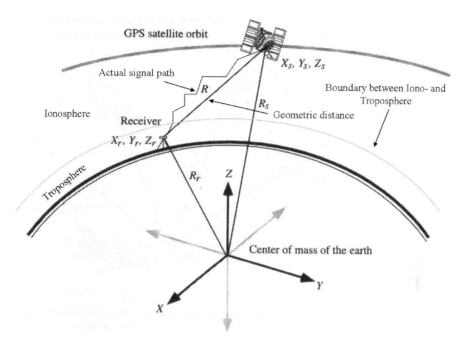

FIGURE 9.4 Atmospheric effects on Global Positioning System (GPS) signal propagation.

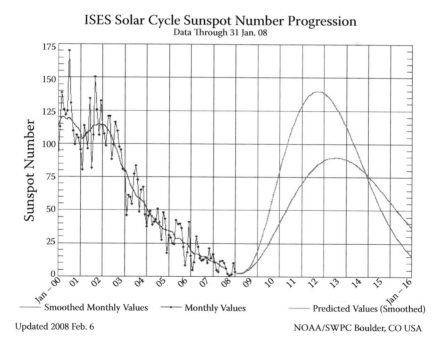

FIGURE 9.5 The progression of Solar Cycle 23 (www.sec.noaa.gov/SolarCycle).

it is accomplished by using empirical models of the tropospheric delay (Hofman-Wellenhof et al., 2001). The total effect in the zenith direction reaches ~2.5 m and increases with the cosecant of the elevation angle up to ~25 m at a 5° elevation angle. The tropospheric delay consists of the dry component (about 90 percent of the total tropospheric refraction), proportional to the density of the gas molecules in the atmosphere, and the wet refractivity due to the polar nature of the water molecules. In general, empirical models, which are functions of temperature, pressure, and relative humidity, can eliminate 90 to 95 percent of the tropospheric effect from the GPS observables.

9.5.1.2 Relativistic Propagation Error

The gravitational field causes space–time curvature of the signal; thus, propagation correction has to be applied to the GPS carrier-phase observable. This effect strongly depends on the geometry between station, satellite, and geocenter, but it is small in magnitude. The maximum effect is 19 mm, and it virtually cancels out in differential GPS (see Section 9.7).

9.5.1.3 Multipath

Multipath is a result of an interaction of the upcoming signal with the objects in the antenna surrounding, such as buildings and rocks. It causes multiple reflections and diffractions, and as a result, the signal arrives at the antenna via direct and indirect paths, increasing the travel time of the signal. These signals interfere with each other, resulting in an error in the measured pseudorange or carrier phase, degrading the positioning accuracy. The magnitude of multipath effect tends to be random and unpredictable, varying with satellite geometry, location, and type of reflective surfaces in the antenna surrounding, and can reach 1 to 5 cm for the carrier phases and 10 to 20 m for the code pseudoranges (Hofman-Wellenhof et al., 2001). Properly designed choke ring antennas can almost entirely eliminate this problem for the surface waves and signals reflected from the ground.

9.5.2 Satellite and Receiver Clock Errors

Even though the satellite onboard clock is a high-accuracy atomic clock, it may have a slight timing error that will affect the GPS observable if not corrected. This error is, however, monitored and estimated by the GPS *Control Segment*; the satellite clock correction (predicted) is provided in the satellite navigation message and can be used to correct the GPS observables. The receiver's built-in clock is not as accurate as the atomic clocks onboard the GPS satellites; therefore, it may be subject to timing error, defined as the difference between the clock's time and the true GPS time. This error is either eliminated by differential (relative) positioning (see Section 9.7.1.2) or must be estimated in the positioning solution if the point positioning mode is used.

It should also be mentioned that the relativistic effect on GPS receiver clock, due to the fact that the receiver is placed in the gravitational field and rotates with Earth, is corrected by the receiver software; it amounts to ~1 ns = 30 cm after 3 hours. Moreover, the constant drift, which is a part of the total correction due to relativistic time difference between the receiver and the satellite, is compensated for before the launch time by reducing the frequency of the satellite clock by 0.00455 Hz from its nominal value of 10.23 MHz. However, the periodic term has to be modeled; for GPS altitude, it has the maximum amplitude of ~30 ns in time or 10 m in distance. The periodic part can be canceled by performing between-stations differencing, but for point positioning it is still problematic if not properly accommodated. For more information about relativistic effects on GPS, see, e.g. (Schwarze et al., 1993; Zhu & Groten, 1988).

9.5.3 Orbital Errors

Orbital errors are also known as broadcast ephemeris errors. These are inaccuracies of the satellite's location (predicted orbit) reported in the broadcast ephemeris (navigation message). The satellite orbits are subject to perturbation forces, such as solar radiation, Earth and planetary gravity fields, tidal effects, relativistic effects, and others, that cause departure from the predefined elliptical orbit. These departures from the predefined orbit are continuously monitored by the *Control Segment*, allowing for accurate orbit prediction that is uploaded to the satellites' onboard computers for real-time broadcasting to the users. A more accurate alternative to the broadcast orbits is provided by the International GPS Service (IGS). Predicted, rapid, and postprocessed (final) GPS orbits are available with accuracy ranging from ~200 cm (predicted) to ~10 cm (rapid), to ~5 cm (final orbits) (http://igscb.jpl.nasa.gov/components/prods.html).

9.5.4 RECEIVER NOISE, INTERCHANNEL BIAS, AND OTHER INSTRUMENTAL BIASES

The most basic kind of noise is thermal noise[*] produced by the movement of the electrons in any material that has temperature above 0 Kelvin (Langley, 1998a, 1998c). The commonly used measure of the received signal strength is the signal-to-noise ratio; in the case of radiofrequency (RF), the most commonly used measure of the signal's strength is the carrier-to-noise-power-density ratio, C/N_0, defined as a ratio of the power level of the signal carrier to the noise power in a 1 Hz bandwidth (Langley, 1998a, 1998c; Van Dierendonck, 1995). C/N_0 is considered a primary parameter in describing the GPS receiver performance, as its value determines the precision of the pseudorange and carrier phase observations. Typical values of C/N_0 for modern high-performance GPS receivers (L1 C/A code) range between 45 and 50 dB-Hz. For example, for a C/N_0 equal to 45 dB-Hz and a signal bandwidth of 0.8 Hz, the root mean square (RMS) tracking error due to receiver thermal noise for the C/A code is 1.04 m, and for high-performance GPS receivers with narrow correlators (Van Dierendonck et al., 1992) with spacing of 0.1 chips, and the same bandwidth and C/N_0, the RMS is only 0.39 cm. The RMS tracking error due to noise for a carrier-tracking loop with a C/N_0 of 45 dB-Hz and a signal bandwidth of 2 Hz is only about 0.2 mm for L1 frequency (Braasch, 1994; Langley, 1998c). In a GPS receiver, the noise translates into an error in range measurement that ranges between 0.1 and 3 m (see Table 9.2). In summary, a good quality receiver should contribute less than 0.5 ms error in interchannel bias and less than 0.2 m in noise.

TABLE 9.2
GPS Error Sources and Their Approximated Magnitudes under SA On/Off

Error Source	P-code		C/A-code	
	SA Off	SA On	SA Off	SA On
Satellite				
Orbit prediction	3–5 m	10–40 m	3–5 m	10–40 m
Clock stability	1–3 m	10–50 m	1–3 m	10–50 m
Signal Propagation				
Ionosphere (two frequencies)	cm–dm	cm–dm	cm–dm	cm–dm
Ionosphere (model)	2–150 m	2–150 m	2–150 m	2–150 m
Troposphere (model)	dm	dm	dm	dm
Multipath effects	1–5 m	1–5 m	5–10 m	5–10 m
Relativistic propagation	~2 cm	~2 cm	~2 cm	~2 cm
Receiver				
Receiver noise	0.1–0.3 m	0.1–0.3 m	0.1–3 m	0.1–3 m
Interchannel bias	0.5 m	0.5 m	0.5 m	0.5 m
Hardware delays	dm–m	dm–m	m	m
Antenna phase center	mm–cm	mm–cm	mm–cm	mm–cm

---

[*] An electrical current generated by the electronsí random motion.

The latest generation of GPS receivers is based on dedicated-channel architecture, where every channel tracks one satellite either on the L1 or L2 frequency. Consequently, even though multichannel receivers are more accurate and less sensitive to loss of signal lock, they can suffer from the interchannel bias. Fortunately, in the modern receiver this bias can be effectively calibrated by the receiver's microprocessor. Any remaining bias can be removed by the differencing techniques (see Section 9.7.1.2).

All transmitted signals are clocked in coincidence with the PRN transitions for the P-signal and occur at the P-signal transition speed. On the L1 channel the data transitions of the two modulating signals (i.e., that containing the P(Y)-code and that containing the C/A-code) is such that the average time difference between the transitions does not exceed 10 nanoseconds (two sigma) (ICD-GPS-200, 1993). Because the GPS *Control Segment* uses GPS P-code measurements to compute the broadcast GPS orbits and clock corrections, there is a possible lack of synchronization between the epochs of the orbits and the C/A code. These instrumental biases (differential code biases, DCBs) are estimated for each day for all GPS satellites as part of the clock estimation procedure with an accuracy of about 0.1 nanoseconds (see, for example, www.aiub.unibe.ch/download/papers/codar_00.pdf; ftp://gage.upc.es/pub/gps_data/GPS_IONO). The P1-C1 DCB is generally on the order of a nanosecond, and P1-P2 can reach up to a few nanoseconds. Similar biases (~1 to 15 nanoseconds in range) exist in GPS receivers; these can be calibrated by the users. If not accounted for in the point positioning procedure, the DCBs may result in a pseudorange error of up to 5 m.

9.5.5 ANTENNA PHASE CENTER LOCATION

The physical (geometric) center of the antenna usually does not coincide with the phase center (the electrical center) of the antenna—a point to which radio signal measurements are referred. Moreover, the phase centers for L1 and L2 signals generally do not coincide, and the location of the electrical phase center can vary with variable azimuth and elevation of the satellites and the intensity of the incoming signal. This effect should not, in general, exceed 1 to 2 cm, and for modern antennas, it reaches only a few millimeters. Using the correct phase center offsets becomes very important when different antenna types are used in a survey. Detailed antenna calibration tables that account for the average spatial relationship between the Antenna Reference Point (ARP), such as Bottom of Antenna Mount (BA) and electrical phase centers, are provided by the National Geodetic Survey (NGS) (www.ngs.noaa.gov/ANTCAL/) and are normally included in any commercial GPS software.

9.5.6 REFERENCE STATION ERROR

This error affects relative positioning in case the reference (base) receiver location has incorrect coordinates in the WGS84 (ITRF2005) reference system. The ITRF (International Terrestrial Reference Frame) is an alternative to WGS84 realization of the global terrestrial reference system that is produced by the International Earth Rotation Service (IERS) based in Paris, France (http://itrf.ensg.ign.fr/). It includes many more stations than the original WGS84. Note that the current implementation of ITRF, labeled ITRF2000, and WGS84 are generally considered to be equivalent. For more information on ITRF, the reader is referred to http://itrf.ensg.ign.fr/ITRF_solutions/index.php.

9.5.7 NUMBER OF VISIBLE SATELLITES AND THEIR GEOMETRY

GPS is a line-of-sight system, where the path of the signal between the satellite and the ground receiver must be clear and unobstructed. Thus, GPS positioning accuracy and reliability strongly depend on the surrounding environment. A minimum of four satellites in view is required to obtain the position fix; however, redundant observations will significantly improve the accuracy. The satellite geometry is also very important, as weak geometry even with a sufficient number of satellites in view may provide a low-accuracy solution. Satellite geometry refers to the relative position of

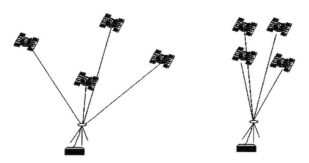

FIGURE 9.6   Good position dilution of precision (PDOP) (left) and bad PDOP (right).

the satellites at any given time with respect to the receiver location. Ideal satellite geometry exists when the satellites are located at wide angles relative to each other and are evenly distributed in azimuth and elevation angle as observed from the user's location. Poor geometry results when the satellites are located in a line or are grouped in clusters which may happen when a part of the sky is obstructed.

The two major factors that reflect the accuracy of GPS are (1) error in the range measurement, σ, and (2) geometric configuration between the receiver and the satellites. The two factors combined define the ultimate positioning error as a product of σ and the geometry factor, PDOP (position dilution of precision)—namely, *standard deviation of 3D positioning* = PDOP $\times$ σ.

The error in range measurement, σ, is called User Equivalent Range Error (UERE) and represents a quality of a single range measurement; it is computed as a square root of a sum of squares of the individual error sources. For SPS, the UERE is around 25 to 33 m under SA and drops to around 5 to 8 m under no SA. For PPS using dual-frequency receivers, the UERE amounts to ~1 to 2 m.

The geometric factor, PDOP, reflects the instantaneous geometry related to a single receiver and is determined by the position of the GPS satellites with respect to the receiver. PDOP can be interpreted as the reciprocal value of the volume of a tetrahedron that is formed by the positions of the satellites and the user. The best geometry corresponds to a large volume and vice versa (see Figure 9.6). Typically, more satellites yield smaller PDOP value, and PDOP of two and less indicates an excellent geometry; PDOP below six refers to a good geometry, and PDOP of seven and above indicates virtually useless data. A geometric factor related to a pair of receivers working in relative (differential) mode (e.g., a base and a rover) is called Relative DOP (RDOP). An example of varying geometry and partial loss of signal lock on GPS positioning accuracy is shown in Figure 9.7, where stand-alone positioning (lower accuracy, top figures) and differential (relative) positioning (higher accuracy, bottom figures) are illustrated. PDOP can be estimated using the approximated location of the user and the satellite broadcast ephemeris included in the satellite navigation message.

9.5.8   Interference and Jamming (Intentional Interference)

Radio interference can, at minimum, reduce the GPS signal's apparent strength (i.e., reduce the signal-to-noise ratio by adding more noise) and, consequently, the accuracy, or, at worst, even block the signal entirely. Medium-level interference may cause frequent losses of lock or cycle slips (a sudden jump in the carrier phase observable by an integer number of cycles), and may render virtually useless data.

9.5.9   Intentional Degradation of the Satellite Signal

Selective Availability (SA) is the DOD policy of denying to nonmilitary GPS users the full accuracy of the system. It is achieved by dithering the satellite clock (called delta process) and degrading the navigation message ephemeris (called epsilon process). The effects of SA (that are highly unpredictable) can be removed with encryption keys or through relative positioning techniques

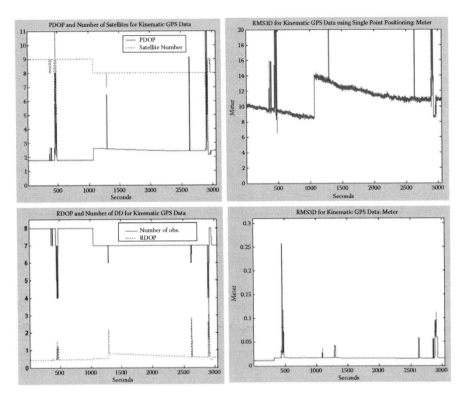

FIGURE 9.7 Position dilution of precision (PDOP) and the number of satellites observed (top left) and the corresponding three-dimensional standard deviation for GPS point positioning with range observation (top right); relative dilution of precision (RDOP) and the number of differential observations (bottom left) and the corresponding three-dimensional standard deviation for GPS relative positioning with carrier phase observations (bottom right).

(see Section 9.7.1.2). SA levels were set to zero on May 2, 2000, at 04:00 UT (Universal Time). For more information on GPS error sources, the reader is referred to Hofman-Wellenhof et al. (2001) and Lachapelle (1990).

9.6 MATHEMATICAL MODELS OF PSEUDORANGE AND CARRIER PHASE OBSERVABLES

Pseudorange is a geometric range between the transmitter and the receiver, distorted by the propagation media and the lack of synchronization between the satellite and the receiver clocks. It is recovered from the measured time difference between the epoch of the signal transmission and the epoch of its reception by the receiver. The actual time measurement is performed with the use of the PRN code. In principle, the receiver and the satellite generate the same PRN sequence. The arriving signal is delayed with respect to the replica generated by the receiver, as it travels ~20,000 km. In order to find how much the satellite's signal is delayed, the receiver-replicated signal is delayed until it falls into synchronization with the incoming signal (it is achieved at the point of maximum correlation between the incoming PRN code and the receiver-generated replica). The amount by which the receiver's version of the signal is delayed is equal to the travel time of the satellite's version (Figure 9.8). The travel time, Δt (~0.06 s), is converted to a range measurement by multiplying it by the speed of light, c.

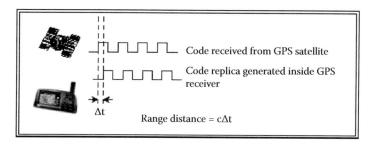

FIGURE 9.8   Principles of pseudorange measurement based on time observation.

There are two types of pseudoranges: C/A-code pseudorange and P-code pseudorange. The precision of the pseudorange measurement is partly determined by the wavelength of the chip in the PRN code. Thus, the shorter the wavelength, the more precise the range measurement would be. Consequently, the P-code range measurement precision (noise) of 10 to 30 cm is about ten times higher than that of the C/A code. Under the Anti-Spoofing policy, the P-code is encrypted to Y-code, as already explained, resulting in more complicated signal recovery on L2 frequency. Because there is no C/A code on L2, signal correlation, as explained above, does not work anymore, as the receiver cannot generate a replica of the unknown Y-code. Consequently, more sophisticated signal tracking techniques must be used (see, for example, Ashjaee, 1993).

The pseudorange observation can be expressed as a function of the unknown receiver coordinates, satellite and receiver clock errors, and the signal propagation errors (Equation (9.1)):

$$P_{r,1}^s = \rho_r^s + \frac{I_r^s}{f_1^2} + T_r^s + c(dt_r - dt^s) + M_{r,1}^s + e_{r,1}^s$$

$$\text{(9.1)}$$

$$P_{r,2}^s = \rho_r^s + \frac{I_r^s}{f_2^2} + T_r^s + c(dt_r - dt^s) + M_{r,2}^s + e_{r,2}^s$$

where

$$\rho_r^s = sqrt\left[\left(X^s - X_r\right)^2 + \left(Y^s - Y_r\right)^2 + \left(Z^s - Z_r\right)^2\right]$$

$$\text{(9.2)}$$

$P_{r,1}^s, P_{r,2}^s$:	Pseudoranges measured between receiver r and satellite s on L1 and L2
ρ_r^s:	Geometric distance between satellite s and receiver r
$\frac{I_r^s}{f_1^2}, \frac{I_r^s}{f_2^2}$:	Range error caused by ionospheric signal delay on L1 and L2
dt_r:	The r-th receiver clock error (unknown)
dt^s:	The s-th satellite clock error (known from the navigation message)
c:	The vacuum speed of light
$M_{r,1}^s, M_{r,2}^s$:	Multipath on pseudorange observables on L1 and L2
T_r^s:	Range error caused by tropospheric delay between satellite s and receiver r (estimated from a model)
$e_{r,1}^s, e_{r,2}^s$:	Measurement noise for pseudorange on L1 and L2
X^s, Y^s, Z^s:	Coordinates of satellite s (known from the navigation message)
X_r, Y_r, Z_r:	Coordinates of receiver r (unknown)
f_1, f_2:	Carrier frequencies of L1 and L2

Carrier phase is defined as a difference between the phase of the incoming carrier signal and the phase of the reference signal generated by the receiver. At the initial epoch of the signal acquisition, the receiver can measure only the fractional phase, so the carrier phase observable contains the initial unknown integer ambiguity, N. Integer ambiguity is a number of full phase cycles between the receiver and the satellite at the starting epoch, which remains constant as long as the signal tracking is continuous. After the initial epoch, the receiver can count the number of integer cycles being tracked. Thus, the carrier phase observable can be expressed as a sum of the fractional part, φ (in cycles), measured with millimeter-level precision, and the integer number of cycles counted since the starting epoch, t_0. The integer ambiguity can be determined using special techniques referred to as ambiguity resolution algorithms. Once the integer ambiguity is resolved, the ambiguous carrier phase observable can be converted to unambiguous range measurement $R = (N + \varphi)\lambda$ by multiplying the sum of the measured phase (in cycles) and the initial integer ambiguity (in cycles) by the corresponding wavelength, λ (see Figure 9.9). It should be noted that starting from epoch t_0, the carrier phase measurement φ will include not only the fractional part of a cycle, but also the number of full cycles since the initial epoch t_0. The phase-range observable, Φ (in meters; Equation (9.3)), equals the sum of R and all the error sources affecting the measurement. This observable is used in the applications where the highest accuracy is required.

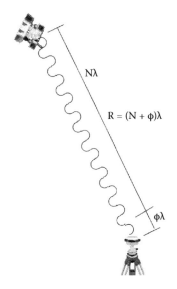

FIGURE 9.9 Carrier phase range measurement.

$$\Phi_{r,1}^s = \rho_r^s - \frac{I_r^s}{f_1^2} + T_r^s + \lambda_1 N_{r,1}^s + c(dt_r - dt^s) + m_{r,1}^s + \varepsilon_{r,1}^s$$

(9.3)

$$\Phi_{r,2}^s = \rho_r^s - \frac{I_r^s}{f_2^2} + T_r^s + \lambda_2 N_{r,2}^s + c(dt_r - dt^s) + m_{r,2}^s + \varepsilon_{r,2}^s$$

where

$\Phi_{r,1}^s, \Phi_{r,2}^s$:   Phase-ranges (in meters) measured between station r and satellite s on L1 and L2

$N_{r,1}^s, N_{r,2}^s$:   Initial integer ambiguities on L1 and L2, corresponding to receiver r and satellite s

$\lambda_1 \approx 19$ cm and $\lambda_2 \approx 24$ cm are wavelengths of L1 and L2

$m_{r,1}^s, m_{r,2}^s$:   Multipath error on carrier phase observables on L1 and L2

$\varepsilon_{r,1}^s, \varepsilon_{r,2}^s$:   Measurement noise for carrier phase observables on L1 and L2

Another observation sometimes provided by GPS receivers, and primarily used in kinematic applications for velocity estimation, is instantaneous Doppler frequency. It is defined as a time change of the phase-range, and thus, if available, it is measured on the code phase (Lachapelle, 1990).

Equation (9.2), which is a nonlinear part of Equation (9.1) and Equation (9.3), requires Taylor series expansion to enable the estimation of the three unknown user coordinates $(X, Y, Z)_r$. Secondary (nuisance) parameters in the above equations are satellite and user receiver clock errors, tropospheric and ionospheric errors, multipath, and integer ambiguities. These are usually removed by differencing mode of GPS data processing (see Section 9.7.1.2), by empirical modeling (troposphere), or by the processing of dual-frequency signals (ionosphere). As already mentioned, ambiguities must be resolved prior to users' position estimation with carrier phase measurements.

Equation (9.1) through Equation (9.3) are parameterized in terms of Cartesian geocentric coordinates X, Y, Z; however, after the positioning solution is obtained, Cartesian coordinates can be converted to geodetic latitude, longitude, and height, which represent an equivalent triplet of coordinates. Because GPS is a geometric system, its coordinates are related to a reference ellipsoid (WGS84 ellipsoid), whose semimajor axis and flattening are needed to convert Cartesian to and from geodetic coordinates (Torge, 1980). Since GPS provides heights above the WGS84 ellipsoid, in order to convert these heights to topographic (orthometric) heights, a geoid undulation (geoid-ellipsoid separation) must be used. National Geospatial-Intelligence Agency (NGA) provides an online service that calculates geoid undulation for a given location using the latest NGA geoid model (http://earth-info.nga.mil/GandG/wgs84/gravitymod/wgs84_180/intptW.html). For more information on geoid and height conversion, the reader is referred to Torge (1980).

9.7 POSITIONING WITH GPS

The main principle behind positioning with GPS is triangulation in space, based on the measurement of a range (pseudorange or phase-range) between the receiver and the satellites (Figure 9.10). Essentially, the problem can be specified as follows: given the position vectors of GPS satellites (such as ρ^s of satellite s in Figure 9.10) tracked by a receiver r, and given a set of range measurements (such as, P_r^s) to these satellites, determine a position vector of the user, ρ_r. A single range measurement to a satellite places the user somewhere on a sphere with a radius equal to the measured range. Three simultaneously measured ranges to three different satellites place the user on the intersection of three spheres, which corresponds to two points in space. One is usually an impossible solution that can be discarded by the receiver. Even though there are three fundamental unknowns (coordinates of the user's receiver), the minimum of four satellites must be simultaneously observed to provide a unique solution in space (see Figure 9.1 in Section 9.2), as explained next.

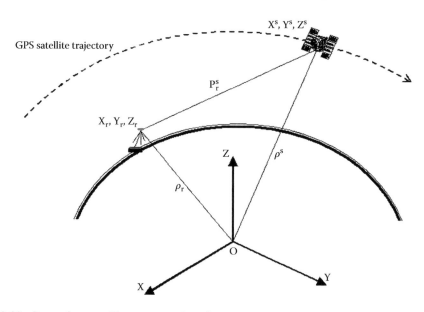

FIGURE 9.10 Range from satellite s to ground receiver r.

As already mentioned, the fundamental GPS observable is the signal travel time between the satellite and the receiver. However, the receiver clock that measures the time is not perfect and may introduce an error to the measured pseudorange (even though we limit our discussion here to pseudoranges, the same applies to the carrier phase measurement that is indirectly related to the signal transit time, as the phase of the received signal can be related to the phase at the epoch of transmission in terms of the signal transit time). Thus, in order to determine the most accurate range, the receiver clock correction must be estimated to bring the receiver clock to synchronization with the satellite clock, and its effect must be removed from the observed range. The synchronization error between satellite and receiver clocks is particularly important given that an error of only 0.1 microsecond in the satellite or receiver clocks results in distance error on the order of 30 m. Hence, the fourth pseudorange measurement is needed, because the total number of unknowns, including the receiver clock, is now four. If more than four satellites are observed, a least-squares solution is employed to derive the optimal solution.

9.7.1   POINT VERSUS RELATIVE POSITIONING

There are two primary GPS positioning modes: point positioning (or absolute positioning) and relative positioning. However, there are several different strategies for GPS data collection and processing, relevant to both positioning modes. In general, GPS can be used in static and kinematic modes, using both pseudorange and carrier phase data. GPS data can be collected and then postprocessed at a later time, or processed in real time, depending on the application and the accuracy requirements. In general, postprocessing in relative mode provides the best accuracy.

9.7.1.1   Point (Absolute) Positioning

In point, or absolute positioning, a single receiver observes pseudoranges to multiple satellites to determine the user's location. For the positioning of the moving receiver, the number of unknowns per epoch equals three receiver coordinates plus a receiver clock correction term. In the static mode with multiple epochs of observations, there are three receiver coordinates and n receiver clock error terms, each corresponding to a separate epoch of observation 1 to n. The satellite geometry and any unmodeled errors will directly affect the accuracy of the absolute positioning.

9.7.1.2   Relative Positioning

The relative positioning technique (also referred to as differencing mode or differential GPS, DGPS) employs at least two receivers: a reference (base) receiver, whose coordinates must be known, and the user's receiver, whose coordinates can be determined relative to the reference receiver. Thus, the major objective of relative positioning is to estimate the 3D baseline vector between the reference receiver and the unknown location. Using the known coordinates of the reference receiver and the estimated ΔX, ΔY, and ΔZ baseline components, the user's receiver coordinates in WGS84 can be readily computed. Naturally, the user's WGS84 coordinates can be further transformed to any selected reference system.

An observable in differencing mode is obtained by differencing the simultaneous measurements to the same satellites observed by the reference and the user receivers (between receiver differencing), or through "between satellite differencing" and "between epoch differencing." The most important advantage of relative positioning is the removal of the systematic error sources (common to the base station and the user or both satellites and epochs of observation) from the observable, leading to the increased positioning accuracy. Because for short to medium baselines (up to ~40 to 60 km) the systematic errors in GPS observables due to troposphere, satellite clock, and broadcast ephemeris errors are of similar magnitude (i.e., they are spatially and temporally correlated), the relative positioning allows for a removal or at least a significant mitigation of these error sources, when the observables are differenced. In addition, for baselines longer than 10 km, the ionosphere-free linear combination must be used (if dual-frequency data are available) to mitigate the effects of the ionosphere (Hofman-Wellenhof et al., 2001). Single-frequency users are limited to short baselines, unless differential corrections are provided via DGPS services, as explained in the next section. In summary, the following are the primary consequences of GPS data differencing:

- Elimination or reduction of several bias errors
- Reduction of data quantity
- Introduction of mathematical correlation among data
- Increase of the noise level of the differenced data

The primary differential modes are (1) single differencing mode, (2) double differencing mode, and (3) triple differencing mode. The differencing can be performed between receivers, between satellites, and between epochs of observations as already mentioned. The single-differenced (between-receiver) measurement, $\Phi_{i,j}^k$, is obtained by differencing two observables to the satellite k, tracked simultaneously by two receivers i (reference) and j (user): $\Phi_{i,j}^k = \Phi_i^k - \Phi_j^k$ (see Figure 9.11). By differencing observables from two receivers, i and j, observing two satellites, k and l, or simply by differencing two single differences to satellites k and l, one arrives at the double-differenced (between-receiver/between-satellite differencing) measurement: $\Phi_{i,j}^{k,l} = \Phi_i^k - \Phi_j^k - \Phi_i^l + \Phi_j^l = \Phi_{i,j}^k - \Phi_{i,j}^l$. Double difference is the most commonly used differential observable. Furthermore, differencing two double differences, separated by the time interval $dt = t_2 - t_1$, renders triple-differenced measurement, $\Phi_{i,j}^{k,l}(dt) = \Phi_{i,j}^{k,l}(t_2) - \Phi_{i,j}^{k,l}(t_1)$, which in case of carrier phase observables effectively cancels the initial ambiguity term. Differencing can be applied to both pseudorange and carrier phase. However, for the best positioning accuracy with carrier phase double differences, the initial ambiguity term should be first resolved and fixed to the integer value. Relative positioning may be performed in static and kinematic modes, in real time (see the next section), or for the highest accuracy, in postprocessing. Table 9.3 shows the error characteristics for between-receiver single and between-receiver/between-satellite double differenced data.

Basic pseudorange and carrier phase Equation (9.1) through Equation (9.3) represent the functional relationships between the true observations and the underlying parameters. However, the observations (and models involved) are not perfect; thus, the functional models require a respective

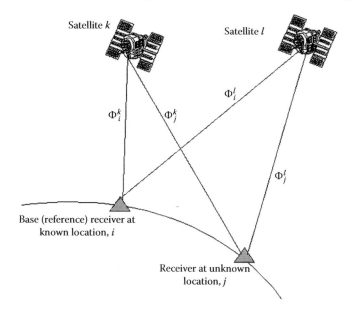

FIGURE 9.11 Between-receiver between satellite phase-range double differencing.

TABLE 9.3

Differencing Modes and Their Error Characteristics

Error Source	Single Difference	Double Difference
Ionosphere	Reduced, depending on the baseline length	Reduced, depending on the baseline length
Troposphere	Reduced, depending on the baseline length	Reduced, depending on the baseline length
Satellite clock	Eliminated	Eliminated
Receiver clock	Present	Eliminated
Broadcast ephemeris	Reduced, depending on the baseline length	Reduced, depending on the baseline length
Ambiguity term	Present	Present
Noise level w.r.t. one-way observable	Increased by $\sqrt{2}$	Increased by 2

stochastic model describing the accuracy of the measurements and its statistical properties. A fundamental part of the stochastic model is a covariance matrix of observations, Σ. The primary least-squares formula representing a general matrix form of a solution of a system of GPS observation equation is shown in Equation (9.4). Note that Equation (9.1) through Equation (9.3) (and in general, any GPS observation equation) are nonlinear; thus, they are linearized before processing through Equation (9.4):

$$\xi = \left(A^T \Sigma^{-1} A \right)^{-1} A^T \Sigma^{-1} y \tag{9.4}$$

where A is a design matrix containing the partial derivatives of the observable with respect to the unknown parameters, ξ; and y is the vector of observations minus "calculated observation" which is computed based on the approximated values of the parameters. For more details on the least-squares solution, see, for example, Strang and Borre (1997).

9.7.2 DGPS Services: An Overview

As explained earlier, the GPS error sources are spatially and temporally correlated for short to medium base–user separation. Thus, if the reference station and satellite coordinates are known (satellite location is known from broadcast ephemeris), the errors in the GPS measurements can be estimated at specified time intervals and made available to the nearby users through a wireless communication as differential corrections. These corrections can be used to remove the errors from the observables collected at the user's (unknown) location (Yunck et al., 1996). This mode of positioning uses DGPS services to mitigate the effects of the measurement errors at the user's location, leading to the increased positioning accuracy in real time. DGPS services are commonly provided by the government, industry, and professional organizations, and enable the users to use only one GPS receiver collecting pseudorange data, while still achieving superior accuracy as compared to the point-positioning mode. Naturally, in order to use a DGPS service, the user must be equipped with additional hardware capable of receiving and processing the differential corrections.

DGPS services normally involve some type of wireless transmission system. They may employ VHF or UHF systems for short ranges, low-frequency transmitters for medium ranges (beacons), or L-band or C-band geostationary satellites for coverage of entire continents, which is called Wide Area DGPS (WADGPS) or Global Satellite Based Augmentation System (GSBAS). WADGPS involves multiple GPS base stations with precisely known locations that track all GPS satellites in view. These data are sent to the master control station (see Figure 9.12) that estimates the errors in the GPS pseudoranges and forms a satellite uplink message that is transmitted to a geo-stationary satellite. This satellite, in turn, broadcasts the information to all users with specialized GPS receivers; the average correction latency is about 5 sec (www.aiub.unibe.ch/download/igsws2004/

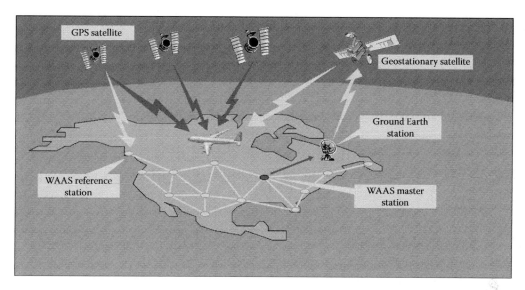

FIGURE 9.12   The Wide Area Augmentation System.

Real_Time_Aspects/TUAM1_Muellerschoen_1.pdf). The satellite uplink message contains GPS differential corrections and GPS satellite health status data. The positioning accuracy of WADGPS ranges from submeter to a few meters, depending on the provider, coverage, number of stations used, error modeling technique, and so forth. DGPS services routinely support the following applications: land survey, offshore positioning, precision agriculture, aerial photogrammetry and LiDAR (light detection and ranging), GIS and asset mapping, machine control, aircraft navigation, and intelligent vehicle highway system (IVHS).

The fundamental concept of WADGPS is the categorization of error sources in the GPS observables. By constructing a model for each error source (e.g., satellite clock, satellite ephemeris, ionospheric delay, and local errors, such as tropospheric delay, multipath, receiver noise, and hardware bias), the system creates a vector correction. This is the distinction between wide area and local area differential corrections. Local area augmentation systems transmit a scalar correction from the reference station to the user for each pseudorange measurement. In contrast, wide area systems transmit the error models to the user, which are then recombined to form a correction for each pseudorange measurement. The benefit of the vector correction is its improved ability to capture the spatial decorrelation of the error sources (http://waas.stanford.edu/tour.html).

9.7.2.1   DGPS Services: Examples

GSBAS, such as OmniSTAR™ (www.omnistar.com) and StarFire™ (www.navcomtech.com/starfire/), are examples of commercial DGPS suppliers, providing worldwide coverage at the submeter to decimeter-level accuracy. For example, the StarFire™ solution represents an advance from the ground-based augmentation systems because it considers each of the GPS satellite signal error sources independently, instead of measuring a total (combined) error or atmospheric error only in the GPS pseudoranges. Examples of government-supported DGPS services include Federal Aviation Administration (FAA)-supported satellite-based Wide Area Augmentation System (WAAS), ground-based DGPS services, referred to as Local Area DGPS (LADGPS), such as U.S. Coast Guard and Canadian Coast Guard services, or FAA-supported Local Area Augmentation System (LAAS). An example of Internet-based WADGPS is the Global Differential GPS (IGDG) provided by NASA JPL, based on the NASA Global GPS Network (GGN) consisting of approximately sixty sites (http://gipsy.jpl.nasa.gov/igdg/). IGDG is designed for dual-frequency users and offers 10 cm horizontal and 20 cm vertical real-time positioning accuracy.

LAAS is installed at individual airports and is effective over just a short range, with accuracy of 1 m or less in all dimensions. The ground equipment includes four reference receivers, a LAAS ground facility, and a VHF data broadcast transmitter. This ground equipment is complemented by LAAS avionics installed on the aircraft. The GPS Reference Receivers and LAAS Ground Facility (or LGF) work together to measure errors in GPS-provided position. LAAS correction message is then sent to a VHF data broadcast (VDB) transmitter. The VDB broadcasts the LAAS signal throughout the LAAS coverage area to avionics in LAAS-equipped aircraft. The LAAS equipment in the aircraft uses the corrections provided on position, velocity, and time to guide the aircraft safely to the runway (http://gps.faa.gov/about/office_org/headquarters_offices/ats/service_units/techops/navservices/nsss/Lass).

WAAS consists of a network of twenty-five ground reference stations and a number of geo-stationary (Inmarsat) satellites broadcasting a signal in the GPS band (Figure 9.12), also providing additional ranging distances that are included in the user positioning solution. The WAAS signals contain information including differential corrections and GPS satellite health status. WAAS has been running 24 hours a day, 7 days a week since early 2000, providing high-integrity navigation signals for nonaviation users such as boaters, precision agriculture, crop dusters, surveyors, vehicle dispatchers and location services, cell phone 911 emergency services, hikers, and other personal recreation uses within the conterminous United States. WAAS was officially commissioned by the FAA for public aviation use on July 10, 2003. The published specifications of WAAS call for 7.6 m in the vertical direction, which corresponds to better than 5 m horizontal accuracy 95 percent of the time (for a single-frequency user receiver), but accuracies around 2 m horizontal RMS have already been demonstrated (http://gps.faa.gov/programs/waas/waas-text.htm; http://www.navcen.uscg.gov/; Lachapelle et al., 2002). Enhanced, dual-frequency WAAS significantly increases the accuracy to approximately 30 to 70 cm in real time.

The GPS and WAAS signals are sent over the same frequency band, with the 50 bps (bits per second) data rate from the normal GPS satellites for information like ephemeris and almanacs, and 500 bps of the raw signal data rate from the WAAS satellite. The increased data rate for WAAS reduces the reliability of the WAAS data transmissions (increased bandwidth reduces SNR). The GPS signals are also slightly stronger than the WAAS signals from the geo-stationary satellites (http://waas.stanford.edu/tour.html).

NAVCEN (www.navcen.uscg.gov/dgps/default.htm) operates the U.S. Coast Guard Maritime Differential GPS (DGPS) Service, consisting of two control centers and over sixty remote broadcast sites, and the developing Nationwide DGPS Service (NDGPS). The Service broadcasts correction signals on marine radio beacon frequencies to improve the accuracy of and integrity to GPS-derived positions. The U.S. Coast Guard DGPS Service guarantees 10 m accuracy, while typical positional error of 1 to 3 m is achieved. October 2007 coverage is presented in Figure 9.13.

9.7.2.2   DGPS Message Format

DGPS receivers support the major international standards for GPS and DGPS (RINEX, RTCM, and NMEA). RINEX (i.e., the Receiver Independent Exchange Format) is an ASCII format, established for an easy exchange of the GPS data collected by different GPS receivers. The format has been optimized for minimum space requirements independent from the number of different observation types of a specific receiver. Three primary types of RINEX files exist: observation data file, navigation message file, and meteorological data file (http://gps.wva.net/html.common/rinex.html#rinex:_the_receiver_independent_exchange_format_version_2.10).

The most widely used international standards for DGPS message format were developed by the Radio Technical Commission for Maritime Services (RTCM), a committee that governs standards for passing data between different equipment used in the Marine Electronics industry. The RTCM Special Committee No. 104 established "Recommended Standards for Differential Navstar GPS Service," dated January 3, 1994, referred to as RTCM SC104, which is a standard format for sending differential

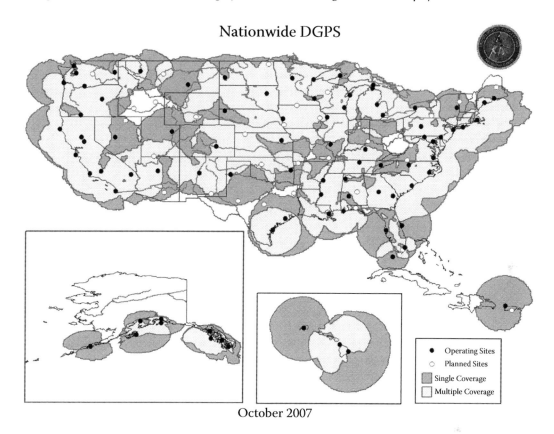

Nationwide DGPS

October 2007

FIGURE 9.13 U.S. differential Global Positioning System (DGPS) coverage. (Courtesy of DGPS System Management Branch, USCG Navigation Center; http://www.navcen.uscg.gov/dgps/coverage/CurrentCoverage.htm.)

correction data to a GPS receiver. The actual format is rather complex and lengthy, but it generally contains the following information: (1) the time of the measurement at the reference station, (2) observed range errors (corrections) for every satellite in view at that reference station, and (3) the range error rate for every satellite in view. RTCM SC104 version 2.0 (RTCM-2.0) format essentially deals with DGPS (code-only) corrections, and RTCM-2.1 incorporates several enhancements, especially for PDGPS (carrier phase) corrections (www.ccg-gcc.gc.ca/dgps/format_e.htm).

The standard message types within RTCM-2.1 are message types 3, 5, 6, 7, 9, 15, and 16. A type 3 message contains information on the identity and surveyed position of the active reference DGPS station. Message type 5 notifies the user equipment suite that a satellite deemed unhealthy by its current navigation message is usable for DGPS navigation. The type 7 message provides information of its broadcasting DGPS station and the other two or three adjacent DGPS stations. The type 6 message is a filler message used only when the reference station has no other message to broadcast. The type 9 message has been selected for broadcasting pseudorange corrections. Two methods of transmitting the type 9 message are possible. The first method of broadcasting GPS pseudorange corrections is based upon "Three-Satellite Type 9 Messages," denoted as "type 9-3" messages. In this method, corrections for all satellites are assigned to either three satellite type 9 messages or to a remainder message of either one or two satellites. The transmission rate could be at either 100 or 200 bps. The second method of broadcasting pseudorange corrections is to broadcast individual type 9 messages for each satellite at a transmission rate of 50 bps. This message is referred to as the "Single Satellite Type 9 Message," denoted as the type 9-1 message. The type 15 message includes atmospheric parameters. The type 16 message can be utilized as a timely supplement to the notice to mariners or shipping, regarding information on the status of the local DGPS service which is not provided in other message types. Additionally, the type 16 message may provide limited information on service outages in adjacent coverage areas or planned outages for scheduled maintenance at any broadcast site. In order to keep data link loading to a minimum, type 16 messages contain only system information that is crucial to the safety of navigation. For more details on the RTCM format and message types, the reader is directed to www.ccg-gcc.gc.ca/dgps/main_e.htm.

The recently updated standards, Version 3.0 RTCM [RTCM, 2004], consist primarily of messages designed to support real-time kinematic (RTK) operations for both GPS and GLONASS (see Section 9.11), including broadcasting of code and carrier phase observables, antenna parameters, and ancillary system parameters by the reference station to the user's location. Unlike the earlier version, this standard does not include tentative messages; it is designed to accommodate modifications to GPS and GLONASS (e.g., new L2C and L5 signals) and to the new systems that are under development (e.g., Galileo, see Section 9.11). In addition, augmentation systems that use geostationary satellites that provide ranging signals and operate in the same frequency bands are now in the implementation stages (RTCM, 2004). The primary reason for this update included the following shortcomings of the earlier versions: (1) parity scheme that uses words with 24 bits of data followed by 6 bits of parity was wasteful of bandwidth; (2) parity was not independent from word to word; (3) with so many bits devoted to parity, the actual integrity of the message was not as high as it should be; and (4) 30-bit words are awkward to handle. Version 3.0 is intended to correct these weaknesses.

Message types contained in the current Version 3.0 standard have been structured in several groups: (1) observations (GPS L1, GPS L1/L2, GLONASS L1, GLONASS L1/L2)—message type 1001–1004, 1009–1012; (2) station coordinates (antenna reference point coordinates and antenna height)—message type 1005–1006; (3) antenna description—message type 1007–1008; and (4) auxiliary operation information—message type 1013 (RTCM, 2004).

Aside from the differentially corrected position coordinates in NMEA (National Marine Electronics Association) format, DGPS receivers might also offer the possibility of storing all of the raw data and correction signals for postprocessing. The raw data at the receiver can generally be stored in either receiver-specific or standard RINEX format. NMEA is an industry association that sets data transmission standards (www.gpsinformation.org/dale/nmea.htm; www.nmea.org) and has developed specifications that define the interface between various pieces of marine electronic equipment, including a set of standard messages defining the possible outputs of a GPS receiver. The idea of NMEA is to send a line of data called a sentence that is totally self-contained and independent from other sentences. There are standard sentences for each device category, and there is also the ability to define proprietary sentences for use by the individual company. All of the standard sentences have a two-letter prefix that defines the device that uses that sentence type (e.g., for GPS receivers the prefix is GP), which is followed by a three-letter sequence that defines the sentence contents (www.gpsinformation.org/dale/nmea.htm).

There are several GP sentences, each one containing some unique data associated with them. They are all in ASCII format and are in the form of comma delimited strings. The character string lengths vary from 30 to 100 characters and are output at the selected intervals. The most common string (or sentence) is called the GGA string that provides essential fix data containing the Time of the Fix, Latitude, Longitude, Height, Number of Satellites used in the fix, DOP, Differential Status, and the Age of the Correction. Other strings may contain Speed, Track, Date, and so forth. NMEA is available in virtually all GPS receivers and is the most commonly used data output format. It is also the format used in most software packages that interface to a GPS receiver.

Some other sentences that have applicability to GPS receivers are as follows:

- AAM—Waypoint Arrival Alarm
- ALM—Almanac data

- APA—Auto Pilot A sentence
- APB—Auto Pilot B sentence
- BOD—Bearing Origin to Destination
- BWC—Bearing using Great Circle route
- DTM—Datum being used
- GGA—Fix information
- GLL—Lat/Lon data
- GSA—Overall Satellite data
- GSV—Detailed Satellite data
- MSK—Send control for a beacon receiver
- MSS—Beacon receiver status information
- RMA—Recommended Loran data
- RMB—Recommended navigation data for GPS
- RMC—Recommended minimum data for GPS
- RTE—Route message
- VTG—Vector track and Speed over the Ground
- WCV—Waypoint closure velocity (Velocity Made Good)
- WPL—Waypoint information
- XTC—Cross track error
- XTE—Measured cross track error
- ZDA—Date and Time
- ZTG—Zulu (UTC) time and time to go (to destination)

The hardware interface for GPS units is designed to meet the NMEA requirements. They are also compatible with most computer serial ports using RS232 protocols; however, strictly speaking, the NMEA standard is not RS232, but rather EIA-422. The interface speed can generally be adjusted (set to 9600 or higher), but the NMEA standard is 4800 baud with 8 bits of data, no parity, and one stop bit (www.gpsinformation.org/dale/nmea.htm).

9.7.3 Network-Based Real-Time Kinematic GPS (RTK GPS)

Another approach, gaining popularity in a number of countries, is to support the users through the local networks of Continuously Operating Reference Stations (CORS) that normally serve a range of applications, especially those requiring high accuracy in postprocessing or in real time (although the real-time support is still limited). Government agencies, such as NGS, Department of Transportation (DOT), or international organizations, such as IGS, deploy and operate these networks. All users typically have free access to the archived data that can be used as a reference (base data) in carrier phase or pseudorange data processing in relative mode. Alternatively, network-based positioning using carrier-phase observations with a single user receiver can be accomplished with the local specialized networks, which can estimate and transmit carrier phase correction (see, e.g., Cannon et al., 2001; Chen, 2000; Dai et al., 2001; Kim and Langley, 2000, Raquet, 1998; Rizos, 2002a; Vollath et al., 2000; Wanninger, 2002).

The long-range RTK technique (defined for 100 to 300 km or longer station separation) is the most challenging GPS data reduction method. As the base-rover separation increases, many distance-dependent biases, such as atmospheric or orbital errors, may become significant even in the relative mode, which complicates the ambiguity resolution process. This, in turn, may seriously corrupt the positioning results, unless these effects are properly accounted for. In general, the success of precise GPS positioning over long baselines depends on the ability to resolve the integer phase ambiguities when short observation time spans are required, which is especially relevant to RTK applications. The distance over which carrier-phase ambiguity can be resolved may be significantly increased by employing a multireference station approach. Over the past few years, the use of the reference station network approach has shown great promise in extending the interreceiver separation. The implementation of multiple reference stations in a permanent array offers several advantages over the standard single-baseline approach. It improves the accuracy of the mobile (user) receiver and makes the results less sensitive to the length of the baselines, at the same time acting as a "filter" for the lower-quality measurements coming occasionally from some stations. Consequently, the precision of the baseline, expressed as its estimated standard deviation, is improved. The most important contribution of the network solution, as compared to the single baseline case, is not, however, the improvement in precision, but rather the improved reliability and availability, as well as the opportunity to model the atmospheric errors (van der Marel, 1998). A commonly used base station separation in the network ranges between a few kilometers to a few hundreds of kilometers, depending on the applications served, coverage (network extent), and the required accuracy.

The network-based RTK concept can be summarized as follows: a network of permanent tracking stations uses GPS data to estimate the ionospheric and tropospheric corrections in real time and broadcasts this information to the users, who require accuracy better than DGPS. Thus, network-based positioning (either RTK or postprocessed) is a three-step procedure: (1) network solution (no user data is involved) to estimate the instantaneous ionospheric corrections and tropospheric delays (tropospheric total zenith delay, TZD) per station, (2) transmission of the corrections to the users (Internet, beacon, mobile phone), and (3) user positioning solution (in relative mode) using data from a single or multiple reference stations together with the atmospheric corrections derived in step (1). Naturally, if RTK mode is used, the user needs a data transmit from a reference station to form a double-difference carrier phase observable for positioning solution (Grejner-Brzezinska et al., 2005a, 2005b; Kashani et al., 2007; Wielgosz et al., 2004, 2005). This approach is similar to the virtual reference station (VRS) concept, but without the actual geometric relocation of the observation to the vicinity of the user receiver (van der Marel, 1998; Wanninger, 2002).

The VRS system uses observations from multiple reference stations to estimate the systematic errors in the data and to create a unique virtual reference station for each user's location. A synthetic data set (at the VRS location) is sent to the user located in the vicinity of VRS and is then used for relative positioning solution. The systematic errors modeled include ionosphere, troposphere, satellite orbit errors, and multipath; data are delivered to the rover in RTCM/CMR+ format. The following infrastructure is required for Trimble VRS™: dual-frequency GPS receivers, PCs, software, and communications links that are permanently or semipermanently stationed to provide continuous data logging, monitoring and data broadcast, rover receiver, rover interface, and office software that handles the data. The following are the basic processing steps (see Figure 9.14 and Figure 9.15):

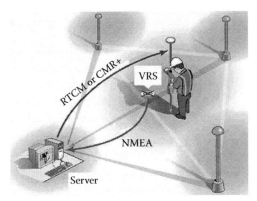

FIGURE 9.14 Virtual reference station (VRS) data flow. (Courtesy Trimble Navigation Ltd.)

- Reference station data streams back to server through LAN, Internet, or radio links
- Roving receiver sends an NMEA string back to the server using cellular modem
- VRS position is established for each user location
- Server uses VRS position to create corrected observables and broadcasts them to the rover
- Rover surveying in normal RTK mode but data are relative to the VRS
- User receiver logs files for postprocessing

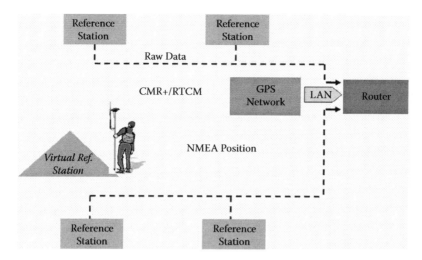

FIGURE 9.15   Basic infrastructure and data flow in the virtual reference station (VRS) network. (Courtesy Trimble Navigation Ltd.)

The primary benefits of the network-based RTK, from the user perspective, are as follows:

- No need to establish base or reference station
- Reduction of time and chance for errors inherent in setting up a base
- Savings in equipment and time cost
- Extension of the operating range with improved initialization and accuracy

The primary applications are monitoring and modeling of the movement of man-made and natural structures using GPS and integrated sensors, and monitoring the integrity of high-order geodetic networks. The market is still expanding as the number of VRS installations is growing nationwide; the evolving applications include oilfield subsidence, dam deformation monitoring, landslide monitoring, volcano monitoring, construction site surveying (Figure 9.16), and other engineering applications (www.trimble.com/vrs.shtml).

9.7.4   Precise Point Positioning

As explained in Sections 9.7.1 through 9.7.3, relative or differential GPS techniques are used to eliminate the atmospheric, receiver-specific, and satellite-specific error sources. A promising alternative to relative positioning and DGPS is precise point positioning (PPP)—an absolute GPS positioning technique with all major error sources carefully modeled (i.e., removed from the direct pseudorange or carrier phase observables). For example, the errors in GPS broadcast ephemeris and clock corrections can be successfully reduced using the precise IGS (http://igscb.jpl.nasa.gov/) orbits and clocks (accuracy of the final IGS orbits and clocks is better than 5 cm and 0.1 ns, respectively). Ionospheric corrections can be provided by IGS or by the local or regional CORS networks, as explained in Section 9.7.3. The influence of the troposphere may be reduced in several ways. The simplest solution is to use the standard atmospheric models; however, the resulting accuracy may not be sufficient. Better results are obtained if meteorological data (temperature, pressure, and relative humidity) are used in the atmospheric models instead of the standard atmosphere. The best results can be achieved if the tropospheric zenith delay corrections are derived from the local or regional CORS networks. The receiver clock correction remains an unknown in the PPP mathematic model, similar to the standard point positioning technique.

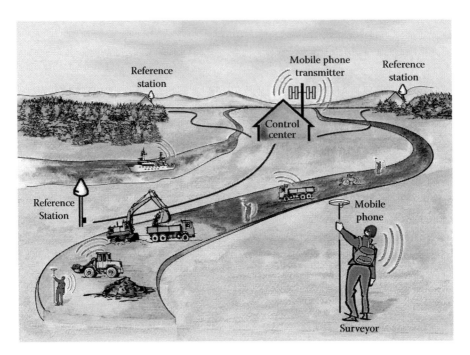

FIGURE 9.16 Virtual reference station (VRS) example setup for construction area. (Courtesy Trimble Navigation Ltd.)

The availability of IGS and CORS networks and orbital, atmospheric, and timing products that are derived from the GPS observables collected by these networks made the PPP technique feasible. The obvious benefits of this positioning approach are as follows:

- Single receiver operation (low cost)
- Can be applied anywhere and anytime (remote areas, space applications, etc.) under different dynamics
- Not limited by a baseline length (no baseline processing)
- Independence on GPS reference stations
- Can be applied for static and kinematic platforms
- Simple processing algorithms

For more details on the PPP error modeling and currently achievable accuracy (subdecimeter to submeter level, depending on the type of observables used, mode of positioning—static versus kinematic—and quality of corrections applied), the reader is referred to the literature (Gao and Shen, 2002; Kouba and Héroux, 2001; Ovstedal, 2002; Wielgosz et al., 2005).

9.8 REAL TIME VERSUS POSTPROCESSING

A brief outline of the most commonly used GPS positioning modes was presented in Section 9.7, with a special emphasis on DGPS services and the growing network-based RTK solutions and applications. In summary, Figure 9.17 provides an overview of the GPS positioning and data-processing strategies. Table 9.4 compares the real-time and postprocessing scenarios, and Table 9.5 compares the accuracy of various DGPS modes with the static and RTK GPS accuracy. Further discussion on GPS accuracy is provided in the next section.

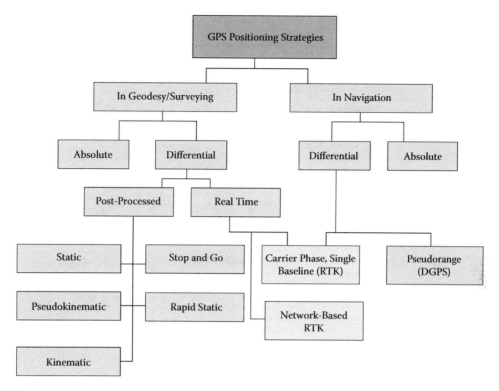

FIGURE 9.17 Global Positioning System (GPS) processing strategies. (Modified from Langley, R. B., *GPS World*, September, pp. 70–74, 1998. With permission.)

TABLE 9.4

Comparison of Real-Time and Postprocessing Scenarios

Positioning Mode/Attribute	Accuracy	Time	Navigation	Cost	Remote Locations	Portability
Postprocessing	Advantage			Advantage		Advantage
Real time		Advantage	Advantage		Advantage	

9.9 HOW ACCURATE IS GPS?

The positioning accuracy of GPS depends on several factors, such as the number and the geometry of the observations collected, the mode of observation (point versus relative positioning), the type of observation used (pseudorange or carrier phase), the measurement model used, and the level of biases and errors affecting the observables. Depending on the design of the GPS receiver and the factors listed above, the positioning accuracy varies from 10 m with SA turned off (about 100 m with SA turned on) for pseudorange point positioning, to better than 1 cm when carrier phases are used in relative positioning mode. In order to obtain better than 10 m accuracy with pseudoranges, differential positioning, DGPS services, or PPP techniques must be employed.

In summary, the current GPS constellation performance significantly exceeds the system design specifications (in point positioning) and can be summarized as follows (SPS, 2001):

- PDOP availability:
    - Requirement—PDOP of 6 or less, 98 percent of the time or better
    - Actual—99.99

TABLE 9.5

Achievable DGPS Accuracy Compared to Static and RTK GPS Accuracy (http://www.smi.com/Downloads/catalogs/2005catalog.pdf; modified here)

Correction Type	Horizontal Accuracy	Vertical Accuracy
Single-frequency WAAS	3–7 m[a]	3–7 m[a]
Dual-frequency WAAS	<50 cm	<70 cm
StarFire((dual frequency)	<10 cm	<15 cm
IGDG (dual frequency)	<50 cm	<70 cm
Static (using NGS CORS)	5 mm[b]	5 mm[b]
Static (using base and rover)	5 mm[b]	5 mm[b]
RTK	1 cm[b]	2 cm[b]
Network-based RTK	5 mm[b]	10 mm[b]

[a] According to the WAAS specifications; however, much better accuracies (<2 m, even up to 30–70 cm) were reported, as explained in Section 9.7.2.1.

[b] Increases with the baseline length.

- Horizontal service availability:
    - Requirement—95 percent threshold of 13 m, 99 percent of the time or better
    - Actual—4.49 m
- Vertical service availability:
    - Requirement—95 percent threshold of 22 m, 99 percent of the time or better
    - Actual—6.43 m
- User range error:
    - Requirement—6 m or less, constellation average
    - Actual—1.47 m

Other factors affecting the GPS positioning accuracy depend on (1) whether the user is stationary or moving (static versus kinematic mode), (2) whether the positioning is performed in real time or in postprocessing, (3) the data reduction algorithm, (4) the degree of redundancy in the solution, and (5) the measurement noise level. The currently achievable GPS accuracies, provided as two-sigma, corresponding to 95 percent confidence level, are summarized in Table 9.6. The lower bound of the relative positioning accuracy listed in Table 9.6 cannot be stated with precision, as it depends on several hardware and environmental factors, as well as the survey geometry, among others (the symbol → indicates the increase of the values listed). Thus, the accuracy levels listed in Table 9.6 should be understood as the best achievable accuracy.

9.10 GPS INSTRUMENTATION

During the past two decades, the civilian as well as military GPS instrumentation evolved through several stages of design and implementation, focusing primarily on achieving an enhanced reliability of positioning and timing, modularization, and miniaturization. In addition, one of the most important aspects, especially for the civilian market, has been the decreased cost of the receivers, as the explosion of GPS applications calls for a variety of low-cost, application-oriented, and reliable equipment. By far, the majority of the receivers manufactured today are of the C/A-code single-frequency type. However, for the high-precision geodetic applications, the dual-frequency solution is standard. Even though the civilian and military receivers, as well as application-oriented instruments, have evolved in different directions, one might pose the following question: *Are*

TABLE 9.6

Currently Achievable GPS Accuracy

	Positioning Mode		
Point Positioning (Pseudorange)		Relative Positioning	
PPS	1–5 m	Static survey (carrier phase)	2 mm (→) plus 1 ppm[a] (up to <0.1 ppm)
SPS, SA off	4–10 m	Kinematic survey (carrier phase)	5 mm (→)
SPS, SA on	0–100 m	DGPS services (pseudorange)	50 cm (→)

[a] ppm, part per million.

Source: From Rizos, C., in *Manual of Geospatial Science and Technology*, Taylor & Francis, London, 2002. With permission.

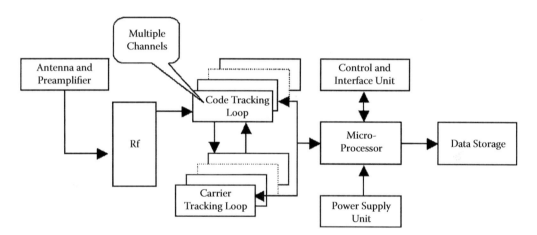

FIGURE 9.18 Basic components of a Global Positioning System (GPS) receiver. (From Grejner-Brzezinska, D. A., in *Manual of Geospatial Science and Technology*, Taylor & Francis, London, 2002. With permission.)

all GPS receivers essentially the same, apart from functionality and user software? The general answer is yes: all GPS receivers support essentially the same functionality blocks, even if their implementation differs for different types of receivers.

The following are the primary components of a generic GPS receiver (Figure 9.18): antenna and preamplifier, radio-frequency (RF) front-end section, a signal tracker block, microprocessor, control/interface unit, data storage device, and power supply (Grejner-Brzezinska, 2002; Langley, 1991; Parkinson and Spilker, 1996). Any GPS receiver must carry out the following tasks:

- Select the satellites to be tracked based on GDOP and Almanac[*]
- Search and acquire each of the GPS satellite signals selected
- Recover navigation data for every satellite
- Track the satellites, measure pseudorange and carrier phase
- Provide position and velocity information

---

[*] A set of parameters included in the GPS satellite navigation message that is used by a receiver to predict the approximate location of a satellite. The almanac contains information on all of the satellites in the constellation.

- Accept user commands and display results via control unit or a PC
- Record the data for postprocessing (optional)
- Transmit the data to another receiver via radio modem for real-time solutions (optional)

An important characteristic of the RF section is the number of channels and hence the number of satellites that can be tracked simultaneously. Older receivers had a limited number of channels (even as little as one), which required sequencing through satellites in order to acquire enough information for 3D positioning. Modern GPS receivers are based on dedicated channel architecture, where every channel tracks one satellite on L1 or L2 frequency.

9.11   GPS MODERNIZATION AND OTHER SATELLITE SYSTEMS

The current plan of GPS modernization is focused on improving the quality of civilian uses of GPS, primarily through the implementation of a new L2C code on L2 frequency and a new civilian signal on L5 frequency of 1227.6 MHz (Spilker and Van Dierendonck, 2001). In addition, a new M-code (encrypted) has been implemented exclusively for military use, ensuring that military and civilian users will have entirely separate signals and codes. Consequently, AS policy will be abandoned. (SA has already been turned off.) The new dual-frequency civilian tracking capability is available on the Block IIR-M GPS satellites (currently, five satellites in orbit, with the most recent launch of Decembr 20, 2007), and more improvements are planned for Block IIF satellites. As of February 2008, the first GPS Block IIF launch was projected by 2010 (http://facility.unauco.org/science_tech_gnss_modernization.html) . The GPS Block IIF spacecraft represents the next-generation GPS system designed to meet both military and civil customer requirements. Among other features, the GPS Block IIF spacecraft will feature extended design life up to 15 years, modular design, 3 m spherical error probable (CEP), options for L5 civil signals, increased autonomy from ground segment, and rapid on-orbit reprogrammability (www.deagel.com/pandora/?p=pm00371004).

In terms of GPS instrumentation, hardware miniaturization, further development of software receivers, improvements in reliability, faster sampling rates, lower noise and more multipath resistance, more real-time operations, and ubiquitous dual-frequency measurement capability are expected (Rizos, 2002c). The current trend of integrating GPS with other sensors such as inertial, vision systems, laser scanners, and pseudolites will continue to serve more specialized, customized applications. It is expected that the increasing number of CORS-based local services will serve real-time users.

Other GNSS that are complementary to GPS are the Russian Glonass system originally developed for the Russian military, and the European Galileo system (civilian), currently under implementation (first satellite was launched on December 28, 2005). Glonass became operational with a full twenty-four-satellite constellation in 1996. However, as of February 2008, the number of operational satellites is sixteen, with the most recent launch of December 25, 2007, where four new Glonass satellites were placed in orbit. Still, the system's long-term stability might be questionable. Nevertheless, GPS/Glonass receivers, which take advantage of the extended constellation created by the additional satellites in view, were developed. For more information on Glonass and Galileo, the reader is referred to http://www.gpsworld.com/; http://www.insidegnss.com/.

9.12   GPS MAPPING PROJECT AND CONNECTION
TO AGRICULTURAL GEOPHYSICS

Modern geophysics uses a variety of sensors and techniques for exploration, probing, and mapping, in applications such as mining/petroleum exploration, environmental studies, engineering investigations, and agricultural field assessments. The geophysical techniques presently used most often for agricultural purposes include resistivity, electromagnetic induction, and ground-penetrating radar.

These three geophysical methods are being employed for soil mapping of salinity, water content, and nutrient levels along with location of claypans and buried farm infrastructure. Other geophysical methods not currently used in agricultural geophysics but having a promising future are, for example, seismic methods, which might be useful for determining shallow subsurface geology, soil engineering properties, and the level of soil compaction; self-potential investigations employed to find leaks in animal waste storage ponds and treatment lagoons; or geomagnetic surveying that can support farm infrastructure mapping. A common denominator for all these geophysical methods and their applications is the need for proper geolocation (geospatial location) of the geophysical quantities being measured. This can be provided by GPS which, depending on the accuracy required, can be used in one of the positioning modes discussed in this chapter. GPS can, in general, provide 3D geolocation of the sensing device as well as a precise timing stamp (1 pps signal), if timing information is required.

One way of defining or classifying geospatial applications that are relevant in the context of this book is to adopt some accuracy requirements, such as follows (Rizos, 2002b):

- Scientific Surveys (category A): better than 1 ppm
  - Precise engineering deformation analysis, geodynamics applications
- Geodetic Surveying (category B): 1 to 10 ppm
  - Geodetic surveys for establishing, maintenance, and densification of geodetic control
- General Surveying (category C): lower than 10 ppm
  - Engineering and cadastral applications
  - Geophysical mapping, and so forth
- Mapping/Geolocation (category D): better than 2 m
  - General-purpose geolocation tasks primarily for GIS data capture, some geophysical mapping

Regardless of the accuracy requirement and GPS technique used, each GPS mapping project consists of five basic steps:

- *Mission Planning*: All the initial preparations that take place before GPS and geophysical sensor data are collected
- *Data Collection*: Collecting GPS and geophysical sensor data in the field
- *Manipulation*: All the processing of GPS data that occurs between the collection period and data analysis, such as the downloading, export, quality control, and processing of GPS data
- *Analysis*: Using GPS data as spatially referenced information in a research problem: here— geolocation of geophysical quantities mapped
- *Application*: Applying the results of the analysis phase in the real world

Stanoikovich and Rizos (2002) and Stewart and Rizos (2002) provide excellent reviews of the GPS survey planning process and carrying out of a mapping task.

GPS receivers most commonly used in agricultural geophysics are Trimble (such as Trimble AgGPS series; www.trimble.com/agriculture.shtml), Topcon receivers (http://www.topcongps.com), Magellan WAAS receivers (www.magellangps.com) and Garmin WAAS products (www.garmin.com). When purchasing a GPS receiver, one of the most important requirements to carefully consider is the accuracy. The accuracy of the receiver has the greatest impact on its cost. Thus, the geolocation accuracy requirement should be a defining factor for the GPS receiver selection. Another important factor is the source of differential correction the receiver can accept (i.e., DGPS radio beacons, commercial satellite differential service providers, or the WAAS correction). Many receivers are designed to use only one correction option. Others, such as, for example, Trimble AgGPS 132 or Topcon GMS-110, allow a choice of the source of differential corrections. Thus, it is crucial to make sure the selected receiver can provide the accuracy needed and can use the correction sources available in the project area. System compatibility with the geophysical sensors is also very important. Two fundamental requirements are (1) sufficient number of ports to connect to the equipment and (2) format of the GPS receiver output that can be supported by the equipment used in geophysical application. Another important consideration when choosing a GPS receiver is its reliability—its ability to withstand the rigors of the application, like high vibration, extreme heat or cold, dust, or moisture. In summary, all factors listed here should be considered carefully and in the context of the application-specific geophysical sensor and GPS requirements.

In conclusion, a few commercial mapping software packages, designed for processing, visualizing, and analyzing large earth science data sets suitable for geophysics applications are listed next. These include, for example, Geosoft products (www.geosoft.com), ESRI GIS and mapping software, such as, ArcGIS, ArcView, or ArcInfo (www.esri.com), SOILTEQ products (www.soilteq.com), or Integrated Mapping Systems Inc. solutions (http://www.mappingsystems.com), just to name a few.

ACKNOWLEDGMENTS

I would like to thank my PhD students, Tae-Suk Bae and Chang-Ki Hong, for their helpful comments on this manuscript. Special thanks go to another PhD student, Yudan Yi, who created Figures 8.6 and 8.8.

REFERENCES

Ashjaee, J., 1993. An analysis of Y-code tracking techniques and associated technologies, *Geodetical Info Magazine*, 7(7), 26–30.

Braasch, M.S., 1994. Isolation of GPS Multipath and Receiver Tracking Errors, *Navigation*, 41(4), 415–434.

Cannon, M.E., Lachapelle, G., Alves, P., Fortes, L.P., Townsend B., 2001. GPS RTK Positioning Using a Regional Reference Network: Theory and Results. Presented at 5th GNSS International Symposium, Seville, Spain, May 8–11 (http://plan.geomatics.ucalgary.ca/papers/01gnssmec.pdf).

Chen, H.Y., 2000. An instantaneous ambiguity resolution procedure suitable for medium-scale GPS reference station networks, *Proceedings of the 13th International Technical Meeting of the Satellite Division of the Institute of Navigation, ION GPS*, Salt Lake City, UT, September 19–22, Institute of Navigation, Fairfax, VA, pp. 1061–1070.

Dai, L., Wang, J., Rizos, C., and Han, S., 2001. Real-time carrier phase ambiguity resolution for GPS/GLONASS reference station networks. *Proceedings of the International Symposium on Kinematic Systems in Geodesy, Geomatics and Navigation* (KIS2001), Banff, Canada, June 5–8, University of Calgary, Calgary, Canada, pp. 475–481.

Gao, Y. and Shen, X., 2002. A new method for carrier phase based precise point positioning, *Navigation*, 49(2), 109–116.

Grejner-Brzezinska D.A., 2002. GPS Instrumentation Issues, Chapter 10 in *Manual of Geospatial Science and Technology*, J. Bossler, J. Jensen, R. McMaster and C. Rizos (editors), Taylor & Francis, London, pp. 127–145.

Grejner-Brzezinska, D.A., Kashani, I., and Wielgosz, P., 2005a. On Accuracy and Reliability of Instantaneous Network RTK as a Function of Network Geometry, Station Separation, and Data Processing Strategy, *GPS Solutions*, 9(3), 179–193.

Grejner-Brzezinska, D.A., Wielgosz, P., Kashani, I., Smith, D.A., Spencer, P.S.J., Robertson, D.S., and Mader, G.L., 2005b. The analysis of the Effects of Different Network-Based Ionosphere Estimation Models on the Rover Positioning Accuracy, *Journal of Global Positioning Systems*, 3(1–2), 115–131.

Hofman-Wellenhof, B., Lichtenegger, H., and Collins, J., 2001. *GPS Theory and Practice*, 5th ed., Springer Verlag Wien, New York, 382 pages.

ICD-GPS-200, Revision C, Initial Release, 1993. (http://www.navcen.uscg.gov/pubs/gps/icd200/icd200cw1234.pdf) last accessed in March 2008.

Kashani, I., Grejner-Brzezinska, D.A., and Wielgosz, P., 2005. Towards instantaneous network-based RTK GPS over 100 km distance, *Navigation* 52(4), 239–245.

Kashani, I., Wielgosz, P., and Grejner-Brzezinska, D.A., 2007. The impact of the ionospheric correction latency on long-baseline instantaneous kinematic GPS positioning, *Survey Review*, 39(305), 238–251.

Kim, D. and Langley, R.B., 2000. GPS ambiguity resolution and validation: Methodologies, trends and issues, *Proceedings of 7th GNSS Workshop and International Symposium on GPS/GNSS*, Seoul, Korea, November 30–December 2, GNSS Technology Council, Seoul, Korea, pp. 213–221.

Klobuchar, J.A., 1987. Ionospheric time-delay algorithm for single-frequency GPS users, *IEEE Transactions on Aerospace and Electronic Systems*, AES-23(3), Aerospace and Electronic Systems Society, pp. 325–331.

Kouba, J. and Héroux, P., 2001. Precise point positioning using IGS orbit and clock products, *GPS Solutions*, 5(2), 12–28.

Lachapelle, G., Ryan, S., and Rizos, C., 2002. Servicing the GPS user, in *Manual of Geospatial Science and Technology*, J. Bossler, J. Jensen, R. McMaster, and C. Rizos (Eds.), Taylor & Francis, London, pp. 201–215.

Lachapelle, G., 1990, GPS observables and error sources for kinematic positioning, in *Kinematic Systems in Geodesy, Surveying, and Remote Sensing*, Schwarz, K.P. and Lachapelle, G. (Eds.), IUGG, Springer-Verlag, pp. 17–26.

Langley, R.B., 1991. The GPS receiver: An introduction, *GPS World*, January, 50–53.

Langley, R.B., 1998a. A primer on GPS antennas, *GPS World*, July, 50–54.

Langley, R.B., 1998b. RTK GPS, *GPS World*, September, 70–74.

Langley, R.B., 1998c. Propagation of the GPS signals, and GPS receivers and the observables, in *GPS for Geodesy*, 2nd ed., Teunissen, P.J. and Kleusberg, A. (Eds.), Springer, pp.112–185.

Ovstedal, O., 2002. Absolute positioning with single-frequency GPS receivers, *GPS Solutions*, 5(4), 33–44.

Parkinson, B.W. and Spilker, J.J. (Eds.), 1996. *Global Positioning System: Theory and Applications*, Vol. 1, American Institute of Aeronautics and Astronautics, Reston, VA.

Raquet, J., 1998. Development of a method for kinematic GPS carrier phase ambiguity resolution using multiple reference receivers. Ph.D. Thesis. UCGE Report Number 20116, University of Calgary. Canada.

RTCM, 2004. RTCM Recommended Standards for Differential GNSS (Global Navigation Satellite Service), Version 3.0; (www.rtcm.org), last accessed March 2008.

Rizos, C., 2002a. Network RTK research and implementation: A geodetic perspective, *Journal of Global Positioning Systems*, 1(2), 144–150.

Rizos, C., 2002b. Introducing the Global Positioning System, in *Manual of Geospatial Science and Technology*, J. Bossler, J. Jensen, R. McMaster. and C. Rizos (Eds.), Taylor & Francis, London, pp. 77–94.

Rizos, C., 2002c. Where do we go from here?, in *Manual of Geospatial Science and Technology*, J. Bossler, J. Jensen, R. McMaster, and C. Rizos (Eds.), Taylor & Francis, London, pp. 216–229.

Schwarze, V.S., Hartmann, T., Lenis, M., and Soffel, M., 1993. Relativistic effects in satellite positioning, *Manuscripta Geodaetica*, 18, 306–316.

Spilker, J.J. and Van Dierendonck, A.J., 2001. Proposed new L5 civil GPS codes, *Navigation*, 48(3), 135–143.

SPS, 2001. *Global Positioning System Standard Positioning Service Performance Standard*, Assistant Secretary of Defense for Command, Control, Communications and Intelligence (http://pnt.gov/public/docs/SPS-2--1-final.pdf).

Stanoikovich, M. and Rizos, C., 2002. Carrying out GPS Surveying/Mapping Task, in *Manual of Geospatial Science and Technology*, J. Bossler, J. Jensen, R. McMaster. and C. Rizos (Eds.), Taylor & Francis, London, pp. 183–199.

Stewart, M. and Rizos, C., 2002. GPS projects: Some planning issues, in *Manual of Geospatial Science and Technology*, J. Bossler, J. Jensen, R. McMaster, and C. Rizos (Eds.), Taylor & Francis, London, pp. 162–181.

Starng, G. and Borre, K., 1997. *Linear Algebra, Geodesy and GPS*, Wellesley-Cambridge Press, Wellesley, MA.

Torge, V., 1980. *Geodesy*, Walter de Gruyter, Berlin, New York.

Van Dierendonck, A.J., 1995. Understanding GPS receiver terminology: A tutorial, *GPS World*, January, 34–44.

Van Dierendonck, A.J., Fenton, A., and Ford, T., 1992. Theory and performance of narrow correlator spacing in a GPS receiver, *Navigation*, 39(3), 265–283.

van der Marel, H., 1998. Virtual GPS reference stations in The Netherlands. *Proceedings, 11th International Technical Meeting of the Satellite Division of the Institute of Navigation, ION GPS,* Nashville, TN, 15–18 September, Institute of Navigation, Fairfax, VA, pp. 49–58.

Vollath, U., Buecherl, A., Landau, H., Pagels, C., and Wagner, B., 2000. Multi-base RTK positioning using virtual reference stations, *Proceedings, 13th International Technical Meeting of the Satellite Division of the Institute of Navigation, ION GPS*, September 19–22, Salt Lake City, UT, pp. 123–131.

Wanninger, L., 2002. Virtual reference stations for centimeter-level kinematic positioning. *Proceedings, 15th International Technical Meeting of the Satellite Division of the Institute of Navigation, ION GPS*, September 24–27, Portland, Oregon, Institute of Navigation, Fairfax, VA, pp. 1400–1407.

Wielgosz, P., Grejner-Brzezinska, D.A., and Kashani, I., 2004. Network approach to precise medium range GPS navigation, *Navigation*, 51(3), pp. 213–220.

Wielgosz, P., Grejner-Brzezinska, D.A., and Kashani, I., 2005. High-Accuracy DGPS and precise point positioning based on Ohio CORS network, *Navigation,* 52(1), 23–28.

Yunck, T.P., Bar-Sever, Y.E., Iijima, B.A., Lichten, S.M., Lindqwister, U.J., Mannucci, A.J., Muellerschoen, R.J., Munson, T.N., Romans, L.J., and Wu., S.C., 1996. A prototype WADGPS system for real time sub-meter positioning worldwide, *Proceedings, 9th International Technical Meeting of the Satellite Division of the Institute of Navigation, ION GPS,* Kansas City, MO, September 17–20, Institute of Navigation, Fairfax, VA, pp. 1819–1826.

Zhu, S.Y. and Groten, E. 1988. Relativistic effects in GPS, *GPS Techniques Applied to Geodesy and Surveying, Lecture Notes in Earth Sciences*, Vol. 19, Springer-Verlag, Berlin, Heidelberg, New York, pp. 41–46.

10 Integration of Geographic Information Systems (GISs) with Agricultural Geophysics

Carolyn J. Merry

CONTENTS

10.1 INTRODUCTION

Proper agricultural management decisions often require the integration and analysis of information from a number of different sources, including measurements obtained from one or more geophysical methods (resistivity, electromagnetic induction, ground-penetrating radar, etc.) along with other farm field geospatial data (topography, crop yields, soil properties, satellite remote sensing imagery, etc.). Integration and analysis of geophysical and nongeophysical information from multiple sources is especially important when employing precision agricultural techniques to separately manage different parts of a farm field. Geographic information systems (GISs) provide a set of tools that are particularly well suited for management, integration, and analysis of multiple geophysical and nongeophysical geospatial data sets. GIS software essentially allows geospatial data to be organized, stored, edited, displayed, and analyzed in an effective and efficient manner.

10.2 COMPONENTS OF A GIS

There are four basic components to a GIS: an input system, a database management system which includes data storage and editing, a data analysis system, and an output system (Figure 10.1). Basic functions within a GIS include data acquisition, data management, data manipulation, data analysis, modeling, and the display of spatial data. The input system to a GIS allows the user several different ways to incorporate data. Data input devices, such as digitizing tables and scanners, allow one to incorporate historical data from existing hard-copy maps. Software utilities are available to import data in various standard formats, such as ASCII, shapefiles, .tif, .img, and .dxf.

The database management system is the core of the GIS. The geospatial data will be in various computer formats in the database—vector data, raster data, shape files, and the attribute information. The database is typically in a standard format, depending on the GIS software. Data storage

199

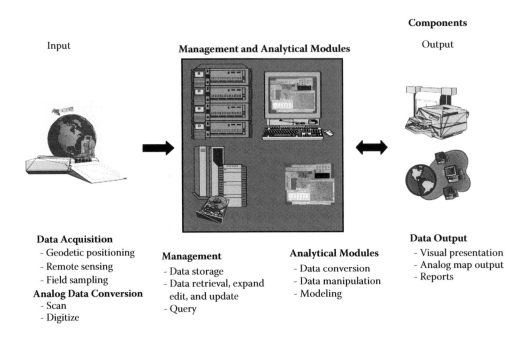

FIGURE 10.1   Components of a geographic information system (GIS).

and editing capabilities provide for a variety of software tools to store and maintain the digital representation of the data sets. The database needs to be kept up-to-date, so editing tools are available to correct and update the data layers. Procedures to join maps and to perform edge matching between adjacent map areas are also available.

Most GISs include a wide range of capabilities for analysis of geospatial data (geostatistics, numerical methods, etc.). The output system is used to create high-quality maps, charts, and statistical summaries from analysis of the data layers residing in the spatial database. There are several public domain and commercially available GIS software packages. These include ArcGIS, Geo-Media, MapInfo, Idrisi, ERDAS, AUTOCAD MAP, MicroImages, and Manifold. Bolstad (2005) provides a brief review of these software packages. There are several basic textbooks that provide excellent descriptions of GIS concepts. These include works by Bolstad (2005), Bernhardsen (2002), Bossler (2002), Chang (2002), Chrisman (2002), DeMers (2005), and Lo and Yeung (2002).

10.3   MAPPING CONCEPTS

A GIS is composed of data layers that represent the different spatial properties of an area. Example data layers that would be important for agricultural geophysics would include data layers of topography, soil type, crop coverage, hydrologic features, land cover type, crop yield, and, of course, the geophysical measurements. New spatial data layers can be derived from existing data layers. This relationship can be defined as

$$\text{NEW Layer} = f \text{ (Existing Data Layers)} \tag{10.1}$$

When using geospatial data models, the spatial properties of each data layer need to be considered. This functionality can be expressed as the following:

$$F (P1, P2, \ldots, Pn) \, x,y,z = f (P1)x,y,z + f (P2)x,y,z + \ldots + f (Pn)x,y,z \tag{10.2}$$

where F ( ) is the model, and P1 … + (Pn) are the parameters of the model. Unlike traditional methods, GIS data are organized based on geographic relationships among the data attributes. Areas that share the same property are grouped into a data layer. The attribute data in GIS layers inherit the spatial properties of the data. Data modeling is the focus when manipulating layers in a GIS. In general, the GIS model can be expressed as the following:

$$F (L1, L2, …, Ln) \ x,y,z = f (L1)x,y,z + f (L2)x,y,z + … + f (Ln)x,y,z \qquad (10.3)$$

where F ( ) is the model, and L1 … + (Ln) are the data layers in the GIS.

Geographic data, such as the topography of an agricultural field, crop yields, soil type, and geophysical measurements, are composed of two types of data: the spatial components (x,y) and the nonspatial attribute data (a). The spatial component stores the position of a feature, and the attribute data store the property or condition of that particular feature. Geospatial data can be categorized into four fundamental types—point, line, polygon, and surface—and are stored as either a vector or a raster format in a GIS (Figure 10.2).

Vector and raster data structures are the two major types of spatial data structures used in a GIS. A vector data structure is an object-based approach to represent the real world and includes points, lines, and polygons as spatial objects. A raster data structure is a field-based approach to represent geographic phenomena that are continuous over a large area.

A vector data format uses a series of points and mathematical functions to describe the shapes and boundaries of features. For example, a line is composed of two or more points. The polygon is represented by a series of points that closes at the original starting point. Currently, two types of vector data models are commonly used in a GIS. These include traditional vector data using Cartesian coordinates (the spaghetti model) and a topological vector model. The primary difference between these two types of vector data models is that the topological vector model adds the topolog-

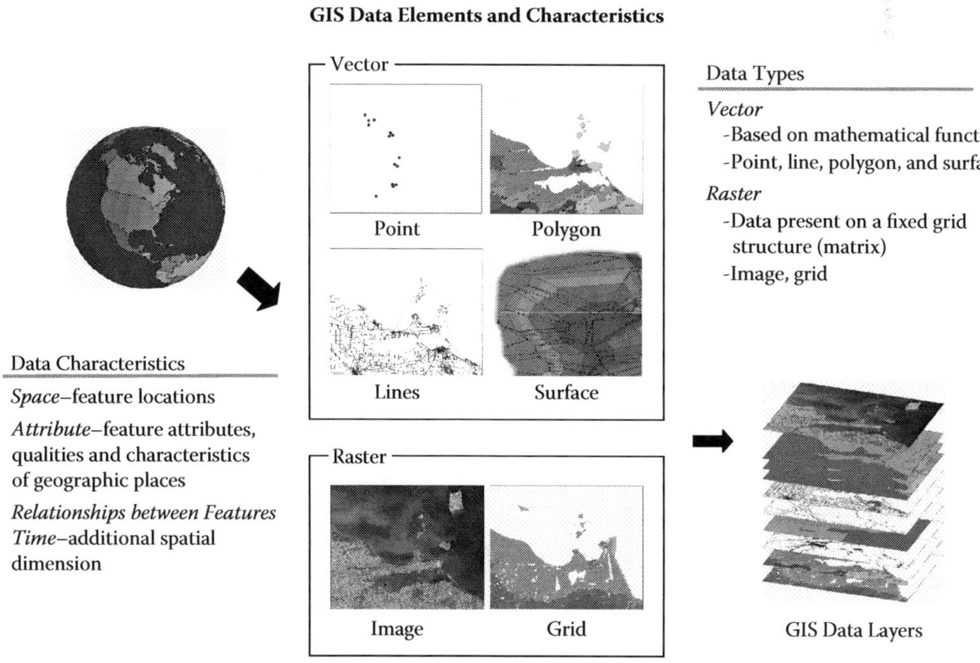

GIS Data Elements and Characteristics

Vector
Point   Polygon
Lines   Surface

Raster
Image   Grid

Data Characteristics

Space–feature locations

Attribute–feature attributes, qualities and characteristics of geographic places

Relationships between Features
Time–additional spatial dimension

Data Types

Vector
-Based on mathematical function
-Point, line, polygon, and surface

Raster
-Data present on a fixed grid structure (matrix)
-Image, grid

GIS Data Layers

FIGURE 10.2   The four fundamental data types—point, line, polygon, and surface—in a geographic information system (GIS) that can be represented by a vector or raster data structure.

ical relationships—adjacency, containment, connectivity—to the basic elements of the vector data. These topological descriptions provide information in addition to the spatial relationships between features. Some common vector data formats include GBF/DIME, TIGER, DLG, AutoCAD DXF, IGDS DGN file, ArcInfo coverage, ArcInfo E00, shapefile, and CGM.

A raster data format is used to store data in a grid cell array. The size of a grid cell is fixed throughout the entire data set with each cell equally spaced. A regular grid is used and can be the shape of a square, rectangle, triangle, or hexagon. Because there is a fixed grid size, the spatial resolution of the raster data layer will be equivalent to the size of the grid cell.

The use of raster and vector data has strengths and weaknesses. The decision to implement a raster or vector data format relies on the convenience of implementation, the type of GIS operations to be performed, the desired scale and accuracy, and the format of the original data sets. One should be familiar with these data types and select an appropriate format. Table 10.1 provides a summary of the advantages and disadvantages of raster and vector data.

Nonspatial attribute data in a GIS can be a set of tables or individual data records from a database. These attribute data provide a description of features. Two methods are used to link spatial and attribute data: attribute relationships and spatial relationships.

TABLE 10.1

Advantages and Disadvantages of Raster and Vector Data

	Advantages	Disadvantages
Vector	More compact data structure	Complex data model
	Topological processing	Difficult to perform overlay processing, can be
	Cartographic quality	computationally complex—involves geometric
	Sophisticated attribute data handling	intersection, topology building, and error checking
	Applications that rely on individual spatial features represented by points, lines, and polygons are much easier (i.e., network analyses that rely on streets as discrete features; land parcel-based applications, such as land title registration and forest resource inventories that rely on linear boundaries)	Difficult presentation of spatial variability Expensive data collection Use of expensive technology
	Mapping applications that rely on linear features, such as roads, streams, coastline, building outlines, and parcel boundaries are clearly defined using coordinates	
Raster	Simple data model	No topological processing
	Use of cheap technology	Limited attribute data handling
	Ease of data collection and data processing of raster data	Less compact data structure
	Ability to represent different types of continuous surfaces (topography, land use/land cover, air quality)	Low cartographic output quality
	Fast computer processing for overlaying operations	Not suitable for applications that rely on individual spatial features represented by points, lines, and polygons (network applications, land parcel-based
	Fast display of surface data	applications)
	Ability to handle very large databases	Raster data processing restricted by the resolution
	Suitable for applications that are difficult or impossible to perform using vector data (hydrologic modeling, portraying spatial processes such as spread of wildfire, movement of pollutants from a point source, growth of settlements)	of the source data

Source: From Lo, C.P., and Yeung, A.K.W., *Concepts and techniques of geographic information systems,* Prentice Hall, Upper Saddle River, NJ, 2002. With permission.

A common identifier—a key—is used to link spatial data with nonspatial attributes. This method is the most widely used and easiest to implement in a GIS. As long as data sets contain a common identifier, data sets can be related to each other. Data can be distributed into several small databases or tables and organized based on convenience. In addition, any data acquired later can be added into the GIS by adding a common identifier.

The second method used to link spatial data and attribute data is based on the spatial properties where both data sets share the same spatial properties. This method is useful for relating different data layers that contain a spatial property but have no common identifier. For example, to map soil types within a crop field requires a spatial link, because both data sets do not have a common identifier. To successfully implement a GIS, one should carefully examine the relationship between both the spatial and attribute data and select an appropriate method to manage the data layers. Figure 10.4 shows a series of GIS data layers created for an agricultural test site in Ohio.

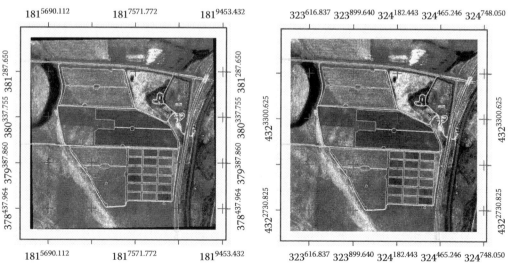

FIGURE 10.4 The Management System Evaluation Area (MSEA) site (Ohio) shown in three different standard map coordinate systems. (Courtesy of S.-S. Lin, pers. comm., 2006.)

10.4 COMMON COORDINATE SYSTEMS

A GIS is used to study various phenomena based on the spatial relationships among features. The location information of every entity in a GIS must be in the same coordinate system. The data layer needs to be registered to the same reference coordinate system. Otherwise, errors will result that will lead to problems in interpretation and analysis during the later stages of GIS operations. Converting the coordinate system of individual data layers to a unified coordinate system is the first priority when developing a GIS database.

Current GIS software has the ability to unify the coordinate system of layers that contain spatial information (i.e., projection-on-the-fly). However, the projection-on-the-fly function still relies on the fact that each individual data layer must have an assigned coordinate system. Due to the differences of vector and raster data structures, the projection-on-the-fly function still has problems when performing GIS operations that involve interactions between raster and vector data. All the data layers in a GIS should be converted to the same coordinate system. This ensures less complications with GIS operations and eliminates data management issues due to different coordinate systems.

The differences between various coordinate systems and map projections should be understood so that a standard spatial reference system can be selected. Although a GIS is operational as long as the data layers share the same spatial reference system, the use of a standard coordinate system is highly recommended. Incompatibility problems that may be associated with an arbitrary coordinate system can be avoided by using a standard coordinate system.

There are two types of coordinate systems used in GIS: a geographic coordinate system and a plane coordinate system. Geographic coordinates are represented as latitude and longitude values. The units used in a geographic coordinate system are decimal degrees. This type of coordinate system is not equally spaced in the x,y direction, which is a result of the different x,y axis associated with the spherical shape of the earth. For the y direction, latitude values range from $-90°$S to $90°$N and for the x direction, the longitude values range from $-180°$W to $180°$E. This type of coordinate system is useful for locating the spatial position of features for a large area that considers the earth's curvature—a spherical coordinate system. The geographic coordinate system has been the primary coordinate system used in navigation and fundamental surveying applications. However, due to the different lengths of a degree for latitude and longitude, Earth features appear elongated in the x direction (longitude) or shortened in the y direction (latitude). This characteristic makes the geographic coordinate system unsuitable for remote sensing imagery that uses a fixed grid size in both the latitude and longitude directions.

A second type of coordinate system, the plane coordinate system (also called rectangular coordinate system), defines the position on a flat map representation instead of the curved surface of the earth. Currently, there are two commonly used plane coordinate systems used in the United States: the Universal Transverse Mercator (UTM) system and the State Plane Coordinate (SPC) system. Each uses different projection methods to project the earth's surface onto a flat surface. The UTM system is more suitable for larger areas (regional scales), whereas the SPC system is more suitable for smaller areas (local scales).

Because the earth is a sphere, and does not have a flat surface, the geographic coordinate system is used as a positioning guide, because it represents the curvature of the earth's surface. Due to the units (decimal degrees) associated with a geographic coordinate system, a transformation is required to convert the decimal degree units into another linear system, such as feet or meters. The technique used to transform the spherical earth into a two-dimensional plane is called a map projection. The control points to support the map projection are called a datum.

The map projection is a mathematical function to transform the curved earth to a flat map, and the datum is the reference point used in the transformation. When using two maps with the same map projection, but with a different datum, the results will not be the same. The map projection and map datum information should be carefully examined before performing a projection conversion.

The National Geodetic Survey developed a SPC system for each state. To convert a geodetic position to plane rectangular coordinates, the point needs to be projected. A mathematical process is used to project points from the earth ellipsoid to an imaginary developable surface, which is a surface that can be unrolled and laid out flat without any distortion of shape or size. Two commonly used map projections are the Lambert conformal conic, where the projection is made onto the surface of an imaginary cone, and the transverse Mercator projection, which uses a cylinder as the developable surface. The Lambert projection is typically used for mapping states that are narrow in the north–south direction, but are longer in the east–west direction. Examples include states like Kentucky, Montana, Pennsylvania, and Tennessee. The transverse Mercator projection is used for mapping states that are narrow in the east–west direction, but are longer in the north–south direction. Examples include states like Illinois, Indiana, and New Jersey. Figure 10.3 shows an agricultural test site in Ohio—the Management System Evaluation Area (MSEA) site located in Piketon—in three different standard map coordinate systems. Another common map projection system is the UTM system that is used worldwide. The UTM system was developed by the military and covers the earth from 84°N latitude to 80°S latitude. There are sixty zones, with each zone 6° wide in longitude. The x,y coordinates are in meters, in northing and easting values.

Selecting an appropriate coordinate system for a GIS requires several considerations, which include the data types, existing coordinate system in the data layers, and future data layers to be generated. The data types will often result in positioning accuracy issues during a coordinate system conversion. For example, converting a raster data layer from one coordinate system to another may require rotating, shifting coordinates, and a resampling process that may severely distort the

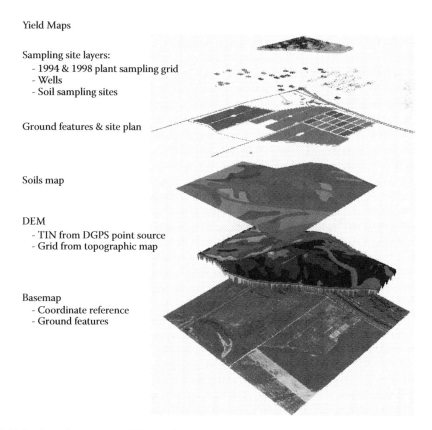

Yield Maps

Sampling site layers:
- 1994 & 1998 plant sampling grid
- Wells
- Soil sampling sites

Ground features & site plan

Soils map

DEM
- TIN from DGPS point source
- Grid from topographic map

Basemap
- Coordinate reference
- Ground features

FIGURE 10.3 Data layers created for use in a geographic information system (GIS) for the Management System Evaluation Area (MSEA) site in Ohio. (Courtesy of S.-S. Lin, pers. comm., 2006.)

original data. The errors that result from coordinate transformation must be considered to ensure the quality of each data layer. Most surveys performed on small areas are based on the assumption that the earth is a planar surface. For large-area surveys, it is necessary to consider the earth's curvature by computing an x,y position in terms of geodetic latitude and longitude. These computations can be quite complicated and lengthy, so a plane rectangular coordinate system is preferable.

Due to the constant improvement of geodetic surveys, the lack of agreement for a standard coordinate system among agencies, and the slow update of the older geodata layers, GIS analysts often encounter spatial data in a variety of projections and coordinate systems. For example, one may obtain a soils map in an Alber's equal area projection, Digital Elevation Model (DEM) data in a UTM coordinate system using the NAD 1927 datum, Global Positioning System (GPS) data of specific points located in a geographic coordinate system, a Landsat Thematic Mapper (TM) image in a UTM coordinate system using the WGS 1984 datum, or aerial photographs with a local coordinate system. Familiarity with the concepts of map projection and datums is necessary to convert data layers with different coordinate systems into a unified standard coordinate system.

10.5 SPATIAL DATABASES

GIS databases are used to organize and process geospatial data layers. The relational database model is the most widely used model in a GIS. The relational database is a collection of tables, known also as relations, which are connected to each other by keys. The row of the table is called a tuple (or record), and the column of the table is called an attribute (or a field) that describes the spatial entity. The attributes can also be called items or variables. The tables are related to each other by using keys. The keys represent one or more attributes whose values uniquely identify a record in a table. Data in the table can be logically joined together using a key that is the common attribute data value between two tables. An example of a relational data model is shown in Figure 10.5. The key connecting zoning and parcel is the zone code, and the key connecting parcel and owner is the PID (parcel ID number). When used together, the keys can relate zoning and owner.

The relational database—simple relational tables—is queried using Structured Query Language (SQL). This is an English-like language that uses logical commands to manipulate the data stored in the relational tables. Examples of logical commands would include restrict, project, union, intersect, difference, product, join, and divide (Bolstad, 2005).

A comprehensive database used in soil surveys is the Soil Survey Geographic (SSURGO) database. The Natural Resources Conservation Service (NRCS) collects SSURGO data from field mapping activities. The data are archived in 7.5 min quadrangles for a soil survey area. Linked to SSURGO is the Map Unit Interpretations Record (MUIR) attribute database, which contains a series of tables of around eighty-eight estimated soil physical and chemical properties, interpretations, and performance data. There are three important tables in the MUIR database: the soil survey

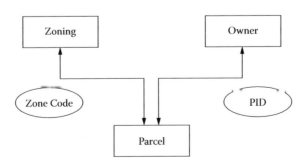

FIGURE 10.5 An example of a relational data model.

area table, the soil survey map unit table, and the map unit components table. Keys are available for each table that relate to the other tables in the MUIR database. Additional information on the MUIR database can be found at http://soildatamart.nrcs.usda.gov/.

10.6   ACQUISITION OF NONGEOPHYSICAL SPATIAL DATA

Geophysical measurements are commonly integrated with additional geospatial infromation in order to make appropriate agricultural management decisions. This additional geospatial information can be collected using field surveying methods, aerial photogrammetry, satellite imagery, and GPS (Global Positioning System). Field surveying is used to determine the positions or coordinates of features observed in the field. Field mapping is typically conducted when detailed surveys are required. Distances and angles are measured with surveying instruments—usually with total stations—to provide detailed x,y,z coordinates of ground features.

Aerial photogrammetry is used to develop topographic maps from a stereopair of aerial photographs. These maps can be quite detailed and at large scale, depending on the aircraft height and the aerial camera used. Topographic maps at a scale of 1:50,000 can be prepared from satellite images acquired by the French SPOT satellite because of its off-nadir viewing capability.

Satellite imagery ranging from Landsat (15, 30 m), to SPOT (5, 10, 20 m), to MODIS (250 m, 500 m, 1 km) is used for representing the earth's surface at various scales. Imagery from these satellite systems covers large areas of the earth. Land cover and land use maps can be prepared using standard image processing techniques. Other geospatial maps, such as NDVI (normalized difference vegetation map), elevation, temperature, soil moisture, snow moisture, and suspended sediment concentration maps, can also be prepared using selected bands from these satellite sensors. Other sensors provide images of the earth's surface in the microwave region at spatial resolutions ranging from 10 m to 100 m. These data are used to represent soil moisture conditions or structural rock features. GPS is used to collect point information of earth features. The data are collected in a format that can be directly input to a GIS database, using the shapefile format for ArcGIS.

10.7   GIS MODELS AND MODELING

GIS modeling uses a process of building models using spatial data. The GIS is a tool that can integrate different data sources, including maps, DEMs, GPS data, images, and tabular data. This makes GIS modeling particularly attractive for exploratory data analysis, data visualization, and database management. The models built with a GIS can be vector or raster based, depending more on the nature of the model, the data sources, and the computing algorithm. The distinction between raster-based or vector-based models does not prevent GIS users from integrating both types of data in the modeling process, because algorithms for converting data types are easily available in a GIS system. GIS modeling can take place within the GIS or may require linking the GIS to other computer programs, such as a statistical analysis package like SAS. Many GIS software programs have analytical functions for modeling.

There are four types of models available in a GIS system: binary models, index models, regression models, and process models (Chang, 2002). A binary model will use a logical expression to select map features from a composite map or from multiple grids, with the output being a binary map (1 [true] that satisfies a logical expression and 0 [false] for map features that do not). An index model uses an index value calculated from a composite map or multiple grids that are used to produce a ranked map. Usually the observed values of each variable on a map are evaluated and given numeric scores, then the relative importance (weighting) of a variable is evaluated against other variables to produce a final ranked map.

A regression model—either linear regression or logistic regression—is used to relate a dependent variable to a number of independent variables through the use of an equation, which can then be used for prediction or estimation. A process model integrates existing knowledge about environmental processes in the real world into a set of relationships and equations to quantify a physical process. The process model can offer both a predictive and an explanatory capability that is inherent in the proposed processes. The output from a process model is typically a set of equations that can be used for predictive purposes.

10.8   SUMMARY

GISs provide the necessary tools to manage, integrate, and analyze geophysical and nongeophysical information for the purpose of improving agricultural practices. In particular, as the employment of precision agriculture techniques continues to grow, there is expected to be an increasing need to input geophysical data into the GIS used to make decisions regarding how to best manage different areas of a farm field for fertilizer or pesticide application, tillage, irrigation, among other applications. Consequently, GIS is likely to play a greater and greater role in enhancing the usefulness of geophysical data collected in agricultural settings.

REFERENCES

Bernhardsen, T., *Geographic information systems—An introduction,* Second edition, John Wiley & Sons, Hoboken, NJ, 2002, 428 p.

Bolstad, P., *GIS fundamentals: A first text on geographic information systems,* Second edition, Eider Press, White Bear Lake, MN, 2005, 543 p.

Bossler, J.B., Ed., *Manual of geospatial science and technology,* Taylor & Francis, New York, 2002, 623 p.

Chang, K.T., *Introduction to geographic information systems,* McGraw-Hill, New York, 2002, 348 p.

Chrisman, N., *Exploring geographic information systems,* Second edition, John Wiley & Sons, Hoboken, NJ, 2002, 305 p.

DeMers, M.N., *Fundamentals of geographic information systems,* Third edition, John Wiley & Sons, Hoboken, NJ, 2005, 468 p.

Lo, C.P., and Yeung, A.K.W., *Concepts and techniques of geographic information systems,* Prentice Hall, Upper Saddle River, NJ, 2002, 492 p.

Section IV

Resistivity and Electromagnetic Induction Case Histories

There has been a long history in agriculture regarding the use of geophysical methods to measure soil electrical conductivity, or its inverse, soil electrical resistivity. Two different geophysical methods, resistivity and electromagnetic induction, are employed for measuring in situ bulk soil electrical conductivity, which is commonly referred to as "apparent soil electrical conductivity" (EC_a). An historical perspective on the agricultural use of geophysical methods to measure soil electrical conductivity can be found in Chapter 1, and to a far greater extent, in Chapter 2. Some theoretical considerations with respect to soil electrical conductivity are discussed in Chapter 4. Detailed descriptions of the resistivity and electromagnetic induction methods can be found in Chapters 5 and 6, respectively.

Initial geophysical EC_a measurement efforts in agriculture were focused largely on soil moisture monitoring and salinity assessment. However, within the last fifteen years, there has been a dramatic increase in the number of different agricultural uses found for geophysical soil electrical conductivity measurement. Many of these more recent agricultural applications for resistivity and electromagnetic induction methods are documented within Section IV of this book. Chapters 11, 12, and 13 detail the spatial correlation between electromagnetic induction EC_a and various soil properties. Capabilities for mapping pesticide partition coefficients with resistivity and electromagnetic induction methods are addressed by the Chapter 14 and Chapter 15 case histories. The employment of resistivity method EC_a mapping to delineate management units, soil drainage classes, and productivity zones within agricultural fields are topics in Chapters 16, 17, and 18. Geophysical EC_a measurement methods have additionally been utilized to monitor nutrient levels (Chapter 19), gauge uniformity and variability between research field plots (Chapter 20), and provide insight on subsurface features and conditions beneath golf course greens and tees (Chapter 21). The Chapter 22 case history compares results between resistivity and electromagnetic induction methods. These Section IV case histories list but a few of the potential geophysical EC_a measurement applications in agriculture, and there are many more new uses likely to be discovered in the near future.

11 Apparent Electrical Conductivity for Delineating Spatial Variability in Soil Properties

Brian J. Wienhold and John W. Doran

CONTENTS

11.1   INTRODUCTION

Soils commonly exhibit spatial variability in inherent soil properties such as texture, depth of topsoil, and organic C content. Inherent soil properties influence many chemical and biological properties that ultimately affect plant growth. Traditional methods of soil sampling (grid sampling, stratified sampling) require large inputs of labor and are expensive. The labor and cost requirements associated with traditional sampling often make it prohibitive to sample at a density that will accurately measure the spatial patterns present in a field. In agricultural fields, crops growing in areas of a field differing in inherent soil properties may differ in yield potential. Because yield varies within a field, the potential exists for improving crop utilization of inputs such as lime and fertilizer by varying application rates based on spatial patterns in inherent soil properties. To realize this potential, more efficient methods for delineating spatial variability in inherent soil properties are needed. Apparent soil electrical conductivity (EC_a) measured using electromagnetic induction combined with Global Positioning System (GPS) has potential for delineating soil variability and as an aid in selecting soil sampling locations (Lesch et al., 1995). A study was conducted to determine the relationship between EC_a and inherent soil properties that potentially affect soil–water dynamics and chemical and biological properties.

11.2   METHODS

This study was conducted in central Nebraska, near Bruning. The 20 ha field was in a soybean (*Glycine max* L.), winter wheat (*Triticum aestivum* L.), sorghum (*Sorghum bicolor* (L.) Moench) rotation and had been cropped to soybeans the previous year. Soils at this site are a Crete silt loam with small areas of Butler silt loam and Hasting silt loam. In June 2003, an EC_a survey was conducted using an

EM-38 (Geonics Ltd., Mississauga, Ontario, Canada) mounted on a nonmetallic cart that was pulled through the field with a four-wheeled all terrain vehicle. The EC_a values were recorded with a data logger every second (1.5 m), and the location was geo-referenced using a Trimble GPS (Trimble Navigation, Sunnyvale, CA) that was differentially corrected to provide an accuracy of <1 m.

Survey data were analyzed using the ESAP-95 software package (Lesch et al., 2000). The ESAP-95 software package assesses the spatial dependency of the data, calculates soil sampling locations that best encompass the variability present in the field, and uses measured soil data from those locations and a stochastic calibration model to predict the spatial pattern of secondary soil properties. Soil properties measured were 1:1 soil:distilled water electrical conductivity ($EC_{1:1}$) and pH (Smith and Doran, 1996), clay content (Kettler et al., 2001), and 2 M KCl extractable NO_3-N (Keeney and Nelson, 1982). The EC_a survey data and the output files of predicted secondary soil properties were used to generate spatial maps by kriging using the GS+ software package (Robertson, 2000).

11.3 RESULTS AND DISCUSSION

Apparent electrical conductivity values ranged from 45.5 to 81.1 mS m^{-1} and were spatially dependent (Figure 11.1). The spatial dependence in EC_a was best fit with a spherical model ($r^2 = 0.91$, residual mean square (RSS) = 26.9). Areas of the field exhibiting high EC_a values were slightly higher in elevation and had been subjected to erosion (tillage, wind, and water), and a portion of the topsoil had been lost. Areas of the field exhibiting low EC_a values tended to be depositional areas of the field.

The correlation between EC_a and $EC_{1:1}$ was not strong ($r^2 = 0.22$), likely because there was a small range of $EC_{1:1}$ values (0.27 to 0.45 dS m^{-1}) in these nonsaline soils. Bulk soil EC_a is affected by a number of soil properties including depth of topsoil, clay content, water content, and salt content (Rhoades and Corwin, 1990; Johnson et al., 2001). Laboratory $EC_{1:1}$ is more strongly correlated with dissolved salts. In spite of the weak correlation between EC_a and $EC_{1:1}$, the predicted $EC_{1:1}$ map (Figure 11.2) is visually similar to that for EC_a (Figure 11.1). The spatial dependence predicted for $EC_{1:1}$ was best fit with a spherical model ($r^2 = 0.91$, RSS = 1.9E^{-8}).

Clay content varied twofold (range 279 to 797 g kg^{-1}) and was strongly correlated with EC_a ($r^2 = 0.92$). Others have also demonstrated a strong correlation between EC_a and clay content (Kitchen et al., 2003). As noted above, high clay content was related to landscape position. Predicted clay content was best described by an exponential model ($r^2 = 0.81$, RSS = 46.7). Predicted clay content was positively correlated with EC_a with areas predicted to have a high clay content (Figure 11.3) being areas exhibiting high EC_a values (Figure 11.1).

Soil NO_3-N values ranged from 0.7 to 24.7 g kg^{-1} and were correlated with EC_a values ($r^2 = 0.86$). In these nonsaline soils, NO_3-N is a major anion, and EC_a has been shown to have great potential in monitoring NO_3-N dynamics during the growing season (Eigenberg et al., 2002). The correlation between predicted NO_3-N and EC_a was negative with low NO_3-N values in areas of the field (Figure 11.4) exhibiting high EC_a

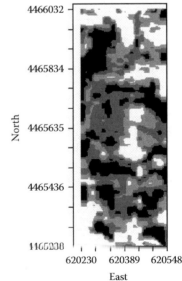

FIGURE 11.1 Spatial map of soil apparent electrical conductivity for the study site near Bruning, NE. ■ >72, ▨ 72 to 68, ▦ 68 to 64, □ <64 mS m^{-1}.

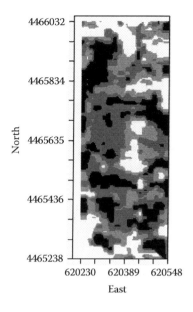

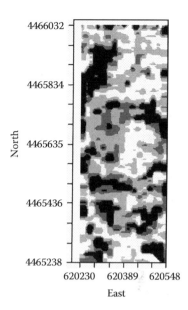

FIGURE 11.2 Spatial map of laboratory measured 1:1 soil electrical conductivity in the 0 to 15 cm depth for the study site near Bruning, NE. ■ >0.40, ■ 0.40 to 0.38, ▨ 0.38 to 0.36, □ <0.36 mS m^{-1}.

FIGURE 11.3 Spatial map of clay content in the 0 to 15 cm depth for the study site near Bruning, NE. ■ >350, ■ 350 to 320, ▨ 320 to 290, □ <290 g kg^{-1}.

(Figure 11.1). Predicted soil NO_3-N was best described by an exponential model ($r^2 = 0.67$, RSS = 25.6).

There was little variation in soil pH (range 5.4 to 5.6), and the correlation with EC_a was very poor ($r^2 = 0.03$). In spite of the poor correlation between EC_a and soil pH, it is apparent in the map of predicted soil pH (Figure 11.5) that higher pH is associated with areas of the field exhibiting high EC_a (Figure 11.1). Johnson et al. (2001) found a stronger relationship between EC_a and pH in

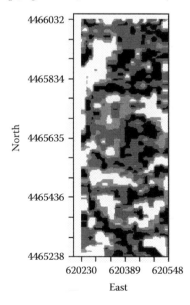

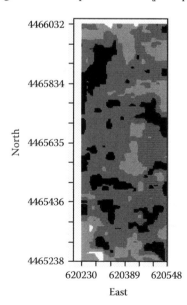

FIGURE 11.4 Spatial map of soil NO_3-N concentration in the 0 to 15 cm depth for the study site near Bruning, NE. ■ >23, ■ 23 to 18, ▨ 18 to 14, □ <14 mg kg^{-1}.

FIGURE 11.5 Spatial map of soil pH in the 0 to 15 cm depth for the study site near Bruning, NE. ■ >5.50, ■ 5.50 to 5.45, ▨ 5.45 to 5.40, □ <5.40.

a semiarid landscape where carbonates were present, and therefore, there was a larger range of pH values. Soil pH was best described by a spherical model ($r^2 = 0.91$, RSS = $1.7E^{-8}$).

Use of statistical methods to direct soil sampling and estimate the spatial pattern exhibited by soil properties will provide additional information that can be used to understand crop production and improve management. Previous research has shown that the relationship between EC_a and crop yield is inconsistent and appears to be dependent on crop, growing season precipitation, and incidence of fallow (Johnson et al., 2003; Kitchen et al., 1999). Maps showing the relationship between EC_a and inherent soil properties will be useful for explaining the interaction between weather, crop, and soils. At this study site, the areas having high clay content tend to be low yielding in years of below-normal precipitation. In years of above average precipitation these areas often exhibit higher yields than depositional areas where drainage is insufficient to prevent short-term waterlogging and conditions favoring denitrification.

Maps delineating the spatial distribution of soil chemical properties have potential for improving the management of inputs such as fertilizer and lime. In this study, parts of the field having high EC_a also had low NO_3-N concentrations. Even though these areas of the field are lower yielding than the field average, they still require fertilizer N. Areas having low EC_a values were areas of the field having high NO_3-N concentrations. Applying additional N to these areas will probably have little effect on yield and will increase the potential for emission of nitrous oxides or leaching of NO_3-N. Because there are so many processes affecting N availability, much work needs to be done to realize the potential for varying fertilizer N inputs.

Lime is another major input for many agricultural systems. In this study, soil pH was below the optimum for crop production. Even though pH was poorly correlated with EC_a there was spatial structure in the map of predicted pH. When lime requirement was determined for the sample sites and a variable rate lime application was compared to a constant application rate, the variable rate was 85 percent that of the constant rate. The variable rate application required 6 Mg less lime and represents a substantial reduction in input costs for this field.

11.4   CONCLUSIONS

This study utilized a statistical approach for directing soil sampling that can be used to characterize the spatial structure in soil properties affecting crop production. An understanding of the spatial structure of soil properties combined with yield maps has been shown to be useful for explaining year-to-year differences in yield patterns (Schepers et al., 2004). The approach presented in this study has potential for reducing the sampling intensity required for characterizing spatial patterns present in a field. Characterizing the spatial variability will be useful in studying dynamic processes affecting water and nutrients. Improved understanding of these processes will allow development of management practices that better match inputs with crop needs.

11.5   AUTHORS' NOTE

Trade or manufacturer's names mentioned do not constitute endorsement, recommendation, or exclusion by the U.S. Department of Agriculture–Agricultural Research Service (USDA-ARS). USDA-ARS, Northern Plains Area, is an equal opportunity/affirmative action employer and all agency services are available without discrimination.

REFERENCES

Eigenberg, R.A. et al. Electrical conductivity monitoring of soil condition and available N with animal manure and a cover crop. *Agric. Ecosyst. Environ.* 88, 183–193, 2002.

Johnson, C.K. et al. Field-scale electrical conductivity mapping for delineating soil condition. *Soil Sci. Soc. Am. J.* 65, 1829–1837, 2001.

Johnson, C.K. et al. Site-specific management zones based on soil electrical conductivity in a semiarid cropping system. *Agron. J.* 95, 303–315, 2003.

Keeney, D.R., and Nelson, D.W. Nitrogen: Inorganic forms, in *Methods of soil analysis* A.L. Page et al. (ed.). Part 2. 2nd Ed. Agron. Monogr. 9. ASA and SSSA, Madison, WI. 1982, Chap. 33.

Kettler, T.A., Doran, J.W., and Gilbert, T.L. A simplified method for soil particle size determination to accompany soil quality analyses. *Soil Sci. Soc. Am. J.* 65, 849–852, 2001.

Kitchen, N.R., Sudduth, K.A., and Drummond, S.T. Soil electrical conductivity as a crop productivity measure for claypan soils. *J. Prod. Agric.* 12, 607–617, 1999.

Kitchen, N.R. et al. Soil electrical conductivity and topography related to yield for three contrasting soil-crop systems. *Agron. J.* 95, 483–495, 2003.

Lesch, S.M., Strauss, D.J., and Rhoades, J.D. Spatial prediction of soil salinity using electromagnetic induction techniques: 2. An efficient spatial sampling algorithm suitable for multiple linear regression model identification and estimation. *Water Resour. Res.* 31, 3387-3398, 1995.

Lesch, S.M., Rhodes, J.D., and Corwin, D.L. ESAP-95 Version 2.01R User Manual and Tutorial Guide. USDA-ARS, George E. Brown, Jr. Salinity Laboratory Research Report No. 146, Riverside, CA, 2000.

Rhoades, J.D. and Corwin, D.L. Soil electrical conductivity: effects of soil properties and application to soil salinity appraisal. *Commun. Soil Sci. Plant Anal.* 21, 837-860, 1990.

Robertson, G.P. GS+: Geostatistics for the Environmental Sciences. Gamma Design Software, Plainwell, MI., 2000.

Schepers, A. et al. Delineation of management zones that characterize spatial variability of soil properties and corn yields across years. *Agron. J.* 96:195–203, 2004.

Smith, J.L., and Doran, J.W. 1996. Measurement and use of pH and electrical conductivity for soil quality analysis, in *Methods for assessing soil quality.* J.W. Doran and A.J. Jones (ed.). SSSA Spec. Publ. 49. SSSA, Madison, WI. 1996, Chap. 10.

12 Dependence of Soil Apparent Electrical Conductivity (EC$_a$) upon Soil Texture and Ignition Loss at Various Depths in Two Morainic Loam Soils in Southeast Norway

Audun Korsaeth

CONTENTS

12.1 INTRODUCTION

The use of sensor techniques provides a time- and cost-effective alternative to traditional soil sampling and laboratory analyses, in order to monitor spatial and temporal soil variation. One sensor-based, noninvasive technique is to measure apparent electrical conductivity (EC$_a$) of a soil profile by means of an electromagnetic (EM) induction approach. The device selected for the EM measurements in this paper was the Geonics EM38 (Mississauga, ON, Canada; www.geonics.com), which is the EM-EC$_a$ sensor most often used in agriculture (Sudduth et al., 2001). The device may be operated in one of two measurement modes that give an effective measuring depth of either 0.75 or 1.5 m.

EM-EC$_a$ has been used successfully as an indirect indicator of important soil physical and chemical properties, such as soil salinity hazard (Williams and Baker, 1982), soil water content (Khakural et al., 1998), and topsoil inorganic N content (Korsaeth, 2005). Whereas these studies focus mainly on soil variation in the lateral plane, others have used the EM-EC$_a$ technique with more emphasis on the vertical soil variation, such as topsoil thickness (Kitchen et al., 1999) and depth of sand deposition (Kitchen et al., 1996).

Fields on morainic soils, which represent a significant proportion (>25 percent) of the areas used for cereal production in Norway, typically show high variability in texture and organic matter, both laterally and vertically. Nevertheless, most interest is normally paid to lateral variation in topsoil properties. Reasons for this are that such variation may be directly observed in the field, and that more quantitative information, such as soil analyses, usually exists for the upper layer than for those below. Using the EM-EC_a technique, however, the effective measuring depth goes much deeper than the topsoil layer. Consequently, subsoil soil properties may affect the EM-EC_a signal significantly. It is thus of interest to explore how soil variation in deeper layers affects such measurements. The objective of this study was to establish relations between measurements of EC_a, soil texture, and ignition loss at different depths on two morainic soils in southeast Norway, in order to evaluate the suitability of the EM-EC_a technique to map the texture and organic matter content of such soils.

12.2   MATERIAL AND METHODS

12.2.1   LOCATIONS

Two locations were selected: a field at Apelsvoll Research Centre (60°42′ N, 10°51′ E, 250 m above sea level) and the long-term experiment at Møystad (60°47′ N, 11°10′ E, 150 m above sea level). The soil at both locations is an imperfectly drained brown earth (Gleyed melanic brunisolls, Canada Soil Survey) with dominantly loam and silty sand textures (see Table 12.1), which is typical for the region.

12.2.2   EM-EC_a DEVICE

The device used to measure EC_a (EM38) has an intercoil spacing of 1 m and may be operated in one of two measurement modes. In vertical mode (coil axes perpendicular to soil surface), the effective measuring depth is approximately 1.5 m, whereas in horizontal mode (coil axes parallel to soil surface), the effective measuring depth is approximately 0.75 m. In the following, measurements of

TABLE 12.1
Mean Soil Properties at Apelsvoll (n = 18) and Møystad (n = 16)*

	Depth (cm)					
	Apelsvoll			Møystad		
Soil Property[a]	0–20	30–50	60–80	0–20	20–40	40–60
Gravel[b]	200 (355)	236 (640)	182 (449)	163 (125)	202 (146)	228 (224)
Coarse sand	78.9 (210)	127 (580)	97.8 (320)	75.6 (40)	83.8 (60)	81.9 (70)
Medium sand	242 (130)	273 (290)	274 (270)	224 (60)	233 (50)	267 (170)
Fine sand	224 (100)	235 (210)	254 (310)	223 (40)	233 (50)	241 (280)
Total sand[b]	544 (210)	631 (100)	628 (530)	558 (170)	544 (210)	554 (220)
Coarse silt	134 (80)	126 (130)	134 (140)	132 (30)	128 (40)	129 (70)
Medium silt	108 (80)	90.6 (110)	88.3 (110)	143 (40)	134 (40)	126 (50)
Fine silt	72.8 (40)	60.0 (70)	56.7 (100)	73.1 (30)	69.4 (30)	63.1 (60)
Total silt[b]	316 (160)	273 (300)	276 (310)	344 (40)	330 (70)	316 (120)
Clay[b]	142 (150)	95.0 (180)	96.7 (260)	121 (120)	121 (120)	118 (110)
Ign. loss	62.2 (51)	37.8 (41)	32.3 (46.7)	50.7 (60)	48.7 (57)	48.0 (57)

*   Ranges in parenthesis
[a]   Units: gravel given in g kg^{-1} bulk soil, and other components in g kg^{-1} fine earth (<2 mm).
[b]   Particle size of gravel: >2000 μm, total sand: 60 to 2000 μm, total silt: 2 to 60 μm and clay: <2 μm.

EC$_a$ in horizontal and vertical modes will be denoted EM$_H$ and EM$_V$, respectively. The device was placed in a plastic cylinder, which was mounted on a plastic sledge and towed behind an off-road vehicle for continuous measurements. A differential Global Positioning System (DGPS) receiver (Star Track GSW 12, Communication Technology GmbH, Germany) was attached on the sledge to deliver the geographical coordinates of the EC$_a$ measurements.

12.2.3   MEASUREMENTS

At Apelsvoll, soil samples were taken on 29 September 2002 at eighteen points from a field that was about to be drained. The sampling points were distributed with a spacing of approximately 15 m along two drainage pipelines (Figure 12.1). The samples were taken at three depths (0 to 20, 30 to 50, and 60 to 80 cm) as the pipes were lain and were analyzed for texture (pipette method and wet sieving) and ignition loss (550°C). At the long-term experiment at Møystad, such data were measured in 1992 at three depths (0 to 20, 20 to 40, and 40 to 60 cm) on sixteen plots. Measurements of EC$_a$ were conducted in both measuring modes (EM$_H$ and EM$_V$) on 30 October 2001 at Møystad and on 19 September 2002 at Apelsvoll.

12.2.4   DATA ANALYSIS AND STATISTICS

The measurements of electrical conductivity were all log-transformed, because they were assumed to be log-normally distributed (Hendrickx et al., 2002). The statistical software package MINITAB (Release 14.13, www.minitab.com) was used for basic statistics (correlation) and linear regression analysis (stepwise selection, $\alpha = 0.15$). The variables were checked for multicolinearity by calculating the variance inflation factors (VIF). The VIF should not exceed 10 in order to avoid serious problems with multicolinearity (Montgomery and Peck, 1992).

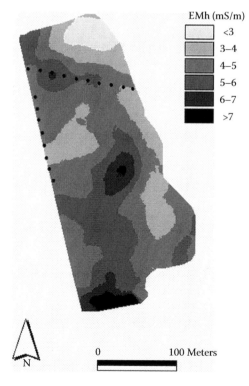

FIGURE 12.1   Interpolated measurements of EM$_H$ at the Apelsvoll field. Black dots indicate soil sample locations.

12.3   RESULTS

At Apelsvoll, measurements of EC_a were correlated with most of the measured soil properties (Table 12.2). The contents of gravel and fine sand were the only properties that did not affect EC_a. Most correlations were found between EC_a and soil properties in the depth layer 30 to 50 cm. The correlations were generally strongest when EC_a was measured in horizontal mode (EM_H). This was especially so in the topsoil layer, where there was no correlation between EM_V and the measured soil properties.

At Møystad, most correlations were found between EC_a and the topsoil properties (Table 12.3). EC_a was unaffected by the contents of fine sand, fine silt, and coarse silt, regardless of depth. As found at Apelsvoll, the topsoil properties were more strongly correlated with EM_H than with EM_V. In contrast to the results at Apelsvoll, however, this trend appeared to change with depth at Møystad, where the soil properties in the deepest layer (40 to 60 cm) were most strongly correlated with EM_V.

At both locations, the strongest correlations were found between EM_H and topsoil clay content. The total content of topsoil sand was the second most important soil property influencing EM_H. The content of clay and total sand were strongly negatively intercorrelated ($|r| > 0.580$) at both locations and at all depths. In contrast to the clay content, the sand content was, as expected, negatively related to EM_H. At Apelsvoll, these two properties were the only ones that had a significant influence on EC_a at all three depths. At Møystad, this applied for ignition loss as well, and here the relative influence of ignition loss appeared to increase with increasing depth.

Even though many of the soil properties were rather strongly correlated with EC_a, only a few were included in the regression models using the stepwise procedure (Table 12.4). Topsoil clay content alone accounted for 57 percent of the variation in measured EM_H at Apelsvoll and 67 percent of that at Møystad. When expanding the models to include all potential predictors, the corresponding degrees of explanations increased up to 72 and 89 percent, respectively. Topsoil clay

TABLE 12.2

Correlation Coefficients between Selected Soil Properties at Three Depths and Apparent Soil Electrical Conductivity (Log-Transformed), Measured in Autumn 2002 at Apelsvoll ($n = 18$) Using EM38 in Either Horizontal (EM_H) or Vertical (EM_V) Modes

Soil Property	Depth (cm)					
	0–20		30–50		60–80	
	log EM_H	log EM_V	log EM_H	log EM_V	log EM_H	log EM_V
Gravel[a]	n.s.	n.s.	n.s.	n.s.	n.s.	n.s.
Coarse sand	n.s.	n.s.	−0.51*	−0.55*	n.s.	n.s.
Medium sand	n.s.	n.s.	−0.47*	n.s.	−0.60**	0.58*
Fine sand	n.s.	n.s.	n.s.	n.s.	n.s.	n.s.
Total sand[a]	−0.63**	n.s.	−0.75***	−0.59**	−0.54*	n.s.
Coarse silt	n.s.	n.s.	0.47*	0.49*	n.s.	n.s.
Medium silt	n.s.	n.s.	0.63**	0.53*	0.49*	n.s.
Fine silt	0.51*	n.s.	0.62**	0.55*	n.s.	n.s.
Total silt[a]	n.s.	n.s.	0.60**	0.56*	n.s.	n.s.
Clay[a]	0.77***	n.s.	0.77***	0.50*	0.50*	n.s.
Ignition loss	0.62**	n.s.	n.s.	n.s.	n.s.	n.s.

Note: * $p < 0.05$; ** $p < 0.01$; *** $p < 0.001$, n.s. not significant.

[a] Particle size of gravel: >2000 µm, total sand: 60 to 2000 µm, total silt: 2 to 60 µm and clay: <2 µm.

TABLE 12.3

Correlation Coefficients between Selected Soil Properties at Three Depths and Apparent Soil Electrical Conductivity (Log-Transformed), Measured in Autumn 2001 at Møystad (n = 16) Using EM38 in Either Horizontal (EM_H) or Vertical (EM_V) Modes

| Soil Property | Depth (cm) | | | | | |
| | 0–20 | | 20–40 | | 40–60 | |
	log EM_H	log EM_V	log EM_H	log EM_V	log EM_H	log EM_V
Gravel[a]	−0.54*	n.s.	−0.60*	n.s.	n.s.	n.s.
Coarse sand	−0.53*	−0.56*	n.s.	n.s.	n.s.	n.s.
Middle sand	−0.52*	n.s.	n.s.	n.s.	n.s.	n.s.
Fine sand	n.s.	n.s.	n.s.	n.s.	n.s.	n.s.
Total sand[a]	−0.73**	−0.57*	−0.53*	−0.60*	n.s.	−0.64**
Coarse silt	n.s.	n.s.	n.s.	n.s.	n.s.	n.s.
Middle silt	n.s.	n.s.	n.s.	n.s.	n.s.	0.52*
Fine silt	n.s.	n.s.	n.s.	n.s.	n.s.	n.s.
Total silt[a]	n.s.	n.s.	n.s.	n.s.	n.s.	0.56*
Clay[a]	0.83***	0.75**	0.77**	0.74**	0.62*	0.67**
Ignition loss	0.49*	n.s.	0.69**	n.s.	0.62*	0.67**

Note: * $p < 0.05$; ** $p < 0.01$; *** $p < 0.001$, n.s. not significant.

[a] Particle size of gravel: >2000 μm, total sand: 60 to 2000 μm, total silt: 2 to 60 μm and clay: <2 μm.

TABLE 12.4

Regression Models from Stepwise Regression Analyses, which Included Only Significant Constants and Predictors[a]; Dependent Variable is Log-Transformed EM_H, and Adjusted Coefficients of Determination (r^2_{adj}) Are Shown

Regression Equations	r^2_{adj}
Apelsvoll	
log $EM_H = 0.426 + 0.018\ Cl_{0-20}$	0.572
log $EM_H = 0.453 + 0.016\ Cl_{0-20} + 0.007\ FSa_{60-80} - 0.006\ MSa_{30-50}$	0.724
Møystad	
log $EM_H = 1.597 + 0.015\ Cl_{0-20}$	0.666
log $EM_H = 1.590 + 0.012\ Cl_{0-20} + 0.011\ IGL_{20-40}$	0.736
log $EM_H = 1.706 + 0.027\ IGL_{20-40} - 0.011\ MSi_{20-40} - 0.011\ CSa_{40-60} +$	0.887
$0.019\ CSi_{0-20} - 0.002\ FSa_{40-60}$	

[a] Predictors: Cl, clay; FSa, fine sand; MSa, medium sand; CSa, coarse sand; MSi, medium silt; CSi, coarse silt; IGL, ignition loss. Subscripts indicate depths in centimeters.

content was included as predictor in four of the five regression models resulting from the stepwise procedure. An interpolated map of measured EM_H at the Apelsvoll location is shown in Figure 12.1. The variation in EM_H within the field was reasonably well represented by the measurement points, but zones with more than 7 mS/m and less than 3 mS/m were not represented among the soil sampling locations.

12.4 DISCUSSION

Apparent electrical conductivity (EC_a) correlated strongly with soil texture (Tables 12.2 and 12.3), as is commonly found in nonsaline soils (Sudduth et al., 2001). This may be partly explained by the exchangeable cations associated with clay minerals, which represent an important pathway for EC in soil (Corwin and Lesch, 2003). Another pathway is through the liquid phase. Because both water content and the amount of exchangeable cations usually decrease with increasing sand content, this may explain the negative relation between sand content and EC_a observed in this and many other studies (e.g., Khakural et al., 1998; Kitchen et al., 1996). The strong, positive correlation between EC_a and ignition loss confirms earlier findings on the same soil type (Korsaeth, 2005). One should, however, take into account that ignition loss and clay content were to some extent positively intercorrelated (data not shown), so that the result may in fact have been due to variation in clay content.

Measured EM_H correlated generally more strongly with the topsoil properties than did EM_V (Table 12.2 and Table 12.3), which agrees with the results presented by Korsaeth (2005). The superiority of EM_H to EM_V in terms of detecting variation in topsoil properties is also reported under other conditions (Boettinger et al., 1997; Khakural et al., 1998). Such a phenomenon is to be expected from the sensitivity functions of EM38 (McNeill, 1980), which show that the relative contribution to the signal from the topsoil is larger for EM_H than for EM_V. With regard to the subsoil, the results at both locations appear to reflect the larger relative weighting of the EM_V signal there. An exception was found in the deepest layer at Apelsvoll (60 to 80 cm), where EM_H was almost as well correlated in this layer as in the topsoil. No obvious explanation for this is apparent.

The regression analyses underlined the importance of topsoil clay content for measured EC_a in such soils (Table 12.4). Nevertheless, the regression models that explained most of the measured EM_H included, at both locations, more subsoil than topsoil properties as predictors (Table 12.4). This shows that both topsoil and subsoil properties must be considered when interpreting soil survey maps made by the EM-EC_a technique.

12.5 CONCLUSIONS

- When using the EM-EC_a technique to map topsoil properties, EM_H is superior to EM_V.
- On typical morainic soils in southeast Norway, topsoil clay, total sand, and ignition loss correlate best with EM_H.
- The EM-EC_a technique is suitable for mapping soil variation on such soils, but both topsoil and subsoil properties should be considered when interpreting such maps.

REFERENCES

Boettinger J.L., Doolittle J.A., West N.E., Bork E.W., and Schupp E.W. 1997. Nondestructive assessment of rangeland soil depth to petrocalcic horizon using electromagnetic induction. *Arid Soil Res Rehabil* 11: 375–390.

Corwin D.L., and Lesch S.M. 2003. Application of Soil Electrical Conductivity to Precision Agriculture: Theory, Principles, and Guidelines. *Agron J* 95: 455–471.

Hendrickx J.M.H., Borchers B., Corwin D.L., Lesch S.M., Hilgendorf A.C., and Schlue J. 2002. Inversion of soil conductivity profiles from electromagnetic induction measurements: Theory and experimental verification. *Soil Sci Soc Am J* 66: 673–685.

Khakural B.R., Robert P.C., and Hugins D.R. 1998. Use of non-contacting electromagnetic inductive method for estimating soil moisture across a landscape. *Commun Soil Sci Plant Anal* 29: 2055–2065.

Kitchen N.R., Sudduth K.A., and Drummond S.T. 1996. Mapping of sand deposition from 1993 midwest floods with electromagnetic induction measurements. *J Soil Water Conserv* 51: 336–340.

Kitchen N.R., Sudduth K.A., and Drummond S.T. 1999. Soil electrical conductivity as a crop productivity measure for claypan soils. *J Prod Agric* 12: 607–617.

Korsaeth A. 2005. Soil apparent electrical conductivity (EC$_a$) as a means of monitoring changes in soil inorganic N on heterogeneous morainic soils in SE Norway during two growing seasons. *Nutr Cycl Agroecosyst* 72: 213–227.

McNeill J.D. 1980. Electromagnetic terrain conductivity measurement at low induction numbers. Tech. Note TN-6. Geonics, ON, Canada.

Montgomery D.C., and Peck E.A. 1992. *Introduction to linear regression analysis.* 2nd edition. John Wiley & Sons, New York.

Sudduth K.A., Drummond S.T., and Kitchen N.R. 2001. Accuracy issues in electromagnetic induction sensing of soil electrical conductivity for precision agriculture. *Comput Electron Agric* 31: 239–264.

Williams B.G., and Baker G.C. 1982. An electromagnetic induction technique for reconnaissance surveys of soil salinity hazards. *Aust J Soil Res* 20: 107–118.

13 Relations between a Commercial Soil Survey Map Based on Soil Apparent Electrical Conductivity (EC$_a$) and Measured Soil Properties on a Morainic Soil in Southeast Norway

Audun Korsaeth, Hugh Riley, Sigrun H. Kværnø, and Live S. Vestgarden

CONTENTS

13.1 INTRODUCTION

The use of noninvasive electromagnetic (EM) induction to measure apparent profile soil electrical conductivity (EC$_a$) provides a time- and cost-effective tool to map within-field variation in soil properties. Apparent electrical conductivity (EC$_a$) shows the depth-weighed summarized effect of all factors influencing EC in soil, and a range of chemical and physical properties have been indirectly determined by the EM-EC$_a$ method, such as soil salinity hazard (Williams and Baker, 1982), soil water content (Sheets and Hendrickx, 1995), claypan thickness (Doolittle et al., 1994), topsoil inorganic N content (Korsaeth, 2005a), and nutrient levels (Heiniger et al., 2003).

Soil mapping based on the EM-EC$_a$ technique is offered commercially in many countries, such as New Zealand (www.nzcpa.com), Denmark (www.gpsagro.dk), Sweden (www.analycen.se), and

Norway (www.planteforsk.no). It is often argued that *a priori* information on soil variation could be useful when deciding where to take soil samples, which in practice are limited in number, due to the high costs of soil analyses. The information provided by such EM-EC$_a$ maps may also be used to divide a field into different management units, where each unit is more homogeneous in terms of soil properties than the field as a whole (Kitchen et al., 1999). Each unit could then be managed individually to adjust for variation between the management units.

The result of a commercial soil survey is normally a color map, showing zones with different EC$_a$. A crucial question is whether the zones on the soil survey maps show substantial differences in soil properties. This question cannot be answered in general, but should be tested for different soil types. In this study we wished to test whether the standard procedure used for commercial soil survey with the EM-EC$_a$ method in Norway may provide maps that show significant differences in soil properties.

13.2 MATERIAL AND METHODS

13.2.1 LOCATION

A 32 ha agricultural field located at Løten in SE Norway (60°50′ N, 11°18′ E, 315 m above sea level), typical for the higher regions around lake Mjøsa, was selected for this case study. The field is characterized by considerable variations in topography, with peaks and hollows spread over the area, with a maximum height difference within the field of about 20 m. The soil, which is of morainic origin, ranges from imperfectly drained brown earth (Gleyed melanic brunisoils, Canada Soil Survey) mainly at the peaks, to humified peaty gley (Terric Humisol, Canada Soil Survey), mainly in the hollows. (See Table 13.1 for mean values and ranges of the soil properties.)

TABLE 13.1
Soil Properties and Their Correlation (*r*) with EM$_H$

Soil Property	Mean	Min	Max	Unit	*n*	*r*	*p*
Ignition loss	115	30.0	390	g kg^{-1}	310	0.830	<0.001
Soil organic matter	96.0	10.0	443	g kg^{-1}	310	0.819	<0.001
pH	7.04	5.50	8.00	g kg^{-1}	310	0.128	0.024
P-AL	0.10	0.02	1.35	g kg^{-1}	310	−0.015	n.s.
K-AL	0.07	0.02	0.15	g kg^{-1}	310	0.155	0.006
Mg-AL	0.11	0.03	0.32	g kg^{-1}	310	0.325	<0.001
Ca-AL	8.09	0.90	51.7	g kg^{-1}	310	0.226	<0.001
K-HNO$_3$	0.29	0.10	0.68	g kg^{-1}	310	−0.226	<0.001
Na-AL	0.02	0.01	0.08	g kg^{-1}	310	0.405	<0.001
Soil water content	248	134	535	g kg^{-1}	154	0.835	<0.001
Coarse sand	63.0	15.0	128	g kg^{-1}	40	−0.520	0.001
Medium sand	164	97.0	267	g kg^{-1}	40	−0.475	0.002
Fine sand	175	127	266	g kg^{-1}	40	−0.506	0.001
Total sand (60–2000 μm)	402	257	634	g kg^{-1}	40	−0.520	<0.001
Silt (2–60 μm)	424	294	634	g kg^{-1}	40	0.636	0.001
Clay (<2 μm)	174	24.0	282	g kg^{-1}	40	0.006	n.s.
K$^+$	0.16	<0.01	0.36	cmol$_c$ kg^{-1}	21	0.174	n.s.
Mg^{2+}	0.78	0.43	1.40	cmol$_c$ kg^{-1}	21	0.662	0.001
Ca^{2+}	31.0	7.25	61.0	cmol$_c$ kg^{-1}	21	0.825	<0.001
Na$^+$	0.04	<0.01	0.07	cmol$_c$ kg^{-1}	21	0.432	n.s.
H$^+$	2.50	<0.01	7.10	cmol$_c$ kg^{-1}	21	−0.127	n.s.
Cation exchange capacity	34.5	12.4	63.7	cmol$_c$ kg^{-1}	21	0.837	<0.001

13.2.2   EM-EC$_A$ Device

The device used to measure EM-EC$_a$ was the Geonics soil conductivity meter EM38 (Mississauga, ON, Canada; www.geonics.com), which is the sensor most often used for such measurements in agriculture (Sudduth et al., 2001). The device may be used in one of two measuring modes, in vertical mode (coil axes perpendicular to soil surface, denoted EM$_V$) or in horizontal mode (coil axes parallel to soil surface, denoted EM$_H$). The respective effective measuring depths are approximately 1.5 m and 0.75 m. When measuring, the device was placed in a plastic cylinder, which was mounted on a plastic sledge and towed behind an off-road vehicle for continuous measurements (ca. 200 measurements ha^{-1}). A differential Global Positioning System (DGPS) receiver (Ceres, CSI Wireless Inc., Canada) was attached on the sledge to deliver the geographical coordinates of the EC$_a$ measurements.

13.2.3   Measurements

On 30 April 2003, 310 soil samples were taken from the topsoil (0 to 20 cm) in a 20 × 20 m grid, covering about 38 percent of the field (12 ha) and analyzed for ignition loss, pH, P-AL (AL denotes the ammonium lactate/acetic acid mixture used for extraction; Égner et al., 1960), K-AL, Mg-AL, Ca-AL, Na-AL, and K-HNO$_3$. A selection of 154 soil samples were additionally analyzed gravimetrically for water content, and mechanical analysis of texture was performed on forty of these samples. From the latter group, twenty-one samples were extracted with 1 M ammonium acetate, and exchangeable Ca^{2+}, K$^+$, Mg^{2+}, and Na$^+$ were analyzed spectrometrically (ICAP 1100, Thermo Jarrell Ash Corp, US). Exchangeable H$^+$ was determined by titration (NaOH) and cation exchange capacity (CEC) calculated as the sum of the five cations. Soil organic matter (SOM) was determined from ignition loss, by correcting for clay content, where the clay content was estimated by the finger method in those samples that lacked ordinary texture analysis. The same day as the soil sampling, EM-EC$_a$ was measured in the field with the EM38 in horizontal mode only. The rationale for using this mode is that the relative contribution to the signal from the topsoil is larger for EM$_H$ than for EM$_V$ (McNeill, 1980), and that EM$_H$ has been found to be superior to EM$_V$ in practice, when using the EM-EC$_a$ technique to map topsoil properties on fairly comparable soils (Korsaeth, 2005a, 2005b).

13.2.4   Statistics and Data Analyses

The EC$_a$ data were not transformed, as this is seldom done in commercial practice. The statistical software package MINITAB (Release 14.13, www.minitab.com) was used for basic statistics (calculating means, standard deviations, and correlations), one-way analysis of variance (ANOVA), and Fisher's least significant difference (LSD) test for comparisons. ArcView GIS (version 3.2, www.esri.com) was used to produce the maps.

Soil properties that showed a reasonably strong correlation with EM$_H$ ($|r| > 0.5$), and from which at least forty observations were available (thus excluding the exchangeable cations), were selected to test the standard procedure used for commercial soil survey. On the basis of measured EC$_a$, the selected data were grouped separately into classes with intervals of 2 mS m^{-1} (groups with less than three observations were excluded). Thereafter, one-way ANOVA and Fisher's LSD test were used to test for differences between groups with respect to each of the selected variables. The rationale for using an interval of 2 mS m^{-1} was that experience from commercial mapping of a range of different soil types in Norway has shown that an interval of this size is normally most suitable.

13.3   RESULTS

Measured EM$_H$ correlated significantly with sixteen of the twenty-two measured soil properties (Table 13.1). Properties with a correlation coefficient (r) larger than 0.8 were ranked in descending order: CEC > soil water content > ignition loss > exchangeable Ca > SOM. All significant

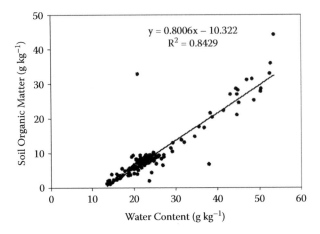

FIGURE 13.1   The relation between soil organic matter (SOM) and soil water content ($n = 154$).

TABLE 13.2
Selected Soil Properties Grouped on the Basis of Measured EM_H,
with Intervals Commonly Used for Soil Survey Maps in Norway*

		Soil Organic Matter			Silt			Total Sand		
		g kg⁻¹			g kg⁻¹			g kg⁻¹		
Class	EM_H Interval mS m⁻¹	n	Mean	Stdv	n	Mean	Stdv	n	Mean	Stdv
1	$0 \geq EM_H < 2$	38	30.3[a]	14.8	5	327[a]	36.7	5	568[a]	48.0
2	$2 \geq EM_H < 4$	140	71.1[b]	28.5	12	372[ab]	69.0	12	463[b]	93.4
3	$4 \geq EM_H < 6$	75	96.6[c]	37.0	11	470[c]	40.7	11	310[c]	55.6
4	$6 \geq EM_H < 8$	18	133[d]	74.8	3	428[bc]	51.4	3	373[bc]	95.0
5	$8 \geq EM_H < 10$	15	190[e]	79.8	3	400[abc]	11.5	3	390[bc]	26.2
6	$10 \geq EM_H < 12$	15	227[f]	88.6	3	490[cd]	124.9	3	335[c]	17.2
7	$12 \geq EM_H < 14$	7	326[g]	68.9	3	576[d]	72.4	3	327[c]	38.6

* Number of observations (n), mean, and standard deviation (Stdv) within each group are shown, and means indicated by the same letter are not significantly different.

correlations were positive, except those between EM_H and the sand fractions and K-HNO₃. Measured EC_a was unaffected by the clay content.

The soil properties selected for testing the commercial soil survey practice were soil water content, ignition loss, soil organic matter, silt content, and all sand fractions except medium sand. The sand fractions were naturally strongly intercorrelated, as were ignition loss and SOM (data not shown). However, ignition loss and SOM also correlated strongly with soil water content, as shown for SOM in Figure 13.1. For convenience, only soil properties having a low intercorrelation are presented in Table 13.2.

Mean SOM increased with each EM_H class up to the highest class included in the test, and there were statistical differences between all seven classes (Table 13.2). The good match between the EC_a zones (based on the commercial procedure) and the SOM content is illustrated in Figure 13.2. For ignition loss and soil water content, there were significant differences between six and five of the

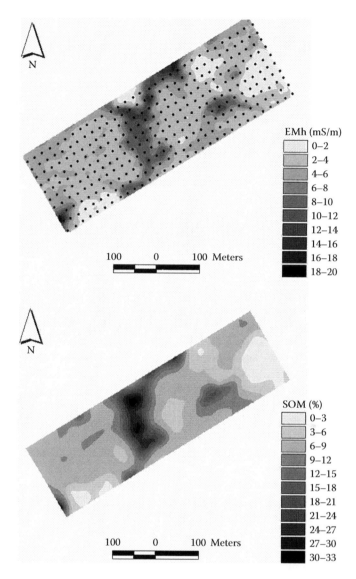

FIGURE 13.2 Map showing EM$_H$, using the standard procedure used in Norway for commercial soil survey with the EM-EC$_a$ method (above), and measured soil organic matter (SOM) content. Soil sample locations are indicated by black dots.

classes, respectively (data not shown). Mean silt content increased from the lowest to the highest class but showed a decrease from class 3 to 5. A similar trend was seen for the sand fractions (for total sand see Table 13.2), but with opposite signs, as the sand content decreased from the lowest to the highest class. Hence, none of the texture fractions were unambiguously reflected by the soil survey map.

13.4 DISCUSSION

There were significant correlations between apparent electrical conductivity (EC$_a$) and all the texture fractions except for clay content. This exception is not typical, because EC$_a$ is normally strongly correlated with clay content on nonsaline soils (Sudduth et al., 2001). Moreover, other findings on morainic soil in the region have shown strong correlations between EC_a and clay content (Korsaeth, 2005a, 2005b). Such high correlations may be partly explained by the exchangeable cations associated with clay minerals, which represent an important pathway for EC in soil (Corwin and Lesch, 2003). In the current study there was a very strong correlation between EC_a and CEC, and particularly, exchangeable Ca affected measured EC_a strongly, but clay content was not correlated with the exchangeable cations (data not shown). CEC was, however, strongly correlated with SOM ($r = 0.918$).

After CEC, soil water content was the most important soil property, influencing measured EC_a. It was quite wet at the time of measuring, as the soil had just thawed after the winter frost, and there had been a period of rain shortly before. A strong correlation between soil water content and EC_a is to be expected, because the liquid phase is one of three pathways for electrical conductivity in soil (Corwin and Lesch, 2003), and the results are in agreement with numerous other reports (e.g., Khakural et al., 1998; Korsaeth, 2005a; Sudduth et al., 2001).

The relation between soil water content and SOM was very high (Figure 13.1). This may be explained by the fact that the peaty subregions within the field were mainly located in the hollows, where the water content is normally higher than on the peaks. High water content is an essential part of peat formation, because it reduces the oxygen availability in the soil and thereby retards the decay of organic matter. A logical consequence of higher content of both water and organic matter in the hollows would be increased EC_a at such locations, as was actually observed. Clay et al. (2001) observed the same phenomenon at four different fields in eastern North Dakota. They attributed the landscape differences in EC_a to water leaching salts from peaks (summit) to hollows, lower water contents in peaks than in the hollows (toeslope areas), and water erosion that transported surface soil from upper to lower areas.

The statistical test of the soil survey map data revealed that there were significantly different contents of SOM between all the zones tested (Table 13.2). This appears reasonable considering the obvious coincident of EM_H and SOM content shown in Figure 13.2. One important soil property, which is highly related to the SOM content, is the mineralization potential of nitrogen from the soil organic matter pool. On the investigated field, the most frequent crop in the rotation is spring barley, and the risk of lodging, due to overfertilization in SOM-rich areas, is considered to be high. Hence, a map showing differences in SOM content would provide useful information for the farmer (e.g., for decisions concerning site-specific fertilization).

Soil water content was more strongly related to EC_a than was SOM (Table 13.1), but the statistical test revealed, nevertheless, a lower coincidence between soil water content and the EC_a-zones on the soil survey map (Table 13.2). This may partly be related to the difference in the number of samples analyzed. All 310 samples were analyzed for SOM, whereas only about half the samples (154) were analyzed for soil water content. The range of the measured SOM was considerably larger than that of the measured soil water content (Table 13.1). The combination of both larger overall variation in the measurements and higher degrees of freedom within the EC_a classes favored SOM in the statistical test.

For silt and sand content, there were differences between the EC_a classes, but they were not as distinct as for SOM. The texture coincided poorly with the EC_a classes, particularly in the range of 4 to 10 mS m^{-1} (Table 13.2). This reflected the fact that soil texture was generally more weakly correlated with EC_a than SOM and soil water content (Table 13.1), and that soil texture had much lower degrees of freedom in the statistical test, due to fewer analyses, compared with SOM.

13.5   CONCLUSIONS

This case study has shown that the standard procedure used in Norway for commercial soil survey with the EM-EC_a method may provide maps that show significant differences in soil properties.

REFERENCES

Clay D.E., Chang J., Malo D.D., Carlson C.G., Reese C., Clay S.A., Ellsbury M., and Berg B. 2001. Factors influencing spatial variability of soil apparent electrical conductivity. *Commun Soil Sci Plant Anal* 32: 2993–3008.

Corwin D.L., and Lesch S.M. 2003. Application of soil electrical conductivity to precision agriculture: Theory, principles, and guidelines. *Agron J* 95: 455-471.

Doolittle J.A., Sudduth K.A., Kitchen N.R., and Indorante S.J. 1994. Estimating depths to claypans using electromagnetic induction methods. *J. Soil Water Conserv.* 49: 572–575.

Égner H., Riehm H., and Domingo W.R. 1960. Untersuchung über die chemische Boden-Analyse als Grundlage für die Beurteilung des Nährstoffzustandes der Boden. Kungl. (Using chemical soil analysis as a basis for assessing soil nutrient analysis.) *Landtbrukshögskolans Annaler* 26: 199–215.

Heiniger R.W., McBride R.G., and Clay D.E. 2003 Using soil electrical conductivity to improve nutrient management. *Agron J* 95: 508–519.

Khakural B.R., Robert P.C., and Hugins D.R. 1998. Use of non-contacting electromagnetic inductive method for estimating soil moisture across a landscape. *Commun Soil Sci Plant Anal* 29: 2055–2065.

Kitchen N.R., Sudduth K.A., and Drummond S.T. 1999. Soil electrical conductivity as a crop productivity measure for claypan soils. *J Prod Agric* 12: 607–617.

Korsaeth A. 2005a. Soil apparent electrical conductivity (EC$_a$) as a means of monitoring changes in soil inorganic N on heterogeneous morainic soils in SE Norway during two growing seasons. *Nutr Cycl Agroecosyst* in print.

Korsaeth A. 2005b. Relations between soil apparent electrical conductivity (EC$_a$), soil texture, ignition loss and depth of soil properties on two morainic loam soils in southeast Norway. [This book].

McNeill J.D. 1980. Electromagnetic terrain conductivity measurement at low induction numbers. Tech. Note TN-6. Geonics, ON, Canada.

Sheets K.R., and Hendrickx J.M.H. 1995. Noninvasive Soil-Water Content Measurement Using Electromagnetic Induction. *Water Resour Res* 31: 2401–2409.

Sudduth K.A., Drummond S.T., and Kitchen N.R. 2001. Accuracy issues in electromagnetic induction sensing of soil electrical conductivity for precision agriculture. *Comput Electron Agric* 31: 239–264.

Williams B.G., and Baker G.C. 1982. An electromagnetic induction technique for reconnaissance surveys of soil salinity hazards. *Aust J Soil Res* 20: 107–118.

14 Mapping Pesticide Partition Coefficients By Electromagnetic Induction

Dan B. Jaynes

CONTENTS

14.1   INTRODUCTION

One of the difficulties in predicting the fate and leaching risk of field-applied pesticides is that their transport properties are affected by soil properties that can vary greatly across the landscape. For example, the affinity of a chemical to sorb to the soil matrix has been shown to vary spatially within a field (Cambardella et al., 1994; Rao et al., 1986; Wood et al., 1987) and even within the same soil map unit (Novak et al., 1997). Yet Loague et al. (1990) found that sorption can be the predominant process contributing to the variability of pesticide mobility across the landscape.

The partitioning of an herbicide between the solution and soil phases is typically represented by Freeze and Cherry (1979):

$$s = K_d c \tag{14.1}$$

where s is the sorbed concentration on the soil (mg kg^{-1}), c is the solution concentration (mg L^{-1}), and K_d is the partition coefficient (L kg^{-1}). Standard methods for measuring K_d (Novak et al., 1994) are time consuming and require costly specialized equipment. Due to the expense and time required to measure partition coefficients for combinations of specific soils and chemicals, the coefficients are often estimated from related soil properties. In particular, organic carbon content is commonly used because this soil fraction strongly interacts with nonionic herbicides (Bailey and White, 1970). The relationship between soil organic carbon mass fraction, SOC (kg kg^{-1}) and K_d can be expressed as

$$K_d = K_{oc} \, \text{SOC} \tag{14.2}$$

where K_{oc} (L kg^{-1}) is the coefficient of proportionality. Although K_d is not perfectly correlated with SOC because of additional sorption to the clay fraction (Laird et al., 1992), the K_{oc} approach has proven to be satisfactory for many purposes (Rao and Davidson, 1980). However, if we wish to characterize the spatial variability of K_d over a field or larger unit, even this approach is cumbersome because measuring SOC for a large number of samples is not a trivial task. Thus, alternative, more easily measured surrogate parameters of both K_d and SOC are needed that can be used at the field scale or larger.

Soil electrical conductivity as measured by electromagnetic induction (EC_a) has been used successfully to map soil characteristics. The electrical conductivity of a soil is determined by a combination of soil water content, dissolved salt content, clay content and mineralogy, and soil temperature (McNeill, 1980). In many fields, a single property (e.g., salinity) is the primary factor directly controlling soil electrical conductivity. Thus, once the correlation between electrical conductivity and this property is established, an EC_a survey can be used to map this soil attribute quickly and cheaply. For example, EC_a measurements have been successfully used to measure soil salinity (Cameron et al., 1981; Lesch et al., 1992; Rhoades and Corwin, 1981) and soil water content (Kachanoski et al., 1988); to map groundwater contaminant plumes associated with elevated chloride, sulfate, and nitrate levels (Drommerhausen et al., 1995; Greenhouse and Slaine, 1983); and measure clay content (Williams and Hoey, 1987).

EC_a measurements have also been used to determine soil and field properties that it cannot measure directly. EC_a has been used to determine soil cation exchange capacity and exchangeable Ca and Mg (McBride et al., 1990), depth to claypans (Doolittle et al., 1994), field-scale leaching rates of solutes (Slavich and Yang, 1990), spatial pattern of groundwater recharge (Cook et al.,1989, 1992), and yield (Jaynes et al., 1995). These studies were successful because the parameter of interest either influenced a soil property (e.g., water content) that affects the EC_a reading directly or because the parameter is associated with pedogenic processes that create properties that affect EC_a.

Given its utility as a surrogate for many important soil and field properties, EC_a may be an easily used, acceptable surrogate for K_d. Data presented in Cambardella et al. (1994); Jaynes et al. (1994), and Novak et al. (1997) will be used to illustrate how EC_a measured by electromagnetic induction can be used to estimate K_d for the herbicide atrazine [6-chloro-N-ethyl-N-(1-methylethyl)-1,3,5-triazine-2,4-diamine across an agricultural field.

14.2   MATERIALS AND METHODS

Measurements were made in a 32 ha field within the Walnut Creek watershed in central Iowa (41°58′ N, 93°43′ W). A detailed description of the soils, geology, and farming practices within the watershed can be found in Hatfield et al. (1999). The landscape within the watershed is characterized by gentle swell-swale relief of several meters (Daniels and Handy, 1966). Surface drainage is poorly developed, resulting in numerous closed depressions or potholes that have been extensively tile drained in the past 100 years.

Soils within the field were formed from till from the most recent substage of the Wisconsin glaciation. The toposequence of soils within the field range from well-drained Clarion loam (fine-loamy, mixed, superactive, mesic Typic Hapludolls), to the somewhat poorly drained Nicollet loam (fine-loamy, mixed, superactive, mesic, Aquic Hapludolls), to the poorly drained Canisteo silty clay loam (fine-loamy, mixed, superactive, calcareous, mesic Typic Endoaquolls) and Harps loam (fine-loamy, mixed, superactive, mesic Typic Calciaquolls), and ending with the very poorly drained Okoboji mucky silty loam (fine, smectitic, mesic Cumulic Vertic Endoaquolls) (Figure 14.1).

Prior to planting in 1992, a grid was laid out across a 250 m by 250 m area in the southern half of the field (Figure 14.1). Detailed information about the grid and soil properties not reported here can be found in Cambardella et al. (1994) and Novak et al. (1997). The grid spacing was 25 m in both the easterly and northerly directions. Additional grid points were established at closer spacings but were not included in this study. Within 1 m of each grid point, three 6-cm diameter soil cores were taken to a depth of 15 cm and composited for analysis. The mass of organic carbon was measured using dry combustion methods with a Carlo-Erba NA1500 NCS elemental analyzer (Haake Buchler Instruments, Paterson, NJ) after carbonates had been removed with $2M$ H_2SO_4.

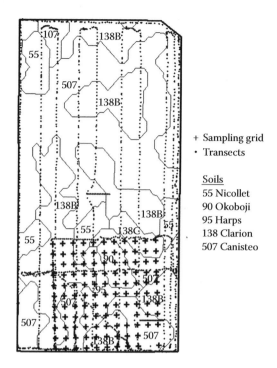

FIGURE 14.1 Grid locations for soil sampling and apparent electrical conductivity, EC_a, measurements, and transects where additional EC_a measurements were made superimposed on the soils map for a 32 ha field.

Atrazine sorption was determined on duplicate 4 g samples of air-dried soil that had passed a 2 mm sieve. Soil was equilibrated with 15 ml of solution containing 1.5 mg L^{-1} atrazine dissolved in $0.01M$ $CaCl_2$ for 72 h. The solution was then centrifuged, and atrazine in the supernatant analyzed with high-performance liquid chromatography (HPLC) as outlined by Novak et al. (1994).

At the same time soil samples were collected, apparent electrical conductivity measurements (EC_a) were made with an EM38 electromagnetic induction meter (Geonics Limited, Ontario, Canada). A single reading was made at each grid point with the meter in the vertical dipole orientation and suspended 20 cm above the soil surface. Readings were taken with the meter suspended above the soil surface rather than on the surface so that the results were directly applicable to a companion transect survey. A vertical rather than horizontal orientation was used so that slight variations in the spacing between the meter and soil surface would have minimal effect on the readings (Rao and Davidson, 1980) during the transect survey. A horizontal orientation may have resulted in values better correlated to the surface soil properties measured for this study. Two days after making EC_a measurements at the grid locations, a transect survey of the entire 32 ha field was made. The EM38 was attached to a wooden boom suspended 20 cm above the ground and pulled by a five-wheel, all-terrain vehicle equipped with a GPS (Ambuel et al., 1991). Electrical conductivity was recorded along with position at 2 s intervals while driving transects across the field (Figure 14.1).

Contour maps of the values collected within the intensive grid area were produced using Surfer software (Golden Software, Inc., Golden, CO) based on a 41 × 41 grid produced by a linear kriging interpolation procedure embedded within the software. A contour map of EC_a from the transect survey of the entire 32 ha field was produced in a similar manner for a 50 × 100 grid using the relationship found between EC_a and K_d.

Descriptive statistics and regressions were computed using standard methods (Draper and Smith, 1966; Snedecor and Cochran, 1967). Correlograms were calculated using the method described in Davis (1973). The spatial structure of the mapped properties was quantified using Moran's I statistic (Moran, 1950; Upton and Fingleton, 1985). Moran's I is similar in concept to correlation and ranges from −1 to 1 where values near −1 indicate that like members of a group are evenly interspersed across the field like the colored squares of a checkerboard, and values near 1 indicate that members are grouped closely together in space. Moran's I was calculated using the Excel 97/2000 Visual Basic routine written by Sawada (1999).

14.3 RESULTS

Measurements at only 117 of 121 grid points were used because standing water in the southeast corner of the field prevented access to three locations and the fourth measurement was lost. The mean K_d for atrazine was 4.94 L kg^{-1} (Table 14.1) and ranged from 1.74 to 10.92 L kg^{-1} with the sample population distribution being positively skewed (mean > median). SOC averaged 0.0280 kg kg^{-1} and varied between 0.0123 and 0.0556 kg kg^{-1}, and the distribution was also positively skewed. Both K_d and SOC were better described by lognormal distributions than normal distributions as determined by the Kolmogorov-Smirnov test and the D'Agostino-Pearson test (Cambardella et al., 1994). EC_a measurements ranged from 20.4 to 65.4 mS^{-1} and averaged 41.0 mS^{-1}. These values are relative, however, and are not direct measures of the true soil electrical conductivity (Lesch et al., 1992). EC_a data were not described well by either normal or lognormal distributions, although the relative difference between the mean and median was less for the untransformed data.

The spatial distribution of each parameter over the intensive-grid area is shown in Figure 14.2. None of the parameters were randomly distributed across the area but were instead grouped into areas of like values. Moran's I statistic for K_d, SOC, and EC_a was 0.52, 0.74, and 0.59, respectively, indicating substantial spatial clustering. Each parameter showed lower values in areas of the field mapped as well-drained Clarion loam and higher values in the area mapped as very poorly drained Okoboji mucky silt loam. This agrees with our hypothesis that all three parameters should be correlated with drainage class.

In addition to similar spatial patterns, the spatial structures of the three parameters were almost identical as determined by similar correlograms (not shown). The correlograms indicated a spatial dependence between measurements for each parameter to distances of about 80 m. The similar spatial distribution leads to a significant simple correlation statistic between each pair with the correlation between EC_a and SOC being the strongest (Table 14.1). Given the similar spatial patterns, spatial structure, and positive correlation between the parameters, it appears likely that K_d can be successfully estimated from either SOC or EC_a. K_{oc} was calculated by regressing the K_d values versus the SOC values using linear least sum-of-squares. The resulting value for K_{oc} was 0.171 L kg^{-1} with a standard error of regression of 1.42 L kg^{-1}.

An analogous regression was performed between K_d and EC_a. However, because the K_d values were lognormally distributed and the EC_a values were better described by a normal distribution, the expression equivalent to K_{oc} is ln K_d = b EC_a or K_d = exp (b EC_a), where b was found by nonlinear

TABLE 14.1
Descriptive Statistics and Correlation Coefficients for 117 Measurements of Apparent Electrical Conductivity (EC_a), Atrazine Partition Coefficient (K_d), and Mass Fraction Organic Carbon (SOC)

Property	Mean	Median	Stan. Dev.	Correlation SOC	Correlation K_d
EC_a (mS m^{-1})	41.0	41.8	13.0	0.765	0.575
K_d (L kg^{-1})	4.94	4.61	1.86	0.686	
SOC (kg kg^{-1})	0.0280	0.0274	0.0100		

regression to be 0.0336. The resulting regression had a standard error of 1.87 L kg^{-1} or about a third larger than for the K_{oc} regression. Estimates of K_d from both equations are plotted versus measured K_d in Figure 14.3. Both estimates fall along the 1:1 line with similar scatter except near high K_d values where SOC gives a better estimate of K_d than does EC_a, which accounts for much of the smaller sum-of-squares.

Perhaps more important than overall agreement between calculated and estimated K_d values is whether or not the calculated values show the same spatial patterns as the measured values. The spatial distributions of K_d calculated from EC_a and SOC are shown in Figure 14.4. Compared to the measured K_d values (Figure 14.2), both calculated distributions are more diffuse and do not reach the maximum values measured within the Okoboji map unit (Figure 14.1). However, both calculated distributions accurately recreate the low values of K_d indicative of the better drained Clarion map units. Underestimations from EC_a data of 3 L kg^{-1} or more are concentrated within the Okoboji map unit and the lower southwest corner of the grid. In contrast, calculations from SOC greatly underestimate K_d only in limited areas along the northern boundary of the grid. Estimates based on EC_a and SOC both overestimate K_d in the center of the intensive grid in an area roughly bounded by Harps soil that has a higher pH than the other soils within the grid (Novak et al., 1997). Atrazine sorption affinity has been shown to decrease with increasing pH (Yamane and Green, 1972). Neither SOC nor EC_a alone captured this pH interaction with atrazine K_d.

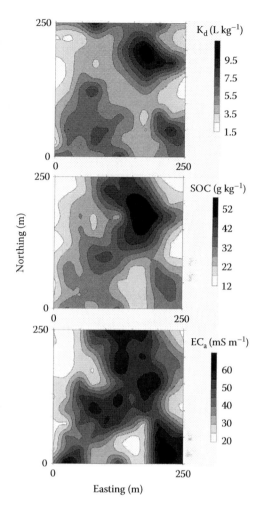

FIGURE 14.2 Spatial distribution of (a) atrazine sorption affinity, K_d, (b) soil organic carbon fraction, SOC, and (c) apparent electrical conductivity, EC_a, over the gridded area.

Overall, EC_a measurements provided reasonable estimates of K_d, although not as accurate as the standard procedure of estimating K_d from SOC. Spatial patterns for estimated K_d values were also similar to measured patterns, although more diffuse, and estimates based on SOC were closer to measured values in the regions of high K_d. The real advantage to using EC_a measurements to estimate K_d, however, is in the speed and ease of making many measurements over a wide area. Measurements of EC_a over the grid were made in about an hour, and SOC determinations took many days of field and laboratory effort.

Once calibrated, maps of K_d estimated from EC_a surveys would be useful in determining the leaching potential of herbicide applications for specific areas in fields. We can illustrate this by mapping K_d across the entire 32 ha field by using the EC_a data collected along the transects and the relationship between EC_a and K_d developed above. Using linear kriging to interpolate EC_a for a regular grid across the field and then applying the nonlinear regression relating K_d to EC_a, we can develop a map of K_d with little extra effort (Figure 14.5). This map clearly shows the spatial pattern of K_d variation across the field and could be used as the basis of computing a leaching risk for atrazine using methods such as those proposed by Loague et al. (1990). Although not as accurate as

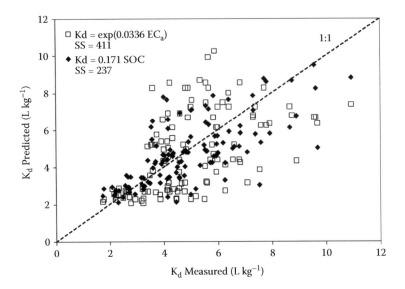

FIGURE 14.3 Measured atrazine K_d versus K_d predicted from apparent electrical conductivity, EC_a, or soil organic carbon fraction, SOC.

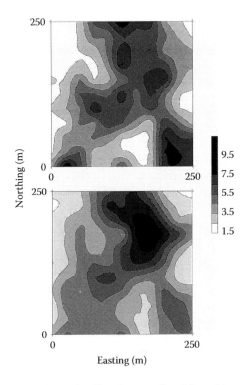

FIGURE 14.4 Spatial distribution of atrazine K_d values predicted from (a) apparent electrical conductivity, EC_a, and (b) soil organic carbon fraction, SOC.

direct measurement of K_d, estimating K_d from EC_a measurements is nearly as accurate as the commonly used method of estimating K_d from SOC with the major advantage that using EC_a to map K_d is a rapid, easy, and inexpensive method once it has been calibrated.

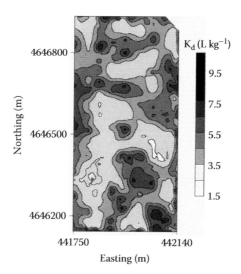

FIGURE 14.5 Map of atrazine K_d values for the entire 32 ha field predicted from apparent electrical conductivity, EC_a.

14.4 AUTHOR'S NOTE

Trade and company names are used for the benefit of readers and do not imply endorsement by the U.S. Department of Agriculture (USDA).

REFERENCES

Ambuel, J., T.S. Colvin, and K. Jeyapalan. 1991. Satellite based positioning system for farm equipment. *SAE Trans., J. Commercial Vehicles,* 100-2:323–329.

Bailey, G.W. and J.L. White. 1970. Factors influencing adsorption, desorption and movement of pesticides in soil. *Res. Rev.* 32:29–92.

Cambardella, C.A., T.B. Moorman, J.M. Novak, T.B. Parkin, D.L. Karlen, R.F. Turco, and A. Konopka. 1994. Field-scale variability of biological, chemical, and physical soil properties in central Iowa soils. *Soil Sci. Soc. Am. J.* 24:36–41.

Cameron, D.R., E. DeJong, D.W.L. Read, and M. Oosterveld. 1981. Mapping salinity using resistivity and electromagnetic inductive techniques. *Can. J. Soil Sci.* 61:67–78.

Cook, P.G., M.W. Hughs, G.R. Walker, and G.B. Allison. 1989. The calibration of frequency-domain electromagnetic induction meters and their possible use in recharge studies. *J. Hydrol.* 107:251–265.

Cook, P.G., G.R. Walker, G. Buselli, I. Potts, and A.R. Dodds. 1992. The application of electromagnetic techniques to groundwater recharge investigations. *J.. Hydrol.* 130:201–229.

Daniels, R.B. and R.L. Handy. 1966. Surficial sediments on undulating Cary ground moraine in central Iowa. Iowa State Univ. Eng. Res. Inst. Tech. Rep. 3, Ames, IA.

Davis, J.C. 1973. *Statistics and data analysis in geology.* John Wiley & Sons, New York.

Doolittle, J.A., K.A. Sudduth, N.R. Kitchen, and S.J. Indorante. 1994. Estimating depths to claypans using electromagnetic induction methods. *J. Soil and Water Cons.* 49:572–575.

Draper, N.R. and H. Smith. 1966. *Applied regression analysis.* John Wiley & Sons, New York.

Drommerhausen, D.J., D.E. Radcliffe, D.E. Brune, and H.D. Gunter. 1995. Electromagnetic conductivity surveys of dairies for groundwater nitrate. *J. Environ. Qual.* 24:1083–1091.

Freeze, R.A. and J.A. Cherry. 1979. *Groundwater.* Prentice Hall, Englewood Cliffs, NJ.

Greenhouse, J.P. and D.D. Slaine. 1983. The use of reconnaissance electromagnetic methods to map contaminant migration. *Ground Water Monitoring Rev.* 3(2):47–59.

Hatfield, J.L., D.B. Jaynes, M.R. Burkart, C.A. Cambardella, T.B. Moorman, J.H. Prueger, and M.A. Smith. 1999. Water quality in Walnut Creek watershed: Setting and farming practices. *J. Environ. Qual.* 28:11–24.

Jaynes, D.B., T.S. Colvin, and J. Ambuel. 1995. Yield mapping by electromagnetic induction. pp. 383–394. In P.C. Robert, R.H. Rust, and W.E. Larson (ed.) *Site-specific management for agricultural systems. Proc. 2nd Inter. Conf.* Minneapolis, MN. 27–30 Mar. 1994. Am. Soc. Am., Inc., Madison, WI.

Jaynes, D.B., J.M. Novak, T.B. Moorman, and C.A. Cambardella. 1994. Estimating herbicide partition coefficients from electromagnetic induction measurements. *J. Environ. Qual.* 24:36–41.

Kachanoski, R.G., E.G. Gregorich, and I.J. Van Wesenbeeck. 1988. Estimating spatial variations of soil water content using noncontacting electromagnetic inductive methods. *Can. J. Soil Sci.* 68:715–722.

Laird, D.A., E. Barriuso, R.H. Dowdy, and W.C. Koskinen. 1992. Adsorption of atrazine on smectites. *Soil Sci. Soc. Am. J.* 56:62–67.

Lesch, S.M., J.D. Rhoades, L.J. Lund, and D.L. Corwin. 1992. Mapping soil salinity using calibrated electromagnetic measurements. *Soil Sci. Soc. Am. J.* 56:540–548.

Loague, K. R.E. Green, T.W. Giambelluca, T.C. Liang, and R.S. Yost. 1990. Impact of uncertainty in soil, climatic, and chemical information in a pesticide leaching assessment. *J. Contaminant Hydrol.* 5:171–194.

McBride, R.A., A.M. Gordon, and S.C. Shrive. 1990. Estimating forest soil quality from terrain measurements of apparent electrical conductivity. *Soil Sci. Soc. Am. J.* 54:290–293.

McNeill, J.D. 1980. Electromagnetic terrain conductivity measurement at low induction numbers. Technical Note TN-6. Geonics Ltd. Mississauga, Ontario, Canada.

Moran, P.A.P. 1950. Notes on continuous stochastic phenomena. *Biometrika* 37:17.

Novak, J.M., T.B. Moorman, and C.A. Cambardella. 1997. Atrazine sorption at the filed scale in relation to soils and landscape position. *J. Environ. Qual.* 26:1271–1277.

Novak, J.M., T.B. Moorman, and D.L. Karlen. 1994. Influence of aggregate size on atrazine sorption kinetics. *J. Agric. Food Chem.* 42:1809–1812.

Rao, P.S.C. and J.M. Davidson. 1980. Estimation of pesticide retention and transformation parameters required in nonpoint source pollution models. pp. 23–67. In M.R. Overcash (ed.) Environmental impact of nonpoint source pollution. Ann Arbor Sci. Publ. Inc.,Ann Arbor, MI.

Rao, P.S.C., K.S.V. Edvardsson, L.T. Ou, R.E. Jessup, P. Nkedi-Kizza, and A.G. Hornsby. 1986. Spatial variability of pesticide sorption and degradation parameters. p. 100–115. In W.Y. Garner, R.C. Honeycutt, and H.N. Nigg (ed.) *Evaluation of pesticides in ground water.* ACS Symp. Ser. 315. ACS, Washington, DC.

Rhoades, J.D. and D.L. Corwin. 1981. Determining soil electrical conductivity—Depth relations using an inductive electromagnetic soil conductivity meter. *Soil Sci. Soc. Am. J.* 45:255–260.

Sawada, M. 1999. ROOKCASE: An Excel 97/2000 Visual Basic (VB) add in for exploring global and local spatial autocorrelation. *Bull. Ecol. Soc. Am.* 80:231–234.

Slavich, P.G. and J. Yang. 1990. Estimation of field scale leaching rates from chloride mass balance and electromagnetic induction measurements. *Irrig. Sci.* 11:7–14.

Snedecor, G.W. and W.G. Cochran. 1967. *Statistical Methods.* The Iowa State University Press, Ames, IA.

Upton, G.J. and B. Fingleton. 1985. *Spatial data analysis by example. Volume 1: Point pattern and quantitative data.* Wiley, New York.

Williams, B.G. and D. Hoey. 1987. The use of electromagnetic induction to detect the spatial variability of the salt and clay contents of soils. *Aust. J. Soil Res.* 25:21–27.

Wood, L.S., H.D. Scott, D.B. Marx, and T.L. Lavy. 1987. Variability in sorption coefficients of metolachlor on a Captina silt loam. *J. Environ. Qual.* 16:251–256.

Yamane, V.K. and R.E. Green. 1972. Adsorption of ametryne and atrazine on an oxisol, montmorillonite, and charcoal in relation to pH and solubility effects. *Soil Sci. Soc. Am. Proc.* 36:58–64.

15 Can Apparent Soil Electrical Conductivity Be Used to Map Soil-Herbicide Partition Coefficients?
A Case Study in Three Colorado Fields

Dale L. Shaner, Hamid J. Farahani, Gerald W. Buchleiter, and Mary K. Brodahl

CONTENTS

15.1 INTRODUCTION

Herbicides are one of the major pesticide inputs in agriculture. In 2002, atrazine, acetochlor, and *S*-metolachlor were applied to the soil on 62, 25, and 15 percent of the treated hectares, respectively (Anonymous, 2003). The activity and fate of soil-applied herbicides are affected by multiple factors including soil organic matter (OM), pH, cation exchange capacity, and texture (Blackshaw et al., 1994). These factors are spatially variable, and herbicide adsorption to soil can differ considerably within fields (Koskinen et al., 1994). The soil adsorption of herbicides determines the bioavailability of the chemical to both weeds and microbes. Liu et al. (2002) found that the efficacy and mineralization of atrazine and alachlor varied across a field depending on soil properties, and weed control was poor where the herbicides dissipated rapidly.

Herbicides could be applied at varying rates within a field depending on differences in soil properties as well as weed populations (Williams et al., 2002), but it is difficult to economically measure the variation in soil properties at the field scale. What is needed is a method that can economically map the heterogeneity of soil properties that are closely related to herbicide behavior within a field. In the absence of salinity, apparent soil electrical conductivity (EC_a) is strongly influenced by soil texture, including clay content, soil water content, and OM.

Jaynes et al. (1995) used EC_a measured by electromagnetic (EM) induction to estimate the fraction of organic carbon (f_{oc}) and herbicide-soil partition coefficient (K_d) for atrazine within a 32 ha field in Iowa. The correlation coefficient between K_d and EC_a was 0.57 and between f_{oc} and EC_a was 0.69. These results suggest that EC_a maps could be used to predict herbicide K_d variability within a field. The objectives of this study were to determine the relationship between EC_a and f_{oc} and between f_{oc} and the K_d of three different herbicides (EPTC, metribuzin, and metolachlor) in soil samples taken from three fields in Colorado and to examine whether such a relationship could be used to map the differences in binding of these herbicides to the soil.

15.2   MATERIALS AND METHODS

15.2.1   Sites

Three fields irrigated by center pivots were used in this project. Two were located near Wiggins, Colorado (designated as Wiggins 1 and Wiggins 2), and one near Yuma, Colorado (designated as Yuma). Wiggins 1 and Wiggins 2 are 71 and 52 ha, respectively, and are located a few kilometers apart. The soils in these two fields include coarse-loamy, mixed, superactive, mesic Ustic Haplargids; mixed, mesic Typic Ustipsamments; mixed, mesic Ustic Torripsamments; and coarse-loamy, mixed, superactive, mesic Aridic Argiustolls. Yuma is 57 ha and includes fine-loamy, mixed, superactive, mesic Pachic Argiustolls; fine-loamy, mixed, superactive, mesic Pachic Argiustolls; and fine-loamy, mixed, superactive, mesic Aridic Argiustolls.

15.2.2 EC_A Measurement and Soil Samples

Measurements of EC_a were taken in spring of 1999 and 2000 using the Veris 3100 Soil EC Mapping System (Veris Technologies, Salina, KS) (Farahani and Buchleiter, 2004). Sample sites were identified in each field by selecting areas from delineated EC_a zones (Farahani and Buchleiter, 2004). Samples were taken from the top 0.3 m in 1999 from Wiggins 1 and Wiggins 2 and in 2000 from Yuma. The number of samples taken were 28, 38, and 20 from Wiggins 1, Wiggins 2, and Yuma, respectively. Samples were air-dried, ground, and sieved through a 2 mm mesh. Soil organic matter was determined by the method of Sherrod et al. (2002). A validation set of ten additional soil samples was taken from each field in spring 2003 (Figure 15.1). The sites were chosen in areas of distinct EC_a zones that had not been previously sampled. The samples were taken and handled as described previously.

15.2.3   Herbicide–Soil Binding

The binding of analytical-grade EPTC, metribuzin, and metolachlor was determined by batch-slurry equilibration with 10 g of soil and 10 mL of 0.05M $CaCl_2$. The herbicides were introduced into the $CaCl_2$ solution at a final concentration of 1 ng/mL and allowed to equilibrate with the soil for 24 h. After equilibration, the system was centrifuged, and the concentration of the herbicide in the 0.05M $CaCl_2$ was partitioned into toluene and was measured by using a gas chromatograph equipped with a mass spectrophotometer.

The herbicide–soil sorption coefficient (K_d) was calculated as

$$K_d = \text{(herbicide sorbed to soil } (\mu g/g))/(\text{herbicide in solution } (\mu g/mL)) \tag{15.1}$$

K_{oc} (the soil organic carbon sorption coefficient) for each herbicide was calculated as

$$K_{oc} = K_d/f_{oc}. \tag{15.2}$$

f_{oc} is the soil organic carbon mass fraction that was measured for each soil sample:

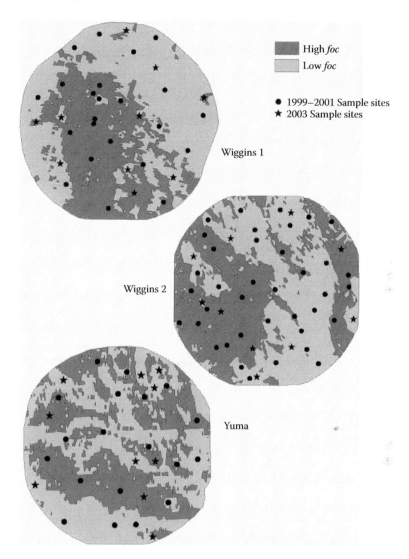

High *foc*
Low *foc*

● 1999–2001 Sample sites
★ 2003 Sample sites

Wiggins 1

Wiggins 2

Yuma

FIGURE 15.1   Map of f_{oc} variability in three fields in Colorado based on regression tree analysis between f_{oc} and EC$_a$.

$$f_{oc} = (\text{Soil Organic Matter} * 0.58) \text{ g/100 g} \tag{15.3}$$

Analyses of the relationships between soil EC$_a$, herbicide K_d values, and various soil properties were done using linear regression functions in SigmaPlot v 7.101 (SPSS Inc., Chicago, IL). For each field, regression tree analysis was performed on the soil EC$_a$ and f_{oc} values using the S-Plus tree analysis tool of S-Plus2000 (Insightful Corp., Seattle, WA) to establish EC$_a$-f_{oc} relationships for mapping f_{oc} classes in the fields. Due to the small size of the data sets for the individual fields, a check of the regression tree results was performed by a boot strapping technique (Hasties et al., 2001).

15.3   RESULTS AND DISCUSSION

There were significant correlations between EC$_a$ and f_{oc} and between f_{oc} and K_d for all three herbicides across all three fields (Table 15.1). These results are similar to those of Jaynes et al. (1995).

TABLE 15.1

Descriptive Statistics and Correlation Coefficients for Combined Samples from All Fields

| | | | | Correlation Coefficients | | |
| | | | | K_d | | |
Property	Mean	SD[a]	fo_c	EPTC	Metribuzin	Metolachlor
EC$_a$ 0–0.3 m (mS/m)	22.1	8.8	0.75[b]	0.57[b]	0.54[b]	0.63[b]
K_d (L/kg) EPTC	0.90	0.32	0.69[b]			
K_d (L/kg) Metribuzin	0.25	0.20	0.63			
K_d (L/kg) Metolachlor	1.20	0.51	0.71[b]			
f_{oc}	0.006	0.002				

[a] Standard deviation.
[b] $p < 0.0001$.

However, the question is: Can these relationships be used to map herbicide–soil binding variability across a field? Each field was mapped into high and low f_{oc} zones (Figure 15.1) based on results from the regression tree classification of the EC$_a$ and f_{oc} relationship. The validation soil data were used to test the resulting maps. The results were good, although not perfect (Table 15.2). As expected, the K_d of the three herbicides followed the f_{oc}. On average, the herbicides bound less tightly to soil samples taken from the low f_{oc} zones compared to soils from the high f_{oc} zones (Table 15.2). These results indicate that if a relationship between f_{oc} and EC$_a$ exists, EC$_a$ maps could be used to map herbicide soil binding variability across a field.

TABLE 15.2

The Fraction of Organic Carbon (f_{oc}) and Herbicide–Soil Binding (K_d) of EPTC, Metribuzin, and Metolachlor in Validation Soil Samples from Three Fields in Colorado

| | | | Yuma | | |
| | | | K_d | | |
Zone	EC$_a$ 0–0.3 m	EPTC	Metribuzin	Metolachlor	f_{oc}
	15.06	0.38	0.04	0.58	0.0058
	22.49	0.52	0.12	0.83	0.0058
Low f_{oc}	18.47	0.52	0.15	0.91	0.0047
	18.19	0.77	0.21	1.19	0.0061
	11.33	0.78	0.29	1.29	0.0055
Avg	**17.11**	**0.59**	**0.16**	**0.96**	**0.0056**
Std. dev.	**4.17**	**0.17**	**0.10**	**0.29**	**0.0005**
	27.03	0.98	0.35	1.44	0.0085
High f_{oc}	40.94	1.11	0.42	1.66	0.0065
	35.89	1.18	0.45	1.86	0.0110
	44.38	1.32	0.64	2.26	0.0113
Avg	**30.48**	**1.15**	**0.47**	**1.81**	**0.0075**
Std. Dev.	**16.09**	**0.14**	**0.13**	**0.35**	**0.0044**

TABLE 15.2 (continued)
The Fraction of Organic Carbon (f_{oc}) and Herbicide–Soil Binding (K_d) of EPTC, Metribuzin, and Metolachlor in Validation Soil Samples from Three Fields in Colorado

Wiggins I

Zone	EC_a 0–0.3 m	EPTC	Metribuzin	Metolachlor	f_{oc}
	7.63	0.44	0.16	0.68	0.0044
	9.65	0.47	0.18	0.71	0.0047
Low f_{oc}	10.45	0.57	0.20	0.88	0.0049
	10.60	0.55	0.17	0.90	0.0049
Avg	**9.58**	**0.51**	**0.18**	**0.79**	**0.0047**
Std. dev.	**1.37**	**0.06**	**0.02**	**0.11**	**0.0002**
	17.50	0.65	0.24	1.08	0.0062
	16.77	0.72	0.24	1.14	0.0060
	11.76	0.88	0.26	1.24	0.0051
High f_{oc}	11.90	0.78	0.22	1.45	0.0051
	21.10	1.32	0.28	1.46	0.0068
	21.67	1.33	0.39	1.85	0.0069
Avg	**16.61**	**0.95**	**0.27**	**1.37**	**0.0060**
Std. dev.	**5.52**	**0.30**	**0.06**	**0.28**	**0.0008**

Wiggins 2

Zone	EC_a 0–.3 m	EPTC	Metribuzin	Metolachlor	f_{oc}
	16.68	0.56	0.21	0.83	0.0040
Low f_{oc}	16.40	0.58	0.18	0.85	0.0052
	13.33	0.60	0.18	0.95	0.0044
	15.16	0.62	0.17	0.95	0.0058
Avg	**15.39**	**0.59**	**0.18**	**0.90**	**0.0048**
Std. dev.	**1.52**	**0.02**	**0.02**	**0.06**	**0.0008**
	21.79	0.70	0.25	1.03	0.0070
	33.70	0.69	0.28	1.19	0.0058
	19.24	0.67	0.38	1.21	0.0042
High $f oc$	18.76	0.52	0.23	1.35	0.0041
	27.84	0.71	$0 \rightarrow 3$	1.35	0.0064
	25.22	0.79	0.34	1.36	0.0062
Avg	**22.77**	**0.68**	**0.30**	**1.25**	**0.0052**
Std. dev.	**4.48**	**0.09**	**0.06**	**0.13**	**0.0013**

REFERENCES

Anonymous, Agricultural Chemical Usage 2002 Field Crops Summary USDA NASS, 2003, 104 p.

Blackshaw, R., Moyer, J.R., and Kozub, G.C., Efficacy of downy brome herbicides as influenced by soil properties, *Can. J. Plant Sci.*, 74, 177, 1994.

Farahani, H.J. and Buchleiter, G.W., Temporal stability of bulk soil electrical conductivity in irrigated sandy soil, *Trans. Am. Soc. Agric. Eng.* 47, 79, 2004.

Hasties, T.R., Tibshirani, R., and Friedman, J., *The Elements of Statistical Learning: Data Mining, Inference, and Prediction*, Springer-Verlag, New York, 2001, 533 p.

Jaynes, D.B. et al., Estimating herbicide partition coefficients from electromagnetic induction measurements, *J. Environ. Qual.*, 24, 36, 1995.

Koskinen, W.C. et al., Spatial variability of herbicide sorption on soil, In *Terrestrial Field Dissipation Studies: Purpose, Design, and Interpretation*, Arthur, E.L., Barefoot, A.C., and Clay, V.E., Eds., ACS Symp. Series 842, Washington, DC, 2003, 88.

Liu, Z., Clay, S.A., and Clay, D.E., Spatial variability of atrazine and alachlor efficacy and mineralization in an eastern South Dakota field, *Weed Sci.*, 50, 662, 2002.

Sherrod, L.A., G. Dunn, G.A. Peterson, and R.L. Kolberg. Inorganic carbon analysis by modified pressure-calcimeter method. *Soil Sci. Soc. Am. J.* 66:299–305, 2002.

Williams, M.M. et al., Spatial inference of herbicide bioavailability using a geographic information system, *Weed Technol.*, 16, 603, 2002.

16 Delineating Site-Specific Management Units Using Geospatial EC_a Measurements

Dennis L. Corwin, Scott M. Lesch, Peter J. Shouse,
Richard Soppe, and James E. Ayars

CONTENTS

16.1 INTRODUCTION

Site-specific crop management (or site-specific management, SSM) is a means of managing the spatial variability of edaphic (i.e., soil related), anthropogenic, topographic, biological, and meteorological factors influencing crop yield. The aim of SSM is to increase crop productivity, sustain the soil–plant environment, optimize inputs, increase profitability, and minimize detrimental environmental impacts. The spatial variability of edaphic factors is a consequence of pedogenic and anthropogenic activities, which produce variation in soil physical and chemical properties within agricultural fields. In the arid southwestern United States, the primary soil properties influencing crop yield are salinity, soil texture and structure, plant-available water, trace elements (particularly B), and ion toxicity from Na^+ and Cl^- (Tanji, 1996).

Bullock and Bullock (2000) indicated that efficient, reliable methods for measuring within-field variations in soil properties are important for precision agriculture. Because apparent soil electrical conductivity (EC_a) is influenced by a variety of soil properties (i.e., salinity, water content, texture, bulk density, organic matter, and temperature) and is a reliable measurement that is easy to take, geospatial measurements of EC_a have become one of the most frequently used measurements to characterize within-field variability for agricultural applications (Corwin and Lesch, 2003). Geospatial measurements of EC_a have been used to characterize spatial variation in soil salinity and nutrients such as NO_3^-, water content, texture-related properties, bulk density–related properties such as compaction, leaching, and organic matter–related properties (Corwin and Lesch, 2005a).

In the past, geo-referenced EC_a measurements have been correlated to associated yield-monitoring data with mixed results (Corwin et al., 2003; Jaynes et al., 1993; Johnson et al., 2001; Kitchen et al., 1999; Sudduth et al., 1995). These mixed results are due, in part, to a misunderstanding of the relationship between EC_a measurements and variations in crop yield. As pointed out by Corwin and Lesch (2003), crop yield inconsistently correlates with EC_a due to the influence of soil properties (e.g., salinity, water content, texture, etc.) that are being measured by EC_a, which may or may not influence crop yield within a particular field, and because a temporal component of yield variability is poorly captured by a state variable such as EC_a. Corwin and Lesch (2005a) provide a recent review of the application of geo-referenced EC_a measurements in agriculture with particular attention to precision agriculture applications.

Site-specific management units (SSMUs) have been proposed as a means of dealing with the spatial variability of edaphic properties influencing crop productivity to achieve the goals of SSM. A SSMU is simply a mapped unit of soil that is managed the same to achieve SSM goals. In a strict sense, the task of delineating SSMUs is extremely complicated because all edaphic, anthropogenic, topographic, biological, and meteorological factors influencing a crop's yield must be considered. One means of simplifying the complexity of delineating SSMUs is to define SSMUs based on a single factor, such as edaphic properties, and determine the extent of the variability of crop yield due to the single factor.

It is hypothesized that in instances where EC_a correlates with crop yield, spatial EC_a information can be used to direct a soil sampling plan that identifies sites that adequately reflect the range and variability of various soil properties thought to influence crop yield. The objective of this study is to utilize an intensive geo-referenced EC_a survey to direct soil sampling and to identify edaphic properties influencing cotton yield, and to use this spatial information to make recommendations for SSM of cotton by delineating SSMUs based solely on edaphic properties influencing cotton yield. This paper draws from previous more detailed work conducted and published by Corwin and colleagues (Corwin and Lesch, 2003, 2005b; Corwin et al., 2003).

16.2 MATERIALS AND METHODS

16.2.1 STUDY SITE

A 32.4-ha field located in the Broadview Water District of the San Joaquin Valley's west side in central California was used as the study site. The soil at the site is a Panoche silty clay (thermic Xerorthents), which is slightly alkaline with good surface and subsurface drainage. The subsoil is thick, friable, calcareous, and easily penetrated by roots and water.

16.2.2 YIELD MONITORING AND EC_A SURVEY

Spatial variation of cotton yield was measured at the study site in August 1999 using a four-row cotton picker equipped with a yield sensor and Global Positioning System (GPS) receiver. The yield sensors measured average seed cotton yield. All subsequent references to cotton yield are with respect to seed cotton yield. A total of 7706 cotton yield readings were collected (Figure 16.1a). Each yield observation represented a total area of approximately 42 m². From August 1999 through March 2000 the field was fallow. On March 2000, an intensive EC_a survey was conducted using mobile fixed-array electrical resistivity equipment developed by Rhoades and colleagues (Carter, 1993; Rhoades, 1992). The fixed-array electrodes were spaced to measure EC_a to a depth of 1.5 m. Over 4000 EC_a measurements were collected (Figure 16.1b).

16.2.3 SAMPLE SITE SELECTION, SOIL SAMPLING, AND SOIL ANALYSES

Data from the EC_a survey were used to direct the selection of sixty sample sites. A spatial statistics software package, ESAP-95 version 2.01, developed by Lesch et al. (2000) was used to determine

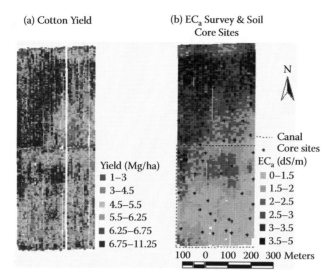

FIGURE 16.1   Maps of (a) cotton yield and (b) EC$_a$ measurements including sixty soil core sites. (Modified from Corwin, D.L., Lesch, S.M., Shouse, P.J., Soppe, R., Ayars, J.E., *Agron. J.*, 95, 352–364, 2003. With permission.)

the sample sites from the EC$_a$ survey data. The software uses a model-based response-surface sampling strategy. The selected sites reflect the observed spatial variability in EC$_a$ while simultaneously maximizing the spatial uniformity of the sampling design across the study area. Figure 16.1b shows the spatial EC$_a$ survey data and the locations of the sixty core sites. Soil samples were collected at 0.3 m increments to a depth of 1.8 m. Soil samples were analyzed for physical and chemical properties thought to influence cotton yield including gravimetric water content (θ_g), bulk density (ρ_b), pH, B, NO$_3$-N, Cl$^-$, electrical conductivity of the saturation extract (EC$_e$), leaching fraction (LF), percentage clay, and saturation percentage (SP). All samples were analyzed for physical and chemical properties following the methods outlined in Agronomy Monograph No. 9 (Page et al., 1982).

16.2.4 STATISTICAL AND SPATIAL ANALYSES

Statistical analysis was conducted using SAS software (SAS, 1999). The statistical analysis consisted of three stages: (1) determination of the correlation between EC$_a$ and cotton yield using data from the sixty sites, (2) exploratory statistical analysis to identify the significant soil properties influencing cotton yield, and (3) development of a crop yield response model based on ordinary least squares regression (OLS) adjusted for spatial autocorrelation with restricted maximum likelihood. Because the location of EC$_a$ and cotton yield measurements did not exactly overlap, ordinary kriging was used to determine the expected cotton yield at the sixty sites.

Spatial analysis was accomplished with a geographic information system (GIS). The commercial GIS software ArcView 3.3 (ESRI 2002) was used to compile, manipulate, organize, and display all spatial data. Delineation of SSMUs was accomplished using the GIS, exploratory statistical analyses, and crop yield response model adjusted for spatial autocorrelation.

16.3 RESULTS AND DISCUSSION

16.3.1 CORRELATION BETWEEN CROP YIELD AND EC$_A$

The correlation of EC$_a$ to yield at the sixty sites was 0.51. The moderate correlation between yield and EC$_a$ suggests that some soil properties influencing EC$_a$ measurements may also influence cotton yield, making an EC$_a$-directed soil sampling strategy a potentially viable approach at this site. The similarity of the spatial distributions of EC_a measurements and cotton yield in Figure 16.1 visually confirms their close relationship.

16.3.2 Exploratory Statistical Analysis

Exploratory statistical analysis was conducted to determine the significant soil properties influencing cotton yield and to establish the general form of the cotton yield response model. The exploratory statistical analysis consisted of three stages: (1) a preliminary multiple linear regression (MLR) analysis, (2) a correlation analysis, and (3) scatter plots of yield versus potentially significant soil properties. Both preliminary MLR and correlation analysis showed that the 0 to 1.5 m soil depth increment resulted in the best correlations and best fit of the data; consequently, the 0 to 1.5 depth increment was considered to correspond to the active root zone. Preliminary MLR analysis indicated that the following soil properties were most significantly related to cotton yield: EC_e, LF, pH, percentage clay, θ_g, and ρ_b. Table 16.1 reveals that the correlation coefficients between EC_a and θ_g, EC_e, B, percentage clay, ρ_b, Cl^-, LF, and SP were significant at the 0.01 level. The correlation coefficients were 0.79, 0.87, 0.88, 0.76, −0.38, 0.61, −0.50, and 0.77, respectively, indicating high correlations between EC_a and the properties of θ_g, EC_e, B, percentage clay, and SP. However, B is a property not measured by EC_a. Rather, the high correlation of B to EC_a is an artifact due to its close correspondence to salinity (i.e., EC_e) stemming from the leaching process. The high correlation of EC_a to both percentage clay and SP is expected because it reflects the influence of texture on the EC_a reading. So, in this particular field, EC_a is highly correlated with salinity, θ_g, and texture. Table 16.1 also indicates the correlation between cotton yield and soil properties, with the highest correlation occurring with salinity (EC_e).

TABLE 16.1
Simple Correlation Coefficients between EC_a and Soil Properties and between Cotton Yield and Soil Properties

Soil Property[a]	Fixed Array EC_a[b]	Cotton Yield[c]
θ_g	0.79**	0.42**
EC_e	0.87**	0.53**
B	0.88**	0.50**
pH	0.33*	−0.01
% clay	0.76**	0.36*
ρ_b	−0.38**	−0.29*
NO_3-N	0.22	−0.03
Cl^-	0.61**	0.25*
LF	−0.50**	−0.49**
SP	0.77**	0.38*

Notes: * Significant at the $P < 0.05$ level; ** significant at the $P < 0.01$ level. θ_g = gravimetric water content; EC_e = electrical conductivity of the saturation extract (dS m^{-1}); LF = leaching fraction; SP = saturation percentage.

[a]   Properties averaged over 0 to 1.5 m.

[b]   Pearson correlation coefficients based on sixty observations.

[c]   Pearson correlation coefficients based on fifty-nine observations.

Source: Modified from Corwin, D.L., Lesch, S.M., Shouse, P.J., Soppe, R., Ayars, J.E., *Agron. J.*, 95, 352–364, 2003. With permission.

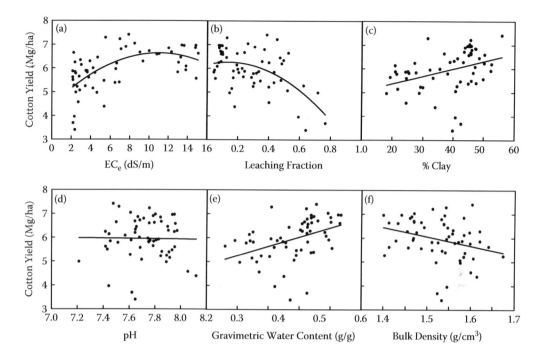

FIGURE 16.2   Scatter plots of soil properties and cotton yield: (a) electrical conductivity of the saturation extract (EC$_e$, dS m^{-1}), (b) leaching fraction, (c) percentage clay, (d) pH, (e) gravimetric water content, and (f) bulk density (Mg m^{-3}). (Taken from Corwin, D.L., Lesch, S.M., Shouse, P.J., Soppe, R., Ayars, J.E., *Agron. J.*, 95, 352–364, 2003. With permission.)

A scatter plot of EC$_e$ and yield indicates a quadratic relationship where yield increases and then decreases (Figure 16.2a). The scatter plot of LF and yield shows a negative, curvilinear relationship (Figure 16.2b). Yield shows a minimal response to LF below 0.4 and falls off rapidly for LF > 0.4. Clay percentage, pH, θ_g, and ρ_b appear to be linearly related to yield to various degrees (Figure 16.2c through Figure 16.2f, respectively). Even though there was clearly no correlation between yield and pH ($r = -0.01$; see Figure 16.2d), pH became significant in the presence of the other variables, which became apparent in both the preliminary MLR analysis and in the final yield response model. Based on the exploratory statistical analysis, it became evident that the general form of the cotton yield response model was

$$Y = \beta_0 + \beta_1(\text{EC}_e) + \beta_2(\text{EC}_e)^2 + \beta_3(\text{LF})^2 + \beta_4(\text{pH}) + \beta_5(\% \text{ clay}) + \beta_6(\theta_g) + \beta_7(\rho_b) + \varepsilon \quad (16.1)$$

where, based on the scatter plots of Figure 16.2, the relationships between cotton yield (Y) and pH, percentage clay, θ_g, and ρ_b are assumed linear; the relationship between yield and EC$_e$ is assumed to be quadratic; the relationship between yield and LF is assumed to be curvilinear; β_0, β_1, β_2, ..., β_7 are the regression model parameters; and ε represents the random error component.

16.3.3 Crop Yield Response Model Development

Ordinary least squares regression based on Equation (16.1) resulted in the following crop yield response model:

$$Y = 20.90 + 0.38(\text{EC}_e) - 0.02(\text{EC}_e)2 - 3.51(\text{LF})2 - 2.22(\text{pH}) + 9.27(\theta_g) + \varepsilon \quad (16.2)$$

where the nonsignificant t test for percentage clay and ρ_b indicated that these soil properties did not contribute to the yield predictions in a statistically meaningful manner and dropped out of the

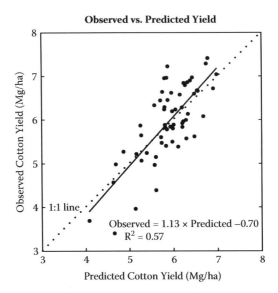

FIGURE 16.3 Observed versus predicted cotton yield estimates using Equation (16.3). Dotted line is a 1:1 relationship. (Taken from Corwin, D.L., Lesch, S.M., Shouse, P.J., Soppe, R., Ayars, J.E., *Agron. J.*, 95, 352–364, 2003. With permission.)

regression model, while all other parameters were significant near or below the 0.05 level. The R^2 value for Equation (16.2) is 0.61, indicating that 61 percent of the estimated spatial yield variation is successfully described by Equation (16.2). However, the residual variogram plot indicates that the errors are spatially correlated, which implies that Equation (16.2) must be adjusted for spatial autocorrelation.

Using a restricted maximum likelihood approach to adjust for spatial autocorrelation, the most robust and parsimonious yield response model for cotton was Equation (16.3):

$$Y = 19.28 + 0.22(EC_e) - 0.02(EC_e)^2 - 4.42(LF)^2 - 1.99(pH) + 6.93(\theta_g) + \varepsilon \qquad (16.3)$$

Figure 16.3 shows the observed versus predicted cotton yield estimates from Equation (16.3). Figure 16.3 suggests that the estimated regression relationship has been reasonably successful at reproducing the predicted yield estimates with an R^2 value of 0.57. Sensitivity analysis reveals that LF is the most significant factor influencing cotton yield with the degree of predicted yield sensitivity to one standard deviation change resulting in a percentage yield reduction for EC_e, LF, pH, and θ_g of 4.6, 9.6, 5.8, and 5.1 percent, respectively. The point of maximum yield with respect to salinity is calculated by setting the first partial derivative of Equation (16.3) to zero with respect to EC_e, which results in a value of 7.17 dS m^{-1}, which is similar to the salinity threshold for cotton of 7.7 dS m^{-1} reported by Maas and Hoffman (1977).

16.4   DELINEATED SITE-SPECIFIC MANAGEMENT UNITS

From Equation (16.3) and scatter plots of cotton yield versus properties (Figure 16.2), management recommendations were made that spatially prescribed what could be done to increase cotton yield at those locations with less than optimal yield. Four recommendations can be made to improve cotton productivity at the study site: (1) reduce the LF in highly leached areas (i.e., areas where LF > 0.5), (2) reduce salinity by increased leaching in areas where the average root zone (0 to 1.5 m) salinity is >7.17 dS m^{-1}, (3) increase the plant-available water in coarse-texture areas by more frequent irrigation, and (4) reduce the pH where pH > 7.9.

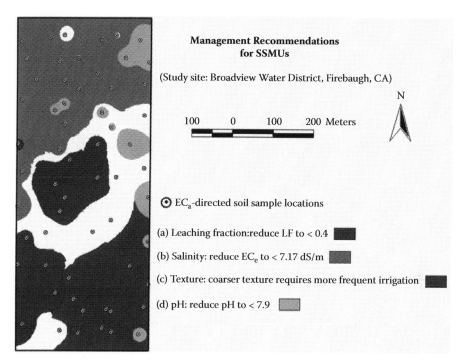

FIGURE 16.4 Site-specific management units for a 32.4 ha cotton field in the Broadview Water District of central California's San Joaquin Valley. Recommendations are associated with the SSMUs for (a) leaching fraction, (b) salinity, (c) texture, and (d) pH. (Taken from Corwin, D.L., Lesch, S.M., *Comput. Electron. Agric.* 46(1–3), 11–43, 2005a.)

Subsequently, Corwin and Lesch (2005a) delineated SSMUs, which are depicted in Figure 16.4. Figure 16.4 indicates the areas pertinent to the above recommendations. All four recommendations can be accomplished by improving water application timing and distribution with variable-rate irrigation technology (Evans, 1997; Perry et al., 2003) and by the precision application of soil amendments. Highly leached zones were delineated where the LF needed to be reduced to <0.5; high-salinity areas were defined where the salinity needed to be reduced below the salinity threshold for cotton, which was established from Equation (16.3) to be EC$_e$ = 7.17 dS m^{-1} for this field; areas of coarse texture were defined that needed more frequent irrigations; and areas were pinpointed where the pH needed to be lowered below a pH of 8 with a soil amendment such as OM. This work brought an added dimension because it delineated within-field units where associated site-specific management recommendations would optimize the yield, but it still falls short of integrating meteorological, economic, and environmental impacts on within-field crop-yield variation. Furthermore, these SSMUs have not been tested to evaluate whether their use would actually increase yield.

In instances where crop yield correlates with EC$_a$, the spatial distribution of EC$_a$ provides a means of directing soil sampling to determine edaphic properties influencing yield. This information provides a basis for delineating SSMUs. The method for delineating SSMUs consists of the following general steps: (1) intensive yield monitoring and EC$_a$ survey; (2) EC$_a$-directed soil sampling; (3) statistical analyses to determine the correlation between EC$_a$-directed soil sampling and crop yield, to identify the significant soil properties influencing crop yield, and to develop a crop yield response model adjusted for spatial autocorrelation; and (4) use of GIS to define SSMUs based on scatter plots and crop yield response model.

Even though EC$_a$-directed soil sampling provides a viable means of identifying some soil properties that influence within-field variation of yield, it is only one piece of a complicated puzzle of interacting factors that result in observed within-field crop variation. Crop yield is influenced by complex interactions of meteorological, biological, anthropogenic, topographic, and edaphic factors. Furthermore, SSM requires more than just a myopic look at crop productivity. It must balance sustainability, profitability, crop productivity and quality, optimization of inputs, and minimization of environmental impacts. Nevertheless, the presented approach is a step forward because it provides spatial information for use in site-specific soil and crop management.

REFERENCES

Bullock, D.S., Bullock, D.G., Economic optimality of input application rates in precision farming. *Prec. Agric.*, 2, 71–101, 2000.

Carter, L. M., Rhoades, J.D., Chesson, J.H., Mechanization of soil salinity assessment for mapping, Proc. 1993 ASAE Winter Meetings, Chicago, IL, 12–17 Dec. 1993, ASAE, St. Joseph, MI, 1993.

Corwin, D.L., Lesch, S.M., Application of soil electrical conductivity to precision agriculture: Theory, principles, and guidelines. *Agron. J.*, 95, 455–471, 2003.

Corwin, D.L., Lesch, S.M., Apparent soil electrical conductivity measurements in agriculture, *Comput. Electron. Agric.* 46(1–3), 11–43, 2005a.

Corwin, D.L., Lesch, S.M., Characterizing soil spatial variability with apparent soil electrical conductivity: I. Survey protocols, *Comput. Electron. Agric.* 46(1–3), 103–133, 2005b.

Corwin, D.L., Lesch, S.M., Shouse, P.J., Soppe, R., Ayars, J.E., Identifying soil properties that influence cotton yield using soil sampling directed by apparent soil electrical conductivity, *Agron. J.*, 95, 352–364, 2003.

ESRI, ArcView 3.3, ESRI, Redlands, CA, 2002.

Evans, R.G., If you build it will it work? Testing precision center-pivot irrigation, *Precision Farming*, 10–17, (August) 1997.

Jaynes, D.B., Colvin, T.S., Ambuel, J., Soil type and crop yield determinations from ground conductivity surveys, ASAE Paper No. 933552, 1993 ASAE Winter Meetings, 14–17 Dec. 1993, Chicago, IL, ASAE, St. Joseph, MI, 1993.

Johnson, C.K., Doran, J.W., Duke, H.R., Weinhold, B.J., Eskridge, K.M., Shanahan, J.F., Field-scale electrical conductivity mapping for delineating soil condition. *Soil Sci. Soc. Am. J.*, 65, 1829–1837, 2001.

Kitchen, N.R., Sudduth, K.A., Drummond, S.T., Soil electrical conductivity as a crop productivity measure for claypan soils, *J. Prod. Agric.*, 12, 607–617, 1999.

Lesch, S.M., Rhoades, J.D., Corwin, D.L., ESAP-95 Version 2.01R: User manual and tutorial guide, Research Rpt. 146, USDA-ARS George E. Brown, Jr. Salinity Laboratory, Riverside, CA, 2000.

Maas, E.V., Hoffman, G.J., Crop salt tolerance—Current assessment, *J. Irrig. Drain. Div., Am. Soc. Civ. Eng.* 103(IR2), 115–134, 1977.

Page, A.L., Miller, R.H., Kenney, D.R., eds., *Methods of Soil Analysis, Part 2—Chemical and Microbiological Properties*, 2nd Ed., Agron. Monogr. No. 9, ASA-CSSA-SSSA, Madison, WI, 1982.

Perry, C., Pocknee, S., Hansen, O., Variable-rate irrigation, *Resource*, 11–12, (January) 2003.

Rhoades, J.D., Instrumental field methods of salinity appraisal, in *Advances in Measurement of Soil Physical Properties: Bring Theory into Practice*, Topp, G.C., Reynolds, W.D., Green, R.E., eds., SSSA Special Publ. No. 30, Soil Science Society of America, Madison, WI, 1992, 231–248.

SAS Institute, 1999, SAS Software, Version 8.2, SAS Institute, Cary, NC.

Sudduth, K.A., Kitchen, N.R., Hughes, D.F., Drummond, S.T., Electromagnetic induction sensing as an indicator or productivity on claypan soils, in *Proc. 2nd International Conference on Site-Specific Management for Agricultural Systems*, Robert, P.C., Rust, R.H., Larson, W.E., eds., ASA-CSSA-SSSA, Madison, WI, 1995, 671–681.

Tanji, K.K., ed., *Agricultural Salinity Assessment and Management*, ASCE, New York, NY, 1996.

17 Mapping of Soil Drainage Classes Using Topographical Data and Soil Electrical Conductivity

A. N. Kravchenko

CONTENTS

17.1   INTRODUCTION

Soil drainage is an important soil property affecting plant growth, water flow, and solute transport in soils. Accurate and inexpensive prediction and mapping of soil drainage classes for agricultural fields or small watersheds are of great importance for both agricultural and environmental management. However, accurate drainage mapping of a particular site is often not possible without taking soil cores—a task that is both time consuming and expensive. Therefore, it would be beneficial to determine other easily measured factors for quantitative prediction of soil drainage on small scales and to estimate potential accuracy of such prediction.

Topography is one of the easily evaluated drainage affecting factors that can greatly facilitate mapping and prediction of soil drainage (Bell et al., 1992, 1994; Troeh, 1964). Another easily measured factor potentially related to drainage is soil electrical conductivity (EC). It can be measured in the field using recently developed fast and nondestructive methods (Doolittle et al., 1994; Kitchen et al., 1999). Electrical conductivity depends on a number of soil physical properties, including soil salinity, soil water, and clay content (Rhoades et al., 1989). Sheets and Hendrickx (1995) found a linear relationship between the electrical conductivity measurements and soil profile water content. Williams and Hoey (1987) observed positive correlations between conductivity and soil clay content. Because both soil profile water contents and clay contents are related to the soil drainage properties, it would be reasonable to expect that electrical conductivity data might be helpful in mapping and predicting soil drainage classes on a field-scale basis.

The objectives of this study were (1) to examine the relationships between soil drainage and various topographical factors and to determine the factors that can be most helpful in predicting soil drainage classes on a field scale using the results from a farm field in Central Illinois, (2) to study the possibility of using soil electrical conductivity data and topographical information for predicting soil drainage classes, and (3) to compare accuracy and efficiency of discriminant analysis and geostatistical procedures in mapping soil drainage classes on an agricultural field scale.

17.2 MATERIALS AND METHODS

17.2.1 SOIL AND TOPOGRAPHICAL DATA

The studied area was a square 20 ha central portion of a 259 ha agricultural field located in Central Illinois. The field has been in a corn and soybean rotation for at least 20 years. Soil sampling was conducted on a semiregular grid, and sample locations are shown in Figure 17.1. A total of 107 soil cores, 100 cm deep, were collected from the studied area.

The drainage class for each sample location was determined based on the visual examination of the cores. Drainage classes were assigned according to the depth to the seasonal high water table, as indicated by the presence of low-chroma (<2) mottles (USDA, 1993). The soil was assigned to the well-drained class (WD) if no mottling was observed in the soil core within a 100 cm depth. Soil with mottling occurring at 50 to 100 cm depth was classified as moderately well drained (MWD), and soils with mottling occurring within 25 to 50 cm and within top 25 cm were classified as somewhat poorly drained (SWPD), and poorly drained (PD), respectively. Out of 107 soil cores, 48 were classified as WD soils, 38 as MWD soils, 15 as SWPD soils, and 6 as PD soils (Figure 17.1). Because the number of PD soil cores was insufficient for meaningful quantitative analysis, they were combined in one class with SWPD soils (SWPD/PD).

Soil electrical conductivity data were collected in fall 1999 using a Veris 3100 sensor cart that operates on the principle of electromagnetic induction (Division of Geoprobe® Systems, Salina, Kansas). Geo-referenced electrical conductivity measurements were taken every 3 to 5 m with the

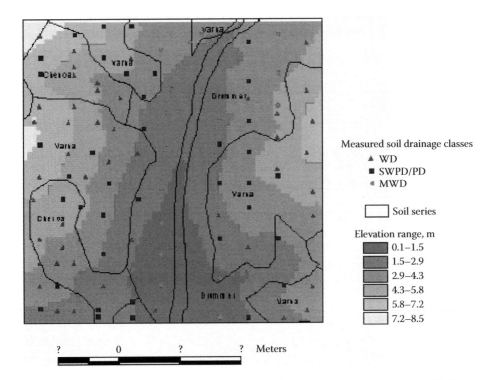

FIGURE 17.1 Locations of the core samples used to determine soil drainage classes along with the soil survey map and elevation map of the studied area. Each sampling site is labeled according to its drainage class.

distance between cart passes about 10 m, resulting in about 6900 EC measurement points for the studied area. Two sets of EC measurements were collected corresponding to depths of approximately 0 to 30 cm (shallow EC) and 0 to 90 cm (deep EC) (Sudduth et al., 1998).

Survey grid GPS (Leica 500 RTK) was used to measure elevation. About 1500 elevation measurements were taken on a semiregular grid with a mean distance between measurements of approximately 10 m. The map of field elevation ranges is shown in Figure 17.1. ArcView Spatial Analyst (Environmental Systems Research Institute, 1996) was used to analyze elevation data and to derive the topographical land features—namely, slope, curvature, and flow accumulation. ArcView was also used to calculate the shortest distance from each of the sampling sites to the drainageway that ran in the middle of the study area (Figure 17.1). This distance was used as one of the variables with potential effect on soil drainage class.

17.2.2 DISCRIMINANT ANALYSIS AND GEOSTATISTICAL PROCEDURES

In this study, discriminant analysis was used to find a decision rule for separating sites with different drainage classes based on the values of the topographical and EC variables measured at each site. The decision rule was first defined based on the measured data, and then the decision rule was applied to predict drainage classes at unsampled sites where measurements of topographical and electrical conductivity variables were available.

Two combinations of the variables were considered. The first combination included topographical variables, such as elevation, slope, curvature, flow accumulation, and distance to a drainageway. The purpose of this combination was to determine how accurate drainage class prediction can be if it is based on topographical data only. The second combination of the variables included all the topographical and electrical conductivity variables to study the possibility of using soil electrical conductivity data with topographical information for predicting soil drainage classes.

Stepwise discriminant procedure, STEPDISC (SAS Institute Inc., 1999) was applied to both combinations of variables to select the variables that had significant influence on the soil drainage class. Only variables significant at 0.15 significance level (Bell et al., 1992) were used in further discriminant analysis. Discriminant analysis was conducted using DISCRIM procedure (SAS Institute Inc., 1999).

Cross-validation was applied to evaluate accuracy of drainage data prediction and classification by discriminant analysis. For cross-validation, each value from the data set was eliminated in turn and, then, estimated using information from the rest of the data (Khattree and Naik, 2000). Posterior probabilities of the three drainage classes obtained for each data point were compared, and the site was assigned a drainage class with the highest posterior probability. Percent of correct drainage class estimates was used to compare different combinations of variables and their effectiveness in predicting drainage classes.

For geostatistical analysis, soil drainage class was treated as a categorical variable. At each data location this variable assumed only one of the three mutually exclusive possible states, corresponding to either WD, MWD, or SWPD/PD drainage classes. Indicator transformation was applied to each drainage class data resulting in three indicator variables. Kriging procedures were used to obtain indicator variable estimates at unsampled locations. The kriging estimate was equivalent to the probability of finding a certain drainage class at this location and assumed a value between 0 and 1. For each drainage class, kriging produced a map of probabilities of finding this drainage class at each particular location, and as in the discriminant analysis, the location was assigned to a drainage class with the highest probability.

Two geostatistical procedures compared in the study included ordinary indicator kriging and soft indicator cokriging (Goovaerts, 1997). Indicator kriging uses the drainage data only; cokriging allows for combining primary drainage data with any available secondary data related to drainage. One of the disadvantages of cokriging is that it becomes extremely cumbersome and time consuming with a large number of secondary variables. In this study, we conducted cokriging with one primary variable (indicator transformed drainage class) and two secondary variables. The variables found to have a significant effect on drainage class during the stepwise discrimination procedure were further used in cokriging estimation. The cross-validation results for kriging and cokriging procedures were further used to compare their accuracy with accuracy of discriminant analysis. Geostatistical analysis including variogram calculation, cross-validation, kriging, and cokriging was performed using the geostatistical software package GSLIB (Deutsch and Journel, 1998).

17.3   RESULTS AND DISCUSSION

Mean values of the variables that were selected as significant during stepwise discriminant procedure ($P = 0.15$) are shown in Table 17.1. When only topographical variables (i.e., elevation, slope, curvature, flow accumulation, and distance to drainageway) were used in the stepwise discriminant procedure, the variables with significant effect on drainage were slope and distance to drainageway. All the other topographical variables were not significant in discriminating between the drainage classes for this field. When all the above listed topographical variables and both shallow and deep EC measurements were used in the stepwise discrimination, only distance to drainageway and deep EC were selected as significant variables.

SWPD/PD soils were located close to the drainageway, at sites with the lowest slope values (Table 17.1). These sites were also characterized by the highest values of deep EC. One of the factors influencing the relationship between EC and soil drainage classes is soil clay content. Soil EC was found to be positively correlated with clay content (Williams and Hoey, 1987), and at the same time, higher clay content is one of the characteristic properties of poorly drained and somewhat poorly drained soils in this study. Analysis of soil texture revealed that SWPD/PD soils had somewhat higher clay contents than MWD and WD soils. The differences in clay contents were most notable at lower parts of the profile at depths 60 to 100 cm, where clay contents of WD, MWD, and SWPD/PD soils were equal to 26.3, 28.2, and 31.3 percent, respectively. Another factor of potential influence on the relationship between EC and soil drainage classes is soil water content. Similar to clay content, soil water content is positively correlated to soil EC, and in the field conditions, poorly drained soils often are characterized by higher water contents compared with MWD and WD soils. Unfortunately, soil water content was not monitored during the EC measurements; hence, it was not possible to separate influences of soil water content and clay content on EC data in this study. Distance to drainageway seemed to be a useful parameter in discriminating between WD and MWD drainage classes in this field. Sites with MWD soils were located generally closer to the drainageway than the sites with WD soils (Table 17.1). Slopes and deep EC values of WD and MWD soils were similar; hence, these parameters were not effective in separating WD and MWD drainage classes using stepwise discriminant procedure.

Cross-validation results for comparing the studied methods of predicting and mapping soil drainage classes are presented in Table 17.2. In discriminant analysis the WD and SWPD/PD soils were predicted most accurately, with about 70 percent of each being correctly estimated based on

---

TABLE 17.1

Mean Values of the Discriminant Variables Selected by Stepwise Discriminant Procedure ($P = 0.15$) for the Three Soil Drainage Classes

Soil Drainage Class	Distance to Drain (m)	Slope (degree)	Deep EC (mS/m)
WD	158	1.46	33.5
MWD	125	1.45	34.0
SWPD/PD	37	1.05	49.5

topography and topography plus electrical conductivity data. Separation of WD and MWD classes was less successful due to the above-mentioned insignificant differences between slope and deep EC values corresponding to WD and MWD drainage classes. Discriminant analysis with topography and EC slightly improved the estimation accuracy comparing with topography only due to better separation of WD and MWD drainage classes.

Cross-validation results of indicator kriging were not as accurate as those from discriminant analysis. Indicator kriging correctly estimated drainage classes at only 56 percent of the sites comparing with more than 60 percent of correct estimations from discriminant analysis. Better results were obtained when information on spatial variability of drainage class data was supplemented by the topographical data in the cokriging procedure. Indicator cokriging with slope and distance to drainageway produced the most accurate drainage class prediction comparing with the other methods (Table 17.2).

The most substantial difference between discriminant analysis and kriging and cokriging for creating maps pertains to the fact that kriging and cokriging procedures are exact estimators, and

TABLE 17.2

Number of Correctly and Incorrectly Classified WD, MWD, and SWPD/PD Drainage Classes and the Total Percent of Correctly Identified Classes from Discriminant Analysis, Indicator Kriging, and Indicator Cokriging; Values in Parentheses Are Percentages of Correctly Estimated Sites for Each Drainage Class

True Drain Class	Classified as			Total % Correct
	WD	**MWD**	**SWPD/PD**	
Discriminant analysis with topographical variables (significant from stepwise procedure are slope and distance to drain)				
WD	33 (69%)	14	1	
MWD	19	17 (45%)	2	61.7
SWPD/PD	1	4	16 (76%)	
Discriminant analysis with topographical and EC variables (significant from stepwise procedure are EC deep and distance to drain)				
WD	36 (75%)	10	2	
MWD	18	17 (45%)	3	63.6
SWPD/PD	1	5	15 (71%)	
Indicator kriging				
WD	31 (65%)	16	1	
MWD	18	15 (39%)	5	56.1
SWPD/PD	3	4	14 (67%)	
Indicator cokriging with slope and distance to drain				
WD	47 (98%)	1	0	
MWD	26	10 (26%)	2	65.4
SWPD/PD	7	1	13 (62%)	
Indicator cokriging with EC deep and distance to drain				
WD	36 (75%)	9	3	
MWD	21	8 (21%)	9	57.9
SWPD/PD	3	0	18 (86%)	

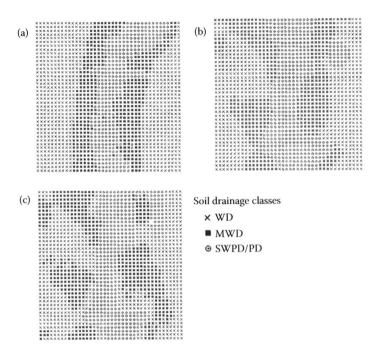

FIGURE 17.2 Drainage maps obtained from (a) discriminant analysis with deep electrical conductivity (EC) and distance to drainageway, (b) indicator kriging based on the measured drainage classes , and (c) indicator cokriging based on the measured drainage classes with slope and distance to drainageway as secondary variable.

discriminant analysis is not. That is, the maps created based on the discriminant analysis ignore the actual data in sampled locations. Hence, indicator kriging and cokriging are preferable compared with discriminant analysis for mapping purposes. Drainage maps obtained from (1) discriminant analysis with deep EC and distance to drainageway, (2) indicator kriging, and (3) indicator cokriging with slope and distance to drainageway are shown in Figure 17.2a through Figure 17.2c, respectively.

17.4 CONCLUSIONS

Discriminant analysis identified terrain slope, deep (0 to 90 cm) EC measurements, and distance to drainageway as the most useful secondary variables for predicting soil drainage classes. The combination of discriminant analysis and geostatistics seems to be the most advantageous approach to drainage class mapping. In this approach, the data are first analyzed with discriminant analysis, and a limited number of variables with significant effect on the drainage classes are selected among all the available topographical and soil variables. Then, the significant variables are used in the cokriging for the most accurate mapping of soil drainage classes.

REFERENCES

Bell, J.C., Cunningham, R.L., and Havens, M.W., Calibration and validation of a soil-landscape model for predicting soil drainage class, *Soil Sci. Soc. Am. J.*, 56, 1860, 1992.

Bell, J.C., Cunningham, R.L., and Havens, M.W., Soil drainage probability mapping using a soil-landscape model, *Soil Sci. Soc. Am. J.*, 58, 464, 1994.

Deutsch, C.V. and Journel, A.G., *Geostatistical Software Library and User's Guide*, Oxford University Press, New York, 1998.

Doolittle, J.A., Sudduth, K.A., Kitchen, N.R., and Indorant, S.J., Estimating depths to claypans using electromagnetic induction methods, *J. Soil Water Conserv.*, 49, 572, 1994.

Environmental Systems Research Institute, Inc., *ArcView Spatial Analyst*, ESRI, Redlands, CA, 1996.

Goovaerts, P., *Geostatistics for Natural Resources Evaluation*, Oxford University Press, New York, 1997.

Khattree, R. and Naik, D.N., *Multivariate data reduction and discrimination with SAS software*, Cary, NC: SAS Institute Inc., 2000.

Kitchen, N.R., Sudduth, K.A., and Drummond, S.T., Soil electrical conductivity as a crop productivity measure for claypan soils, *J. Prod. Agric.*, 12, 607, 1999.

Rhoades, J.D., Manteghi, N.A., Shouse, P.J., and Alves, W.J., Soil electrical conductivity and soil salinity: New formulations and calibrations, *Soil Sci. Soc. Am. J.*, 53, 433, 1989.

Sheets, K.R. and Hendrickx, J.M.H., Noninvasive soil water content measurement using electromagnetic induction, *Water Resours. Res.*, 31, 2401, 1995.

Sudduth, K.A., Kitchen, N.R., and Drummond, S.T., Soil conductivity sensing on claypan soils: Comparison of electromagnetic induction and direct methods, in *Proc. 4th Int. Conf. on Precision Ag.*, 979, 1998.

Troeh, F.R., Landform parameters correlated to soil drainage, *Soil Sci. Soc. Am. J.*, 28, 808, 1964.

USDA. *Soil Survey Manual*, Soil Survey Division Staff, Soil Conservation Service, U.S. Department of Agriculture Handbook 18, 1993.

Williams, B.G. and Hoey, D., The use of electromagnetic induction to detect the spatial variability of the salt and clay contents of soils, *Aust. J. Soil Res.*, 25, 21, 1987.

18 Productivity Zones Based on Bulk Soil Electrical Conductivity
Applications for Dryland Agriculture and Research

Cinthia K. Johnson, Rhae A. Drijber, Brian J. Wienhold, and John W. Doran

CONTENTS

18.1   INTRODUCTION

Soil is inherently variable in the physical, chemical, and biological properties that determine yield potential, a fact that complicates the identification and implementation of sustainable management practices. Historically, a lack of means to delineate and address within-field heterogeneity has forced farmers to operate at the large scale, using the field as a primary management unit. Soils are sampled and inputs applied to target average requirements across fields, resulting in underperforming yields in some areas and wasted inputs in others. Increasingly, farmers seek to manage land at a smaller level of resolution (site-specific management) to improve economic and ecological outcomes.

Paradoxically, soil heterogeneity has the opposite impact on agronomic research. Scientists have traditionally relied upon small or plot-scale experiments as a means to control spatial variability and reduce experimental error. Yet, the importance of research conducted at a level of scale so divergent from that of production agriculture is sometimes questioned. Conclusions based upon data extrapolated beyond the spatial context of sampling assume the scale independence of patterns and processes (Wiens, 1989), an erroneous assumption (Cambardella et al., 1994). For this reason, growing numbers of scientists and farmers promote large-resolution (field-scale) experiments to address

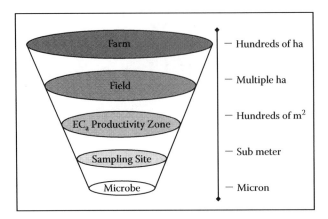

FIGURE 18.1   The EC_a-based productivity zone functions as a pivotal point through which measurements spanning the hierarchy of experimental scale can be related.

questions of operational scale and advance research relevance (Drinkwater, 2002; Rzewnicki, 1991). Field-scale research permits study of intact agroecosystems wherein the interrelationships between soil edaphic properties, biological processes, and management can be studied for yield, farm economic, pest, and on- and off-site environmental impacts. However, two critical issues must be resolved if field-scale investigations are to gain widespread acceptance. Innovative methods are needed to obtain reasonable estimates of experimental error in lieu of replication to and relate data collected at small (microbial), medium (soil physical/chemical), and large scales (crop yields and farm economics).

Cost-effective technologies capable of producing maps that define within-field areas of similar production potential, termed "productivity zones" (Kitchen et al., 2005), may benefit both production agriculture and agronomic research. Productivity zones could function as intermediate units of reference to permit shifts in operational scale (Figure 18.1)—that is, smaller for farmers and larger for researchers—to better address sustainability issues. Apparent soil electrical conductivity (EC_a) may be one option for mapping productivity zones.

In individual fields, the magnitude and spatial patterns of mapped EC_a are largely driven by one or two soil characteristics including salinity, clay type/percentage, bulk density, water content, or temperature (Rhoades et al., 1989). Crop yields are not always correlated with EC_a because correlation requires that the same soil factors control both EC_a and yield. For this reason, the EC_a-yield relationship varies among reports (Corwin et al., 2003a; Jaynes et al., 1993; Johnson et al., 2001; Kitchen et al., 1999).

Research in a dryland system in semiarid northeastern Colorado was designed to evaluate the effectiveness of EC_a-based productivity zones as a framework for addressing key sustainability issues in production agriculture and research. Specifically, EC_a zones were considered as a basis for (1) zone soil sampling and monitoring the impact of management on soil ecological trends; (2) site-specific application of inputs (i.e., fertilizer, herbicide, and seed); (3) linking small- (microbial-scale), medium- (sampling site), and large-scale (field and farm) economic and ecological measurements; and (4) designing and statistically evaluating field-scale experiments to circumvent traditional replication.

18.2   MATERIALS AND METHODS

All experiments were conducted at the farmer-owned and -managed Farm-Scale Intensive Cropping Study (FICS) near Sterling, Colorado (Johnson et al., 2001) (Figure 18.2). This dryland experiment encompasses a contiguous section of farmland ($\approx$250 ha) managed for nearly seventy years in a traditional winter wheat (*Triticum aestivum* L.)—fallow rotation using conventional tillage.

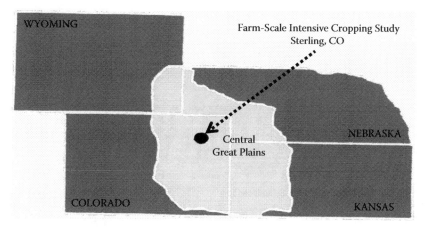

FIGURE 18.2 Site map for the Farm-Scale Intensive Cropping Study in northeast Colorado, located in the heart of the Central Great Plains.

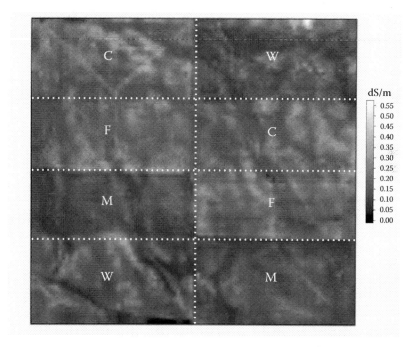

FIGURE 18.3 Farm-Scale Intensive Cropping Study: experimental layout and electrical conductivity map. W = winter wheat, C = corn, M = millet, and F = fallow.

In 1999, it was converted to an intensified no-till winter wheat to corn (*Zea Mays* L.) to proso millet (*Panicum miliaceum* L.) to fallow rotation. Crop treatments were assigned to each of the eight ≈31 ha fields within the site (Figure 18.3) so that each phase of the 4-year rotation was present in two replicates each year.

In 1999, the experimental site was EC_a mapped (0 to 30 and 0 to 90 cm depths) by direct contact, using a Veris 3100 Sensor Cart (Veris Technologies, a division of Geoprobe Systems, Salina, KS) (Figure 18.3), and separated into four zones (ranges) of EC_a: low, medium low, medium high, and high (Figure 18.4 and Table 18.1). This was done by individually interpolating EC_a field maps (0 to 30 cm) by inverse-distance weighting and spatially clustering the interpolated data into twelve classes

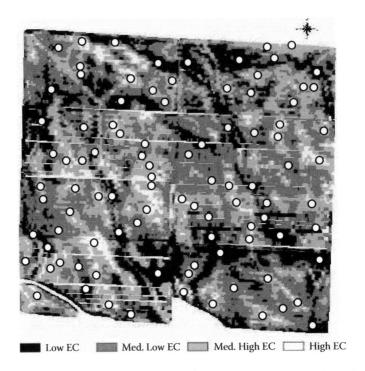

FIGURE 18.4   Farm-Scale Intensive Cropping Study EC_a zone map and soil sampling scheme.

TABLE 18.1

Case Study VI: Partial Listing of Soil Properties (0 to 30 cm) Including within-Electrical Conductivity (EC_a) Zone Means and Significance across Crop Treatments, and Correlation (*r*) with Measured EC_a

	EC_a Ranges dS m^{-1}	Productivity-Associated Factors						Erosion-Associated Factors		
		Water Content kg kg^{-1}	SOM ¶ Mg ha^{-1}	Total C Mg ha^{-1}	Total N Mg ha^{-1}	P ¶ kg ha^{-1}	PMN ¶ kg ha^{-1}	Bulk Density g cm^{-3}	Clay %	pH
EC_a zone		*	**	**	**	**	*	†	*	**
Low	0.00–0.17	0.207	124.8	43.8	4.08	111.8	86.4	1.32	22.8	6.33
Medium low	0.12–0.23	0.187	115.9	35.2	3.45	69.2	67.0	1.39	24.3	6.42
Medium high	0.14–0.29	0.185	110.4	32.2	3.09	27.8	59.3	1.39	27.3	6.72
High	0.18–0.78	0.178	112.6	32.7	3.10	26.7	54.4	1.42	28.1	6.92
r values ($p < 0.001$)		−0.33	−0.34	−0.36	−0.38	−0.58	−0.50	0.49	0.50	0.37

¶ SOM = total soil organic matter; P = extractable P; PMN = potentially-mineralizable NH4-N.

†, *, ** Comparisons of ECa class treatments are significant at the 0.10, 0.05, and 0.01 levels, respectively.

of EC_a using unsupervised classification (ERDAS, 1997). These twelve classes were then recoded (combined) to form four productivity zones. Twelve geo-referenced sampling sites were selected in each of the eight fields, three per EC_a zone, for a total of ninety-six sites across the section.

Data collected spanned both spatial and biological scales (Figure 18.1). Soil samples were taken (0 to 7.5 and 0 to 30 cm depths) using the EC_a-based sampling scheme, and a portion of each sample was separated into 1 to 2 mm aggregates (significant to soil structure). Microbial-scale measures of the presence and activity of vesicular-arbuscular mycorrhizal (VAM) fungi were made on this fraction. These organisms were selected for study because of their contributions to both crop productivity and soil conservation (aggregation). Measurements were made using $C16:1(cis)11$ fatty acid methyl ester biomarker, immunoreactive total glomalin immunoassay, and wet aggregate stability methods. Sampling-site scale analyses of nineteen whole-soil physical and chemical characteristics and surface residue were also performed, and field-scale Global Positioning System (GPS)-referenced yield maps were collected for winter wheat and corn in 1999 and 2000.

Using geographic information system (GIS) technology, microbial, soil, crop yield, and EC_a data layers were superimposed to create a "grid stack" wherein data at identical geo-referenced positions within the various layers could be exported in spreadsheet format for comparison and statistical evaluation. Microbial, soil, and crop yield data were appraised for significant relationships to EC_a and EC_a zones using regression and analysis of variance (ANOVA) for a randomized complete block strip-split plot design with crop and EC_a zones as treatment factors. All statistical analyses were conducted using SAS (SAS Inst., Cary, NC).

Geo-aligned data layers were analyzed in phases to address the various sustainability issues outlined in the research objectives. Each phase built upon the findings of the previous phases to expand linkages between EC_a and EC_a productivity zones, factors underlying soil production potential (edaphic, crop residue, and biological components), and crop yields. Experimental phases are individually addressed in the five sections below; associated publications are cited in the section headings for those interested in more detailed information.

18.3 RESULTS AND DISCUSSION

18.3.1 PHASE I: THE EC_A MAP (JOHNSON ET AL., 2001)

Although soil heterogeneity arises from interactions among the soil-forming factors (Jenny, 1941), it is confounded by management such as tillage and fertilization (Bouma and Finke, 1993). At the FICS, it is interesting to note that mapped EC_a reflects both historical and recent management. The V-shaped patterns in the four corner fields of the section are consistent with the plow path followed in the 1930s when the site was farmed as two half sections (Figure 18.3 and Figure 18.4). In addition, varying mean levels of conductivity, for the eight fields in the study, distinguish among different crops grown the year before EC_a mapping.

18.3.2 PHASE II: EC_A VERSUS SOIL EDAPHIC PROPERTIES (JOHNSON ET AL., 2001)

Sampling-site scale analyses of surface residue and nineteen soil physical and chemical parameters (0 to 7.5 and 0 to 30 cm) were compared with EC_a maps (0 to 30 cm) to "ground-truth" EC_a and EC_a zones. Except for NO_3- and NH_4-N, which exhibited a narrow range of variability across the site, all residue and soil parameters were significantly different among EC_a productivity zones at one or both depths of analysis (0 to 7.5 and 0 to 30 cm). Surface residue mass and soil properties related to yield potential were negatively correlated with EC_a, and properties associated with soil erosion were positively correlated (Table 18.1). These relationships are a function of the calcareous soils at the FICS, where clay content and $CaCO_3$ salts dominate measured EC_a (0 to 30 cm). The loss of topsoil in less-productive eroded areas of each field has exposed underlying clay and $CaCO_3$ horizons, which increase EC_a. These horizons are also characterized by associated elevations in bulk density and pH, soil properties positively correlated with EC_a and negatively correlated with productivity.

At the FICS, strong correlations were not found between EC_a and individual soil properties at point sources. This is because EC_a integrates multiple soil properties, wherein changes in one may be buffered by corresponding changes in another. In this semiarid environment, EC_a is most useful for delimiting overall soil productivity and for defining distinct zones of within-field yield potential. Therefore, soil and residue sampling based upon EC_a productivity zones appears to be a useful basis for (1) zone soil sampling, (2) tracking the temporal impact of farm management on soil productivity, and (3) assessing soil parameters to calculate fertilizer and herbicide inputs in site-specific management.

18.3.3 PHASE III: EC_A VERSUS CROP YIELDS (JOHNSON ET AL., 2003B)

Field-scale measures of crop yield were evaluated for significant relationships to both EC_a and the sampling-site scale soil properties integrated by EC_a (Phase II) using two years of geo-referenced yield maps for wheat and corn. Winter wheat yields were negatively correlated with EC_a (0 to 30 cm) and positively correlated with soil properties, indicative of production potential. Correlations between EC_a and yield have been corroborated by investigators in other regions (Corwin et al., 2003b; Kitchen et al., 2003). A wheat-yield response curve for 1999, a high-yielding year, revealed a boundary line of maximum yield that decreased with increasing EC_a (Figure 18.5), an effective basis for identifying yield goals.

Although crop yield maps provide the most realistic picture of yield heterogeneity, they integrate all factors driving crop yields, including weather variations. Productivity zones, based on EC_a, distinguish only the soil-based factors underlying yield heterogeneity that can be managed. These zones offer a basis for three key aspects of site-specific management: (1) yield goal determination, (2) soil sampling to assess residual nutrients and soil attributes affecting herbicide efficacy, and (3) prescription maps for metering fertilizer, pesticide, and seed inputs.

No consistent associations were found between EC_a (0 to 30 cm) and corn yields probably due to high drought stress in corn during the 2-year study. However, both wheat and corn yields were positively correlated with EC_a (0 to 90 cm), a reversal of the negative relationship between EC_a (0 to 30 cm) and wheat yield. This is likely due to differences in the impact of soil clay content on

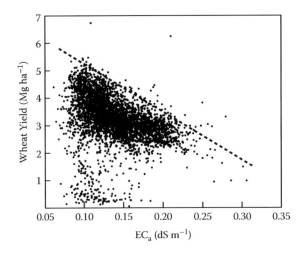

FIGURE 18.5 Scatter plot of 1999 winter-wheat yield as a function of apparent soil electrical conductivity (EC_a) measured at approximately 0 to 30 cm soil depth. The red line is a "boundary line" of maximum potential yield defined as yield points falling at the ninetieth percentile of yield frequency for each 0.01 increment of EC_a.

measured EC_a with depth. Although clay content controls EC_a (0 to 30 cm), soil water and salts (NO_3 and NH_4) may drive EC_a (0 to 90 cm) and yield. Hence, the positive correlation.

18.3.4   Phase IV: EC_A versus Microbial-Scale Measurements (Johnson et al., 2004)

Soil edaphic and biological (community composition, diversity, and activity) variability occurs across multiple levels of scale. It is useful to evaluate this variability, as it relates to sustainable management within the context of a farm. This can be conceptualized as originating with the microbe (micron scale) and continuing upward to sampling site, within-field, field, and farm levels (multiple-ha scale) (Figure 18.1). *Microbial-scale* analyses of VAM fungi were evaluated for significant correlations with EC_a, and with *sampling-site scale* soil properties and *field-scale* yield data integrated by EC_a (Phases II and III). Concentrations of glomalin were negatively correlated with EC_a and positively correlated with soil quality and winter wheat yields, across crop treatments. The C16 biomarker and wet aggregate stability were different among crop treatments as fallow < wheat < corn < millet and were negatively correlated with EC_a in the fallow treatments (effect of crop removed). Due to significant partitioning of these *microbial-scale* measures among EC_a productivity zones, they can be linked to both soil chemical and physical characteristics and crop yields. Thus, EC_a zones offer a pivotal point of reference through which *microbial-*, *sampling-site-*, and *field-scale* data can be related.

18.3.5   Phase V: EC_A for Experimental Design and Analysis (Johnson et al., 2003a, 2005)

In classic experimentation, small plots are arranged in a randomized complete block design where blocks serve to increase precision by reducing experimental error due to soil heterogeneity. Blocks are placed in areas of similar production potential that have been identified by analyzing soil samples for yield-significant properties. Topography, soil fertility, and soil series exemplify traditional blocking factors, and EC_a productivity zones were examined for this purpose at the FICS. Figure 18.6 illustrates the relationship between EC_a productivity zones and plot-scale blocking. The 32 ha field shown on the left is separated into four classes of EC_a (a), three of which form blocks in the traditional plot-scale experiment set in a randomized complete block design (b). Because blocks are homogeneous, plots need not be adjacent but could be placed anywhere in field (a) within assigned blocks. Thus, the entire 32 ha field can be conceptualized as an enlarged version of the plot-scale experiment, where variance across EC_a-delineated productivity zones is equivalent to experimental error in the plot-scale experiment.

To test this, soil and residue data from the FICS were compared with those taken from a nearby plot-scale experiment (Peterson et al., 1993). For each soil and residue measurement, within-field

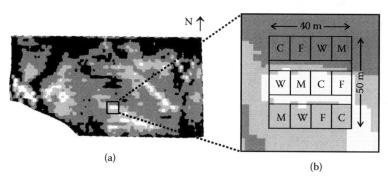

(a)   (b)

FIGURE 18.6 Relationship between apparent soil electrical conductivity (EC_a) productivity zones and plot-scale blocking. (a) An EC_a-zone map of a $\approx$31 ha field at the Farm-Scale Intensive Cropping Study and (b) a typical plot-scale experiment identified within the field using EC_a productivity zones as a basis for blocking.

mean square (MS) error at the FICS was calculated and compared with MS error derived from blocking in the plot-scale experiment. Experimental errors were similar, indicating that field-scale EC_a-classified variability effectively estimated soil heterogeneity partitioned by plot-scale blocking. These findings were corroborated at a second and disparate site in central California, an irrigated system with saline soils (Johnson et al., 2005).

Comparisons were also made between EC_a-delineated estimates of experimental error and experimental error derived from replication using multiple soil properties measured at the FICS. Again, within-field variance was an effective measure of experimental error for most of the nineteen parameters evaluated. These findings indicate that, for some experimental objectives, within-field variance may serve as a surrogate for replication.

In locations where soil factors contributing to EC_a are also yield limiting, EC_a productivity zones can be used to design and place plot-scale experiments. They can also be used to design and evaluate field-scale experiments, functioning as an alternative to replication and blocking. This is appropriate because EC_a productivity zones are related to outcome (crop yield) differences expected in the absence of treatments, the rationale for blocking.

18.4   CONCLUSIONS

Although soil heterogeneity complicates the identification and implementation of sustainable management practices, the ability to delineate this variability offers unique opportunities for both production agriculture and research. In this dryland experiment, EC_a effectively mapped soil heterogeneity, depicting a gradient of within-field productivity useful for delineating zones of similar production potential. Practical applications for EC_a-based productivity zones include zone sampling and site-specific nutrient management. Zone sampling is superior to traditional random sampling because it reduces standard error and decreases the number of samples required to evaluate a field. These advantages are particularly significant in semiarid regions where intensive grid sampling is cost prohibitive in the predominately large, low-input, dryland farms (McCann et al., 1996).

In high-rainfall regions where both drought and excessive precipitation are yield limiting, inconsistent relationships have been found between EC_a and yield across years (Jaynes et al., 1993; Kitchen et al., 1999). Yet, in semiarid and arid regions yield reductions from excessive precipitation are rare, which may make EC_a a more reliable predictor of yield. Consistency of spatial patterns (Sudduth et al., 2000; Veris Technologies, 2001) and yield relationships (Johnson et al., 2003b; Corwin et al., 2003b) across years reinforce EC_a productivity zones as a useful basis for site-specific management in these regions.

Continued field-scale research is required for EC_a productivity-zone-based site-specific nutrient management to determine optimal fertilizer rates for each zone, evaluate soil and crop ecological response, and compare economic return to that of uniform fertilizer management. It may be possible to "fine-tune" productivity zones by using different classification methods or depths of EC_a measurement, or by integrating yield, topographical, or soil color maps. Other factors affecting farm economics and crop production that are potentially delineated by EC_a productivity zones should also be investigated, including grain quality and pest populations (weed, insect, and disease agents).

Beyond practical utility for production agriculture, EC_a productivity zones have important applications in agronomic research. As farmers increasingly move toward management of the soil resource in space and time, new options are needed for experimental design and evaluation if agronomic research is to address relevant issues at relevant levels of scale. Research at the FICS indicates that EC_a productivity zones offer such an option.

First, EC_a-productivity zones serve as a pivotal point of reference, a means to integrate and compare data collected at different biological, spatial, and temporal scales. The framework provided by these zones can be used to monitor management-induced trends in soil quality at the field scale (Corwin et al., 2005). It allows the linkage of microbial-scale findings to farm-scale economic and ecological outcomes in intact agroecosystems. This may advance farm management as a tool for manipulating microbial populations (biological control) to improve system productivity and sustainability.

Second, the relationship between plot-scale blocking and experimental error derived from EC_a-delineated productivity zones supports an alternative experimental design. In nonreplicated field-scale experiments where EC_a and yield are correlated, EC_a-delineated within-field error can be used to estimate plot-scale experimental error. By roughly evaluating treatments in this manner, research questions can be identified that require further study in traditional plot-scale experiments. Beginning investigations at the field scale, and then moving to plot or laboratory studies, represents a reverse in research direction that may improve research relevance and foster farmer involvement and adoption of sustainable management practices.

REFERENCES

Bouma, J., Finke, P.A. 1993. Origin and nature of soil resource variability. p. 3–4. In P.C. Robert, R.H. Rust, W.E. Larson (ed.) *Soil Specific Crop Management*, ASA-CSSA-SSSA, Madison, WI.

Cambardella, C.A., Moorman, T.B., Novak, J.M., Parkin, T.B., Karlen, D.L., Turco, R.F., Konopka, A.E. 1994. Field-scale variability of soil properties in central Iowa Soils. *Soil Sci. Soc. Am. J.* 58, 1502–1511.

Corwin, D.L., Kaffka, S.R., Hopmans, J.W., Mori, Y., van Groenigen, J.W., van Kessel, C., Lesch, S.M., Oster, J.D. 2003a. Assessment and field-scale mapping of soil quality properties of a saline-sodic soil. *Geoderma*. 1052, 1–29.

Corwin, D.L., Lesch, S.M., Shouse, P.J., Soppe, R., Ayars J.E. 2003b. Identifying soil properties that influence cotton yield using soil sampling directed by apparent soil electrical conductivity. *Agron. J.* 95, 352–364.

Corwin, D.L. Lesch, S.M., Oster, J.D., Kaffka, S.R. 2005. Monitoring spatio-temporal changes in soil quality through soil sampling directed by apparent management-induced electrical conductivity. *Geoderma*. 131, 369–387.

Drinkwater, L.E. 2002. Cropping systems research: Reconsidering agricultural experimental approaches. *Horttechnology*. 12, 355–361.

ERDAS. 1997. ERDAS Field Guide. pp. 225–232. ERDAS, Inc., Atlanta, GA.

Jaynes, D.B., Colvin, T.S., Ambuel, J. 1993. Soil type and crop yield determinations from ground conductivity surveys. Paper no. 933552. Am. Soc. Agric. Eng. St. Joseph, MI: ASAE, Winter 1993.

Jenny, H. 1941. Factors of soil formation—A system of quantitative pedogenesis—A review. McGraw-Hill, New York.

Johnson, C.K., Doran, J.W., Duke, H.R., Wienhold, B.J., Eskridge, K.M., Shanahan, J.F. 2001. Field-scale electrical conductivity mapping for delineating soil condition. *Soil Sci. Soc. Amer. J.* 65, 1829–1837.

Johnson, C.K., Eskridge, K.M., Wienhold, B.J., Doran, J.W., Peterson, G.A., Buchleiter, G.W. 2003a. Using electrical conductivity classification and within-field variability to design field-scale research. *Agron. J.* 95, 602–613.

Johnson, C.K., Mortensen, D.A., Wienhold, B.J., Shanahan, J.F., Doran, J.W. 2003b. Site-specific management zones based on soil electrical conductivity in a semiarid cropping system. *Agron. J.* 95, 303–315.

Johnson, C.K., Wienhold, B.J., Doran, J.W, Drijber, R.A., Wright, S.F. 2004. Linking microbial-scale findings to farm-scale outcomes. *Precision Agric.* 5, 311–328.

Johnson, C.K., Eskridge, K.M., Corwin, D.L. 2005. Apparent soil electrical conductivity: Applications for designing and evaluating field-scale experiments. *Comput. Electron. Agric.* 46, 181–202.

Kitchen, N.R., Sudduth, K.A., Drummond, S.T. 1999. Soil electrical conductivity as a crop productivity measure for claypan soils. *J. Prod. Agric.* 12, 607–617.

Kitchen, N.R., Drummond, S.T., Lund, E.D., Sudduth, K.A., Buchleiter, G.W. 2003. Soil electrical conductivity and topography related to yield for three contrasting soil-crop systems. *Agron. J.* 95, 483–495.

Kitchen, N.R., Sudduth, K.A., Myers, D.B., Drummond, S.T., Hong, S.Y. 2005. Delineation of productivity zones on claypan soil fields using bulk soil electrical conductivity. *Comput. Electron. Agric.* 46, 385–308.

McCann, B.L., Pennock, D.J., van Kessel, C., Walley, F.L. 1996. The development of management units for site-specific farming. pp. 296–302. In *Proc. 3rd Int. Conf. on Precision Agriculture*, St. Paul, MN, 23–26 June 1996, ASA, CSSA, SSSA, Madison, WI.

Peterson, G.A., Westfall, D.G., Cole, C.V. 1993. Agroecosystem approach to soil and crop management research. *Soil Sci. Soc. Amer. J.* 57, 1354–1360.

Rhoades, J.D., Manteghi, N.A., Shouse, P.J., Alves, W.J. 1989. Soil electrical conductivity and soil salinity: New formulations and calibrations. *Soil Sci. Soc. Am. J.* 53, 433–439.

Rzewnicki, P. 1991. Farmer's perceptions of experiment station research, demonstrations, and on-farm research in agronomy. *J. Agron. Educ.* 20, 31–36.

SAS Institute. 1997. SAS/STAT Software: Changes and enhancements through Release 6.12. SAS Inst., Cary, NC.

Sudduth, K.A., Drummond, S.T., Kitchen, N.R. 2000. Measuring and interpreting soil electrical conductivity for precision agriculture. pp. 1444–1451. In *Proc. 2nd Int. Conf. Geospatial Information in Agriculture and Forestry.* Lake Buena Vista, FL, 10–12 June.

Veris Technologies. 2004. [Online]. [2 p.] Available at www.veristech.com [cited 31 July 2004; verified 5 February 2005]. Veris Technologies, Salina, KS.

Wiens, J.A. 1989. Spatial scaling in ecology. *Func. Ecol.* 3, 385–397.

19 Four-Year Summary of the Use of Soil Conductivity as a Measure of Soil and Crop Status

Roger A. Eigenberg, John A. Nienaber, Bryan L. Woodbury, and Richard B. Ferguson

CONTENTS

19.1   INTRODUCTION

Sustainable agriculture requires innovative and practical tools to optimize farm economics, conserve soil organic matter, and minimize negative environmental impacts (Johnson et al., 2003). Electromagnetic induction soil conductivity sensors may provide one such tool. Electromagnetic techniques are well suited for mapping soil conductivity to depths useful for agriculturalists (McNeill, 1990). Electrical conductivity (EC) methods have been shown to be sensitive to high nutrient levels (Eigenberg et al., 1996, 2000) and have been used to detect ionic concentrations on or near the soil surface resulting from field application of cattle feedlot manure. EC has generally been associated with determining soil salinity; however, EC also can serve as a measure of soluble nutrients (Smith and Doran, 1996) for both cations and anions, and is useful in monitoring the mineralization of organic matter in soil (De Neve et al., 2000). Doran et al. (1996) demonstrated the predictive capability of soil conductivity to estimate soil nitrate.

The objective of this work was to determine the utility of electromagnetic induction (EMI) methods in evaluating the agronomic effectiveness and environmental consequences of N fertilization for varying rates of compost, manure, and commercial fertilizer with the use of cover crops. Additionally, sequential EMI surveys were examined as a tool in monitoring N cycle dynamics in a corn silage research field.

19.2   METHODS

19.2.1   Site Description

A center-pivot irrigated field (244 m × 244 m) of silage corn (*Zea mays* L.), located at the U.S. Meat Animal Research Center (USMARC), served as a comparison site for various manure and compost application rates for replacement of commercial fertilizer, with the same treatment assigned to field plots for 10 consecutive years. The soil series at this site is a Crete silt loam (fine, montmorillonitic, mesic Pachic Argiustolls), 0 to 1 percent slope.

19.2.2   Field Operations on the Research Cornfield

The study site was laid out as a split-plot design (Figure 19.1), with four replications of the main plot of cover crop (+CC) versus no cover (−CC) (the cover crop was a winter wheat [*Secale cereale* L.] no-till drilled following silage harvest). Subplot application to treatment strips (6.1 m, eight corn rows wide) was made with two manure sources: beef feedlot manure and composted beef feedlot manure. Applications were made each spring according to two strategies: (1) to approximately supply the total crop demand for N (Ferguson et al., 2003), denoted MN and CN for manure and compost, respectively, and (2) to supply the approximate crop demand of P (Ferguson et al., 2003), denoted MP and CP for manure and compost, respectively. Treatments MP and CP each had sufficient carryover phosphorus after 1998 so that no manure or compost was applied to these treatment strips for the 2000 to 2003 study.

19.2.3   Equipment

Two different magnetic dipole soil conductivity meters were used in this study: (1) an EM-38, manufactured by Geonics Ltd., Mississauga, Ontario, Canada (2000 and 2001 seasons) and (2) a Dualem-2 manufactured by Dualem Inc., Milton, Ontario, Canada (2002 and 2003 seasons). The EM-38 was operated horizontally and had a response that varied with depth in the soil, yielding EC_a that was centered at a depth of about 0.75 m. The Dualem-2 operates in the horizontal and vertical dipole modes simultaneously, with EC_a centered at 1.5 and 3.0 m, respectively. Only the horizontal response of each instrument is reported in this study and is designated EC_a. The EMI instruments were mounted on a plastic sled and transported through the field, pulled by an all-terrain vehicle (ATV), or by hand when the corn became too tall for the ATV. Surveys were planned on weekly intervals throughout the corn growing season for all 4 years. A Trimble PRO-XR GPS (global positioning satellite) unit was used to obtain positional data.

Soil conductivity responds to soil temperature (McKenzie et al., 1989). An array of thermocouple temperature probes was buried at the cornfield site prior to the 2002 growing season. Probes were installed at incremental depths to 315 cm. The thermal profile was used with known Dualem-2 response (Dualem, www.dualem.com) curves to provide an effective soil temperature correction for each survey date (McKenzie et al., 1989). The soil temperatures demonstrated predictable temperature patterns that allowed temperature corrections to be estimated for the Geonics instrument for the 2000 and 2001 season. Precipitation events were recorded at the cornfield site using a tipping bucket rain gauge and a portable event recorder.

Soil cores were taken on survey dates throughout the growing season with a hand probe to a depth of 0 to 30 cm. The cores were analyzed by a local commercial soil testing laboratory to

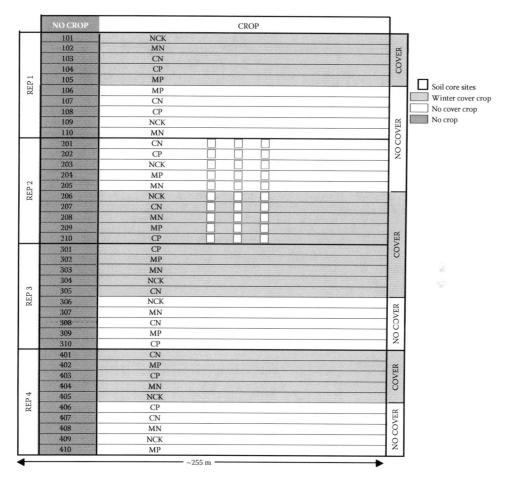

FIGURE 19.1   Split-plot design with four replicates of main plot (cover and no-cover) and subplots (manure and compost applied at either the N or P needs of the silage corn crop, as well as commercial fertilizer). Soil cores were taken in replicate 2. Treatments are designated: manure at N rate (MN), compost at N rate (CN), manure at P rate (MP), compost at P rate (CP), and a commercial fertilizer check (NCK); cover (+CC) and no cover (−CC). (Reproduced with permission from the Soil Science Society of America.)

determine NO_3-N, soil pH, electrical conductivity on 1:1 soil to water extracts, and soil moisture content. Bulk density measurements and soil moisture were used to determine soil water-filled pore space (WFPS).

19.3   RESULTS AND DISCUSSION

19.3.1 EC_A MAPS AND PLOTS

The number of EC_a survey maps generated during the growing seasons of 2000, 2001, 2002, and 2003 were 29, 26, 25, and 34, respectively. A representative EC_a image of the cornfield is shown in Figure 19.2 as generated on June 17, 2002 (DOY 168). When the series of images for a growing season are viewed in sequence, the maps illustrate field dynamics with overall EC_a values rising uniformly with time (images not shown). The image dynamics are more evident in the mean values extracted from the survey data and illustrated in Figure 19.3.

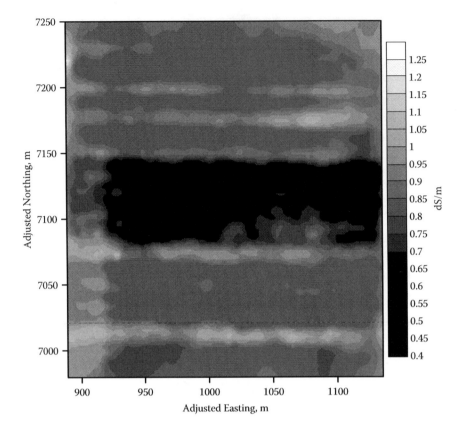

FIGURE 19.2   Representative apparent soil conductivity (EC_a) map as generated on June 17, 2002.

19.3.1.1   Main Plot—Cover Crop

The effect of the main plot treatment, cover crop, is shown in Figure 19.3. Each of the 4 years shows a pattern of –CC EC_a values higher at the start of the season, then converging during mid-season, and diverging again after establishment of the cover crop in the fall. Figure 19.3 indicates convergence (data averaged across subtreatments and replicates) ($P > 0.05$) of –CC, as compared to +CC.

Ferguson et al. (2003) reported that use of a winter cover crop was effective in reducing NO_3 levels at depths below 0.5 m since 1998. Nitrate levels from 0.5 m to the surface tend to be higher under the cover crop, indicating that the cover crop is immobilizing NO_3-N in the upper portion of the soil profile. Statistical analysis supports an association of EC_a and NO_3-N (see Section 19.3.2). Sequential EC_a graphs bolster the view that the immobilized NO_3-N is released when the +CC and –CC curves converge. Each of the 4 years investigated shows a convergence within the crop growing season. The convergence indicates the nitrogen is associated with the organic matter of the cover crop until microbial activity converts the organic N to inorganic NO_3-N.

19.3.1.2   Subplot—Organic and Commercial Amendments

Treatments MP and CP each had sufficient carryover phosphorus after 1998 so that no manure or compost was applied to these treatment strips in subsequent years, including the 2000 to 2003 study; they became essentially equivalent to the commercial fertilizer check (NCK) treatment. Treatments MN and CN were the only treatments receiving manure and compost application. Each of the 4 years showed the EC_a of MN and CN to be significantly greater than the EC_a of NCK through major portions of the growing season (Eigenberg et al., 2005).

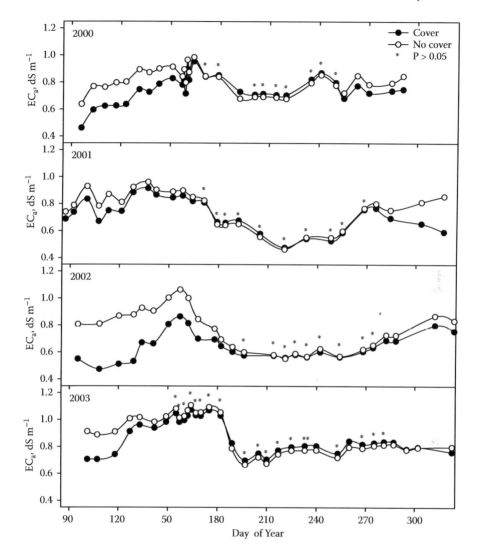

FIGURE 19.3   Apparent soil conductivity (EC_a) for the main treatment of cover and no cover during 2000, 2001, 2002, and 2003. (Reproduced with permission from the Soil Science Society of America.)

19.3.2 Statistical Analysis

Three primary dynamic quantities in a cornfield are the soil temperature, soil moisture, and nutrient levels. The EC_a data for 2000 to 2002 have been corrected to the equivalent temperature of 25°C (McKenzie et al., 1989). The remaining dynamics of soil moisture and nutrient levels (NO_3-N being a primary contributor) are examined in the sections that follow.

19.3.2.1 Main Plots—Cover Crop

Cover crop is the main plot of the split-plot design. Survey EC_a temperature corrected data were collated with all treatments identified and all survey dates concatenated for each year. Soil data were included in the data file and associated with EC_a of replicate 2. A standardized estimate was generated that partitions the percent contribution between the values of soil moisture and NO_3-N toward the variability of EC_a across survey dates; this test was done for temperature-corrected EC_a data for 2000 to 2003, with results shown in Table 19.1. The results indicated that for the –CC treatments,

Year	NO$_3$-N		WFPS		Treatment
	%	**P<**	**%**	**P<**	
2000	81.2	0.004	18.8	0.12	+CC
2001	22.9	0.10	77.1	0.004	+CC
2002	53.0	0.005	47.0	0.007	+CC
2003	29.3	0.006	70.7	0.0001	+CC
2000	97.7	0.023	2.3	0.707	–CC
2001	98.4	0.0001	1.6	0.42	–CC
2002	97.0	0.0003	3.0	0.46	–CC
2003	66.3	0.0001	33.7	0.0001	–CC

* Standard regression estimates shown for 2000–2003 comparing the relative contribution of NO$_3$-N and wfps (water-filled pore space) in explaining the variation in EC$_a$ (all data from Rep 2). Comparisons are made to temperature-corrected EC$_a$ values. (Reproduced with permission from the Soil Science Society of America.)

NO$_3$-N contributed 98, 98, 97, and 66 percent to the variability between water-filled pore space (WFPS) and NO$_3$-N for 2000, 2001, 2002, and 2003, respectively. The +CC treatment indicates NO$_3$-N accounts for 81, 23, 53, and 29 percent of the variability between WFPS and NO$_3$-N. Standardized estimate differences for –CC and +CC treatments can be attributed to the dry years, in which the surveys occurred. The average annual precipitation for the South-Central Nebraska region is 690 mm; the rainfall for the 4-year study was 541, 524, 374, and 492 mm. Below-normal precipitation characterized the 4-year period and is reflected in the standard estimate values for the +CC treatment where both water and nitrate dynamics reflect crop uptake. Analyzing all data from 2000 through 2003 shows that NO$_3$-N dynamics accounted for 79 percent of the –CC variability between WFPS and NO$_3$-N. Additionally, comparisons were made between NO$_3$-N and WFPS differences as contributors to the EMI differences; nitrate contributed 79, 98, 93, and 98 percent of the variability for years 2000, 2001, 2002, and 2003, respectively, and 86 percent for all 4 years combined. The primary contributor to the differences between EMI –CC and +CC over the growing seasons was nitrate level differences. The methodology demonstrated nitrate as a dynamic player in the crop production cycle, and EMI as a viable tool for observing NO$_3$-N dynamics in a crop production system.

19.3.2.2   No-Cover, All Treatments

Table 19.1 indicates that NO$_3$-N is significant and the primary contributor to the –CC profile weighted soil electrical conductivity plot dynamics for all 4 years. Plot treatment means distinctive shapes (Figure 19.4) can be interpreted from the perspective of NO$_3$-N as responsible for key features. Every plot begins with manure or compost being applied early, followed by an EC$_a$ value that gradually increases as the season progresses. Planting dates varied from DOY 114 to 137 and did not have an immediate impact on EC$_a$. The crop produces a visible change in EC$_a$ approximately 50 days after planting, as the crop achieves about 30 cm height (30 cm is approximately the v6-v9 stage, a time of increasing N uptake) (Ritchie et al., 1986); the apparent conductivity makes a noticeable downturn that lasts until about the time the corn silks. The downturn in EC$_a$ corresponds to the time of maximum nutrient uptake; a time when NO$_3$-N was rapidly being removed from the soil. Silk stage occurs at physiological maturity; it is also the point at which the EC$_a$ curve levels out until harvest, when EC$_a$ begins a gradual increase, until the end of the season. Once maturity is achieved, very little nutrient uptake is occurring and EC$_a$ is relatively stable. After harvest, mineralization increases NO$_3$-N concentration and EC$_a$ increases in proportion.

19.4   CONCLUSIONS

A 4-year study, which included field measurement of EC$_a$, identified the effects of manure, compost, fertilizer, and cover crop on EC$_a$ values. Compost and manure applied at the N rate resulted in consistently higher conductivity and available N, when compared to the commercial fertilizer and

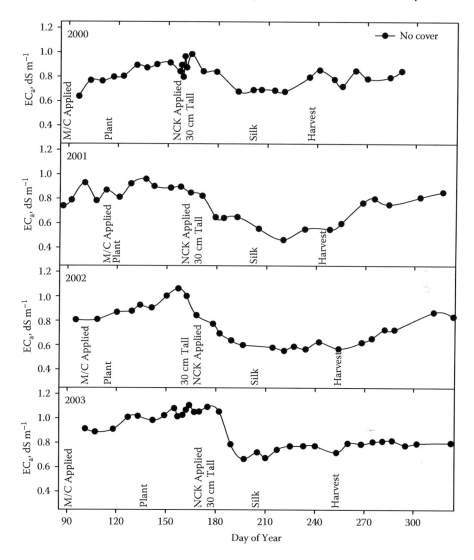

FIGURE 19.4   Distinctive shapes of the apparent soil conductivity (EC_a) mean values shown with chronological events for each season. (Reproduced with permission from the Soil Science Society of America.)

manure and compost at the phosphorus rate (which had not been applied since 1998). Sequential measurements of profile weighted EC_a effectively identified the dynamic changes in available soil N, as affected by animal manure and N fertilizer treatments, during the corn growing season. The sequential measurements also clearly identified the effectiveness of cover crops in minimizing levels of available soil N before and after the corn-growing season, when soluble N is most subject to loss. Ferguson et al. (2003) reported that use of a winter cover crop was effective in reducing NO_3-N levels at depths below 0.5 m. Nitrate levels from 0.5 m to the surface tended to be higher under the cover crop, indicating that the cover crop is releasing NO_3-N in the upper portion of the soil profile. Sequential EC_a graphs indicate that the immobilized NO_3-N is released when the +CC and –CC curves converge. This 4-year study supports the initial findings of a 1999 study that soil conductivity appears to be a reliable indicator of soluble N gains and losses in the soil under study, and may serve as a measure of N sufficiency for corn early in the growing season, as well as an indicator of N surplus after harvest when N is prone to loss from leaching and runoff.

REFERENCES

De Neve, S. et al., Using time domain reflectometry for monitoring mineralization of nitrogen from soil organic matter, *Eur. J. Soil Sci.*, 51, 295, 2000.

Doran, J.W. et al., Influence of cropping/tillage management on soil fertility and quality of former CRP land in the central High Plains, in *Great Plains Soil Fertility Conf. Proc., Potash & Phosphate Institute,* Vol. 6, Havlin, J.L., Ed., Manhattan, KS, 1996, 205.

Eigenberg, R.A., Korthals, R.L., and Nienaber J.A., Electromagnetic survey methods applied to agricultural waste sites, ASAE Paper No. 963014, Am. Soc. Agric. Eng., St. Joseph, MI, 1996.

Eigenberg, R.A. et al., Soil conductivity maps for monitoring temporal changes in an agronomic field, in *Proc. 8th Int. Symp. Animal Agri. Food Proc. Wastes (ISAAFPW)*, Moore, J., Ed., Am. Soc. Agric. Eng., St. Joseph, MI, 2000, 249.

Eigenberg, R.A., et al., Soil conductivity as a measure of soil and crop status—A four year summary. *Soil Sci. Soc. Am. J.*, 70, 1600–1611, 2006.

Ferguson, R.B. et al., Long-term effects of beef feedlot manure application on soil properties and accumulation and transport of nutrients, in *Proc. 9th Int. Symp. Animal, Agric. Food Proc. Wastes (ISAAFPW)*, Raleigh, NC, October 11–14, 2003, Am. Soc. Agric. Eng., St. Joseph, MI, 1.

Johnson, C.K. et al., Status of soil electrical conductivity studies by central state researchers, ASAE Paper No. 032339, Am. Soc. Agric. Eng., St. Joseph, MI, 2003.

McKenzie, R.C., Chomistek, W., and Clark, N.F., Conversion of electromagnetic induction readings to saturated paste extract values in soils for different temperature, texture, and moisture conditions. *Can. J. Soil Sci.*, 69, 25, 1989.

McNeill, J.D., Use of electromagnetic methods for groundwater studies, in *Investigations in Geophysics* No. 5, Ward, S.H., Ed., Geotechnical and Environmental Geophysics, Society of Exploration Geophysicists, Tulsa, OK, 1990, 191.

Ritchie, S.W., Hanway, J.J., and Benson, G.O., How a corn plant develops, *Special Report No. 48*. Herman, J., Ed., Iowa State Univ. Sci. Technol., Coop. Ext. Serv., Ames, IA, 1986.

Smith, J.L. and Doran, J.W., Measurement and use of pH and electrical conductivity for soil quality analysis, in *Methods for Assessing Soil Quality,* SSSA Spec. Publ. 49, Doran, J.W. and Jones, A.J., Eds., Soil Sci. Soc. Am, Madison, WI, 1996, 169.

20 Soil EC$_a$ Mapping Helps Establish Research Plots at a New Research and Extension Center

Hamid J. Farahani, David A. Claypool, R. P. Kelli Belden, Larry C. Munn, and Robert A. Flynn

CONTENTS

20.1 INTRODUCTION

The establishment of a new agricultural research and extension center is a rare event, full of opportunities for success or failure. The University of Wyoming Agricultural Experiment Station purchased land in 2003 to launch a new research and extension center to replace older facilities at Torrington and Archer, Wyoming. The Sustainable Agriculture Research and Extension Center (SAREC), which consists of approximately 149 ha of irrigated and 617 ha of dryland farmland, 775 ha of pasture, and 24 ha of farmstead, is located west of the town of Lingle in Goshen County in southeast Wyoming and along the North Platte River (Figure 20.1). SAREC is a field research and extension station for investigating various aspects of crop and livestock production including integrated crop–livestock systems (Claypool et al., 2004a).

With the creation of the new research and extension center, the College of Agriculture at the University of Wyoming was presented with a unique opportunity to develop baseline soils and yield maps before research plots are established (Belden et al., 2005). Understanding and documenting the status of the soil resource before initiating research offers a unique opportunity to quantify changes in soil quality. Knowledge of soil resource variability also aids researchers with the spatial placement of their plots, particularly if soil variance is to be kept at a minimum.

Two advances in agricultural sensing and mapping technology—on-the-go mapping of yield and apparent soil electrical conductivity (EC$_a$)—provided the Agricultural Experiment Station with an exciting set of tools with which to plan and manage SAREC. After the land was purchased and in cooperation with USDA-ARS Water Management Research unit in Fort Collins, Colorado, the Agriculture Experiment Station allocated resources to obtain important baseline data: digital imagery, yield, EC$_a$, and elevation maps, and grid soil sampling and analysis. As discussed by many in this book, geospatial EC$_a$ mapping provides the simplest and most rapid assessment of soil

FIGURE 20.1   Aerial image of the irrigated farmland at SAREC.

variability. Spatial data will help future management and design of research projects. Existing EC$_a$ and yield data will be analyzed to identify possible problem areas. The integration of this baseline information means that much more will be known about the soils at SAREC before research begins than during the establishment of any previous Wyoming research and extension center. In this case study, we present a summary of the baseline data and briefly discuss their uses.

20.2 METHODS

Baseline data were collected as follows. In late summer 2003 and spring 2004, measurements of soil EC$_a$ were taken using the Veris 3100 Soil Mapping System (Veris Technologies, Salina, KS). The Veris unit has six coulter electrodes mounted on an implement that is pulled by a pickup truck. The unit simultaneously measured EC$_a$ at two soil depths, referred to as "shallow" (0 to 0.3 m) and "deep" (0 to 0.9 m) EC$_a$ readings, and recorded their spatial coordinates using a Global Positioning System (GPS) unit (Lund et al., 1999). Maneuvering speeds through the field averaged 11 km/hr with measurements taken every second on 12 m swath widths, corresponding to an average 3 m spacing between measurements in the direction of travel. Figure 20.2a shows the raw EC$_a$ data for the entire irrigated area at SAREC. The irrigated fields were named after states whose shape they most resembled. A total of 47,500 EC$_a$ readings were collected in the irrigated fields, representing approximately 330 data points per ha (or one point every 30 m^2). This level of detail is helpful because many research studies occupy much less than 1 ha, but published soil surveys typically do not map areas smaller than 0.5 ha.

Soils were sampled on a 1 ha grid the same day EC$_a$ measurements were taken to characterize soil variability. Sample depths were 0 to 0.3, 0.3 to 0.6, and 0.6 to 0.9 m. Soil water content, paste pH, saturated paste electrical conductivity (EC), particle size distribution, and organic matter were determined using standard procedures.

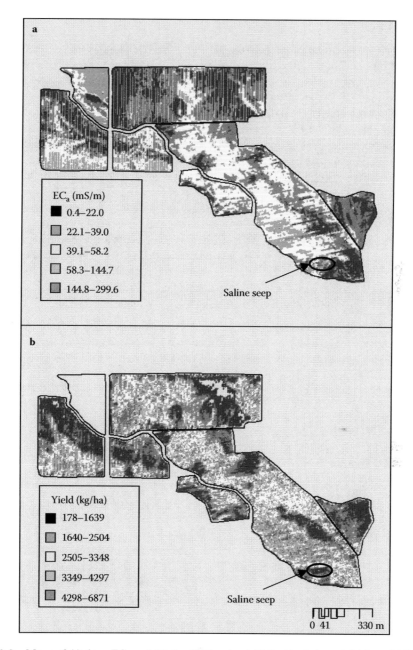

FIGURE 20.2 Maps of (a) deep EC$_a$ and (b) the 2003 oats yield for the irrigated fields at SAREC. (Both maps are classified according to natural breaks.)

Spatial variation of oats yield was measured across the irrigated fields in 2003 using a combine with a yield monitor that measured grain yield on a 9.1 m swath and recorded the spatial coordinates (Figure 20.2b). A total of 59,600 yield readings were collected in the irrigated fields (except no crop was planted in the Illinois field), representing approximately 380 data points per hectare (or one point every 26.3 m^2). All spatial data were organized and mapped using ArcGIS 9.0 (ESRI, Redlands, CA).

20.3 RESULTS

Even though oats are not a typical crop in the North Platte River Valley, they were planted on the entire irrigated farmland (except the Illinois field) in spring 2003 because of their sensitivity to the kind of soil variability that also directly impacts yield variability. Visual comparison of the yield and EC_a maps (Figure 20.2) shows similarities in some location-specific patterns (a positive correlation between the two), implying that the yield response was soil related.

Results from the yield and EC_a mapping for the irrigated fields at SAREC are summarized in Table 20.1. Mean yield and coefficient of variation (CV) across all fields were 2206 kg/ha (ranging from 167 to 6379 kg/ha) and 39 percent, respectively. The yield map obtained from the yield monitor data identifies areas of high and low yield potential (Figure 20.2b). As shown, yield variability was significant even within individual fields as small as New Hampshire (7.5 ha). Mean EC_a for the top 0.9 m soil and across all fields was 34.8 mS/m (milli Siemen per meter), ranging from 0.4 to 300 mS/m. Regardless of the size of the field, soil EC_a showed high variability (Table 20.1), with an overall CV of 56 percent. The highest EC_a values were observed in Illinois (mean of 50.6 mS/m), which also had the highest clay (21 percent) and organic matter (1.46 percent) contents. As expected, the lowest EC_a values in Montana (mean of 22.7 mS/m) corresponded to the lowest clay (14 percent) and organic matter (1 percent) contents. The California field contains a saline seep (saturated paste EC = 1420 mS/m) near the southern tip of the field, clearly identified by the EC_a mapping with

TABLE 20.1
Summary Statistics of Oats Yield and EC_a Measurements across Irrigated Fields at SAREC

Irrigated Field	Field Area (ha)	Oats Grain Yield (kg/ha)				
		Mean	Std	Min	Max	CV (%)
New Hampshire	7.5	2019	720	167	6199	36
Triangle	7.6	1655	510	168	5575	31
Illinois	8.0	—	—	—	—	—
Iowa	9.9	1875	740	175	5795	39
Idaho	19.3	2150	1050	171	5465	49
Montana	37.8	2682	1091	204	6386	41
California	54.4	2855	917	182	6379	32
Total area	145					
Mean		2206	838	178	5967	38

		Deep (Top 0.9 m Soil) EC_a (mS/m)				
		Mean	Std	Min	Max	CV(%)
New Hampshire	7.5	36.3	15.7	0.5	99.3	43
Triangle	7.6	23.1	8.7	3.9	54.4	38
Illinois	8.0	50.6	21.0	2.5	123.4	42
Iowa	9.9	43.2	20.6	0.4	108.0	48
Idaho	19.3	34.3	20.9	0.4	123.4	61
Montana	37.8	22.7	12.8	0.4	82.8	56
California	54.4	38.0	17.7	0.7	299.6	47
Total area	145					
Mean		34.8	19.4	0.4	299.6	56

Notes: Std, standard deviation; Min, minimum; Max, maximum; CV, coefficient of variation; N, number of observations.

elevated EC$_a$ values greater than 200 mS/m and corresponding low yield values below 600 kg/ha (Figure 20.2).

The map produced from the EC$_a$ data (Figure 20.2a) is much more detailed than a standard soil survey. This information was used to explore variability within potential research plots to be laid out across the Montana field underneath a newly installed linear move sprinkler system. Figure 20.3a shows the eight east–west tire tracks for the linear system. The Montana field was divided into nine plots (85 m wide and 378 m long) called M1 to M9, with each plot further divided into seven subplots. Mean soil EC$_a$ and yield values from each of the nine Montana plots are shown in Figure 20.3b, with their corresponding CV values shown in Figure 20.3c. Two important observations may be made from Figure 20.3.

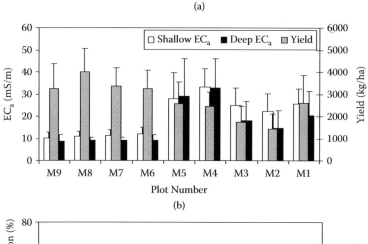

(a)

(b)

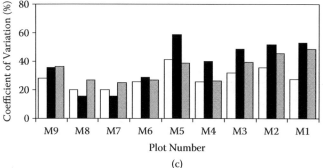

(c)

FIGURE 20.3   (a) Map of research plots and subplots in the irrigated Montana field, (b) mean EC$_a$ and yield (with error bars), and (c) coefficient of variation (CV) for each of the nine research plots across the Montana field.

First, plots M6 to M9 are the most productive (highest yield) but have the lowest EC_a values. In other words, soil EC_a and yield correlation is not necessarily positive. Generally speaking, a yield map represents a crop response surface that integrates the effect of a host of influences including soil, water, nutrients, pests, climate, and management. Because of that, a yield response map could be specific to a growing season and may change over time. On the other hand, a soil EC_a map is strictly a soil phenomenon, a reflection of highly complex interactions of soil physical and chemical properties of texture (for instance, clay content), cation exchange capacity, organic matter, and soil water. The main distinction between a soil EC_a and a yield map is that the former is known to be temporally stable (Farahani and Buchleiter, 2004), providing a useful base map for multiple years.

Second, Figure 20.3c shows that within-plot variability was as pronounced as within-field and within-farmland variability. Figure 20.3 is an attempt to illustrate an example of the usefulness of baseline EC_a and yield information. This information can aid researchers in placing their particular study across a desirable level of soil and productivity variance. As shown in Figure 20.3c, variability ranged from less than 20 percent CV to about 60 percent in Montana plots. Choice of plots with least soil variance is offered by plots M6 to M9. For small plot research, the baseline information may be further refined to infer variability within the subplots.

20.4 CONCLUDING REMARKS

Precision agriculture technology is providing a unique opportunity to create soil maps that are sufficiently detailed to direct research planning at SAREC. In addition to the example given in this article, a few other opportunities to apply the baseline data to practical field problems have already occurred. For instance, the nature of soil variability maps as depicted by EC_a was found useful to (1) the planners with the selection and placement of two newly installed sprinkler irrigation systems for field research (Claypool et al., 2004b), and (2) Belden et al. (2005) who explored the causes of apparent Fe chlorosis exhibited by grain sorghum (*Sorghum bicolor* L. Moench). Areas of the field were symptom free, but there were several patches that exhibited both extreme symptoms and elevated EC_a values. Soil EC_a results are also being correlated with important soil properties such as soil texture, which directly affects water-holding capacity—an important property in both irrigated and dryland soils. The EC_a map should help guide direct sampling and sensor placement.

REFERENCES

Belden, R.P.K., Claypool, D.A., and Farahani, H.J., Spatial variability of soil properties at the University of Wyoming Sustainable Agricultural Research and Extension Center, In *Proc. Western Nutrient Management Conference,* Vol. 6, Salt Lake City, UT, 2005.

Claypool, D. et al., Precision agriculture technology to plan and manage a new research and extension center, In *Proc. Seventh International Conference on Precision Agriculture and Other Precision Resources Management,* July 25–28, Minneapolis, MN, 2004a.

Claypool, D. et al., Modern tools help establish a new agricultural research and extension center, *Reflections,* College of Agriculture, University of Wyoming, Laramie, June 2004, 55, 2004b.

Farahani, H.J. and Buchleiter, G.W., Temporal stability of soil electrical conductivity in irrigated sandy fields in Colorado, *Trans. ASAE,* 47, 79, 2004.

Lund, E.D., Christy, C.D., and Drummond, P.E., Practical applications of soil electrical conductivity mapping, In *Precision Agriculture 99: Proc. 2nd European Conference on Precision Agriculture,* Stafford, J.V., Ed., BIOS Scientific Publishers, Oxford, U.K., 1999, 771.

21 Mapping Golf-Course Features with Electromagnetic Induction, with Examples from Dublin, Ohio

Richard S. Taylor

CONTENTS

21.1   INTRODUCTION

Electromagnetic induction (EMI) is a convenient technique for mapping apparent electrical conductivity over large areas and varied terrain. In recent years, the resolution of EMI instrumentation has improved such that the technique can enable the detailed analysis of spatial and temporal changes in conductivity. High-resolution EMI surveys are suitable for mapping features of the scale typical of golf-course greens and tees. Such surveys can check the depth and extent of the materials used to construct golf-course features, and may be of some use in monitoring the response of features to environmental fluctuations and managerial practices.

This note briefly reviews the measurement of apparent electrical conductivity by EMI, compares the resolution of an available instrument to that required for mapping features at golf courses, and presents several examples from test surveys in Dublin, Ohio.

21.2   APPARENT CONDUCTIVITY BY EMI

An electromagnetic (EM) array of a transmitter and receiver provides the standard means of measuring apparent conductivity by EMI. The EM array is defined by the separation between the transmitter and receiver and the relative orientation of these components. The most popular arrays for measuring apparent conductivity have less than 5 m separation, and their orientations are termed horizontal coplanar (HCP) and perpendicular (PRP).

The apparent electrical conductivity measured by EMI at some height above a layered earth is influenced by the thickness of the "air layer" between the EM array and the earth, and the thickness

FIGURE 21.1   The 1 m horizontal coplanar/perpendicular (HCP/PRP) instrument in sled.

and conductivity of the various layers in the earth. Where the operating frequency of the EM array is sufficiently low (Wait, 1962), the sensitivity of the array to a layer that occupies a given interval of depth beneath the array is a simple function of depth. Consequently, the contribution of each layer to the apparent conductivity is the product of the sensitivity to the layer times the conductivity of the material in the layer, and apparent conductivity is the sum of these contributions. For the PRP array, Wait's criterion for low-frequency (Wait, 1962) (i.e., low induction number [LIN]) operation is

$$|(i\sigma\mu\omega)^{1/2}\rho| < 0.5 \tag{21.1}$$

where i is the square root of -1, σ is the electrical conductivity of the material below the EM array, μ is the magnetic permeability of the material below the EM array, ω is the angular frequency of operation of the EM array, and ρ is the horizontal distance between the transmitter and receiver. For the HCP array, this inequality should be less than 0.16.

Figure 21.1 shows an EM instrument, mounted in a sled, that incorporates both the HCP and the PRP arrays. The transmitter and receiver separation (ρ) is 1 m, and the operating frequency ($\omega/2\pi$) is 9 kHz. For this instrument (H/P-1), the LIN criterion is satisfied where the conductivity of the earth does not exceed 0.36 S/m for the HCP array, and 3.5 S/m for the PRP array, assuming the earth has the permeability of free space. For comparisons pertinent to golf-course applications, the conductivity of native soil seldom exceeds 0.2 S/m, and the conductivity of sand is usually less than 0.02 S/m (20 mS/m). In environmental conditions typical of golf courses, the H/P-1 pictured in Figure 21.1 has an accuracy of about 1 mS/m.

21.3   RESOLUTION OF MEASUREMENT

In addition to conductivity accuracy, spatial resolution must be finer than the features of interest. The functions of depth sensitivity for the HCP and PRP arrays indicate the depth resolution of an LIN instrument. For depth expressed in terms of ρ, the cumulative depth sensitivity of the HCP array is

$$C_H = 1 - 1/(4\,\rho^2 + 1)^{1/2} \tag{21.2}$$

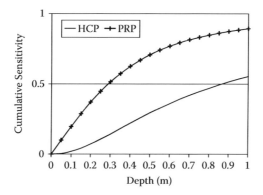

FIGURE 21.2 The 1 m horizontal coplanar/perpendicular (HCP/PRP) depth sensitivities.

and the cumulative depth sensitivity of the PRP array is

$$C_P = 2 \; \rho/(4 \; \rho^2 + 1)^{1/2} \tag{21.3}$$

Figure 21.2 graphs these functions for arrays with ρ of 1 m.

Note that there is an appreciable difference in the functions. For example, about half the sensitivity of the PRP array accumulates within a depth of about 0.3 m, and half the sensitivity of the HCP array accumulates within a depth of about 0.9 m. As these depths are similar to the depths of golf-course features, an H/P-1 at ground level appears to be suitable for mapping such features. The difference in cumulative sensitivity is the basis for resolving changes in conductivity with depth, as the 1 m PRP array is relatively sensitive to material to a depth of about 0.5 m, and the 1 m HCP array is relatively sensitive to material between the depths of about 0.3 and 1.5 m.

21.4   GOLF-COURSE GREEN MODEL

A model based on the sensitivity functions can predict the apparent conductivities that will be measured over a feature of interest (e.g., a golf-course green). The modeling exercise can help assess the potential effectiveness of EMI for a given application. Figure 21.3 shows the apparent conductivities that the H/P-1 would measure over a model golf-green that is composed of materials of given conductivity. The model contains a layer of 10 mS/m material, representing sand, over 50 mS/m material, which represents the native soil. The thickness of the sand layer is zero at stations 0 and 20, 0.3 m at stations 3 through 8 and 12 through 17, and 0.5 m at station 10.

Although the native soil extends to the surface at stations 0 and 20, the HCP apparent conductivity will be a fraction of a mS/m and the PRP apparent conductivity will be a few mS/m less than the native-soil value of 50. This results from the fact that when the H/P-1 rests on the ground, its arrays are about 5 cm above the ground due to their positioning in the instrument.

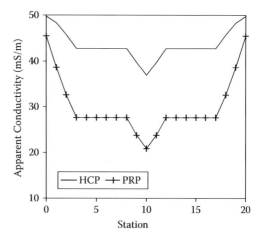

FIGURE 21.3   Model responses for a golf-course green.

The diminished values of apparent conductivity reflect the sensitivity of the arrays this thickness of nonconductive air (inside the instrument) in the sampled depth.

Where the H/P-1 lies above 0.3 m layer of sand, the HCP apparent conductivity will decrease to 43 mS/m; the greater sensitivity of the PRP array through this depth will decrease its apparent conductivity to 28 mS/m, about midway between the actual conductivities of the sand and native soil. Where the sand thickness increases to 0.5 m, the HCP apparent conductivity will decrease to 37 mS/m, and the PRP apparent conductivity will decrease to 21 mS/m.

The modeling suggests that sand-related changes in 1 m HCP apparent conductivity will be significantly above the level of environmental noise, but undefined local changes in the conductivity of the native soil will likely add significant uncertainty to the interpretation of actual HCP measurements. Where the sand is 0.3 m thick, for example, modeling predicts that the HCP apparent conductivity will fluctuate between 26 and 59 mS/m, where the conductivity of the native soil fluctuates between 30 and 70 mS/m. The sand-related changes to 1 m PRP apparent conductivity will be about twice those of the HCP array, and the PRP measurements will be less affected by sub-sand fluctuations in native-soil conductivity. With greater sensitivity to depths typical of the sand beneath golf greens, the PRP apparent conductivity should be effective at mapping the extent of the sand, and perhaps fluctuations in the thickness and moisture of the sand.

21.5   SURVEY EXAMPLES

Surveys at two golf courses in Dublin, Ohio, tested the effectiveness of EMI for golf-course applications. At the Golf Course of Dublin (GCD), measurements were made over a practice green constructed using the California method. At the Muirfield Village Golf Course (MVGC), measurements were made at the seventeenth hole, over both the USGA-method green and the professional tee.

At each site, the H/P-1 was pulled in a sled at walking speed along east–west traverses. Measurements were at 1/8-second intervals, which yielded about eight measurements per meter along the survey traverse. Closely spaced measurements are essential for mapping golf-course features that might include an irregular boundary between sand layers and native soil. The survey traverses were visually controlled according to markers at each end of the traverse, and the nominal spacing between traverses was 1 m.

Figure 21.4 presents profiles of apparent conductivity from traverse 10 N, through the center of the GCD California-method practice green. The apparent conductivities for both HCP and PRP are around 55 mS/m at the ends of the traverse, indicating that native soil of about this conductivity extends to the surface. Over an apparent transition zone of about 2 m (between eastings 1 to 3 and 20 to 18), the PRP apparent conductivity decreases from native-soil levels to about 30 mS/m. This decrease is consistent with a thickening to about 0.3 m of a subturf layer of sand, with conductivity of about 10 mS/m. A decrease in PRP apparent conductivity of a few mS/m at easting 11, and perhaps at easting 6, may indicate thickening of the sand or other coarse material coincident with drainage structures.

Figure 21.5 and Figure 21.6 are contour maps of the H/P-1 apparent conductivities from the traverses on the GCD green. The PRP contours of Figure 21.5 reveal a rectangular area of apparent conductivities below 34 mS/m, enclosed by apparent conductivities in the 40s and 50s of mS/m. In this and subsequent maps,

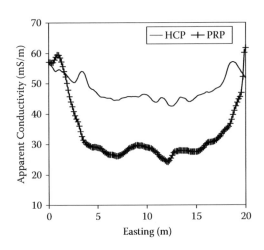

FIGURE 21.4   Traverse 10 N at the Golf Course of Dublin (GCD).

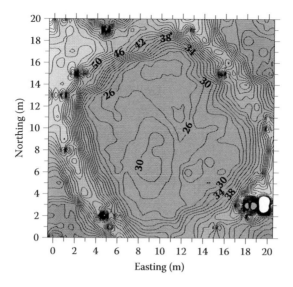

FIGURE 21.5   Perpendicular (PRP) electrical conductivity contours at the Golf Course of Dublin (GCD).

strong and local fluctuations in apparent conductivity are caused by buried metal of watering systems and other utilities (e.g., at 18 E, 3 N).

The area of low conductivity shows the extent of the sand under the green. Within the area, a north–south trend of slightly lower conductivity lies roughly at 11 E. This may indicate a drainage trench that might have been cut into the native soil at the base of the sand and filled with resistive material. The PRP contours suggest drainage structures that extend northwest and northeast from the central portion of the feature at 11 E. Drainage of the area is from north to south, and the lower conductivities in the northern portion of the green suggest that this portion of the sand is the driest. The HCP contours of Figure 21.6 also indicate the extent of the sand but with significantly less contrast in conductivity, as predicted by the previous

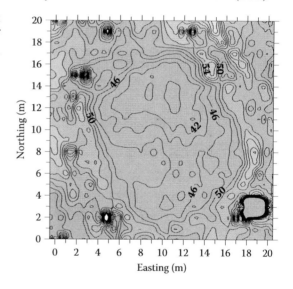

FIGURE 21.6   Horizontal coplanar (HCP) electrical conductivity contours at the Golf Course of Dublin (GCD).

modeling of responses. The HCP measurements also are lowest in the northern portion of the green, indicating that core of the conductivity decrease is near the base of the sand, a depth at which the HCP array has comparable sensitivity to the PRP array.

Figure 21.7 and Figure 21.8 contain contours of the H/P-1 apparent conductivities from the traverses at the MVGC green. The PRP measurements of Figure 21.7 sharply define an area of low conductivity that shows the limit of the sand beneath the green, except on the northern portion of the eastern border where the traverse lines ended on the green. Note that the contours at the edge of the sand are generally closer together than those at the GCD California-method green, in keeping with the vertical interface between sand and soil specified for USGA-method greens. A curved trend of slightly lower conductivity extends from (6 E, 5 N) to (15 E, 21 N) and coincides with the location of the main drain beneath the green. Trends that branch to the west and south from (10 E, 17 N) may

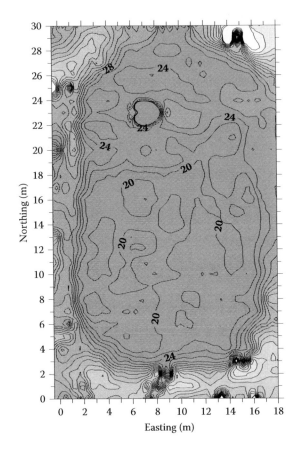

FIGURE 21.7 Perpendicular (PRP) electrical conductivity contours at the Muirfield Village Golf Course
MVGC green.

indicate secondary drains. Drainage is to the northeast; the southwestern portion of the green is the
least conductive, suggesting that it is also driest.

The HCP contours of Figure 21.8 are similar to the PRP contours, but with less contrast and
spatial detail.

Measurements at the MVGC tee (Figure 21.9 and Figure 21.10) were made along north–south
traverses. Unlike the results at the greens, the PRP contours of Figure 21.9 do not show a regular
area of low conductivity, which suggests that the tee has no sand-layer beneath it. Instead, PRP
conductivities decrease below 30 mS/m toward the southern edge of the tee. A corresponding but
subtler decrease in HCP conductivities (Figure 21.10) suggests that the soil texture of the uppermost
0.3 m near the southern edge of the tee is relatively coarse.

Toward the northern edge and the northwest corner of the tee, HCP conductivities rise to their
highest values, and PRP values remain more moderate. In these areas, the diverging conductivities
reveal that the earth beneath 0.5 m depth is increasingly conductive.

The PRP conductivities and, to a lesser extent, HCP conductivities show a linear low extending
along 7 N between 7 E and 14 E. A drainage trench runs under the tee at about 7 N, and the conductivity low may indicate the local effectiveness of the trench in draining the soil.

21.6 CONCLUSIONS

A 1 m EM instrument that incorporates HCP and PRP arrays, with 1 mS/m accuracy, can provide
ground-level measurements with sufficient resolution for mapping features on golf courses. Where

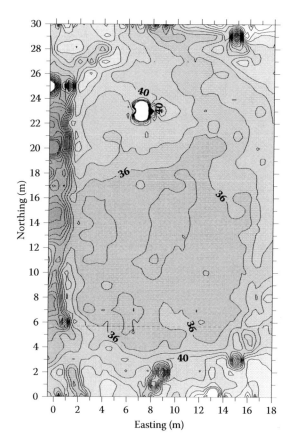

FIGURE 21.8 Horizontal coplanar (HCP) electrical conductivity contours at the Muirfield Village Golf Course (MVGC) green.

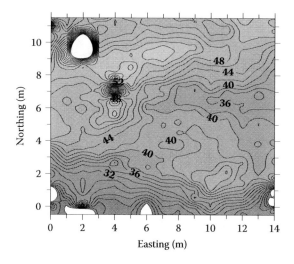

FIGURE 21.9 Perpendicular (PRP) electrical conductivity contours at the Muirfield Village Golf Course (MVGC) tee.

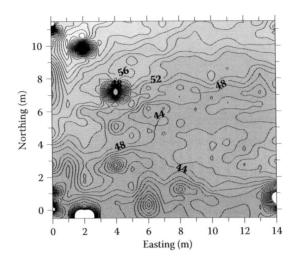

FIGURE 21.10   Horizontal coplanar (HCP) electrical conductivity contours at the Muirfield Village Golf Course (MVGC) tee.

the conductivity of native soil is several times that of the sand used beneath golf greens, modeling results suggest EMI can indicate the lateral extent of the sand, as well as limited information about the shape of the sand and soil interface. Survey tests at two golf courses in Dublin, Ohio, show that the practical application of the 1 m instrument is consistent with the capabilities suggested by the modeling. Further studies are required to determine if EMI can monitor changes in soil moisture and chemistry, and what role EMI surveys might play in golf-course management.

21.7   ACKNOWLEDGMENTS

The author thanks Dr. Barry Allred for proposing and organizing the survey tests, staff at the U.S. Department of Agriculture–Agricultural Research Service–Soil Drainage Research Unit (USDA-ARS-SDRU) and Ohio State University for help with the surveys, and the superintendents at the Golf Course of Dublin and Muirfield Village Golf Course for convenient access to the test sites.

REFERENCES

Wait, James R., 1962, A note on the Electromagnetic Response of a Stratified Earth: *Geophysics*, 27, 382–385.

22 Comparison of Geoelectrical Instruments in Different Soilscapes

Robin Gebbers and Erika Lück

CONTENTS

22.1   INTRODUCTION

There are a number of instruments available for kinematic soil mapping by geoelectrical methods. Thus, a potential user might ask him- or herself, which is the appropriate instrument, what does it measure, and how accurate can the measurement be? What is measured is largely influenced by the underlying measuring principle. From theory, we know that the depth response curves of geoelectrical measurements could vary depending on the layering of conductive and resistive structures (Dabas and Tabbagh, 2003). It is assumed that the galvanic contact resistivity (GCR) and the electromagnetic induction (EMI) methods react different on soil layering. Most of the instruments used for soil mapping in agriculture are based on one of these two principles. A third method, based on capacitively coupled resistivity, has rarely been used in agricultural practice up to now. Thus, we will focus on GCR and EMI instruments.

The measurement error (accuracy and precision) could depend on both the instrument's design and the measuring principle. An example for the importance of the design of an instrument is temperature compensation. All geoelectrical measurements are known to be subjected to temperature drift. A part of this drift can be suppressed by the construction of the instrument. The measuring principle can influence precision (repeatability) as follows: EMI instruments react on metal objects (e.g., pulling vehicle) and ambient electrical interference to a considerable extent (Sudduth et al., 2001), and GCR may have problems under dry soil conditions due to high contact resistivity.

The objective of this study was to compare measurements of different instruments in different soilscapes under field conditions. We want to find out how depth of investigation is influenced by soil variability, which is inherent in different soilscapes. Additionally, precision of the instruments is tested.

22.2 MATERIALS AND METHODS

On selected transects, we collected data with mobile instruments and with a stationary GeoTom 200/100 RES/IP multielectrode array (GeoLog, Germany) as the reference method. The multielectrode array consists of up to 100 electrodes that are evenly placed on a line (spacing 50 cm), connector cables, the measuring instrument, and a mobile computer, which runs the GeoTom controller software. The electrodes were switched in a Wenner configuration. During the measurements, electrodes were systematically activated to obtain readings from different sets. Wider electrode spacings can be related to larger depths of investigation. Data were collected from eight depths according to eight electrode spacings from 0.5 m to 4.0 m. Estimated depths of the maximum of the depth response curve for each electrode spacing are presented in Table 22.1. The GeoTom was chosen as the reference because its design and the stationary measurement provide maximum control during operation. Measurements are repeated automatically until they are stable; the software indicates unusually high and low values. The ER_a (apparent soil resistivity) profile can be visually inspected in real time, and electrodes can be easily checked and adjusted if necessary.

Measurements with borehole methods can be regarded as reliable as well—especially where the soil is layered—because they are able to detect soil EC_a in situ. Due to several drawbacks, like availability or durability of the equipment, slow performance, and restriction to moist soil conditions only, we decided not to refer to borehole methods as the standard. Another reason is that surface methods can be compared much more easily among each other. Relating surface methods to borehole methods requires statistical processing of the data, which can be quite demanding and introduces a source of error as well. Hence, borehole methods were only compared with readings of the multielectrode array.

22.2.1 SITE CHARACTERISTICS

All test sites are located in Germany and named after the nearby towns:

- *Bornim*: Dystric cambisol on old glacial deposits, sand to loamy sand, gentle slope, recultivated land (meadow), transect length 159 m. In addition to the high natural variability, soil heterogeneity was increased by human activities (soil compaction, gravel, water pipe). See Dabas et al. (2004) for details.
- *Beckum*: Cambisol-Rendzina on loess and limestone, gradual transition from sandy loam to loam (loam at the bottom of the profile), gentle slope, arable land, transect length 162 m.
- *Golzow*: Fluvisoil on fluvial sediments, strong contrasts between sand and silty loam in horizontal and vertical direction (loam above sand), flat terrain, arable land, transect length 162 m.
- *Kassow*: Cambisol-luvisol on deposits from the last ice age, sand to loamy sand, dynamic relief, set aside area, transect length 265 m.

The resistivity sections obtained by the GeoTom are shown in Figure 22.1.

22.2.2 INSTRUMENTS

- *EM38, EM38-DD, and a prototype EM38-MK2 (Geonics Ltd., Canada)*: These instruments are based on the EMI method, measuring one or two depths at a time. The EM38 maintains 1 m intercoil spacing and operates at 14.6 kHz (see McNeill, 1980; Sudduth et al., 2001 for more details). The EM38-DD consists of two EM38s, which are arranged in the horizontal and vertical dipole orientations, recording data simultaneously. The EM38-MK2 is a prototype specially constructed for the Technical University of Munich. It is built of two EMI instruments with coil spacings of 0.5 m and 1 m operating at about 40.4 kHz (Geonics, 2002). The EM38-MK2 is intended to measure EC_a at very shallow

TABLE 22.1

Kendall's Tau Correlations between the GeoTom Multielectrode Array (Raw Data) and Instruments for Continuous EC_a (ER_a) Mapping on Transects in Four Different Soilscapes

Field	Bornim				Kassow				Golzow				Beckum			
GeoTom level	1	2	3	4	1	2	3	4	1	2	3	4	1	2	3	4
Electrode spacing [m]	0.50	1.00	1.50	2.00	0.50	1.00	1.50	2.00	0.50	1.00	1.50	2.00	0.50	1.00	1.50	2.00
Depth of maximum response [m]	−0.26	−0.51	−0.77	−1.02	−0.26	−0.51	−0.77	−1.02	−0.26	−0.51	−0.77	−1.02	−0.26	−0.51	−0.77	−1.02
ARP 1	**0.423**	0.317	0.241	0.204	**0.805**	0.749	0.660	0.596	**0.787**	0.776	0.752	0.738	**0.841**	0.800	0.803	0.815
ARP 2	0.503	**0.671**	0.635	0.581	0.811	**0.815**	0.743	0.673	0.773	**0.806**	0.799	0.782	**0.863**	0.837	0.842	0.855
ARP 3	0.358	0.656	**0.743**	0.727	0.721	0.814	**0.832**	0.782	0.754	0.800	**0.807**	0.801	0.863	0.860	**0.875**	0.872
Veris 1	**0.422**	0.190	0.084	0.027	**0.815**	0.720	0.637	0.572	**0.759**	0.710	0.678	0.674	**0.867**	0.813	0.817	0.826
Veris 2	0.584	**0.614**	0.523	0.452	0.683	0.756	**0.763**	0.744	0.833	**0.852**	0.819	0.794	**0.851**	0.823	0.828	0.836
EM38 H	0.339	0.648	**0.694**	0.673		**0.811**	0.808	0.802	0.805	0.817	0.843	0.864	**0.875**			
EM38 V	0.202	0.535	0.606	**0.617**	0.707	0.764	**0.796**	0.788	0.788	0.834	0.843	**0.850**	0.853	0.883	0.895	**0.905**
CM138 H	−0.164	−0.148	−0.125	−0.108	CM138 V best run				**0.718**	0.674	0.659	0.665				
CM138 V	0.196	0.362	**0.386**	0.366	CM138 V worst run				0.435	0.431	0.440	**0.445**				

Notes: GCR electrode spacing: ARP 1 = 0.5 m, ARP 2 = 1 m, ARP 3 = 2 m, Veris 1 = 0.24 m, Veris 2 = 0.72 m; EMI mode: H = horizontal, V = vertical. Coefficients in boldface indicate best correlations with the reference at each site and each instrument layer/mode. Depth of maximum response was approximated by the inversion software RES2DINV®.

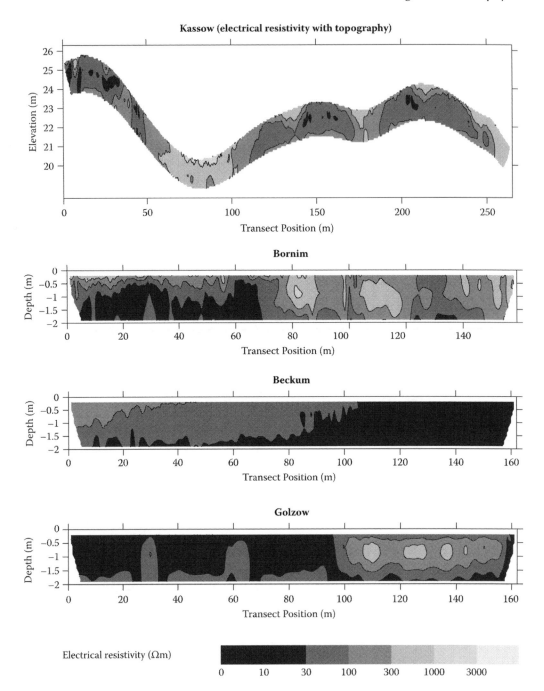

FIGURE 22.1 Depth profiles of electrical resistivity from the four test sites representing different soilscapes. Inversion results from measurements with the GeoTom multielectrode array.

depths. To avoid interference, the paired instruments of the EM38-DD and the EM38-MK2 work at different frequencies (Geonics Ltd., personal communications).

- *CM-138 (Gf Instruments, Czech Republic)*: The CM-138 is an EMI instrument and works at one depth at a time. According to the operation manual, it has very similar parameters as the EM38 (1 m coil spacing, 14.406 kHz).

- *Veris 3100 (Veris Technologies, USA)*: The Veris 3100 is a GCR-based instrument working at two depths. The rolling electrodes (smooth coulters) are arranged in a line with distances of about 24 and 72 cm.
- *ARP-03 (Geocarta SA, France)*: The ARP-03 is a GCR instrument working at three depths. Rolling electrodes (coulters with spikes) are arranged in a trapezoid pattern. Current electrodes have a spacing of 1 m; potential electrodes have spacings of 0.5, 1, and 2 m (Dabas, 2002). Frequency is about 225 Hz; voltage can be varied between 0 and 200 V depending on contact resistance conditions of a site (Geocarta SA, personal communications).
- *Earth Resistivity Meter 14.01 (Eijkelkamp, The Netherlands)*: The Earth Resistivity Meter (ERM) is a GCR instrument for borehole measurements (direct push method). Four electrodes in the sensor tip are arranged in a Wenner array. They sense approximately 80 cm³ of soil volume. Maximum depth of investigation is 1 m with a vertical resolution of about 5 cm. The probe has to be operated manually.

The GeoTom, the EM38, the EM38-MK2, and the ERM measurements were geo-referenced by a measurement tape. Readings of all the other instruments were geo-referenced by differential Global Positioning System (DGPS). Because of limited availability of the instruments and restrictions due to soil conditions, measurements were not always possible at the same time and at the same site with all instruments.

22.2.3 DATA ANALYSIS

We compared all surface methods with the raw data of the GeoTom. To avoid problems due to nonlinearity, outliers, and deviations from normal distribution, we chose Kendall's tau (τ) rank correlation coefficient to express similarities between the measurements. The τ values generally tend to be lower when compared with Pearson's r correlation coefficient. Levels of significance are not reported because all correlations were significant ($p < 0.05$ in all cases). Although readings were not exactly colocated, we calculated τ for pairs of points within a specific distance (lag). To correct GPS errors and offsets, positions are shifted along the transects until τ is maximum. For GPS geo-referenced measurements, lag width was set to 2 m and the shift increment to 0.5 m. For tape measurements, the lag width was set to 1 m and the shift increment was 0.2 m.

Inversion is used to visualize the transects to compare the GeoTom with the borehole method and to quantify spatial variability. Inversion is a procedure that tries to estimate ER_a for distinct, nonoverlapping layers, excluding the influences of the soil volumes that are on top. Calculations were made with RES2DINV (Geotomo software, Malaysia).

22.3 RESULTS

Comparison of the ERM borehole instrument with the inverted GeoTom data is shown in Figure 22.2. The correlations for each single test site were 0.559 at Bornim, 0.709 at Golzow, and 0.896 at Kassow. No measurements with the ERM were made at Beckum. The τ correlation over all fields was 0.703 (Pearson's r is 0.871 for log-log transformed values). This comparison confirms the reliability of the GeoTom surface readings.

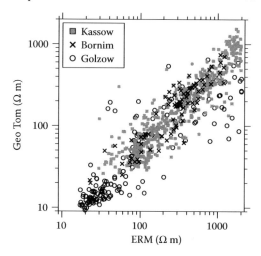

FIGURE 22.2 Log-log scatterplot of inverted surface measurements by the GeoTom versus borehole readings by the Earth Resistivity Meter (ERM).

Correlation results for ARP03, Veris 3100, EM38 (EM38-DD), and CM-138 are presented in Table 22.1. EM38-DD and EM38-MK2 were only tested at Bornim. The results for the EM38-DD are presented in Table 22.1; they are similar to correlations obtained with an EM38 in two runs using the horizontal and the vertical mode. The horizontal mode output of the EM38-MK2 can be related best to GeoTom level 2 ($\tau = 0.606$), the vertical mode output can be related to GeoTom 3 with $\tau = 0.666$. Repeated runs yielded a correlation of 0.721 (horizontal) and 0.797 (vertical).

Repeated runs with the ARP in Golzow achieved a τ of 0.839, 0.874, and 0.864 for the first, second, and third levels. Repeated runs with the CM-138 at Golzow produced a τ of 0.794 in the best case and 0.598 in the worst case. Two runs with Veris in Bornim correlate with a τ of 0.854 (level 1) and 0.843 (level 2).

22.4   DISCUSSION

In addition to the accuracy of the instruments and positioning errors, site conditions can influence correlations. They may have been variations between the different dates of measurements. In our case, repeated runs at the same date show correlations that are not substantially better than correlations between data collected at different dates. To explain this, we can think of a number of reasons. First, the statistical procedure we have chosen is robust. Minor deviations and outliers have little influence on τ, and nonlinearities of relationships have no effect at all as long as monotonically increasing or decreasing models can describe them. For example, soil temperature is not influencing τ because of the monotonic increase of EC_a with rising temperature. Another reason can be the positive correlation of some of the most important factors influencing soil EC_a on nonsaline mineral soils—namely, soil texture and water content (Lück and Eisenreich, 2001). Under the temperate climatic conditions of Germany, in addition to marshland, soil salinity is not an important factor. As long as mineral soils are not affected by spatiotemporal variations of groundwater level, lateral water flow or diverging water uptake by crops, and soil water content largely depends on soil texture. On these soils, organic matter content and cation exchange capacity can be correlated with soil texture as well (Lück and Eisenreich, 2001; Rogasik et al., 2002). Stability of EC_a patterns over time was also observed by other authors (Sudduth et al., 2003).

Comparing the different soilscapes, correlations were best when the rate of fluctuation in EC_a was low in the horizontal and vertical directions (Beckum). Vice versa, Bornim with its extremely heterogeneous soil, shows the worst correlation. When we try to compare correlations of different instruments, it is very important to bear in mind how electrode spacing influences the strength of correlation. None of the GCR instruments had exactly the same spacing as the GeoTom (even the ARP deviates because of the trapezoid arrangement of its electrodes). Best correlations were usually found when electrode configuration (or equivalent coil configuration with EMI) comes *close* to the electrode configuration of the GeoTom. Correlations could be better when configurations are matching exactly (e.g., Veris level 1 will show a higher correlation with GeoTom spacing between 20 and 30 cm). Therefore, one should not easily rate an instrument as superior by this kind of comparison. In our tests, most instruments produce good correlations regarding electrode spacing, depth of investigation, and soil variability. The CM-138 clearly is an exception. Correlations are generally lower and sometimes inconsistent. Repeated runs show low accuracy. There seem to be a number of reasons for this behavior; temperature influences the CM-138 substantially, it shows erratic drifts during stationary measurements, and it has very low dynamics in signal response (Dabas et al., 2004). These problems remained even after the instrument was returned to the manufacturer. The EM38 is subject to ambient influences and drifts as well (Sudduth et al., 2001), but it was much more reliable than the CM-138 (Dabas et al., 2004). To improve temperature stability of the EM38, Geonics Ltd. has been offering an update since 2004.

In most cases, ARP level 1, 2, and 3 correlated best with GeoTom level 1, 2, and 3. Veris levels 1 and 2 roughly coincide with GeoTom levels 1 and 2. EM38 (-DD) in the vertical mode can be related to GeoTom level 4. Two exceptions probably can be treated as "outliers": Veris level 2 does not show the best correlation with GeoTom level 2 in Kassow. This may be due to poor soil contact as a thick turf covered the soil surface. Also in Kassow there is an "irregular" best correlation of EM38 V mode with GeoTom level 3. We cannot explain this at the moment.

Correlation generally increased with depth of investigation. This can be due to a smoothing effect, which improves correlation. Because wider electrode spacing is integrating larger soil volumes, small-scale variability is smoothed out.

Comparing the correlations of EM38 horizontal mode in Beckum, Bornim, and Golzow, it is remarkable that highest τ in Beckum was found at GeoTom level 4, and τ in Golzow was highest at level 1 and in Bornim at level 3. We explain this by higher sensitivity to conductive structures which is characteristic for EMI methods (Dabas and Tabbagh, 2003). Golzow has layered soils where sometimes loam covers sand. In these cases, the EM38 reacts preferentially to the conductive structures in the topsoil, and there is less response to signals from deeper layers. Beckum has opposite conditions: highly conductive clay appears regularly at the bottom of the soil profile. It is interesting to find out whether the DC methods show an opposite behavior—that is, higher sensitivity to resistive structures. The results from Beckum seem to confirm this assumption. ARP 2 and Veris 2 showed the highest correlations with GeoTom 1 in Beckum, but correlations were better with Geo-Tom 2 at all the other sites. This explains the observations from earlier studies (Suddath et al., 2003), where correlation between Veris 3100 and EM38 was influenced by soil profile layering.

Evaluating depth sensitivity, it turns out that correlation over different GeoTom levels changes more gradually with the EMI methods than with the GCR methods. This means that EMI methods have less distinct depth sensitivity. Dabas and Tabbagh (2003) give theoretical reasons for this behavior.

22.5   CONCLUSIONS

The theoretical assumption that depth sensitivity is influenced by layering conductive and resistive structures was confirmed by our field studies. The GCR and the EMI method react different on layering. The EMI method is largely influenced by conductive layers, and the GCR method is more influenced by resistive layers.

With respect to the dominant rooting zone, which is located within 60 cm depth in the case of cereals, the Veris 3100 and the ARP03 are more suitable to obtain information from shallow depths than the EM38. When the EM38 is used, one should consider the horizontal instead of the vertical mode to investigate more shallow soil horizons.

The EM38-MK2 is an interesting prototype and shows how the EMI method is suitable for mapping the topsoil and the rooting zone. The EM38-DD and the EM38-MK2 actually consist of two measurement units, it is remarkable that both instruments are not influencing each other significantly and that they are clearly providing readings from different depths.

REFERENCES

Dabas, M., Apport de l'ARP (résitité électrique) à la connaissance des sols agricoles, in *Actes des 7èmes Journées Nationales de l'Etude des Sols, Octobre 22-24, 2002*, Baize, D. and Duval, O., Eds., afes, Orléans, France, 2002, 267.

Dabas, M. and Tabbagh, A., A Comparison of EMI and DC methods used in soil mapping—theoretical considerations for Precision Agriculture, in *Prec. Agric., Proc. 4th Eur. Conf. Prec. Agric.*, Stafford J. and Werner, A. Eds., Wageningen Academic Publishers, The Netherlands, 2003, 121.

Dabas, M. et al., A comparison of different sensors for soil mapping: The ATB case study, in *Proc. 7th Int. Conf. Prec. Agric.*, ASA/CSSA/SSSA, Madison, WI, 2004, CD-ROM.

Geonics, *EM38-MK2 Ground Conductivity Meter Operating Manual,* Geonics Ltd, Mississauga, Ontario, Canada, 2002.

Lück, E. and Eisenreich, M., Electrical conductivity mapping for precision agriculture, in *Proc. 3rd Eur. Conf. Prec. Agric.*, Blackmore, S. and Grenier, C., Eds., Agro, Montpellier, France, 2001, 425.

McNeill, J.D., *Electromagnetic Terrain Conductivity Measurement at Low Induction Numbers*, Tech. Note TN-6, Geonics Ltd, Mississauga, Ontario, Canada, 1980.

Rogasik, J. et al., Relations between soil fertility and plant yield influencing variable rate fertilization, in *Proc. Int. Conf. Long-Term Exp., Sustainable Agric. Rural Develop., Vol. 1*, Debrecen, Hungary, 2002, 65.

Sudduth, K.A., Drummond, S.T., and Kitchen, N.R., Accuracy issues in electromagnetic induction sensing of soil electrical conductivity for precision agriculture, *Comput. Electron. Agric.* 31, 239, 2001.

Sudduth, K.A. et al., Comparison of electromagnetic induction and direct sensing of soil electrical conductivity, *Agron. J.*, 95, 472, 2003.

Section V

Ground-Penetrating Radar Case Histories

Although probably not employed quite to the extent of soil electrical conductivity measurement methods, such as resistivity and electromagnetic induction, there have been a substantial number of agricultural applications found for ground-penetrating radar (GPR). An historical perspective on the use of GPR for agricultural purposes and a description of the method can be found in Chapters 1, 3 and 7. Section V includes case histories describing just a few GPR agricultural applications.

One important recent development has been use of GPR to map the soil volumetric water content. Spatial maps of soil water content can be valuable for assessing soil drainage conditions and scheduling irrigation events. The Chapter 23, 24, and 25 case histories cover several aspects regarding the use of GPR to measure the volumetric water content in soil. GPR has also been used in agricultural settings (including golf courses) to delineate various subsurface features. The case histories in Chapters 26, 27, 28, and 29 provide examples for employing GPR to acquire information on subsurface features such as the orientation of preferential water flow pathways, soil layer depths and thicknesses, and the locations of buried drainage pipes. Plant root biomass determination using GPR is a fairly new area of investigation having significant implications for forestry along with planting, harvesting, and yield mapping of root crops like potatoes and sugar beets. The last case history in Chapter 30 is focused on GPR root biomass determination within forest environments. As indicated by the Section V case histories, there are certainly a variety of potential agricultural GPR applications, and there are many more GPR uses that could not be included in this section due to book length considerations.

23 GPR Surveys across a Prototype Surface Barrier to Determine Temporal and Spatial Variations in Soil Moisture Content

William P. Clement and Andy L. Ward

CONTENTS

23.1   INTRODUCTION

Measurement of the electromagnetic (EM) properties of the subsurface can provide estimates of important hydrological parameters such as porosity in the saturated zone and soil moisture content in the vadose zone. Importantly, EM methods can be deployed across the ground surface and are thus a noninvasive method to sample the subsurface. Ground-penetrating radar (GPR), a high-frequency EM method, acquires data quickly and at high spatial densities to provide a detailed distribution of the desired property.

Many experiments have used GPR to test the validity of using radar energy to map soil moisture content (Berktold et al., 1998; Chanzy et al., 1996; Charlton, 2000; Du and Rummel, 1994:, Greaves et al., 1996; Huisman et al., 2001; Lesmes et al., 1999; van Overmeeren et al., 1997; Weiler et al., 1998). Most of these experiments were small, test-of-concept surveys. Grote et al. (2002) used GPR to monitor the volumetric water content in soils for a highway construction and maintenance application. Grote et al. (2003) and Hubbard et al. (2002) have recently used GPR to map soil moisture content across a vineyard. GPR measurements are converted to soil moisture content and have shown promising results for measuring soil moisture content.

We present data collected at the U.S. Department of Energy's Hanford Site in southeastern Washington where a study was conducted to assess the viability of GPR to monitor changes in the soil moisture distribution in the engineered barrier. At Hanford, surface barriers are being tested

FIGURE 23.1 View of the prototype surface barrier looking northeast. The surface is covered with sage-brush planted in rows. The basalt riprap is seen on the left side of the photograph. In the background is an irrigation system.

to determine their ability to minimize infiltration into buried wastes (Ward and Gee, 1997). The surface of the barrier was constructed with a 2 percent slope and was revegetated with a mixture of rabbit brush, sage brush, and native grasses (Figure 23.1). The barrier is above grade with a steep (2:1) rip-rap protective side slope to the east and a more gentle (10:1) gravel side slope to the west. Irrigation equipment was located at the north end of the site for the March and May surveys. The barrier is engineered to store up to 600 mm of water in the winter and release this water by evapo-transpiration over the rest of the year while limiting recharge to 0.5 mm/yr or less. The upper 2.45 m of the surface barrier consists of a 1 m thick layer of silt loam with 15 percent pea gravel; a 1 m thick layer of silt loam; a 0.15 m sand filter; and a 0.3 m thick gravel filter underlain by asphalt. Our task was to evaluate the potential for using GPR to determine the temporal and spatial changes in the soil moisture content and water storage in this upper ~2.5 m zone above the asphalt. Although the research at Hanford does not involve a precision agriculture application, the goals of the project, to map the changes in soil moisture, are similar to the goals for precision agriculture.

23.2 METHODS (GPR SURVEYS)

GPR sends radar frequency EM energy into the ground through a transmitting antenna. This energy is recorded at a receiving antenna placed near the transmitter. For the data used in this study, 100 MHz antennas were used. The sample interval in time was 0.8 ns and 500 samples were acquired for each trace for a recording window of 400 ns. We stacked (summed each time sample) the data 64 times for the March data and 32 times for the rest of the data. The reduced number of stacks increased the acquisition rate, yet did not deteriorate the data quality.

Two acquisition steps are necessary to conduct this type of survey. The first step is using the common midpoint (CMP) method to determine the optimal antenna separation or offset. The CMP method acquires data such that the midpoint between the antennas is the same. The two antennas are started close to each other, then each antenna is moved away from the other at a set increment. The antennas are stepped away until the data contain easily identified air wave, ground wave, and reflection events. In our experiment, the optimal offset is chosen as the distance at which the air wave and ground wave are sufficiently separated in time so that they do not interfere with each other (Figure 23.2).

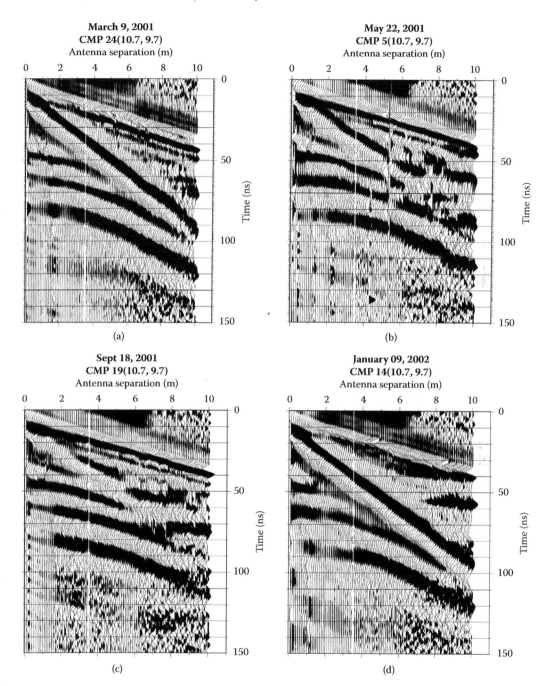

FIGURE 23.2 Common midpoint (CMP) surveys from the same location. The CMPs were acquired in (a) March, (b) May, (c) September, and (d) January. Note the changes in the ground wave and reflection character from March to January. The vertical white line in each plot shows the optimal antenna separation.

The second step acquires the wide-offset reflection (WOR) data. These profiles are denoted as WOR profiles because the offset between the antennas is much wider than is usually used in standard GPR reflection profiles. The method uses an optimal antenna separation derived from CMP surveys to reliably identify the air wave and the direct ground arrival times.

Wide Offset Reflection Acquisition

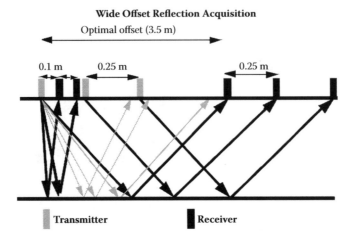

FIGURE 23.3 Acquisition geometry for the wide-offset reflection (WOR) profiles. The transmitter antenna is held stationary and the receiver antenna is moved away at 0.1 m increments until the optimum offset is obtained. This stage is the walkaway stage. The antennas are then moved together in 0.25 m increments across the profile.

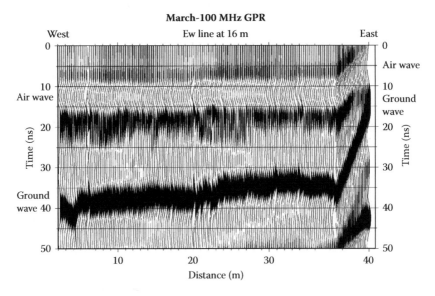

FIGURE 23.4 An east–west wide-offset reflection (WOR) profile along the prototype surface barrier. The walkaway start to the survey is at the east end of the profile. The air and ground wave are labeled. The polarity of the ground wave is opposite to the polarity of the air wave. Note the different length scale between about 38.5 m and 40 m. This scale change is to more easily see the walkaway section of the plot.

The WOR survey is a two-stage process. The first stage of the WOR survey starts with the transmitter and receiver close together, then one antenna is moved away from the other in small increments until the predetermined optimal offset is reached (Figure 23.1). This first stage of acquisition is referred to as a walkaway survey because one antenna is "walked away" from the other. Once the antenna separation in the walkaway stage reaches the optimal offset, both antennas are moved together at a constant step size. Figure 23.4 shows data from a WOR survey. The easily identified air and ground wave in the walkaway section of the profile can be correlated with the same events in the constant, wide-offset section. Without the walkaway start, these events could easily be misidentified. For this experiment, the antenna were located 0.1 m apart at the start of the walkaway stage, then the receiving antenna only was moved 0.1 m until the antenna separation became 3.5 m. At this point, both antennas were moved 0.25 m per trace, keeping a constant antenna separation of 3.5 m.

To determine the soil moisture content, two different radar events are used. The ground-coupled air wave travels directly between the transmitter and the receiver. This event is the first arrival and has the EM velocity of air. The direct ground wave travels along the surface. The EM velocity from this event corresponds to the EM velocity of the ground. We use the travel times of these two phases and the distance between the antennas to determine the EM velocity. The EM velocity character-izes the material through which the energy propagates. For our purposes, soil consists of a matrix and pore space. This pore space is either filled with water or air. Water has a slow EM velocity (0.033 m/ns), and air has a fast EM velocity (0.3 m/ns). In most areas, the material does not change in terms of composition or structure over the short time intervals used in these types of experiments. Thus, changes in soil moisture at the prototype surface barrier cause changes in EM velocity. Slower measured EM velocities indicate a higher amount of water in the pore space.

The method indirectly infers the dielectric constant of the material through which the energy propagates. The dielectric constant or dielectric permittivity represents the ability of a material to polarize or store energy through separation of bound charges. The dielectric constant (κ) can be computed from the EM velocity (v):

$$\sqrt{\kappa} = \frac{c}{v} \tag{23.1}$$

where c is the EM velocity of light (0.3 m/ns).

Water has a high dielectric constant of about 80. Air has a dielectric constant of 1. Dry soil materials and sediments have dielectric constants between 3 and 10. Clays and silts may have a dielectric constant as high as about 30 to 40. The large dielectric constant difference between water and air enables mapping of changes in water content across a survey.

Soil moisture content can be derived from the EM velocity of the soil using Equation (23.1) to convert first to dielectric constant. Mixing laws based on the amounts of the constituent materials present are used to convert dielectric constant to soil moisture content (Knoll et al., 1995). The dielectric constants can also be converted to soil moisture content using established petrophysical relationships such as Topp's equation (Topp et al., 1980):

$$\theta = -0.053 + .0292\kappa - 0.00055\kappa^2 + 0.000004.3\kappa^3 \tag{23.2}$$

where θ is the water content.

We analyzed the GPR profiles using a method that identifies changes in the arrival time of a known radar event and then converts this time to EM velocity (Berktold et al., 1998; Du and Rum-mel, 1994). This method is not widely applied in GPR surveys, but offers great potential to provide spatially densely sampled EM velocity measurements that can be converted to the desired param-eters, such as soil moisture content.

23.3 DATA AND ANALYSIS

We conducted 40 m long profiles to observe spatial changes in the GPR character across the proto-type surface barrier. We also visited the prototype surface barrier site four times, March, May, and September 2001, and January 2002, to determine the change in soil moisture during the different seasons of the year. At Hanford, winters are cool and wet accounting for most of the precipitation, and the summers are warm and dry. This testing methodology allowed us to observe changes in EM energy travel times that we could relate to changes in soil moisture content.

23.3.1 FINDING THE OPTIMAL OFFSET

We acquired several CMP surveys at the beginning of the March field experiment to determine the optimal offset to collect the WOR data (Figure 23.2). The optimal offset was based on the time separation of the air and ground waves. We wanted to avoid interference between the two phases so we could accurately pick the ground wave arrival time and amplitude. If the separation is too close, the air wave will interfere with the later arriving ground wave and potentially cause a mis-interpretation of the arrival time and the amplitude. From the March CMPs, we chose an optimal antenna separation of 3.5 m. The character of the GPR data changed substantially during the May and September field experiments. Fortunately, the 3.5 m offset still allowed picking of the ground wave. Attenuation of the wave's amplitude due to larger separation would have made picking the ground wave unreliable.

Processing of the GPR data consisted of a few standard procedures. We removed low-frequency noise due to the electronics in the radar unit from the data. For the CMP analysis, we bandpass filtered the data between 25 and 200 MHz to increase the signal-to-noise ratio of the arrivals. The WOR data were not filtered. For plotting the images, we used a gain function to emphasize the later arriving events (AGC) with a 25 ns window. Figure 23.2 compares CMPs from the same location for the four acquisition dates. For the March and January data, the ground wave is a strong event. The May data shows a weaker, less-extensive ground wave. (We have displayed the highest-quality CMP survey from the May profiles). The September CMPs show weak ground wave arrivals. In the March and January CMPs, the ground wave projects to arrive at 10 m antenna separation at approximately the same time (~86 ns for March; ~88 ns for January). Thus, their slopes are nearly equal, indicating that the EM velocity of the ground wave is about the same for each date. In May, the ground wave projects to arrive at ~75 ns at 10 m antenna separation. This earlier arrival time indicates that the EM velocity is faster in May.

23.3.2 WIDE-OFFSET SURVEYS

The GPR data were acquired with the same parameters for each survey, except the stacking change mentioned earlier. Each survey was started with a walkaway to help reliably identify the air and ground waves (Figure 23.4). By starting the survey with the walkaway geometry, we can more confidently pick the air and ground waves from the slope and the time intercept of the phases (Figure 23.5 and Figure 23.6). In GPR data, the input waveform has a central, large amplitude peak flanked by two smaller peaks. We picked the central peak for analysis. In addition, the air wave has the opposite polarity from the ground wave providing more confidence in the event picking (Du and Rummel, 1994). Thus, in the presented data, the air wave has a negative (white) amplitude, whereas the ground wave has a positive (black) amplitude. The surface spatial location of these events is known, so we can reference our findings to the ground location.

Two aspects of the character of the WOR GPR surveys are easily observed in Figure 23.5 and Figure 23.6. First, the ground wave is a strong-amplitude, coherent event in the March and January data. The ground wave in the May and September data is weaker in amplitude, more difficult to see, and not as coherent. Fortunately, the walkaway start of each survey makes picking the ground wave more reliable. Second, the arrival time of the ground wave varies between about 35 and 40 ns in the March and January data but occurs at about 30 ns in the May and September data. The earlier ground wave arrival indicates that the EM velocity is faster in May and September compared to March and January.

The GPR data indicate changes in the radar response along the profile and throughout the year. The changes in radar response along the profile are indicative of spatial differences in water content. Because of the specifications used to minimize variability in the silt-loam layer, spatial differences in moisture are mostly due to differences in evapotranspiration. Temporal changes in the radar character are due to changes in soil moisture content resulting from decreasing precipitation and high

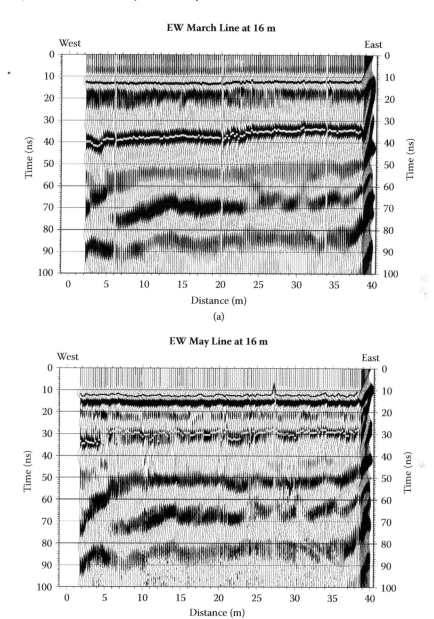

FIGURE 23.5   The wide-offset reflection (WOR) data from (a) March and (b) May. Note the weaker ground wave amplitudes and the earlier arrival times in the May profile compared to the March profile. The air wave picks are marked by the black line; the ground wave picks are marked by the light gray line. Bad picks on the right side (walkaway stage) of the sections are due to a poorly defined picking window and are ignored in the interpretation.

rates of evapotranspiration in the summer months. During the winter months, higher precipitation and reduced evapotranspiration cause an increase in soil moisture.

The WOR GPR data analysis consists of picking the arrival times of the air and ground waves. To quantify changes in soil moisture content, we first determine the dielectric constant of the material sampled by the radar energy. The EM velocity of the air and ground waves is simply the distance between the antennas divided by the travel time between antennas. The EM velocity is then

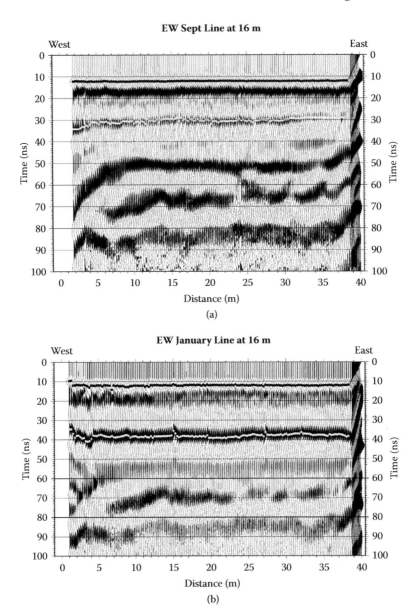

FIGURE 23.6 Similar plots to Figure 23.5, showing data from (a) September and (b) January. Note the weaker ground wave energy and the earlier arrival time of the September profile. The September data are more similar to the data from May. The January data have similar arrival times and amplitudes compared to the March data. Annotations are the same as Figure 23.5.

converted to the dielectric constant by Equation (23.3). A more direct way to determine the dielectric constant is to use the arrival times of the air and ground arrivals without first converting to velocity. The square root of the dielectric constant can be computed from the air and ground wave travel time picks (Huisman et al., 2001):

$$\sqrt{\kappa} = \frac{c\left(t_{ground} - t_{air}\right) + x}{x} \qquad (23.3)$$

where c is again the EM velocity in air, x is antenna separation (3.5 m), t_{ground} is the arrival time of the ground wave, and t_{air} is the arrival time of the air wave.

The dielectric constant estimate represents the dielectric constant in the material sampled by the radar wave. Through experiments, Berktold et al. (1998) determined that the ground wave samples below the surface to a depth between one-half to one wavelength. The wavelength (λ) is computed from the frequency (f) and the velocity of the phase by

$$\lambda = \frac{v}{f} \tag{23.4}$$

In the GPR data, the antenna frequency is 100 MHz and the EM velocity is about 0.12 m/ns. Thus, the dielectric constant is for the material to a depth of about 0.6 to 1.2 m.

Figure 23.7 shows the changes in EM velocity along the GPR profile over the course of the experiment. To more easily see the trends in the values, we applied a five-point running average to smooth the values. In March, the EM velocity is about 0.09 to 0.095 m/ns on the west side of the profile, then increases to about 0.105 m/ns on the east side. The EM velocity increases significantly in May to EM velocities of about 0.12 m/ns, then remains about the same or slightly slower in September. The increase in EM velocity from west to east is not observed in the May data but appears in the September data. This change in trend from May to September may suggest greater evapotranspiration on the west side of the prototype surface barrier. In January, the EM velocity decreased to about 0.09 to 0.095 m/ns along the profile.

Figure 23.8 shows the soil moisture content estimates derived from Topp's equation (Equation (23.2)). Again, we applied a five-point running average to smooth the values. The graph shows reciprocal trends from Figure 23.7; high soil moisture content in March and January, and low moisture content in May and September. From the plot, we infer that the soil dries out from March to May. The soil moisture content remains about the same through September indicating a lack of precipitation. By January, the soil moisture has increased to the highest values observed. Between January and March, the soil has again started the cyclical drying process.

We also briefly investigated the effects of soil moisture on amplitude. Du and Rummel (1994) note that the amplitude of the ground wave increases as $\sqrt{\kappa}$ relative to that of the air wave. Thus, the ground wave is better observed in wet soils compared to dry soils. Reviewing Figure 23.5 and Figure 23.6, the ground wave amplitudes weaken in the drier spring and summer months relative to

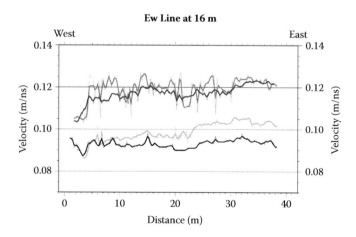

FIGURE 23.7 Velocity changes along the profile from March to January. The thick lines are the five-point smoothed average of the thin lines. The March data are displayed in light gray, May is gray, September is dark gray, and January is black.

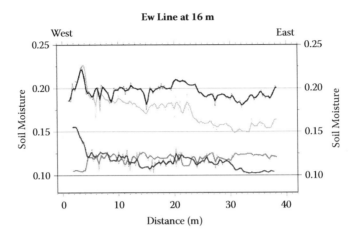

FIGURE 23.8 Soil moisture changes over time along the profile. As in Figure 23.7, the thick lines are the five-point smoothed average of the thin lines. The March data are displayed in light gray, May is gray, September is dark gray, and January is black.

the winter months. The amplitudes from the GPR are influenced by many variable factors, including the battery strength and coupling between the antenna and the ground. By normalizing the ground wave amplitude with the air wave amplitude, we can remove coupling effects and other instrument fluctuations in the ground wave amplitudes between the surveys. Again, the changes in amplitude are due to changes in the moisture conditions. Our preliminary analysis indicates that a correlation exists between high amplitude and wetter soil.

23.4 CONCLUSIONS

GPR has great potential to observe temporal and spatial changes in the soil moisture content of engineered barriers. Changes in the GPR character were easily observed over the course of this experiment. The changes correspond to EM velocity changes that indicate the soil moisture varies in an expected way. Amplitude analysis also indicates the soil moisture changes similar to the results from the EM velocity analysis. Because the EM velocity changes are large between air and water, GPR soil moisture estimates will probably be reliable. Further work is needed to calibrate the EM velocity–soil moisture content relation to specific sites. We also need to test the accuracy of quantitative estimates of the soil moisture using this method. The method can be made more convenient by developing equipment for fast data acquisition and interpretation.

GPR proved successful at imaging changes in soil moisture along the profile and over the year. The spatial changes in soil moisture from May to September may indicate greater evapotranspiration on the west side of the prototype surface barrier. The seasonal GPR changes correlate with the seasonal changes in soil moisture storage in the prototype barrier.

Using GPR to determine soil moisture content has many advantages over traditional methods: (1) the cost and speed of data acquisition is relatively inexpensive; (2) the large spatial sampling density provides greater coverage; and (3) the method is nonintrusive. GPR is a promising technique to determine the three-dimensional distribution of the soil moisture content in the subsurface.

23.5 ACKNOWLEDGMENTS

The U.S. Department of Energy EM-50 Technical Task Plan RL21SS20, "The Application of Electromagnetic Techniques to Cover Performance Monitoring," funded this work. Battelle operates the Pacific Northwest National Laboratory for the U.S. Department of Energy under Contract DE-AC05-76RL01830.

REFERENCES

Berktold, A., K. G. Wollny, and H. Alstetter, Subsurface moisture determination with the ground wave of GPR, *Proceedings from GPR '98, 7th International Conference*, May 27–30, 675–680, 1998.

Chanzy, A., A. Tarussov, A. Judge, and F. Bonn, Soil water content determination using a digital ground-penetrating radar, *Soil Science Society of America Journal*, 60, 1318–1326, 1996.

Charlton, M., Small scale soil-moisture variability estimated using ground penetrating radar, *SPIE Proceedings Series*, 4084, 798–804, 2000.

Du, S., and P. Rummel, Reconnaissance studies of moisture in the subsurface with GPR, Proceedings from GPR '94 5th International Conference, June 12–16, 1241-1248, 1994

Greaves, R. J., D. P. Lesmes, and M. N. Toksoz, Velocity variations and water content estimates from multi-offset, ground-penetrating radar, *Geophysics*, 61, 683–695, 1996.

Grote, K., S. Hubbard, and Y. Rubin, GPR monitoring of volumetric water content in soils applied to highway construction and maintenance, *The Leading Edge*, 21, 482–485; 504, 2002.

Grote, K., S. Hubbard, and Y. Rubin, Field-scale estimation of volumetric water content using ground-penetrating radar ground wave techniques, 39, 1321, doi:10.1029/2003WR002045, 2003.

Hubbard, S., K. Grote, and Y. Rubin, Mapping the volumetric soil water content of a California vineyard using high-frequency GPR ground wave data, *The Leading Edge*, 21, 552–559, 2002.

Huisman, J. A., C. Sperl, W. Bouten, J. M. Verstraten, Soil water content measurements at different scales: Accuracy of time domain reflectometry and ground-penetrating radar, *Journal of Hydrology*, 245, 48–58, 2001.

Knoll, M., R. Knight, and E. Brown, Can accurate estimates of permeability be obtained from measurements of dielectric properties?, *SAGEEP95*, Environmental and Engineering Geopohysical Society, Denver, Colorado, 25–36, 1995.

Lesmes, D. P., R. J. Herbstzuber, D. Wertz, Terrain permittivity mapping: GPR measurements of near-surface soil moisture, *SAGEEP99*, Environmental and Engineering Geopohysical Society, Denver, Colorado, 575–582, 1999.

Topp, G. C., J. L. Davis, and A. P. Annan, Electromagnetic determination of soil water content: Measurements in coaxial transmission lines, *Water Resources Research*, 16, 574–582,1980.

van Overmeeren, R. A., S. V. Sariowan, J. C. Gehrels, Ground penetrating radar for determining volumetric soil water content: Results of comparative measurements at two test sites, *Journal of Hydrology*, 197, 316–338, 1997.

Ward, A.L., and G.W. Gee, Performance and water balance evaluation of a field-scale surface barrier, *Journal of Environmental Quality*, 2, 694–705, 1997.

Weiler, K. W., T. S. Steenhuis, J. Boll, and K.-J. S. Kung, Comparison of ground penetrating radar and time-domain reflectometry as soil water sensors, *Soil Science Society of America Journal*, 62, 1237–1239, 1998.

24 Soil Water Content Measurement Using the Ground-Penetrating Radar Surface Reflectivity Method

J. David Redman

CONTENTS

24.1   INTRODUCTION

Ground-penetrating radar (GPR) has been used extensively in the borehole, multi-offset subsurface reflection and multi-offset ground wave configurations to measure soil water content (Davis and Annan, 2002; Huisman et al., 2003). These GPR methods determine soil water content from a measurement of the time required for an EM wave to travel a known distance within the subsurface. This case study discusses the GPR surface reflectivity method used for the determination of soil water content from measurements of the electromagnetic (EM) reflection coefficient at the air–ground interface, using a GPR elevated ~1 m above the surface.

The time domain reflectometry (TDR) method is, at present, one of the most widely used methods for measuring soil water content. Unfortunately, the TDR method cannot be used to cover large areas efficiently and provides only local point measurements covering a relatively small surface area (~0.005 m^2). For many agricultural applications, the ability to efficiently map soil water content over large areas at a resolution of ~1 to 10 m^2 would improve irrigation practices and reduce water requirements. The surface reflectivity method could provide this capability by providing rapid coverage of relatively large areas, because the GPR can be vehicle mounted and is not in contact with the ground. The method also provides better surface coverage by giving average water content measurements over typical surface areas of ~1 m^2.

Previous field studies have demonstrated that the method has good potential (Chanzy et al., 1996; Redman et al., 2002; Serbin and Or, 2002). It has been shown that the water content measured with the GPR surface reflectivity method was more variable than and, in some cases, was substantially different from the TDR-determined water content (Redman et al., 2002). These effects were attributed to surface scattering, vertical stratification of the water content, and other spatial variability in water content. Numerical modeling of a two-layer soil water content profile demonstrated that the measured water content could, depending on the thickness of the upper layer, be overestimated for a wet layer over a dry layer and underestimated for a dry layer over a wet layer (Redman et al., 2003). This case study will outline the surface reflectivity methodology, present field data examples, and the results of numerical modeling on the effects of surface roughness on GPR measured soil water content.

24.2 METHODOLOGY

The GPR, consisting of a transmitter and receiver, is elevated above the surface at a typical height of ~1 m as shown in Figure 24.1. A vehicle-mounted implementation of a surface reflectivity system (Figure 24.2) shows the GPR mounted to the left of the vehicle, with GPS providing positioning information. The transmitter emits a short EM pulse. Part of the energy is transmitted into the soil and part is reflected back into the air. The reflected pulse amplitude is measured in the receiver.

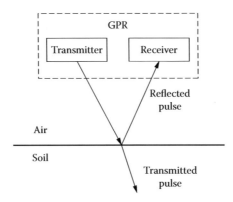

FIGURE 24.1 Ground-penetrating radar (GPR) configuration for surface reflectivity method.

FIGURE 24.2 Vehicle-mounted ground-penetrating radar (GPR) surface reflectivity system for acquiring soil water content measurements. GPR (Noggin 500) is shown on the left side of the vehicle. GPS provides positioning information.

For the incident pulse at the air–soil interface, the reflection coefficient R is determined by the contrast in relative dielectric permittivity (K) between the air (K = 1), and the soil (K = K$_s$):

$$R = \frac{1-\sqrt{K_s}}{1+\sqrt{K_s}} \qquad (24.1)$$

It is assumed that the conductivity is sufficiently small to be ignored, that the surface is flat, and that the soil properties are homogeneous. The above relationship can be rearranged to determine the soil permittivity from a measured reflection coefficient. In practice, the GPR system is calibrated with metal plate target with a known reflection coefficient of −1. The reflection coefficient can be measured by calibrating the GPR system with a sufficiently large metal plate (dimensions > Fresnel zone diameter) placed at the surface (Redman et al., 2002). The magnitude of the reflection coefficient is the ratio of the amplitude A$_r$ of the reflected wavelet from the soil surface to the amplitude A$_m$ of a wavelet measured at the same elevation over a metal plate target with a reflection coefficient of −1. Using Equation (24.1), K$_s$ can be determined from the ratio of the amplitude of the wavelet (A$_r$) reflected from the surface and from the amplitude of the wavelet (A$_m$) reflected from a metal plate:

$$K_s = \left(\frac{1+\dfrac{A_r}{A_m}}{1-\dfrac{A_r}{A_m}} \right)^2 \qquad (24.2)$$

The soil water content θ is then estimated using a suitable mixing formula. For the data presented in this case study, the empirical relationship of Topp et al. (1980), was used:

$$\theta = -0.053 + 0.0292 K_s - 5.5 \times 10^{-4} K_s^2 + 4.3 \times 10^{-6} K_s^3 \qquad (24.3)$$

24.3   SAMPLING VOLUME

The electrical properties of the near surface, directly below the GPR antennas, determine the measured surface reflectivity. This region or sampling volume for the method can be roughly approximated by a vertical cylindrical zone extending from the surface to some defined sampling depth. For GPR antennas at height h above the surface, with a wavelength λ in air at their center frequency, the diameter D of the circular sampling zone at the surface (the Fresnel zone) is given by Equation (24.4). For a 500 MHz center frequency GPR at a height of 1 m, the diameter D is ~1.1 m:

$$D = \sqrt{\frac{\lambda^2}{4} + 2\lambda h} \qquad (24.4)$$

The sampling depth is controlled by the subsurface EM attenuation and velocity. Numerical modeling has shown that for a 500 MHz GPR system, this sampling depth is ~0.09 m for a wet soil with a volumetric water content of 20 percent and ~0.18 m for a dry soil with a volumetric water content of 5 percent (Redman et al., 2003). The sampling depth is inversely proportional to the center frequency of the GPR system, allowing lower-frequency GPR antennas to sample more deeply.

24.4   FIELD MEASUREMENTS

Using the methodology described previously, surface reflectivity measurements were collected over a number of sites (Redman et al., 2002, 2003). These examples demonstrate the variability often

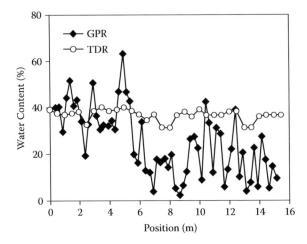

FIGURE 24.3   Soil water content profile from grass (0 to 6 m) onto cornfield (6 to 16 m).

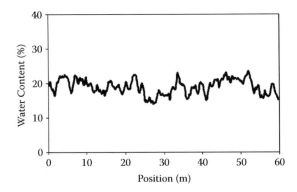

FIGURE 24.4   Example soil water content profile on grass field collected with vehicle-mounted ground-penetrating radar (GPR) system shown in Figure 24.2.

observed in the surface reflectivity data. At the first site, a profile was collected starting on a relatively flat grass area and running onto a corn field perpendicular to the corn rows. The corn had been harvested at this time, so only the short stubble remained. The data, at 0.25 m sampling intervals, were collected with a Sensors & Software pulseEKKO 1000 system using antennas with a center frequency of 450 MHz, separated by 0.75 m and elevated 1.3 m above the surface. A TDR profile of water content was measured along the same profile using 20 cm long probes. The soil water content measured with the surface reflectivity method (Figure 24.3) is clearly more variable than the TDR results and over the corn field is lower, on average, than the TDR.

In another example data set, a profile was acquired on a grass field using a Sensors & Software, Noggin 500 GPR. This GPR has a center frequency 500 MHz, and an antenna separation of 0.15 m. Data were collected continuously with a vehicle-mounted system. GPS provided positioning information resulting in a spatial sampling interval of ~0.15 m. Water contents ranging from 14 to 18 percent were measured at five locations on this field. The variabilities in water content observed in the GPR-based measurements (Figure 24.4) are higher than suggested by the limited number of TDR data.

These example data sets and others that have been acquired demonstrate that there is significant apparent spatial variability in the measured soil water content and, in some cases, significantly different water contents from those measured with TDR. Redman et al. (2002, 2003) attributed these observed differences to surface scattering, lateral variations in the depth dependence of the water content, or other spatial variability in the water content. This was demonstrated with numerical modeling that showed the GPR measured water content can be strongly influenced by stratification of the water content distribution.

24.5   SCATTERING FROM SURFACE ROUGHNESS

As discussed, surface scattering has also been assumed to be a contributor to the observed differences between water content measured with TDR and the GPR surface reflectivity method. Numerical modeling was performed to demonstrate the importance of this effect. Modeling software (GPRMAX2D) developed by Giannopoulos (1997) was used to compute the model response. This software uses a finite difference time domain (FDTD) two-dimensional (2D) algorithm. The standard Ricker wavelet was used to describe the source pulse shape. In this 2D model, all structures extend an infinite length perpendicular to the model plane. An example model (Figure 24.5a) was used to simulate the response for the data example presented in Figure 24.3, the profile from the grass surface onto the corn field. In the model, the grass is perfectly flat and the corn rows are modeled as half cylinders (diameter 0.3 m) spaced every 76 cm, the spacing of the corn rows. The modeled media has a water content of 40 percent corresponding to a relative dielectric permittivity of 25. The modeled GPR system is elevated 1.3 m above the surface, and the pulse has a center frequency of 450 MHz.

The results (Figure 24.5b) for this simple model demonstrate that the scattering effects in the corn field are substantial and consistent with the variability observed in the water content measurements in the field example (Figure 24.3). This example presents one of the worst-case effects of scattering. The results of field measurements for a grass field (Figure 24.4) are more typical of the type of variation that can be attributed to scattering.

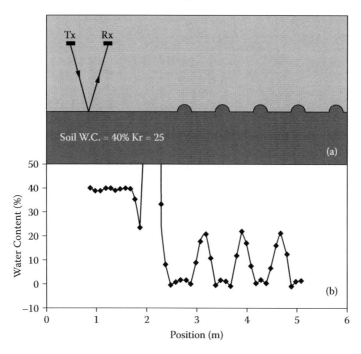

FIGURE 24.5   Two-dimensional model (a) for the case of a grass field abutting a cornfield and modeled water content measurements (b). The water content of the modeled media is 40 percent, giving a relative dielectric permittivity of 25. The water content at 2 m is not shown because the high reflection coefficient from scattering at this position in the model gives unrealistic values.

24.6 CONCLUSIONS

The GPR surface reflectivity method can be used to measure soil water content subject to the limitations imposed by scattering and stratification of water content. The effective sampling depth can be increased and the scattering effects reduced by using a GPR with a lower center frequency. Elevating the antenna reduces the water content variability related to scattering, but the measured values will be less than for a surface without scattering effects. The simplistic interpretation model used for estimating water content may be enhanced to accommodate scattering and stratification effects. At this time, robust and practical processes to do this do not exist and are the subject of further research.

REFERENCES

Chanzy, A., Tarussov, A., Judge, A., and Bonn, F., Soil water content determination using a digital ground-penetrating radar, *Soil Sci. Soc. Am. J.*, 60, 1318, 1996.

Davis, J.L. and Annan, A.P., Ground penetrating radar to measure soil water content, in *Methods of Soil Analysis: Physical Properties*, 5, Dane J.H., and Topp G.C., Eds., Soil Science Society of America Book Series, 446, 2002, chap. 3.1.3.5.

Giannopoulos, A., The investigation of transmission-line matrix and finite-difference time domain method for the forward problem of ground penetrating radar. D. Phil. thesis, Department of Electronics, University of York, UK, 1997.

Huisman, J.A., Hubbard, S.S., Redman, J.D., and Annan, A.P., Measuring soil water content with ground penetrating radar: A review, *Vadose Zone J.*, 2, 476, 2003.

Redman, J.D., Davis, J.L., Galagedara, L.W., and Parkin, G.W., Field studies of GPR air launched surface reflectivity measurements of soil water content, in *Proc. of the 9th International Conference on Ground Penetrating Radar*, Santa Barbara, California, 2002, 156.

Redman, J.D., Galagedara, L.W., and, Parkin, G.W., Measuring soil water content with the ground penetrating radar surface reflectivity method: Effects of spatial variability, Paper 032276, in *Proc. ASAE 2003 Annual International Meeting*, Las Vegas, NV, 2003.

Serbin, G. and Or, D., Near-surface soil water content measurements using horn antenna radar: Methodology and overview, *Vadose Zone J.*, 2, 500, 2002.

Topp, G.C., Davis, J.L., and Annan, A.P., Electromagnetic determination of soil water content: Measurements in coaxial transmission lines, *Water Resour. Res.*, 16, 3, 574, 1980.

25 Assessing Spatial and Temporal Soil Water Content Variation with GPR

Bernd Lennartz, Britta Schmalz, Derk Wachsmuth, and Heiner Stoffregen

CONTENTS

25.1 INTRODUCTION

The soil water content is an important variable in soil physics and agricultural production. At small scales, TDR (time domain reflectometry) probes can measure the water content and especially changes of water content with very high accuracy (Jacobsen and Schjønning, 1993; Nissen et al., 1999; Roth et al., 1990). But there is still a lack of methods suitable for larger areas and measurements of heterogeneity of soil water content. Using TDR, a high number of probes have to be installed, which leads to considerable cost and work. To determine small-scale heterogeneities, the TDR also has the disadvantage of disturbing the area by installing probes. For these cases, ground-penetrating radar (GPR) is an alternative measuring device. With respect to water content variability analysis, two different procedures based on GPR measurements can be distinguished.

Various authors use tomographic or multi-offset radar methods to determine the velocity of an electromagnetic wave through the subsurface. The dielectric constant can be calculated approximately from the velocity. Water content is then usually derived from the dielectric constant using the popular empirical Topp equation (Topp et al., 1980) or mixture formulae (i.e., Roth et al., 1990) which describe the relationships between dielectric and hydraulic parameters. Using this procedure, Hubbard et al. (1997), Parkin et al. (2000), Greaves et al. (1996), and Sénéchal et al. (2000a) investigated the use of radar data for estimating subsurface water content.

Another method of estimating water content distributions from GPR data is to look for a variable that statistically or geostatistically describes this distribution. In this case, the analysis of radar data is used for characterizing the heterogeneity of the subsurface. To calculate any kind of distribution from radar data, the recorded radar traces are analyzed using different attributes. This method is analogous to seismic trace analyses. Chen and Sidney (1997) gave an overview of seismic attributes that are specific measurements of geometric, kinematic, dynamic, or statistical features derived from the recorded data. Using this procedure for electromagnetic applications, Sénéchal et al. (2000b) gave a complex interpretation of a three-dimensional GPR data set using attributes calculated from amplitude analysis of reflected radar waves. They got an understanding of the lateral continuities and discontinuities of the reflectors, the geometry of structures, and their dynamic characteristics. Knight et al. (1997) used the amplitude values recorded in the radar traces for a geostatistical analysis of the GPR data. The spatial variation in dielectric properties in the subsurface was closely related to the spatial variation in grain size. The geostatistical analysis captured information about the spatial distribution of the dominant sedimentological features. Rea and Knight (1998) found agreement using geostatistical analysis of a digitized photograph and the radar data (amplitudes). The GPR data imaged the spatial distribution of lithologies and could be used to quantify the correlation structure of the sedimentary unit. They also observed agreement between the spatial variation in dielectric properties in the subsurface and the spatial variation of the grain size, both indicating the heterogeneity of the subsurface. The authors hypothesized that information extracted from the GPR data can be used to describe spatial variability of hydraulic properties. Also Tercier et al. (2000) found that geostatistical analysis of GPR data gave an effective way of quantifying the correlation structure of the two-dimensional GPR image. Charlton (2000) used GPR techniques for spatially distributed measurements of volumetric soil moisture. He found significant relationships between maximum amplitude and moisture content, indicating the potential of GPR for a quantitative assessment of soil moisture at different depths.

This study summarizes and completes—by showing new data—two earlier experimental works on GPR application in well-defined systems (Schmalz et al., 2002, 2003; Stoffregen et al., 2002). At first, we focused on the general suitability of the GPR technique to depict temporal soil water content changes by analyzing the electromagnetic wave propagation. A more detailed analysis of the GPR signal was performed in a second step in order to get more insight into the spatial heterogeneity of the soil moisture distribution. We selected certain attributes of the individual radar traces, analyzed them statistically, and compared the computations to soil water content distributions as derived from hydraulic simulation studies. It is important to note that the GPR technique can be used to obtain different kinds of information. At first, the velocity analysis of the magnetic wave can be applied to obtain absolute soil water content information as an integral over the soil depth which is penetrated by the waves. At second, certain attributes of the GPR signal can be analyzed in order to obtain depth-resolved information of the underground.

25.2 MATERIAL AND METHODS

25.2.1 GPR Technique

The propagation velocity of GPR electromagnetic waves is determined by the dielectric constant or permittivity of the medium to be investigated, which on its part is a function of the water content. The dielectric constant ε for water is 80, for various soil and geological materials 5 to 15, and 1 for air. Measurements of the wave propagation velocity can be taken to determine soil water contents using various functional relationships between the two quantities (e.g., Topp et al., 1980). The GPR technique is a nondestructive measurement, which is an advantage in comparison to other electromagnetic wave-based methodologies, such as the TDR technique.

In principle, all kinds of GPR measurements require one transmitting and one receiving antenna. In this study, the fixed offset (FO) method in which the distance between the two antennas is held constant was used. Although a 1 GHz antenna (MALÅ Geoscience, Malå, Sweden) was found to be suitable for the investigated lysimeters, we preferred a 500 MHz antenna (SIR-10A from Geophysical Survey Systems Inc., Salem, NH) for the large sand tank. The measurements were recorded at twelve times during the vegetation period from March to September in the lysimeter study in order to identify absolute soil water content changes. In the sand tank study, where we were aiming at depicting the proceeding water front following single irrigation events, GPR data were acquired at 15 to 20 min intervals.

25.2.1.1 Migration of Radargrams

The processing steps applied to the acquired GPR data were first counting back the applied field gain curve, then normalizing in space to 1 cm trace interval, and afterward reapplying a realistic gain curve taking the attenuation of electromagnetic waves in a midelectroconductive environment into account. No filtering was applied to the data beneath a low-cut filter during acquisition for signal stability.

In order to obtain depth profiles, the GPR signal originally acquired in the time domain has to be converted to a depth section. This procedure called "migration" is a common process in seismic data processing. During the migration, the data are not simply rescaled, but, for example, effects of the acquisition in the time domain such as diffraction hyperbolas at edges or isolated bodies are taken into analysis. In this case, a classic Stolt migration algorithm was used to convert the time section into a depth section (Stolt, 1978). The procedure is based on a basic velocity model that could not be computed directly from the acquired database. However, from the independent TDR measurements in eight depths, it was possible to set up a bedded velocity model neglecting strong variations in the horizontal plane. Some artifacts may be introduced using such a simplified model, but these are not critical compared with analysis on those arising from time-domain data.

25.2.1.2 Normalized Maximum Reflection Amplitude Analysis (NMRA)

The energy transmitted into the soil will be, in part, reflected when contrasts in soil permittivity (e.g., soil water contents) are encountered. Figure 25.1 shows a theoretical derived GPR radargram with a fixed antenna separation (single offset) for a layered soil profile with soil water content differences. The first reflection from the top is the direct wave, and the second indicates the different reflection amplitudes with changing water content.

The maximum reflection amplitude is computed by (a) selecting an interval of interest and (b) calculating the absolute value of the maximal amplitude. The values can be presented in normalized form by dividing the selected value with the maximum value of the considered interval (NMRA). In this study, NMRA values derived from one soil depth (from migrated radargrams) were statistically analyzed and compared to soil water content distributions.

25.2.2 Experimental Setup

Two setups were used to generate the experimental database. In both cases, a 2 m thick sand body was investigated. Although a lysimeter stand with a natural vegetation cover subjected to atmospheric conditions served for absolute water content investigation on a seasonal scale, we used a sheltered large physical sand tank (Schmalz et al., 2003) for temporal and spatial high-resolution measurements.

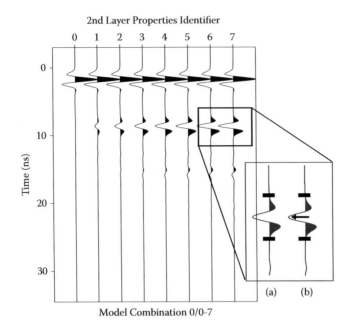

FIGURE 25.1 Idealized ground-penetrating radar (GPR) transect measured with a fixed separation as obtained from a layered soil profile with contrasting soil water contents. The soil water content in the second soil layer is increasing from the left to the right. The first reflection from the top is the direct or air wave, and the second indicates the different reflection amplitudes with changing water content. The normalized maximum reflection amplitude (NMRA) is computed by (i) selection of an interval of interest, (ii) picking the absolute value of the maximal amplitude, and (iii) normalizing the value with the maximum value of the considered transect and soil depth.

25.2.2.1 Lysimeter Stand

The lysimeter station consists of twelve lysimeter cylinders, each with an area of 1 m² (Ø = 113 cm). The two cylinders considered in this study stand on a platform scale that is connected to an electronic scale. The changes in water content can be measured with an accuracy of 100 g, which corresponds to 0.1 mm of the area or 0.000067 m³/m³ of the volume of the lysimeter. The temporal solution of the measurement is 15 min. An artificial groundwater table was established at 2.10 m with the help of a suction system.

25.2.2.2 Large Physical Sand Tank

Infiltration experiments were carried out using a large physical sand model having a base of 5 m × 3 m and a surface of 6 m × 5.6 m and containing three sloped side walls. The chosen construction with three sloped side walls resulted from statical constraints. All soil hydraulic measuring devices were installed from the vertical wall. The tank was filled with a 2 m layer of homogeneous sand (Hagrey et al., 1999). Measurements of the pressure head (two vertical tensiometer profiles) and the water content (one vertical TDR profile) were conducted at eight depths allowing the derivation of θ(h) relationships at 20, 40, 60, 100, 120, 140, 160, and 180 cm soil depth. An automatic irrigation device ensured the homogeneous distribution of the irrigation water over the soil surface. In this study, we consider an irrigation event of 287 mm applied within 14 h, which produced a total discharge of 197 mm over a 14-day period.

25.2.3   SOIL WATER CONTENT PROFILES

As a comparison bases for the GPR signal analysis, we numerically generated two-dimensional soil water content profiles assuming a heterogeneous soil water distribution according to results from earlier studies on homogeneous sand packages (Schmalz et al., 2003). Numerical experiments were conducted using the Hydrus-2D software package of Simunek et al. (1996) to obtain two-dimensional views of the infiltration and redistribution process. The Windows-based Hydrus-2D package solves the Richards equation for variably saturated flow numerically, assuming applicability of the van Genuchten-Mualem soil hydraulic functions.

Because of the geometric features of the sand tank, only flow in one particular cross section was considered. To run Hydrus-2D, we implemented a relatively fine numerical mesh involving 4488 nodes depicting the geometry of the sand tank. An atmospheric boundary condition accounting for infiltration and evaporation was imposed at the soil surface, whereas a seepage face was used at the bottom boundary between the sand and a gravel drainage layer.

A heterogeneous soil water distribution could be depicted by assuming a spatially noncorrelated distribution of the $\theta(h)$ and $K(h)$ relations as measured in situ in eight depths with soil hydraulic probes. The saturated hydraulic conductivity was assessed from the grain distribution using Hazen's empirical formula (Hölting, 1996) and was found to vary from 0.3 to 0.47 mh^{-1} over the entire profile.

25.3   RESULTS

25.3.1   WATER FRONT PROPAGATION

The processed GPR profiles (prior migration) allowed the clear identification of the penetrating water front in infiltration experiments, especially when difference radargrams of two subsequent time steps were computed (Figure 25.2a and Figure 25.2b). Difference radargrams are generated by

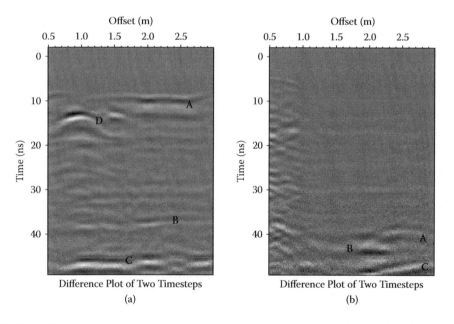

FIGURE 25.2   Difference radargrams (nonmigrated) of the radargrams 3 and 2.45 (a) and 13 and 12.45 (b) hours after onset of irrigation. A marks the water front, B the electrode grid in 2 m depth, C the concrete bottom in 2.5 m depth, and D is a highlighted artefact of a diffraction hyperbola caused by an electrode for geoelectric measurements not mentioned in this paper.

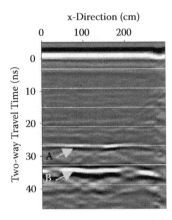

x-Direction (cm)

FIGURE 25.3 Nonmigrated radargram as recorded prior infiltration. Arrows indicate reflectors (A = electrode grid; B = concrete bottom). The radargram revealed the layered structure of the sand body.

directly subtracting the corresponding amplitudes of all radar traces (250 per time step) from the two treated time steps (Δt = 15 min). The result of such a computation must be read from top to the lower limit. Regions that appear rather uniform are unchanged from time step to time step. In contrast, certain reflection patterns (A in Figure 25.2) have changed with time, indicating the moving water front. Most of the visible reflections are artefacts or altered characteristics of very dominant reflections (e.g., concrete base of lysimeter). The nonmigrated radargram showed the construction basis of the lysimeter (concrete) at 2.50 m (Figure 25.3). Furthermore, a layered structure of the sand became visible which was possible due to the filling procedure of the lysimeter (layer-wise filling and compaction).

25.3.2 Seasonal Soil Water Content Changes

The weighed lysimeters exhibited a change of soil water content over the entire 2 m profile of 140 mm over the investigation period (1 year; Figure 25.4a). It has to be noted that no depth-resolved water content distributions were registered. The greatest water content changes occurred during the vegetation period from March to September due to enhanced evaporation. Correspondingly, the reflection times of the electromagnetic waves reduced from 25 to 15 ns (Figure 25.4b).

The relation between the soil water storage change and the change in travel times is depicted in Figure 25.5. The scattering is higher in the dry range and lower in the wet range. This is mainly due to the lower signal-to-noise ratio under the drier condition, when it was uncertain to determine the reflection time. The permittivity was calculated from the reflection time, and the soil water content was estimated using the lysimeter data. The results of the measured relation between the relative dielectric constant and the water content are shown in Figure 25.6. The following calibration curve was fitted to the data:

$$\theta = 0.0245 \, \varepsilon - 0.0304 \tag{25.1}$$

The standard deviation between the fitted curve and data is 15 mm (= 0.01 m³/m³). Using only data of the wetter range, where the scattering of the GPR measurements was less pronounced, the standard deviation reduces to 4 mm (= 0.00026 m³/m³).

The results were compared to three calibration curves developed for TDR (Figure 25.6):

Topp et al. (1980):

$$\theta = -0.053 + 0.0293\varepsilon - 0.00055\varepsilon^2 + 0.0000043\varepsilon^3 \tag{25.2}$$

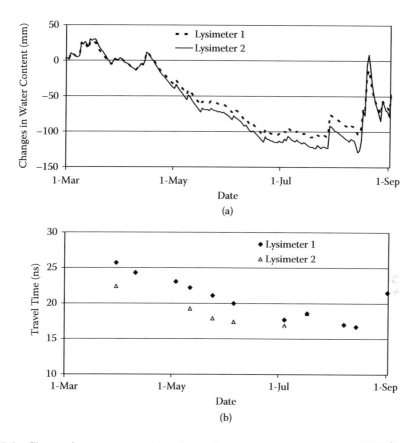

FIGURE 25.4   Changes in water content (a) and according ground-penetrating radar (GPR) reflection times (b) over the 6-month investigation period.

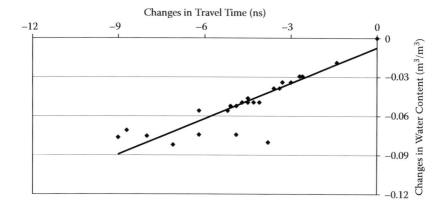

FIGURE 25.5   Correlation between changes in water content and reflection time.

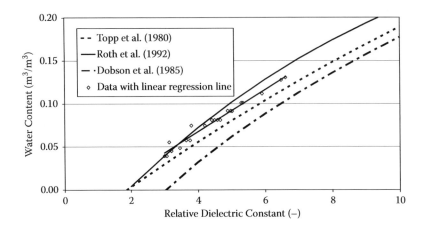

FIGURE 25.6 The measured relative dielectric constant and the soil water content compared to different calibration models.

Roth et al. (1992):

$$\theta = -0.078 + 0.0448\varepsilon - 0.00195\varepsilon^2 + 0.0000361\varepsilon^3 \tag{25.3}$$

Dobson et al. (1985) and Huisman (2002):

$$\theta = \frac{1}{\sqrt{\varepsilon_w - 1}}\sqrt{\varepsilon} - \frac{(1-n)\sqrt{\varepsilon_s} - n}{\sqrt{\varepsilon_w - 1}} \tag{25.4}$$

where ε is the bulk permittivity of a soil–water–air system (assuming that the permittivity of air equals one), ε_w and ε_s are the permittivities of water (81 at 20°) and soil particles (assumed as 5), respectively, and n is the soil porosity. It has to be emphasized that the absolute value of water content was estimated from field capacity in spring. Therefore, absolute values may be erroneous to a certain extent, although the measurements of the water content changes as estimated from lysimeter balance readings are accurate. The equation of Topp et al. (1980) was found to represent the data best (Figure 25.6).

25.3.3 SPATIAL SOIL WATER CONTENT VARIATIONS

25.3.3.1 Dye Experiment

The spatial variations of the soil water content and resulting flux fields were qualitatively investigated using dye tracer tests. Due to the destructive nature of the experiment, the tests had to be conducted at the end of our investigations. The experimental design was chosen according to earlier investigations from Flury et al. (1994): 50 mm of a 4 gl⁻¹ Brilliant Blue (Vitasyn Blau AE85) solution was applied in 4 h. The treated area of 4 m² was horizontally, layer-wise (20 cm) excavated.

Although emphasis was laid on the construction of a *homogeneous* sand tank, the dye penetrated the substrate irregularly (Figure 25.7). The finger-like flow patterns indicate an heterogeneous flow field with mobile and immobile soil water regions. The dye pattern supported the need to set up a heterogeneous numerical model for comparison purposes.

25.3.3.2 Maximum Reflection Amplitude

The general suitability of the maximum reflection amplitude as an indicator for the water content variation and thereby of the variability of the hydraulic functions was shown by Schmalz et al.

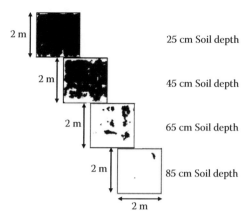

FIGURE 25.7 Stained sand soil profile (top view). The dye pattern indicates a finger-like flow scenario.

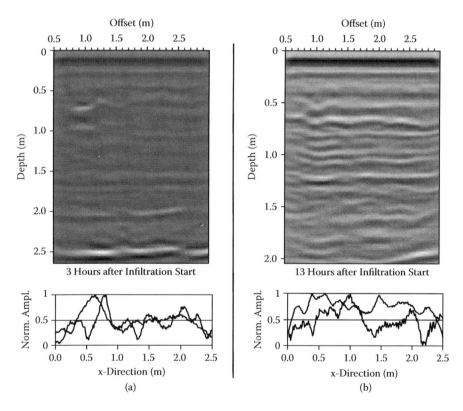

FIGURE 25.8 Migrated radargrams 3 (a) and 13 h (b) after onset of irrigation (top) and NMRA distribution (bottom) in two depth intervals: 30 to 40 cm (thin line) and 120 to 130 cm (bold line).

(2002) in a numerical experiment. In this case, the method was used to evaluate possible soil water content variations in two soil depths (35 and 125 cm) 3 and 13 h after onset of irrigation. The analysis was not meant to directly derive absolute soil water content values but to reveal soil moisture heterogeneities in space. Figure 25.8a and Figure 25.8b show the migrated radargrams (conversion of the travel time axis to distance) and the computed NMRA at two depths over the irrigated transect. Three hours after onset of irrigation, the highest values occurred on the left-hand side of the sand tank, whereas no clear trend can be depicted from the radargram as derived 13 h after the start of irrigation.

25.3.3.3 Simulated Soil Water Content Profiles

Spatially distributed soil physical parameter measurements indicated nonuniform depth-dependent retention characteristics. Optimized van Genuchten parameter values revealed a variation of the saturated water content, θ_s, from 0.27 to 0.36, and the residual water content, θ_r, was found to be 0 in all cases. The range of observed soil hydraulic functions was used to generate layered and heterogeneous flux fields assuming no spatial autocorrelation for any parameter within one soil layer because no spatial autodependency could be observed from spatial highly resolved bulk density measurements. Next to a layer-wise consideration of soil physical properties, we set up a numerical model in which the independently determined hydraulic properties were randomly distributed over the numerical mesh holding 4488 nodes. In Figure 25.9 and Figure 25.10 we present the variations of the soil water content over the entire soil profile and within two selected soil layers. Only in the heterogeneous case could significant volumetric moisture variations of 0.15 cm³ cm⁻³ be observed.

In a next step the soil water content values along with the NMRA signals of two selected soil depths were statistically analyzed and box plots of the distributions were computed (Figure 25.11). It is obvious that there are similarities in the general appearance of the statistical distributions only between the water content variations and the NMRA distribution assuming a heterogeneous distribution of the soil hydraulic properties (middle graphs in Figure 25.11). When a layer-wise orientation of the soil physical properties is considered in the soil water status simulations (top graphs in Figure 25.11), the resulting statistical distribution shows no accordance to the NMRA distribution. Although it is not possible to directly derive absolute soil water content values from the NMRA signal, its statistical distribution might be helpful in generating heterogeneous flux fields.

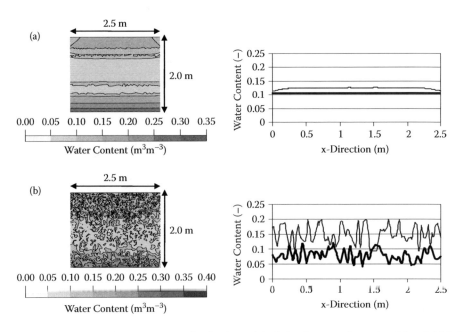

FIGURE 25.9 Simulated soil water content distributions of the 2.5 × 2 m two-dimensional flux field (left) and within two depths: 35 cm (thin line) and 125 cm (bold line) 3 h after onset of irrigation. (a) Shows water content assuming a layered structure and (b) resulted from a random distribution of the soil hydraulic properties.

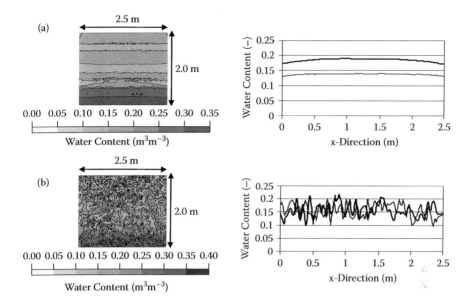

FIGURE 25.10 Simulated soil water content distributions of the 2.5×2 m two-dimensional flux field (left) and within two depths: 35 cm (thin line) and 125 cm (bold line) 13 h after onset of irrigation. (a) Shows water content assuming a layered structure and (b) resulted from a random distribution of the soil hydraulic properties.

25.4 CONCLUSIONS

In sandy soils as they have been investigated in this study, GPR is a valuable technique for mapping soil water content at an intermediate scale in between point (TDR) and area measurements (remote sensing). The observed strong correlation between the change of the reflection time of the electromagnetic wave and the change in soil water content as determined from gravimetrical measurements ($r^2 = 0.82$) confirmed the accuracy of the GPR technique. The popular Topp equation proved suitable for the registration of the seasonal water status development in the top soil. However, the analysis of the velocity of the electromagnetic wave alone does not provide depth-dependent information about the soil water regime. A more detailed inspection of the radargram, especially of the reflection amplitudes, is necessary when information about a change of soil structure or soil water content with depth is needed. In addition, the statistical analysis of the reflection amplitudes or other parameters obtainable from the GPR measurements might be a way to reveal the heterogeneity of the soil water content and the related water flux field.

First results presented in this study indicated that the assumption of a heterogeneous distribution of soil hydraulic properties coincides better with the geophysical data than a generated homogeneous case (homogeneous soil properties within a single soil layer). It can be concluded that even in homogeneous sand profiles, local variabilities dominate water flow which confirms previous findings from other studies in which, for instance, the dye tracer technique has been used to identify flow patterns (Flury et al., 1994). The variances of the soil hydraulic and geophysical parameters differed, although in general the heterogeneous case performed better than the homogeneous case.

The next step essential for evaluating the potential of the method would be using the analyzed variance of GPR amplitudes for each depth interval directly in the numerical models for simulating water flow. Comparisons of water discharge rates derived from numerical models based on soil hydraulic parameter distributions determined from the GPR analysis with real measured data will show if a better agreement of model and experiment can be reached. We expect a better description of the water flux variation by using the information about soil heterogeneity derived from GPR measurements.

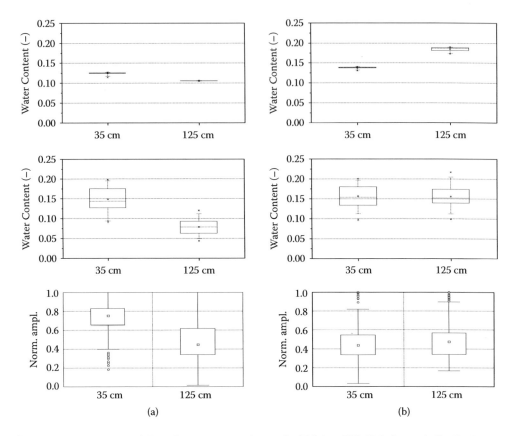

FIGURE 25.11 Box plots of the soil water content (top and middle) and NMRA (bottom) distribution at the two selected soil depths 3 (a) and 13 h (b) after onset of irrigation. The top two graphs characterize the soil water distribution assuming a layer-wise orientation of the soil hydraulic properties, and the middle graphs resulted from a heterogeneous distribution of soil properties.

REFERENCES

Charlton, M., 2000. Small scale soil-moisture variability estimated using ground penetrating radar. Extended abstract, 8th International Conference on Ground Penetrating Radar. Gold Coast/Australia, 23–26.05.2000.

Chen, Q., and Sidney, S., 1997. Seismic attribute technology for reservoir forecasting and monitoring. *The Leading Edge*: 445–456.

Dobson, M.C., Ulaby, F.T., Hallikainen, M.T., and El-Rayes, M.A., 1985. Microwave dielectric behavior of wet soil. Part II. Dielectric mixing models. *IEEE Trans. Geosci. Rem. Sensing GE-23*, 35–46.

Flury, M., Flühler, H., Jury, W.A., and Leuenberger, J., 1994. Susceptibility of soils to preferential flow of water: A field study. *Water Resour. Res.*, 30: 1945–1954.

Greaves, R.J., Lesmes, D.P., Lee, J.M., and Toksöz, M.N., 1996. Velocity variations and water content estimated from multi-offset, ground-penetrating radar. *Geophysics*, 61: 683–695.

Hagrey, S.A., Schubert-Klempnauer, T., Wachsmuth, D., Michaelsen, J., and Meissner, R., 1999. Preferential flow: First results of a full-scale flow model. *Geophys. J. Int.*, 138: 643–654.

Hölting, D., 1996. Hydrogeologie. Einführung in die allgemeine und angewandte Hydrogeologie. Enke, Stuttgart, p. 441.

Hubbard, S.S., Peterson, J.E., Majer, E.L., Zawislanski, P.T., Williams, K.H., Roberts, J., and Wobber, F., 1997. Estimation of permeable pathways and water content using tomographic radar data. *The Leading Edge*: 1623–1628.

Huisman, J.A., 2002. Measuring soil water content with time domain reflectometry and ground-penetrating radar. Dissertation, University Amsterdam, p. 157.

Jacobsen, O.H., and Schjønning, P., 1993. Field evaluation of time domain reflectometry for soil water measurements. *J. Hydrol.*, 151: 159–172.

Knight, R., Tercier, P., and Jol, H., 1997. The role of ground penetrating radar and geostatistics in reservoir description. *The Leading Edge*: 1576–1582.

Nissen, H.H., Moldrup, P., De Jonge, L.W., and Jacobsen, O.H., 1999. Time domain reflectometry coil probe measurements of water content during fingered flow. *Soil Sci. Soc. Am. J.*, 63: 493–500.

Parkin, G., Redman, D., Von Bertoldi, P., and Zhang, Z., 2000. Measurement of soil water content below a wastewater trench using ground-penetrating radar. *Water Resour. Res.*, 36: 2147–2154.

Rea, J., and Knight, R., 1998. Geostatistical analysis of ground-penetrating radar data: A means of describing spatial variation in the subsurface. *Water Resour. Res.*, 34: 329–339.

Roth, C.H., Malicki, M.A., and Plagge, R., 1992. Empirical evaluation of the relationship between soil dielectric constant and volumetric water content as the basis for calibrating soil moisture measurements by TDR. *Soil Sci.* 43: 1–13.

Roth, K., Schulin, R., Flühler, H., and Attinger, W., 1990. Calibration of time domain reflectometry for water content measurement using a composite dielectric approach. *Water Resour. Res.*, 26: 2267–2273.

Schmalz, B., Lennartz, B., and Wachsmuth, D., 2002. Analyses of soil water content variations and GPR attribute distributions. *J. Hydrol.*, 267: 217–226.

Schmalz, B., Lennartz, B., and Van Genuchten, M. Th., 2003. Analysis of unsaturated water flow in a large sand tank. *Soil Sci.*, 168: 3–14.

Sénéchal, P., Perroud, H., and Garambios, S., 2000a. Geometrical and physical parameters comparison between GPR data and other geophysical data. Extended abstract, 8th International Conference on Ground Penetrating Radar. Gold Coast/Australia, 23–26.05.2000.

Sénéchal, P., Perroud, H., and Sénéchal, G., 2000b. Interpretation of reflection attributes in a 3-D GPR survey at Vallée d'Ossau, western Pyrenees, France. *Geophysics*, 65: 1435–1445.

Simunek, J., Sejna, M., and Van Genuchten, M. Th., 1996. Hydrus-2D, Simulating water flow and solute transport in two-dimensional variably saturated media. User's Manual. U.S. Salinity Lab., USDA/ARS, Riverside, California.

Stoffregen, H., Yaramanci, U., Zenker, T., and Wessolek, G., 2002. Accuracy of soil water content measurements using ground penetrating radar: Comparison of ground penetrating radar and lysimeter data. *J. Hydrol.*, 267: 201–206.

Stolt, R.H., 1978. Migration by Fourier transform. *Geophysics, Soc. of Expl. Geophys.*, 43: 23–48.

Tercier, P., Knight, R., and Jol, H., 2000. A comparison of the correlation structure in GPR images of deltaic and barrier-spit depositional environments. *Geophysics*, 65: 1142–1153.

Topp, G.C., Davis, J.L., and Annan, A.P., 1980. Electromagnetic determination of soil water content. Measurements in coaxial transmission lines. *Water Resour. Res.*, 16: 572–582.

26 Ground-Penetrating Radar Mapping of Near-Surface Preferential Flow

Robert S. Freeland

CONTENTS

26.1 INTRODUCTION

A formidable challenge is determining the preferential-flow pathways of near-surface water. A traditional approach involves installing monitoring-well networks, which may be inadequate due to the invasive act of coring and subsequent sampling. Physical probing into the subsurface may compromise delicate pathway integrity and distort matrix flow. Furthermore, improper or insufficient well-layout patterns will produce spatial sampling errors, as the true morphological complexity of the subsurface may have been misperceived or oversimplified.

Spatial sampling error is reduced by increasing spatial sampling frequency. One technology capable of increasing spatial resolution is generally referred to as ground-penetrating radar (GPR), which uses ground-injected electromagnetic impulses. This nonintrusive technology can scan the subsurface in near spatial continuum, thereby yielding significantly higher spatial sampling resolutions than that of point-specific wells.

GPR's basic operating procedure is by towing a ground-coupled antenna across the surface. A basic GPR-antenna configuration contains both a transmitter and a receiver antenna, with its electronics continuously firing electromagnetic impulses into the subsurface and receiving the ensuing echoes off subsurface strata (Figure 26.1). Assigning color values to the digitized samples of the reflected waveform produces a pixel matrix of one column by N-rows of digitized samples (i.e., $N =$ 512, 1024, 2048, etc.). Columns are vertically stacked at approximately 60 Hz to form a scrolling pseudo-image of the subsurface profile. When time stacked, the series of the returning waveforms will profile the substrata as an image scrolling across the display as the antenna travels across the surface (Figure 26.2). Called a radargram, this is a composite display exhibiting line profiles for

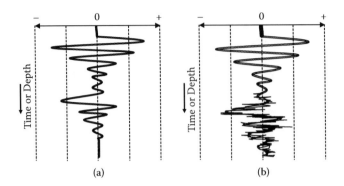

FIGURE 26.1   (a) Example reflected waveform that is digitized over time (increasing depth). Gain is ramped considerably with return time (depth), to expand the amplitude of the waveform's finale such that the peak-to-peak cycles are accentuated. (b) Waveform dances erratically when passing over saturated soil.

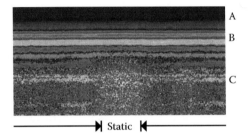

FIGURE 26.2   Radargram taken along the top of a 4 m high levee illustrating its saturated and leakage region "STATIC" where horizons "A" is the well-drained upper profile along the levee crown, "B" the drying front within the levee, and "C" the original soil surface prior to levee construction.

continuous soil horizon boundaries, and inverted hyperbolic-shaped curves for buried point objects such as pipes and rocks.

An electromagnetic impulse will partially reflect off any sharp boundary separating differing dielectric properties. The sharper the dielectric contrast, the sharper the "echo." Within soil and bedrock strata, a fraction of the impulse's energy will continue downward to be reflected off subsequent dielectric boundaries. All of the reflections ensuing from a single impulse will compose the echoed waveform (Figure 26.1). The travel time of each returning waveform cycle (peak-to-peak) is a function of the reflector depth and the overall dielectric of the transmission media. Naturally, the resulting waveform's signal strength will exponentially degrade with depth (or increasing return time) due to this repetitive echoing process as less of the impulse's energy progresses into deeper strata, until the tail of the returning waveform is completely absorbed at some depth range by soil conductivity. Therefore, considerable ramped gain must be applied to the returning waveform. If the depth range is extended beyond the antenna's normal probing capability for the site's soil conditions, static will uniformly appear within the lower extents of the radargram because only background noise is being amplified.

26.1.1   RADARGRAM CHARACTERISTICS OF SATURATED SOIL

A thin layer of perched moisture atop a distinct textural boundary of two soil horizons offers a highly reflective plane, which is easily identifiable within the radargram as dry soil has a much lower dielectric constant than wet soil. Due to this interface, GPR is an excellent tool for soil horizon profiling. Furthermore, a sharp dielectric interface is also supplied by a rapidly progressing wetting front through dry soil. If not for interactions of the capillary fringe, the horizontal boundary between wet and dry soil would always offer an excellent radar reflective plane. For example, a falling or retreating free-water edge is much more difficult (if not impossible) to delineate with GPR than a stagnant or rapidly rising water table due to the more gradual, elongated dielectric change remaining above the saturation zone, as residual free moisture is still draining. Fortunately, a large body of saturated soil does have a differentiating signature from nearby drier soil, due to it being a much more electrically conductive medium. The wetter region, being a much higher conductive body than the surrounding drier soil, significantly weakens the return signal, and thus less energy is reflected back to the surface. Extremely high gain, when applied to a very weak (or no) signal, will highlight the naturally occurring background noise, giving a characteristic telltale static pattern. When observing this return waveform in oscilloscope format, the later-returning component (or the lower tail finale) of the waveform, which is normally somewhat steady in its structured movements (Figure 26.1a), will dance erratically when passing over saturated soil (Figure 26.1b). Within the radargram, adjacent strong horizontal reflectors at the same depth as that of the saturation zone will become masked or obliterated in static as the antenna passes over a large saturated region, only to reappear when reentering a drier region (Figure 26.2). A static pattern that suddenly appears and masks adjacent strong horizontal reflectors (or a strong uniformity signal) suggests passing over a highly conductive wetter zone. In this instance, a rectangle of static will fill the radargram from the saturated zone downward in its entirety.

Subsurface morphology and environmental conditions are widely variable. Subsurface anomalies can generate reflections that mimic the signature of saturated soil, as can also the instrument settings and the individual survey protocol technique of the operator. Surface observations, site experience, and calibrations along with limited ground truth probing can assist substantiating that the GPR signatures are indeed originating from saturated soil.

Employing GPR to trace diminutive pathways of preferential flow is therefore very challenging, as the channels conducting water primarily do so under seasonal wet soil conditions, having also saturated the surrounding soil, and are thereby masked. One approach is to map the channels directly by artificially applying a significant quantity of water to the subsurface during very dry soil conditions, and to track the distinct wetting front as it progresses outward beneath the surface.

26.2 CASE HISTORY

This case study highlights mapping subsurface lateral preferential flow in Major Land Resource Area 134 (MLRA 134—Southern Mississippi Valley Silty Uplands), which extends along the Mississippi River from southern Illinois to northern Louisiana. These highly productive agricultural lands are primarily in row-crop production and represent thousands of hectares that formed in the loess-covered Tertiary-aged Claiborne and Wilcox geologic formations (Hardeman, 1966). At one research site, an interface between the loess (wind-blown silt) and the alluvial clay covering the underlying paleosol (an ancient seabed of deep stratified sands) forms a distinct textural discontinuity for perching water at a below-surface depth of approximately 2 m. The objective of the project was to develop a noninvasive tool that illustrates the rate and direction of the preferential flow that perches and moves atop this interface.

26.2.1 SURVEY PROTOCOL

Conducting a high-resolution GPR survey generates copious data. Gigabytes of GPR data per hectare must be recorded while traveling the lengthy, closely spaced traverses that are required in detecting preferential flow pathways. Thus, it is implausible to employ high-resolution GPR mapping over entire watersheds if relying upon manual radargram interpretations (Figure 26.3). For this reason, we employ a neural-network (NN) classifier to automatically segment patterns based upon

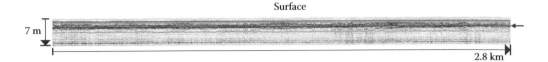

FIGURE 26.3 Raw ground-penetrating radar (GPR) data of a single spiral track unscrolled (2.5 km long by 8 m deep). Arrow at right margin points to a water perching horizon 2 m beneath surface of which Figure 26.7 and Figure 26.8 are top-down views.

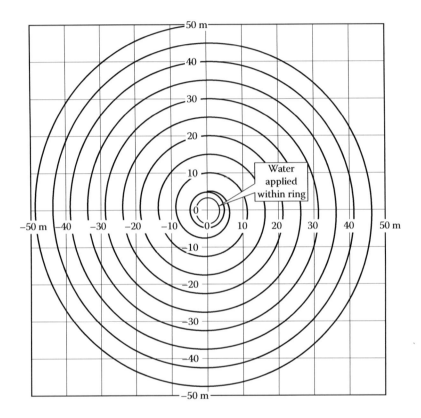

FIGURE 26.4 Course guidance layout design showing flag placement, top-down view.

Odhiambo et al. (2004) and to spatially map these similarity identifiers by corresponding geospatial position.

Figure 26.4 illustrates the survey layout of one such spiral plot, central to which a double-ring infiltrometer is installed. The operator of the mobile GPR platform follows an outwardly spiraling traverse from the infiltrometer as guided by survey flags and tracked in real time by differential GPS (DGPS) (Figure 26.5). Tests are conducted during seasonal extended dry soil conditions such that baseline "dry" condition radar imagery can be registered. After which, water application to the infiltrometer begins at controlled application rates (Figure 26.6). The traverse taken by the mobile platform about the spiral is continuously traveled until a saturated equilibrium state within the sub-surface is detected (Figure 26.7).

26.2.2 DATA PROCESSING

Field data sets are stored in two linkable files: one a traditional GPR-format binary matrix of digitized waveform scans (scans x samples), and a second GPS data file containing 1 sec interval

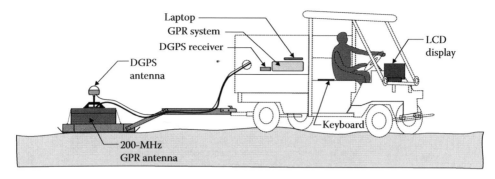

FIGURE 26.5 The University of Tennessee Mobile GPR system. (From Freeland, R. S., R. E. Yoder, J. T. Ammons, and L. L. Leonard, 2002, *Appl. Eng. in Agric..* 18:647–650. With permission.)

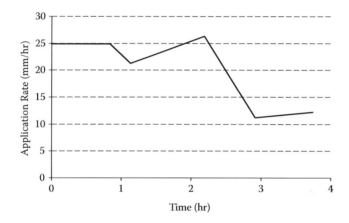

FIGURE 26.6 Water application rate (mm/hr) within 8 m diameter inner ring.

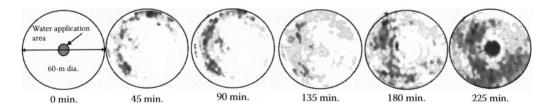

FIGURE 26.7 Top-down view of an ≈5 cm thick horizontal slice immediately atop the alluvium interface depicting movement from the initial state (a) in 45 min sequential steps (b–f). Plot diameter is 60 m.

geospatial coordinates (latitude, longitude, elevation) with indexing pointers to their corresponding scans within the GPR file. The spherical-projected GPS coordinate units (degrees) are transformed to a flat-surface projection coordinate system (feet, meters), either State Plane or Universal Transverse Mercator (UTM). The locations of each scan recorded between the GPS 1 sec position updates are calculated using a linear series best-fit trend Microsoft EXCEL function (Microsoft Corp., Redmond, WA). Thus, each GPR sample value has corresponding three-dimensional spatial coordinates for kriging. For any constant depth value, a horizontal plane can be interpolated and saved in graphical file format. Multiple horizontal planes can be imported into a number of commercial three-dimensional visualization packages (Figure 26.8) (e.g., open source MicroView, GE Health-Care, Waukesha, WI). Stacked in sequence, the horizontal planes form a three-dimensional block.

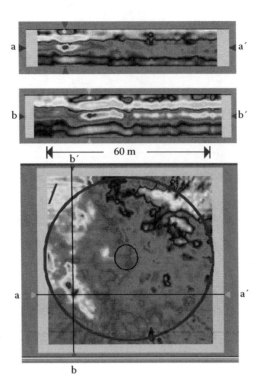

FIGURE 26.8 Three-dimensional image slice of radargram data taken prior to water application. Crosshair is at an alluvium interface fissure.

Noted is the fact that upper strata will distort and mask features of the lower strata, resulting in three-dimensional renderings that may not approach the physical realism typical of modern medical imaging technologies.

Higher-end raster graphical editors allow importing of GPR data for better visualizations and graphical manipulations. Returning GPR waveforms are sampled as 8-, 16-, or 32-bit binary data and recorded in matrix format. For converting to an image graphic, Figure 26.9a shows Adobe Photoshop CS2 (Adobe Systems, Inc., San Jose, CA) RAW File Open As window, illustrating importing

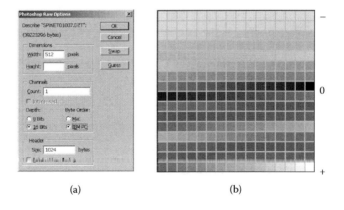

(a)   (b)

FIGURE 26.9 (a) Adobe Photoshop FILE OPEN AS window for importing RAW GPR data, and (b) example INDEXED Color ramp for applying 256-color levels. (Shown is its conversion to 8-bit gray scale to illustrate resolution importance of 256 color scale versus 256 different levels of gray.)

a GSSI-format (Geophysical Survey Systems, Inc., North Salem, NH) data file having a 1024-byte header, with 512 16-bit samples of each waveform. The number of waveforms stored in the data file can be manually calculated, or can easily be inserted automatically by pressing the "GUESS" button. Once opened, the gray scale image can be rotated, zoomed, and assigned to any desired color scheme by converting the file format to INDEXED COLOR, and loading a color table such as a 256-level ramped-color scheme (Figure 26.9b). Although Figure 26.9b in its original form is a ramped color scale, its presentation in gray scale illustrates the importance of color in data resolving. The human eye can resolve thousands of different colors, but only a few dozen shades of gray.

26.3 RESULTS AND DISCUSSION

Figure 26.7 presents seven sequential frames (a through f) illustrating GPR signal change intensities (frames b through f) from the dry initial survey (frame a) to near horizon saturation (frame f) in a top-down view of the 60 m diameter test plot. The depicted image slice is a ≈5 cm thick layer positioned immediately atop the alluvium interface, which was geospatially interpolated using MATLAB (The MathWorks, Inc., Natick, MA). This interface horizon within the raw radargram data of Figure 26.3 that perches water is identified by the right margin arrow. The frames of Figure 26.7 are not radargrams but are intensity maps depicting radargram pattern dynamic change from the initial dry state (frame a) in 45 min increments. To produce these images, the water application rate to the inner 8 m diameter ring was continuous at ≈25 mm/hr for the first 2 hours, and once water ponding within the ring occurred, was reduced to ≈12 mm/hr to maintain constant-state surface ponding for the remainder of the test (Figure 26.6).

In frames b through f of Figure 26.7, a meandering movement of water atop the soil interface can be visualized leaving the central profile beneath the infiltrometer progressing westward. The profile appears almost entirely inundated after 2 hr (Figure 26.7, frame f). The brighter reds (lighter grays) within Figure 26.7 are indicative of the wetting front, as this is where the sharpest change in dielectric occurs resulting in the greatest radar reflection. The light green (uniform darker gray haze) represents saturated soil, as it is a more uniform dielectric that is less reflective. At this site, water applied at the surface infiltrated 2 m in depth and traveled 30 m laterally in less than 45 min.

Figure 26.8 illustrates an image slice (top-down) taken from the three-dimensional radargram data prior to water application. The individual two-dimensional slices were interpolated using Geostatistical Analyst in ESRI ArcMap 9 (ESRI, Redlands, CA). This is the initial state data of Figure 26.6, frame a. The crosshair points to one of the apparent fissures within the alluvium interface, which are permitting drainage into lower strata.

26.4 CONCLUSION

A unique survey protocol was developed that quantifies near-surface preferential flow using GPR, GPS, and a NN pattern classifier. The survey field protocol consists of a mobile GPR system that repeatedly spirals outward along a prescribed course, gathering subsurface data continuously for several hours. After first establishing dry initial-state pattern signatures, metered irrigation begins within a centrally located water-ponding ring. Following vertical infiltration, the water percolates downward and radiates outward from the central application point throughout the subsurface along preferential flow pathways, and the wetting fronts are highlighted by GPR. A radargram NN pattern classifier rapidly segments the profiles by similarities. Within the sequential radargrams during water application, only those specific geo-referenced locales that reveal pattern shifts from the initial dry state are recognized as being excited by water flow phenomena.

Pattern shifts have been found to highlight the preferential flow channels that occur within the patterns of highly complex radargrams. Sample data are provided showing surface-applied water moving beneath the surface, water applied at 25 mm/hr that infiltrated 2 m to a perching horizon, and traveling at lateral velocities of ≈40 m/hr.

26.5 ACKNOWLEDGMENTS

The authors gratefully acknowledge the support of The Trustees of the Hobart Ames Foundation, and the staff of Ames Plantation, Grand Junction, Tennessee.

REFERENCES

Freeland, R. S., R. E. Yoder, J. T. Ammons, and L. L. Leonard. 2002b. Integration of real-time global positioning with ground-penetrating radar surveys. *Appl. Eng. in Agric.*. 18(5):647–650.

Hardeman, W. D. 1966. Geologic Map of Tennessee (West Sheet). State of Tennessee Department of Conservation, Division of Geology, Nashville, TN.

Odhiambo, L. O., R. S. Freeland, and R. E. Yoder. 2004. Soil Characterization using Textural Features Extracted from GPR Data. Paper number 0422108. 2004 ASAE/CSAE Annual Meeting. Ottawa, Ontario, Canada, 1–4 August 2004.

27 An Application of Ground-Penetrating Radar in Golf Course Management

R. Boniak, S.-K. Chong, S. J. Indorante, and J. A. Doolittle

CONTENTS

27.1 INTRODUCTION

Recently, golf became one of the most popular and rapid growing sports in the United States. According to the American Golf Foundation in 2003, there were approximately 17,000 golf courses in the United States. The key to operating a successful golf course is having healthy greens. As stated by O. J. Noer (Tadge, 1980), the two most important ingredients for building and maintaining a successful golf course are "common sense and drainage." Poor soil drainage results in anaerobic conditions. An anaerobic green is more susceptible to disease development and can induce formation of a black layer that impedes turf root development (Bengeyfield, 1976; Chong et al., 2003; Schwartzkopf, 1975). Poor-quality greens always result in poor performance and playability of the field and will eventually jeopardize golf course income. In order to have a healthy green, both irrigation and drainage are of vital importance in golf course management (Chong et al., 2004). Therefore, in order to attract and keep players, the turf must be kept in excellent condition. In other words, a proper functioning drainage and irrigation system is critically important to the success of a golf course.

Over time, drainage and irrigation systems can fail or become plugged due to improper construction or management. Unfortunately, many drainage and irrigation system maps are neither available nor correctly marked, which makes the problem hard to fix. Locating faulty drainage or irrigation pipe can be laborious and time consuming, particularly for those nonmetal drainage lines. Currently, the ground crews lack a viable way to efficiently identify the exact location of these underground features for making necessary repairs.

Ground-penetrating radar (GPR) is a noninvasive geophysical tool for locating subsurface features. It was commercially developed in the mid 1970s. It is primarily used for imaging near-surface features such as buried artifacts (Conyers and Goodman, 1997), drains (Chow and Rees, 1989), irrigation pipes (Vellidis et al., 1990), utility cables (Annan et al., 1984; Morey, 1974), land mines, and human remains. Highway officials also commonly use it to determine roadbed integrity. In addition, GPR has been used to monitor wetting fronts through surface layers (Vellidis et al., 1990), detect perched water tables (Collins and Doolittle, 1987), and chart subsurface soil horizons and layers (Asmussen et al., 1986; Collins and Doolittle, 1987; Mokema et al., 1990; Raper et al., 1990). There has been recent work performed on mapping the water content of soils (Huisman and Bouten, 2002; van Overmeeren et al., 1997; Weiler et al., 1998). The use of Global Positioning System (GPS) and GPR technology for three-dimensional mapping of soil is a recent innovation (Tischler et al., 2002). GPR has been used to test and characterize agriculture contamination transport (Sénéchal, 2002). Recently (Chong et al., 2000), GPR helped characterize the thickness of the sandy rooting mixture in golf greens, locate the drainage pipes, detect areas of surface compaction, and identify areas of concentrated subsurface wetness.

27.2 MATERIALS AND METHODS

27.2.1 DESCRIPTION OF THE SELECTED GOLF GREENS

The GPR was used to locate drainage tile on two different greens. These two golf greens were constructed by different styles. Figure 27.1 shows golf green No. 3, with an area of about 500 m², at the Stone Creek golf course (SCGC), Makanda, IL. This green was constructed following the U.S. Golf Association (USGA) recommendations. The USGA green consisted of a 30 cm thick layer of sand mix overlying a 10 cm layer of gravel. The drainage tile, surrounded with gravel, was installed beneath the gravel layer. Figure 27.2 shows green No. 2, approximately 200 m², at the Hickory Ridge golf course (HRGC), Carbondale, IL. The green was constructed following the California-style recommendations. The California green is similar to the USGA green but without the 10 cm layer of gravel. Even though the drainage tile was surrounded with gravel, the sand mix was placed immediately on top of the native soil. The drain tile often used in golf greens is a 10 cm diameter corrugated plastic pipe. The layout of the drainage tile was often installed either with a herringbone or gridiron

FIGURE 27.1   Green No. 3 located at the Stone Creek golf course, Makanda, IL.

FIGURE 27.2   Green No. 2 located at the Hickory Ridge golf course, Carbondale, IL.

pattern. Spacing between the lateral lines was about 3 to 5 m (USGA Green Section Record, 1993). The blueprints of the drainage system of both greens were available for comparison.

27.2.2 SELECTION OF THE ANTENNA

In mapping the golf green drainage system, a Subsurface Interface Radar Model SIR 2000 (manufactured by Geophysical Survey Systems Inc., Salem, NH) was used. Both 400 and 900 MHz antenna were tested (Boniak et al., 2003). The 400 MHz antenna was good for a deeper soil but provided less resolution. The 900 MHz antenna had a higher resolution for a shallow soil that would seem ideal for the golf green. Prior to measurement, the GPR was calibrated and standardized to select the best viewing parameter settings. The adjustment included maximum depth range, position of ground surface reflection, and proper signal amplification. In the calibration, the depth of the image was set to a subsurface truth object such as the gravel layer. It is best not to apply frequency filters in the field because it may remove critical data that may be needed later.

27.2.3 SITE PREPARATION PRIOR TO THE MEASUREMENT

A two-person team worked together in conducting the experiment. Prior to scanning, a 1 × 1 m grid pattern was overlaid on the entire green. In the establishment of the grid pattern, the sprinkler heads around the green were used as reference points. The sprinkler heads were selected because they are stationary and easily located. If sprinkler heads are not readily available, stakes may put into the ground for permanent reference points. The grids were flagged at every meter on both sides of the green (Figure 27.3). Due to the irregular shape of each green, marking the green boundary is critical for identifying the location of the drainage tile.

27.2.4 MEASUREMENT

Initially, a 900 MHz antenna was used because of its ability to examine shallow subsurface features with high resolution. However, a large amount of unexpected noise appeared in the image. This might be attributed to the granular fertilizer applied to the green earlier. High salt index in the

FIGURE 27.3   Grid layout.

fertilizer may possibly cause some distortion as well. The 400 MHz antenna was used in the same area, but very little noise was found. In order to prevent the interference by the applied chemicals, the 400 MHz antenna was used for the entire study. Data were collected at each flagged line and marked at every meter point. The speed of pulling the antenna would be best if the scanning speed can be maintained constant.

At the Stone Creek Golf Course (SCGC) site, the team took about 45 minutes to complete the entire measurement, including flagging and scanning. For the green at Hickory Ridge Golf Course (HRGC), the experiment was finished within 30 minutes. On both greens, it took longer to overlay and flag the sampling pattern than to scan the site.

27.2.5   DATA PROCESSING

In the analysis, the software RADAN (Geophysical Survey Systems, Inc.. Salem, NH) was used for analyzing the image data. The first step in processing was to correct the horizontal scale by normalizing the data. After removing all pauses or speed changes between markers, the data were filtered to remove any unnecessary noise or clutter and keep the needed reflection signal. The last major step was to migrate the data to display the hyperbola image of the drain tile. The processed two-dimensional image provided the location of the drain tile. The data could be expressed as a two- or three-dimensional image.

27.3   RESULTS AND DISCUSSION

In Figure 27.4, the vertical scale has been exaggerated and the horizontal scale has been compressed. The horizontal scale represents units of distance traveled along the traverse; the vertical scale represents depth. In Figure 27.4, the upper-most interface (upper marked line) represents reflections from the turf surface. The turf surface appears as a series of continuous, parallel bands. At 1 m, this interface appears higher and discontinued on the radar profile. However, on the actual surface of the green, everything appears to be level across the entire green. The apparent rise in the soil surface at 1 m is attributed to a localized acceleration in the velocity of radar signal propagation through this portion of the sandy rooting mixture. Golfers typically exit the green in this area, which caused

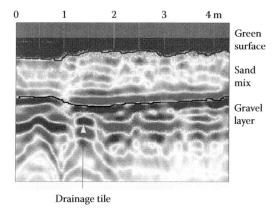

FIGURE 27.4   Radar profile at Hickory Ridge golf course, Carbondale, IL.

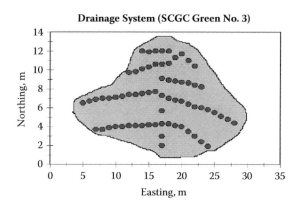

FIGURE 27.5   Map of Stone Creek golf green.

this discontinuity. Because of this steady and focused traffic, the sandy rooting mixture has been compacted over time. Differences in soil compaction affect both the soil's bulk density and moisture content, and in turn, this affects the propagation velocity of the radar signals. Figure 27.4 also shows a discontinuity in shading in the gravel layer between 1 and 1.2 m. (Gravel layer top coincides with lower marked line.) A reverse in shading from light to dark was found between 1.3 and 1.5 m. The reverse in gray scale shading indicated an object appeared in this section of the green. In this case, it was a drainage tile. Figure 27.5 and Figure 27.6 show the drainage pattern beneath Green No. 2 and No. 3 at the HRGC and SCGC, respectively. The results of drainage tile layout matched designs shown in the blueprint when the greens were installed.

The RADAN quick draw software was utilized to plot the three-dimensional map showing the exact location of the tiles. An example of the three-dimensional map from the green at HRGC was constructed to trace a main pipe with one lateral (Figure 27.7). The result was used to compare the drainage system in the blueprint provided by the respective golf courses. Again, the map obtained by the GPR matched the layout shown in the blueprint. Both radar map and blueprint were taken to the field to verify the exact location of the tiles by probing. Both main and lateral lines were accurately located.

Using the GPR for locating and mapping drain tiles beneath golf greens can be very effective. This technology can be used to assist superintendents in locating drain tiles beneath golf greens. However, flagging the green is time consuming; therefore, more work needs to be done in this area.

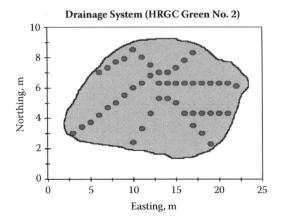

FIGURE 27.6 Map of Hickory Ridge golf green.

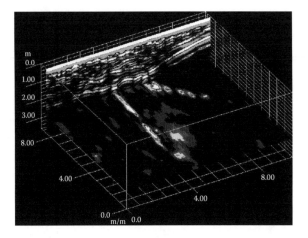

FIGURE 27.7 The three-dimensional image of the main and lateral drainage lines.

It is suggested that a GPS with high accuracy could be used for geo-referencing each location in the process of scanning the green.

27.4 CONCLUSION

The golf green has the perfect condition for the use of GPR to detect underground features. Because of its uniform profile, the response of the electrical permittivity of the soil profile is very consistent. As the electromagnetic wave is transmitted, the velocity of the signal stays unchanged until it detects the layer of gravel or a drain tile. The distinct changes in underground features provide definite reflection. Much of the wave traveling below the clay and gravel interface tends to be scattered, so this interface can be easily identified.

Using the GPR for locating and mapping drain tiles beneath golf greens can be very effective. The noninvasive technology not only can provide accurate and quick answers to superintendents who need to locate drain tiles, it can also detect rooting depth in golf greens. However, in the process of measurement, flagging the green was rather time consuming. It is suggested that a high-accuracy GPS could be used for geo-referencing the measuring of the boundary of the green and scanning location.

The procedure of mapping the golf green drainage system can be summarized as follows:

1. Identify at least two reference points.
2. Create a 1 × 1 m grid system on the golf green.
3. Delineate the boundary of the green.
4. Calibrate the radar.
5. Scan the green.
6. Process data and create scanned image of drain tiles.
7. Plot golf green map with drainage system.

This technology can directly benefit the golf course and the superintendent, and it can indirectly benefit natural resources and the environment. Golf has grown in popularity in recent years. The need to keep courses in pristine condition to attract golfers often fosters practices and management techniques that create environmental hazards. Water leaving the root zone of many golf greens can carry environmentally harmful nitrates or pesticides that, if not redirected, can be harmful to a nontarget population. It is crucial to have this water properly drain into designated areas. Many golf courses are making considerable efforts to be more environmentally conscious and are adopting Audubon regions in proximity of urban dwellings. This GPR technology can provide golf course managers with the critical information they need to make good and environmentally friendly decisions. This technology can also be used on agricultural fields with drain tile and irrigation to examine similar factors.

REFERENCES

Annan, A.P., Davis, J.L. and Vaughn C.J. 1984. Radar mapping of buried pipes and cables. Technical note 1. A-Cubed Inc., Mississauga, Ontario, Canada.

Asmussen, L.E., Perkins, H.F. and Allison, H.D. 1986. Subsurface descriptions by ground-penetrating radar for watershed delineations. *Research Bulletin 340.* Georgia Agricultural Experiment Stations, University of Georgia, Athens.

Bengeyfield, W.H. 1976. Drainage "So easy it's difficult." *USGA Green Section,* January, 1976.

Boniak, R., Chong, S.-K., Indorante, S.J., and Doolittle, J.A. 2002. Mapping golf green drainage and subsurface features using ground penetrating radar. *Ninth International Conference on Ground Penetrating Radar* (Edited by S.K. Koppenjan, and H. Lee) *Proceedings of SPIE,* Vol. 4758: 477–481.

Chong, S.-K., Boniak, R., and Indorante, S. 2003. How do soils breathe? *Golf Course Magazine.* 71:181–183.

Chong, S.-K., Boniak, R., Indorante, S., Ok, C.-H., and Buschschulte, D. 2004. CO_2 content in golf green rhizosphere. *Crop Sci.,* 44: 1337–1340.

Chong, S.K., Doolittle, J., Indorante, S., Renfro, K., and Buck, P. 2000. Investigating without excavating. *Golf Course Manage.* 68: 56–59.

Chow, T.L. and Rees, H.W. 1989. Identification of subsurface drain locations with ground penetrating radar. *Canadian J. Soil Sci.* 69: 223–234.

Collins, M.E. and Doolittle, J.A. 1987. Using ground-penetrating radar to study soil microvariability. *Soil Sci. Soc. of Am. J.* 51: 491–493.

Conyers, L.B. and Goodman, D. 1997. *Ground penetrating radar: An introduction for archeologists.* Alta Mira Press, Walnut Creek, CA.

Huisman, J.A. and Bouten, W. 2002. Mapping surface soil water content with the ground wave of the ground penetrating radar. *Proceedings of the 9th International Conference on Ground Penetrating Radar,* (S.K. Koppenjan and H. Lee, Eds.) Proceedings of SPIE, 162–169.

Morey, R.M. 1974. Continuous subsurface profiling by impulse radar. In: *Proceedings, ASCE Engineering Foundation Conference on Subsurface Exploration for Underground Excavations and Heavy Construction,* Am. Soc. of Civ. Engineers, Henniker, NH, 212–232.

Mokma, D.L., Schaetzel, R.J., Doolittle, J.A., and Johnson, E.P. 1990. Ground penetrating radar study of orstein continuity in some Michigan haplaquods. *Soil Sci. Soc. of Am. J.* 54: 936–938.

Raper, R.L., Asmussen, L.E., and Powell, J.B. 1990. Sensing hardpan depth with ground-penetrating radar. *Trans. ASAE,* 33: 41–46.

Schwartzkopf, C. 1975. Drainage: Why and how. *USGA Green Sect. Rec.,* 13: 1–3.

Senechal, P., Perroud, H., and Bourg, A.C.M. 2002. Characterization of agricultural contaminant transport using ground-penetrating radar and electrical data, *Proceedings of the 9th International Conference on Ground Penetrating Radar,* 460–465.

Tadge, C.H. 1980. Drainage is important to turfgrass management. *USGA Green Sect. Rec.,* 18: 17–19.

Tischler, M., Collins, M.E., and Grunwald, S. 2002. Integration of ground penetrating radar data, Global Positioning Systems and geographic information systems to create three dimensional soil models, *Proceedings of the 9th International Conference on Ground Penetrating Radar,* 313–316.

USGA Green Section Record. 1993. USGA Recommendations for putting green construction. March/April 1993. USGA, Golf House, Far Hills, NJ.

van Overmeeren, R.A., Sariowan, S.V., and Gehrels, J.C. 1997. Ground penetrating radar for determining volumetric soil water content; results of comparative measure of two sites, *J. Hydrology,* 197: 316–338.

Vellidis, G., Smith, M.C., Thomas, D.L., and Asmussen, L.E. 1990. Detecting wetting front movement in a sandy soil with ground-penetrating radar, *Trans. ASAE,* 33: 1867–1874.

Weiler, K.W., Steenhuis, T.S., Boll, J., and Kung, K.-J.S. 1998. Comparison of ground-penetrating radar and time domain reflectometry as soil water sensors, *Soil Sci. Soc. of Am. J.,* 62: 1237–1239.

28 Ground-Penetrating Radar Investigation of a Golf Course Green

Computer Processing and Field Survey Setup Considerations

Barry J. Allred, Edward L. McCoy, and J. David Redman

CONTENTS

28.1   INTRODUCTION

The golf industry in the United States has experienced extraordinary growth since 1950. The number of golfers during the period of 1950 to 1998 increased from approximately 3 million to 27 million (National Golf Foundation and McKinsey & Company, 1999). From 1950 to 2000, the number of golf courses increased by 12,000 (Beard, 2002). By the year 2000, there were over 15,000 golf course facilities throughout the country (National Golf Foundation, 2001). Most of these facilities have nine or eighteen holes, but there are a substantial proportion of facilities that have 27, 36, 45, or more holes. In regard to golf course maintenance costs, the estimate for the amount spent in 1996 alone was over $4.5 billion (Beard, 2002).

Keeping these facilities in good condition requires continual maintenance and occasional remodeling of the various parts that make up a golf course, especially the greens. The most important subsurface features for golf course greens are the constructed soil layers and drainage system infrastructure. There are two commonly used ways of constructing a golf course green: the California Method and the USGA (U.S. Golf Association) Method (Hurdzan, 1996). The investigation described in this case study focused on a USGA Method green because of its greater complexity compared to a California Method green. Recommendations for the USGA Method green call for a 0.3 m layer of sandy material directly below the surface that is, in turn, underlain by a 0.1 m gravel layer resting on native soil into which gravel backfilled trenches have been cut containing 0.1 m diameter drainage pipe (Figure 28.1). The drainage pipe trenches that are cut into the native soil are typically 0.2 m deep and 0.25 m wide. The lateral edges of the sand and gravel layers within a USGA Method green are vertical.

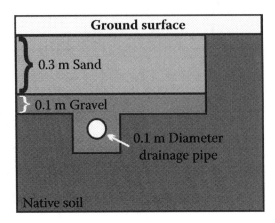

FIGURE 28.1   A U.S. Golf Association (USGA) Method green.

For either the USGA Method green or California Method green, rectangular or herringbone patterns are typically used for placement of the drainage pipe system. With the rectangular pattern, the drainage pipe laterals merge with the main conveyance pipe at an angle of 90 degrees. With the herringbone pattern, the drainage pipe laterals merge with the main conveyance pipe at an angle less than 90 degrees. The spacing distance between the drainage pipe laterals within a green is usually between 3 and 5 m (Boniak et al., 2002).

Ground-penetrating radar (GPR) has the potential to provide valuable information on constructed soil layer thicknesses and depths and buried drainage pipe positions within a golf course green environment. However, to date, there has been very little research conducted on golf course applications of GPR. Boniak et al. (2002) tested GPR at two golf course greens in southern Illinois, and were successful in mapping subsurface features (soil layers and drainage pipes) using a system with 400 MHz center frequency antennas. In this same study, a GPR system with 900 MHz center frequency antennas was also tested but without success due to high signal attenuation that was attributed to recent fertilizer application. A feasibility study by Allred et al. (2005) found GPR antenna center frequencies of 250, 500, and 1000 MHz to work equally well for mapping subsurface drainage pipe systems on a golf course tee and two golf course greens. For producing GPR profiles showing constructed soil layer thicknesses and depths and drainage pipe positions, 900 and 1000 MHz center frequency antennas provided the best data (Allred et al., 2005).

Because of the time and effort devoted to the upkeep of greens, detailed information on features beneath a green's surface will undoubtedly be helpful to the golf course superintendents and architects involved with green maintenance and remodeling activities. When original design plans have been lost, which is often the case, or for the purpose of quality control after new construction, a nondestructive technique such as GPR may be the most viable alternative for determining the thicknesses and depths of constructed soil layers or drainage pipe locations within a golf course green environment. Using GPR to its fullest advantage in this regard requires careful attention to computer-processing procedures and the field survey setup. Consequently, this study had two goals: first to determine the appropriate computer-processing procedures for generating GPR profiles or time-slice amplitude maps of golf course greens, and second, to establish the most effective operational setup for collecting GPR data on a golf course green.

28.2   MATERIALS AND METHODS

The GPR method has been described in detail within Chapter 7 of this book. To summarize, the GPR method involves directing an electromagnetic radio energy (radar) pulse into the subsurface, followed by measurement of the elapsed time taken by the signal as it travels downward from the transmitting antenna, partially reflects off a buried feature, and is eventually returned to the surface, where it is picked up by a receiving antenna. Reflections from different depths produce a signal trace, which is a function of returning radar wave amplitude (and hence, energy) versus two-way travel time. Differences in the dielectric constant across a discontinuity govern the amount of reflected radar energy returning to the surface. In this study, these discontinuities include the boundaries between soil layers along with those interfaces between the soil material surrounding the drainage pipe, the drainage pipe, and the air and water inside the drainage pipe.

It was previously stated that due to its greater overall design complexity compared to a California Method green, the analysis approach employed in this investigation, in order to be more rigorous, focused efforts only on a USGA Method green. The data used for analysis in this investigation was therefore obtained on the seventeenth hole USGA Method green at the Muirfield Village Golf Club in Dublin, Ohio. The GPR system used to collect the data was a Sensors & Software Inc. Noggin^{plus} (Mississauga, Ontario, Canada) with 1000 MHz center frequency transmitter and receiver antennas that were separated by 0.076 m. For data collection, the distance between measurement points along a transect was 0.05 m, eight signal traces were averaged at each point location, and for each signal trace, a 0.1 nanosecond (ns) sampling interval was employed. GPR measurements were collected along two sets of parallel transects oriented perpendicular to one another and forming a rectangular grid covering the green. The spacing distance between adjacent GPR measurement lines was 1 m. These transect measurements were then used to produce GPR images of the soil profile and time-slice amplitude maps.

As discussed in Chapter 7, GPR profiles are constructed by plotting side by side the sequential signal traces collected along a line of measurement. GPR profiles represent the amount of reflected radar energy returning to the surface from different depths beneath the line along which data were collected. The golf course green GPR profiles will be most useful for determining thicknesses and depths of constructed soil layers and horizontal and vertical positions of individual drainage pipes. Information from all of the measurement transects is employed to generate the GPR time-slice amplitude maps, which represent the amount of reflected radar energy returning to the surface over an area from a specified interval of two-way travel time (or depth). The golf course green GPR time-slice amplitude maps have their greatest potential for use in determining the pattern of the drainage pipe network that is present.

Two Sensors & Software Inc. computer software packages were employed to process the GPR data: EKKO View Deluxe for the profiles and EKKO Mapper for the time-slice amplitude maps. The computer-processing procedures evaluated in this study, with respect to producing GPR profiles and time-slice amplitude maps, are listed as follows with corresponding abbreviations to be used throughout the remainder of this case history and a short description of what each procedure does:

1. *Signal Saturation Correction Filter (SSCF)*—This filter removes slowly decaying low-frequency noise introduced by factors related to transmitting and receiving antenna proximity and electrical properties of the ground.
2. *Signal Gain Functions*—Gain functions amplify the radar signal strength.
   a. *Constant Gain Function (CGF)*—For this gain function, the signal trace amplitudes are multiplied by a constant factor, thereby enhancing the signal while preserving relative amplitude variations.
   b. *Automatic Gain Control Function (AGCF)*—This gain function equalizes amplitudes along the radar signal trace.
3. *Signal Trace Enveloping (STE)*—The purpose of this process is to convert the wavelets composing a radar signal trace from ones having positive and negative components to ones that are mono-pulse and all positive. The overall process tends to simplify the data, in turn making interpretation easier in some instances.
4. *Spatial Background Subtraction Filter (SBSF)*—For this filter, a set number of sequential signal traces (which are a subset of all the signal traces collected along a particular line)

are averaged, and the resultant is then subtracted from the signal trace in the center of the sequence. Next, using the same number of sequential signal traces, the process is moved over along the line by one signal trace and then repeated. The overall effect of the filter is to remove the radar signal response from flat-lying features.

5. *2-D Migration (2DM)*—A synthetic aperture image reconstruction process is applied to the GPR data set. The process tends to focus scattered signals, in particular, collapsing hyperbolic responses into localized point responses. (Note that hyperbolic GPR responses have an upside-down U-shaped pattern, are often referred to as "reflection hyperbolas," and for this study, are typically associated with buried drainage pipes.)

Issues related to setting up a GPR golf course green survey need to be addressed before data collection efforts can begin, in order to obtain the best-quality time-slice amplitude maps. Particularly important are decisions regarding whether a unidirectional or bidirectional survey is to be conducted and the spacing distance to use between lines along which GPR measurements are obtained. Because GPR data at the Muirfield Village seventeenth hole green were collected bidirectionally (two sets of parallel measurement transects oriented perpendicular to one another) with a spacing distance of 1 m between adjacent measurement lines, subsets of the total data set can be utilized to simulate the impact of different field survey setups. The different field survey setups to be evaluated include (1) a bidirectional survey with a 1 m spacing distance between adjacent measurement lines, (2) a bidirectional survey with a 2 m spacing distance between adjacent measurement lines, (3) a bidirectional survey with a 3 m spacing distance between adjacent measurement lines, (4) a unidirectional survey based on one set of parallel northwest–southeast transects with a 1 m spacing distance between adjacent measurement lines, and finally, (5) a second unidirectional survey based on the other set of parallel transects, in this case southwest–northeast, with a 1 m spacing distance between adjacent measurement lines.

28.3   RESULTS AND DISCUSSION

Constructed soil layers and drainage pipes within a golf course green will reflect radar energy, and with proper computer processing, have a response that can be clearly exhibited on GPR profiles. Figure 28.2 provides some typical examples for the computer processing of golf course green GPR profiles generated with EKKO View Deluxe. The Figure 28.2 profiles represent the application of different computer-processing procedures to the same set of raw 1000 MHz GPR data collected along a line oriented southwest to northeast on the Muirfield Village Golf Club seventeenth hole USGA Method green. For each Figure 28.2 GPR profile, the left vertical axis represents the two-way radar signal travel time, the right vertical axis represents depth in meters, and the bottom axis is distance in meters along the line of measurement. The raw data are presented in Figure 28.2a. The Figure 28.2a GPR responses to the bottoms of the constructed sand and gravel layers are barely visible, and reflection hyperbolas representing buried drainage pipes cannot be detected. The Figure 28.2b GPR profile processing sequence began with a signal saturation correction filter (SSCF) followed by a constant gain function (CGF) with a factor of 25. The bottom of the constructed sand layer (indicated by the downward pointing arrows), the bottom of the constructed gravel layer (indicated by the upward pointing arrows), and three upside-down U-shaped drainage pipe reflection hyperbolas (one is highlighted by an oval-shaped gray line) are all clearly visible. The actual position for the top of a drainage pipe coincides with the apex position of its reflection

FIGURE 28.2   (*see facing page*) Ground-penetrating radar (GPR) profile computer-processing results, (a) raw data, (b) signal saturation correction filter (SSCF) and constant gain function (CGF), (c) SSCF and automatic gain control function (AGCF), (d) SSCF, CGF, signal trace enveloping (STE), and 2-D migration (2DM), and (e) SSCF, CGF, and spatial background subtraction filter (SBSF).

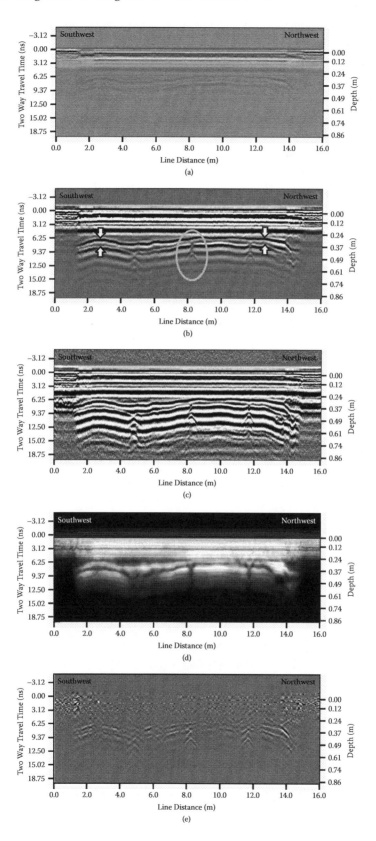

hyperbola response. Interpretation of the Figure 28.2b profile was aided substantially by modeling synthetic GPR profiles with GprMax2D (Giannopoulos, 2003).

The Figure 28.2c GPR profile processing sequence began with the SSCF followed by an automatic gain control function (AGCF). The AGCF, by equalizing amplitudes along the signal trace, oftentimes enhances unwanted signal or "noise." For golf course greens, much of this noise is due to "multiples." Multiples occur when the radar wave travel path from the transmitting antenna, downward into the subsurface, and back up to the receiver antenna involves reflections off of more than one dielectric constant discontinuity interface. The direct, single reflection radar wave travel path to and from a particular subsurface interface will always be shorter than any travel path for a multiple involving this same subsurface interface. Multiples produced by reflections off of two or more shallower interfaces can become a problem then they interfere with the single reflection primary response of a deeper feature. Because only a portion of the incoming radar wave is reflected at an interface, multiples are oftentimes, but not always, somewhat weaker compared to the primary GPR responses produced by direct, single reflection radar wave travel paths. By using an AGCF, multiples that appear in some manner to involve the constructed soil layers are amplified, thereby making overall interpretation of the Figure 28.2c GPR profile more difficult with regard to constructed soil layer thicknesses; however, the three drainage pipe reflection hyperbolas do show up quite well.

The Figure 28.2d GPR profile processing sequence began with the SSCF, then a CGF (factor = 25), followed by signal trace enveloping (STE), and finally, 2-D migration (2DM). The Figure 28.2d computer-processing sequence does not produce a GPR profile that is as easy to interpret as the one for Figure 28.2b. The Figure 28.2e GPR profile processing sequence began with the SSCF, then a CGF (factor = 25), followed by a spatial background subtraction filter (SBSF). The SBSF used for the Figure 28.2e GPR profile employed a seven signal trace moving average sequence. The SBSF removed most of the responses to the constructed soil layers, thereby more completely isolating the drainage pipe reflection hyperbolas. All in all, the results presented in Figure 28.2 indicate that for the purpose of determining constructed soil layer thickness and depth and drainage pipe positions, a computer-processing sequence of a SSCF and a CGF is the best alternative.

Again, drainage pipes tend to reflect radar energy, and with proper computer processing are displayed as lighter shaded linear features on a GPR time-slice amplitude map. Figure 28.3 provides some typical examples for computer processing of golf course green time-slice amplitude maps. These time-slice amplitude maps represent a two-way travel time interval of 9 to 15 ns or a depth interval of 0.38 to 0.68 m. It should be noted that the EKKO Mapper software used to produce the GPR time-slice amplitude maps did not allow input of signal gain functions, because in doing so, the results tend to become distorted. The Figure 28.3 time-slice amplitude maps represent the application of different computer-processing procedures to the same set of raw 1000 MHz antenna center frequency GPR data (bidirectional and 1 m spacing between lines of measurement) collected on the Muirfield Village Golf Club seventeenth hole USGA Method green. For each Figure 28.3 GPR time-slice amplitude map, both the left axis and the bottom axis provide a distance scale in meters. The raw data are presented in Figure 28.3a. Although the areal extent of the golf course green shows up well, there are almost no indications in Figure 28.3a of the subsurface drainage pipe network that is present.

The Figure 28.3b GPR time-slice amplitude map processing sequence began with the signal saturation correction filter (SSCF) followed by signal trace enveloping (STE). Besides clearly depicting the areal extent of the golf course green, the Figure 28.3b time-slice amplitude map begins to provide some very subtle evidence of the drainage pipe system present. The Figure 28.3c GPR time-slice amplitude map processing sequence began with the SSCF, then 2-D migration (2DM), followed by STE. The Figure 28.3c time-slice amplitude map exhibits some definite, although still subtle, indications of the drainage pipe network that is present. Again, in Figure 28.3c, the areal extent of the golf course green is quite apparent.

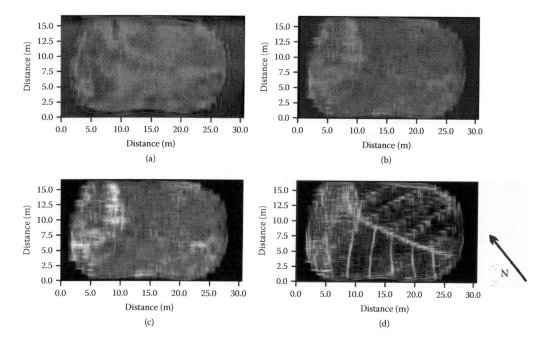

FIGURE 28.3 Ground-penetrating radar (GPR) 9 to 15 ns (0.38 to 0.68 m depth interval) time-slice ampli-tude map computer-processing results: (a) raw data; (b) signal saturation correction filter (SSCF) and signal trace enveloping (STE); (c) SSCF, 2-D migration (2DM), and STE; and (d) SSCF, spatial background subtraction filter (SBSF), 2DM, and STE.

The Figure 28.3d GPR time-slice amplitude map processing sequence began with the SSCF, then a spatial background subtraction filter (SBSF), followed by 2DM, and finally, STE. Not only is the areal extent of the golf course green obvious, the herringbone subsurface drainage pipe pattern is easily distinguishable. The SBSF used for the Figure 28.3d GPR time-slice amplitude map employed a seven signal trace moving average sequence. The SBSF helped substantially by removing most of the constructed soil layer response, thereby enhancing drainage pipe signals. Consequently, for mapping golf course green subsurface drainage pipe systems, a computer-processing sequence of a SSCF, a SBSF, 2DM, and STE is the best choice.

Figure 28.4 depicts the effects of different golf course green GPR field survey setups in rela-tion to a time-slice amplitude map (1000 MHz center antenna frequency data) depiction of the subsurface drainage system present. For reference, Figure 28.4a is an interpreted schematic map showing the Muirfield Village Golf Club seventeenth hole USGA Method green areal extent (gray shaded area) and drain line locations (solid black lines for strong evidence and dashed black lines for weaker evidence). This schematic map is based on 250, 500, and 1000 MHz center antenna fre-quency GPR data collected in July 2004.

All Figure 28.4 time-slice amplitude maps were produced using a computer-processing sequence of a SSCF, a SBSF, 2DM, and STE. The Figure 28.4b time-slice amplitude map is based on a bidi-rectional survey (two sets of parallel measurement transects oriented perpendicular to one another) with a 1 m spacing distance between adjacent measurement lines. The complete subsurface drain-age pipe network shows up very clearly in Figure 28.4b, given the possible exception of the far left portion of the map. The Figure 28.4c time-slice amplitude map is based on a bidirectional survey with a 2 m spacing distance between adjacent measurement lines. The main conveyance line and the lateral drain lines in the lower half of the Figure 28.4c time-slice amplitude map still show up well, but the rest of the subsurface drainage system is not nearly as visible. The Figure 28.4d time-slice amplitude map is based on a bidirectional survey with a 3 m spacing distance between

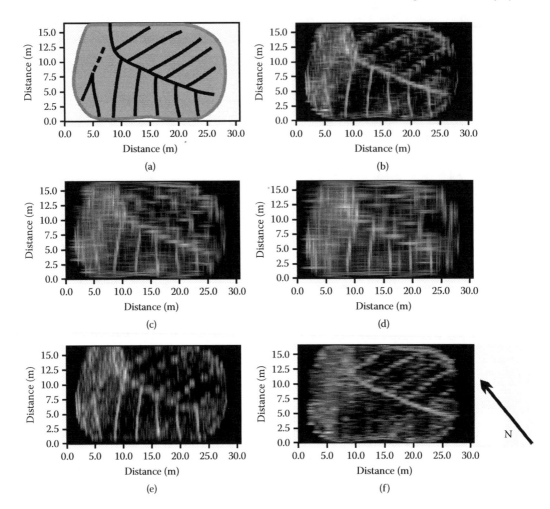

FIGURE 28.4 Ground-penetrating radar (GPR) 9 to 15 ns (0.38 to 0.68 m depth interval) time-slice amplitude map comparison of different field survey setups: (a) interpreted schematic map showing golf course green areal extent (gray shaded area) and drain line locations (solid black lines for strong evidence and dashed black lines for weaker evidence); (b) bidirectional survey with a 1 m spacing distance between adjacent measurement lines; (c) bidirectional survey with a 2 m spacing distance between adjacent measurement lines; (d) bidirectional survey with a 3 m spacing distance between adjacent measurement lines; (e) unidirectional survey based on one set of parallel northwest–southeast transects with a 1 m spacing distance between adjacent measurement lines; and (f) second unidirectional survey based on the other set of parallel southwest–northeast transects with a 1 m spacing distance between adjacent measurement lines.

adjacent measurement lines. The main conveyance line and the lateral drain lines in the lower half of the Figure 28.4d time-slice amplitude map are still evident but not as distinct as in Figure 28.4c; however, the rest of the subsurface drainage system is not interpretable. It is obvious from the results presented in Figure 28.4b through Figure 28.4d that the spacing distance between adjacent measurement transects should be no greater than 1 m in order to effectively resolve a golf course green subsurface drainage system on a GPR time-slice amplitude map.

The Figure 28.4e time-slice amplitude map is based on a unidirectional survey (one set of parallel northwest–southeast transects) with a 1 m spacing distance between adjacent measurement lines. Because the orientation of the measurement transects for Figure 28.4e is essentially perpendicular to the orientation of the drainage pipe laterals shown in the bottom half of the map, these drainage pipe laterals show up extremely well. There are plenty of indications regarding the main conveyance line and the drainage pipe laterals in the upper half of the Figure 28.4e time-slice amplitude map, but they are not nearly as discernable as those in the lower half of the map. The Figure 28.4f time-slice amplitude map is based on a unidirectional survey (one set of parallel southwest–northeast transects) with a 1 m spacing distance between adjacent measurement lines. The drainage pipe laterals in the bottom of the Figure 28.4f time-slice amplitude map do not show up well at all, because the orientation of the measurement transects is essentially parallel to the orientation of the drainage pipe laterals in this part of the green. When they are oriented in the same direction, a GPR measurement line has to follow along directly over top of the drain line in order to detect it. Given circumstances in which they are parallel, if the GPR measurement line and drain line are offset from one another by a small amount, then the drain line will not be identified. One interesting aspect of the Figure 28.4f unidirectional survey GPR time-slice amplitude map is that the main conveyance line and the drainage pipe laterals in the upper half of the map show up remarkably clear, perhaps even better than in Figure 28.4b, which was based on a bidirectional survey. Consequently, a bidirectional GPR survey for a golf course green is required in order to have the best chance of detecting all the drainage pipes present. Although, plotting out the bidirectional and both unidirectional GPR time-slice amplitude maps and then comparing the three is probably worthwhile with respect to getting as good of an interpretation as possible for the overall golf course green subsurface drainage pipe network.

28.4   SUMMARY

GPR can provide valuable information on subsurface features within a golf course green. Getting the most out of a golf course green GPR study requires careful consideration of computer-processing procedures and the field survey setup. Analysis of 1000 MHz center antenna frequency GPR data from a USGA Method green indicate some useful guidelines on GPR computer-processing procedures and the field survey setup:

1. A computer-processing sequence of a signal saturation correction filter (SSCF) and a constant gain function (CGF) is the best alternative for the purpose of producing a GPR profile showing golf course green constructed soil layer thickness and depth and individual drainage pipe positions.
2. A computer-processing sequence of a SSCF, a spatial background subtraction filter (SBSF), 2-D migration (2DM), and signal trace enveloping (STE) is the best choice for GPR time-slice amplitude mapping of a golf course green subsurface drainage pipe system.
3. With regard to field survey setup, a bidirectional GPR survey (two sets of parallel measurement transect oriented perpendicular to one another) with a spacing distance between adjacent lines of measurement no larger than 1 m is the proper approach for complete mapping of a golf course green subsurface drainage system.

REFERENCES

Allred, B.J., D. Redman, E.L. McCoy, and R.S. Taylor. 2005. Golf course applications of near-surface geophysical methods: A case study. *Journal of Environmental and Engineering Geophysics.* v. 10, pp. 1–19.
Beard, J.B. 2002. *Turf Management for Golf Courses.* Ann Arbor Press. Chelsea, MI.
Boniak, R., S.K. Chong, S.J. Indorante, and J.A. Doolittle. 2002. Mapping golf course green drainage systems and subsurface features using ground penetrating radar. In *Proceedings of SPIE, Vol. 4758, Ninth International Conference on Ground Penetrating Radar.* pp. 477–481. S.K. Koppenjan and H. Lee, editors. April 29–May 2, 2002. Santa Barbara, CA. SPIE. Bellingham, WA.
Giannopoulos, A. 2003. *Modeling Ground Penetrating Radar Using GprMax.* www.see.ed.ac.uk/~agianno/GprMax/.

Hurdzan, M.J. 1996. *Golf Course Architecture: Design, Construction & Restoration.* Sleeping Bear Press. Chelsea, MI.

National Golf Foundation and McKinsey & Company. 1999. *A Strategic Perspective on the Future of Golf.* National Golf Foundation. Jupiter, FL.

National Golf Foundation. 2001. *Golf Facilities in the U.S.* National Golf Foundation. Jupiter, FL.

29 Agricultural Drainage Pipe Detection Using Ground-Penetrating Radar

Barry J. Allred and Jeffrey J. Daniels

CONTENTS

29.1 INTRODUCTION

A 1985 U.S. Department of Agriculture (USDA) Economic Research Service economic survey showed that the states in the Midwest United States (Illinois, Indiana, Iowa, Ohio, Minnesota, Michigan, Missouri, and Wisconsin) had by that year approximately 12.5 million hectares that contained subsurface drainage systems (USDA Economic Research Service, 1987). Cropland constituted by far the large majority of this acreage. The same economic survey estimated the 1985 on-farm replacement cost for these cropland subsurface drainage systems to be $18 billion. Today, this subsurface drainage infrastructure would be worth $35 billion based on a 1986 to 2008 average yearly consumer price index inflation rate of 3 percent, and this total does not include the extensive amount of drainage pipe that has been installed in the past 20 years. The magnitude of the area involved along with infrastructure costs indicate how crucial subsurface drainage is to the Midwest U.S. farm economy, without which, excess soil water could not be removed, in turn making current levels of crop production impossible to achieve.

Figure 29.1 is a schematic illustrating drainage pipe placement within the soil profile typical of agricultural fields in Ohio. Prior to the 1960s, agricultural drainage pipe was constructed primarily of clay tile and was then superseded by corrugated plastic tubing (CPT), which is today the material still used in drainage pipe fabrication (Schwab et al., 1981). The drainage pipe diameter is most commonly 10 cm. The pipe is emplaced at the bottom of a trench, which is then backfilled. The trench is typically 0.3 to 0.5 m wide with its bottom depth ranging between 0.5 and 1 m. Modern drain line installation equipment often produces a trench that is wider at the bottom than the top. The water table can be either above or below the drainage pipe depending on the amount of recent rainfall and the mode of operation for the subsurface drainage system (uncontrolled drainage, controlled drainage, or subirrigation). The surface tilled zone is commonly less than 0.3 m, assuming a no-till management strategy has not been adopted.

Increasing the efficiency of soil water removal on farmland that already contains a functioning subsurface drainage system often requires reducing the average spacing distance between drain lines. This is typically accomplished by installing new drain lines between the older ones. By keeping the older drain lines intact, less new drainage pipe is needed, thereby substantially reducing costs to farmers. However, before this approach can be attempted, the older drain lines need to be located.

Finding buried drainage pipe is not an easy task, especially for drain lines installed more than a generation ago. Often, records have been lost, and the only outward appearance of the subsurface drainage system is a single pipe outlet extending into a water conveyance channel. From this, little can be deduced about the network pattern used in drainage pipe placement. Without records that show precise locations, finding a drain line with heavy trenching equipment causes pipe damage requiring costly repairs, and the alternative of using a handheld tile probe rod is extremely tedious at best. Zucker and Brown (1998) indicate

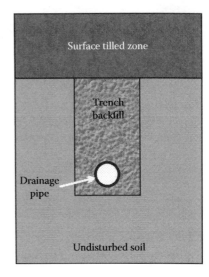

FIGURE 29.1   The drainage pipe position within the soil profile.

that satellite or airborne remote sensing technologies show some promise, but these methods are only applicable during certain times of the year and under limited site conditions.

Consequently, there is definitely a need to find better ways of effectively and efficiently locating buried agricultural drainage pipe. Ground-penetrating radar (GPR) may provide the solution to this problem. Promising results have been achieved using GPR to find buried plastic and metal utility pipelines (Hayakawa and Kawanaka, 1998; LaFaleche et al., 1991; Wensink et al., 1991). However, to date, investigation of GPR drainage pipe detection capabilities has been limited, especially for the Midwest U.S. Chow and Rees (1989) demonstrated the use of GPR to locate subsurface agricultural drainage pipes in the Maritime Provinces of Canada, and Boniak et al. (2002) showed that GPR could be employed to find drainage pipe beneath golf course greens. Allred et al. (2004, 2005) evaluated GPR drainage pipe detection within agricultural field settings typical of the Midwest U.S. The results obtained by Allred et al. (2004, 2005) regarding impacts of antenna frequency, soil hydologic conditions, pipe construction material, and drain line orientation on GPR drainage pipe detection are summarized within this case history along with an assessment of overall GPR drainage pipe detection effectiveness based on data collected at fourteen test plots throughout Ohio.

29.2   MATERIALS AND METHODS

The GPR method has been described in detail in Chapter 7 of this book. To summarize, the GPR method involves directing an electromagnetic radio energy (radar) pulse into the subsurface, followed by measurement of the elapsed time taken by the signal as it travels downward from the transmitting antenna, partially reflects off a buried feature, and is eventually returned to the surface, where it is picked up by a receiving antenna. Reflections from different depths produce a signal trace, which is a function of returning radar wave amplitude (and hence, energy) versus two-way travel time. Antenna frequency and factors influencing soil electrical conductivity, such as shallow hydrologic conditions, clay content, and salinity, all have a substantial influence on the distance beneath the surface to which the radar signal penetrates. The nature of the dielectric constant difference across a discontinuity governs the radar reflection coefficient (see Chapter 7), which, in turn, determines the amount of reflected radar energy returning to the surface. For this study, these discontinuities include those interfaces between with the soil material surrounding the drainage pipe, the drainage pipe, and the air and water inside the drainage pipe.

There are two obvious dielectric constant discontinuities. One is the interface between the drainage pipe and surrounding soil material and the second is the interface between the drainage pipe and the air and water within it (Figure 29.1). However, if the drainage pipe wall thickness is small relative to the radar pulse wavelength (definitely the case for corrugated plastic tubing and likewise for clay tile), then as a result of constructive or destructive interference between the radar pulses reflected off the outer and inner walls of the pipe, the effective GPR reflection response essentially becomes governed by the dielectric constant values of the surrounding soil material and the air and water inside the pipe.

The dielectric constant ranges in value from 1 for air to 80 for water with dry soil closer to the lower end of this range, ~5 to 15, and very moist or saturated soils near the middle of the range, ~30 to 40 (Conyers and Goodman, 1997; Reynolds, 1997; Sharma, 1997; Sutinen, 1992). The dielectric constant, ε, of soil material is directly dependent on the volumetric moisture content, θ. A relationship between ε and θ was empirically developed by Sutinen (1992) for glacial materials, similar to those found in the Midwest United States, and is expressed as follows:

$$\varepsilon = 3.2 + 35.4(\theta) + 101.7(\theta^2) - 63(\theta^3) \tag{29.1}$$

The GPR unit used predominantly for this research was the Sensors & Software Inc. Noggin[plus] with 250 MHz center frequency antennas. In order to investigate the effect of different antenna frequencies on drainage pipe detection, other Sensors & Software Inc. GPR systems were tested in a more limited manner. These included a Noggin[plus] unit employing 500 MHz center frequency antennas and a pulseEKKO 100A unit equipped with 100 MHz center frequency antennas.

For data collection during this project, the distance between measurement points along a transect was 0.05 m, and thirty-two signal traces were averaged at each point location. For each test plot, GPR measurements were typically collected along two sets of parallel transects oriented perpendicular to one another and forming a rectangular grid covering the test plot. The spacing distance between adjacent GPR measurement lines was usually 1.5 m. These transect measurements were then used to produce GPR images of the soil profile and time-slice amplitude maps showing drainage pipe patterns.

As discussed in Chapter 7, GPR profiles are constructed by plotting side by side the sequential signal traces collected along a line of measurement. GPR profiles represent the amount of reflected radar energy returning to the surface from different depths beneath the line along which data were collected. The vertical axis on a GPR profile is given in two-way radar signal travel time unless soil water content or soil dielectric constant data are available to allow converting the two-way travel time values to depth values. Information from all of the measurement transects at a test plot were integrated to generate the GPR time-slice amplitude maps, which represent the amount of reflected radar energy returning to the surface over an area from a specified interval of two-way travel time (or depth).

A certain amount of computer processing was employed to produce the GPR profiles and amplitude maps presented in this case history. Computer processing was essential in order to enhance the GPR drainage pipe response embedded in the raw data. The computer-processing steps for generating GPR profiles included a signal saturation correction filter and a spreading and exponential compensation gain function. For GPR time-slice amplitude maps, a signal saturation correction filter, two-dimensional migration, signal trace enveloping, a high-frequency noise filter, and a spatial background subtraction filter were all used.

The impacts of antenna frequency (100, 250, and 500 MHz), soil hydrologic conditions, pipe construction material, and drain line orientation on GPR detection of agricultural drainage pipes were evaluated at one test plot. This test plot was built specifically for the overall GPR drainage pipe detection project and is located behind the ElectroScience Laboratory (ESL) at Ohio State University in Columbus, Ohio. The surface soil (2.5 to 15 cm depth) texture at the ESL site, as determined

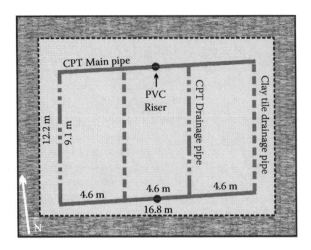

FIGURE 29.2 The ElectroScience Laboratory (ESL) test plot utilized for studying ground-penetrating radar (GPR) drainage pipe detection. GPR surveys were conducted within the dashed boundary.

by grain size analysis (Wray, 1986), is classified as silty clay. Figure 29.2 is a schematic showing the layout of the ESL test plot, which was constructed with both clay tile and CPT drainage pipe placed in 0.5 m wide trenches. Due to land slope, depth to the drainage pipe system on its northwest corner was 1 m, and 0.6 m on the southeast corner. Shortly following backfill of the trenches where the 10 cm diameter pipes were placed, the test plot was tilled down to a depth of 20 cm so that typical agricultural field conditions could be replicated. Two 10 cm diameter riser pipes (Figure 29.2) connect the buried drainage pipe system to the surface, thereby allowing a shallow water table to be maintained at any desired level through use of a water supply hose connected to a Hudson valve suspended inside one of the riser pipes. The overall effectiveness of GPR for detecting buried agricultural drainage pipe was evaluated at fourteen test plots (including the one at ESL), which ranged in size from 200 to 12,000 m², had different soil textures, and were located throughout central, southwest, and northwest Ohio.

29.3 RESULTS AND DISCUSSION

The radar signal penetration depth beneath the surface is governed to a large degree by the GPR antenna center frequency. Radar signal penetration depth decreases as the GPR center frequency is increased. Furthermore, as GPR center frequency increases, object resolution improves (better imaging of smaller subsurface objects). Consequently, determining the proper GPR antenna frequency based on a buried target's depth and size is extremely important.

Figure 29.3 shows the GPR response based on different antenna frequencies. The Electro-Science Laboratory (ESL) test plot data for Figure 29.3 were collected under dry surface conditions with the water table below the drainage pipes (based on field observations and measured monitoring well water levels). Figure 29.3a and Figure 29.3b were obtained with a Noggin[plus] unit and 500 MHz center frequency antennas. Figure 29.3c and Figure 29.3d were obtained with a Noggin[plus] unit and 250 MHz center frequency antennas. Figure 29.3e and Figure 29.3f were obtained with a pulseEKKO 100A unit and 100 MHz center frequency antennas.

The Figure 29.3a, Figure 29.3c, and Figure 29.3e GPR profiles were produced from measurements obtained along the same line, which was oriented perpendicular to the four north–south trending clay tile and corrugated plastic tubing (CPT) drainage pipes at the ESL test plot (Figure 29.2). An upside-down U-shaped feature called a "reflection hyperbola" is the typical drainage pipe response depicted on a GPR profile generated from data collected along a transect oriented perpendicular to the drain lines. This reflection hyperbola response to the buried drainage pipes is shown

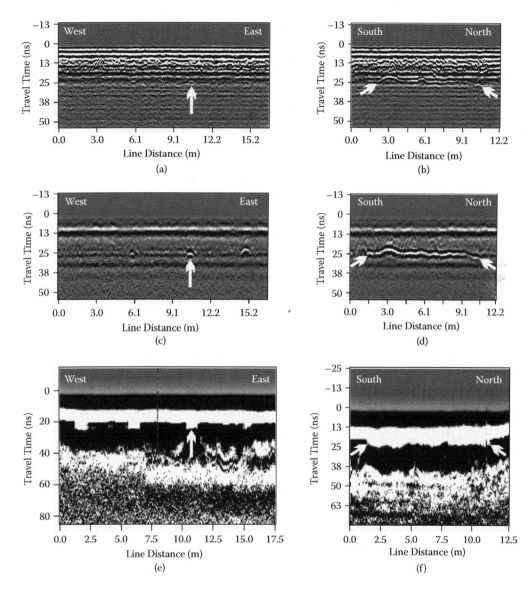

FIGURE 29.3 Ground-penetrating radar (GPR) profiles from ElectroScience Laboratory (ESL) test plot showing the drainage pipe detection effect of antenna frequency; (a) and (b) 500 MHz center frequency antennas, (c) and (d) 250 MHz center frequency antennas, and (e) and (f) 100 MHz center frequency antennas.

to a greater or lesser extent in Figure 29.3a and Figure 29.3c with reflection hyperbola apexes (top of pipe position) found at travel times between 20 and 25 ns. The Noggin^{plus} unit with 500 MHz center frequency antennas barely detected three of the four pipes, of which one is highlighted with an upward pointing arrow (Figure 29.3a). The Noggin^{plus} unit with 250 MHz center frequency antennas detected all four pipes (Figure 29.3c), one of which is highlighted with an upward pointing arrow; although, the response to the one farthest west was fairly subdued. The pulseEKKO 100A unit with 100 MHz center frequency antennas detected all four pipes (Figure 29.3e); however, the GPR response to the buried drainage pipes was not the typical reflection hyperbola expected, but rather a rectangular block-like extension of the top white band down into the black band beneath it (one is highlighted with an upward pointing arrow). The difference in GPR drainage pipe response of the pulseEKKO 100A unit with 100 MHz center frequency antennas compared to the other two systems is at present unclear but may be related to reduced object size resolution and interference with radar pulses traveling directly through the air and along the ground surface.

The Figure 29.3b, Figure 29.3d, and Figure 29.3f GPR profiles were produced using measurements obtained from the same ESL test plot line, which was oriented directly along the trend of one of the buried north–south CPT drainage pipes (the second drain line from the east in Figure 29.2). The typical GPR response in this scenario is a banded linear feature representing the buried drainage pipe. The position of the top of the banded feature corresponds to the top of the buried drain line. The banded linear feature highlighted by the arrows is somewhat subtle on the profile generated using a Noggin[plus] unit with 500 MHz center frequency antennas (Figure 29.3b). In comparison, the banded linear feature for the buried north–south drainage pipe shows up quite well (highlighted by the arrows) on the Figure 29.3d profile generated from data collected using a Noggin[plus] unit with 250 MHz center frequency antennas. Figure 29.3d also shows a strong reflection hyperbola on the south end of the banded feature and a subtle one on the north end, both of which represent the CPT main pipe connected at each end of the drainage line (see Figure 29.2). Finally, the response of the pulseEKKO 100A unit with 100 MHz center frequency antennas was again different. Instead of a distinct banded linear feature, the arrow highlighted drain line position is shown by a long rectangular extension of the top white band down into the black band directly beneath it (Figure 29.3f).

Choosing the proper antenna frequency based on the subsurface depth and size (diameter) of the drainage pipe is an extremely important consideration. Overall, taking into account different drain line orientations with respect to a GPR transect, the 250 MHz center frequency antennas appeared to work best for detecting buried agricultural drainage pipe at depths of up to 1 m. For larger diameter pipes at greater depth, perhaps antennas with a 100 MHz center frequency are the best option.

As the water content of a soil increases, so too does its electrical conductivity, and as soil electrical conductivity increases, radar signal penetration depth decreases. However, with regard to drainage pipe detection, this adverse GPR impact due to increased soil wetness, could potentially be offset if there is a greater amount of radar energy reflected from the drainage pipe due to the nature of the dielectric constant contrast between the soil outside the pipe and the air and water inside the pipe. Large rainfall events in the Midwest United States increase wetness within the soil profile by increasing water contents near the surface and sometimes causing a rise in the shallow water table.

The influence of shallow hydrologic conditions on GPR drainage pipe detection is displayed in Figure 29.4. The data for Figure 29.4 were collected using a Noggin[plus] unit with 250 MHz center frequency antennas. The Figure 29.4a and Figure 29.4d GPR profiles were generated from measurements obtained along one line, which was oriented perpendicular to the four clay tile and CPT north–south trending drainage pipes at the ESL test plot (Figure 29.2). The Figure 29.4b and Figure 29.4e GPR profiles were produced from data obtained from one ESL test plot line measurement transect, which was oriented directly along trend over a buried north–south CPT drainage pipe (second drain line from the east in Figure 29.2). The ESL test plot GPR amplitude maps, Figure 29.4c and Figure 29.4f, represent the amount of reflected radar energy from a 15 ns time window bracketing the drainage pipe positions. Lighter shaded linear features shown on the GPR time-slice amplitude maps indicate drainage pipe patterns.

Figure 29.4a through Figure 29.4c correspond to shallow hydrologic conditions with a wet surface from a recent (<18 h) rainfall of 7.8 mm and a water table raised 0.5 m above the drainage pipes. Figure 29.4d through Figure 29.4f correspond to shallow hydrologic conditions with a very moist soil profile and pipes totally drained of water. The shallow hydrologic conditions for Figure 29.4d through Figure 29.4f were obtained by continually pumping water from the drainage pipes and lowering the water table for 24 h prior to the GPR field survey. Essentially, Figure 29.4a, Figure 29.4b, and Figure 29.4c are representative of shallow hydrologic conditions with a wet soil profile and water-filled pipes, and Figure 29.4d through Figure 29.4f are representative of shallow hydrologic conditions with a wet soil profile and air-filled pipes.

Figure 29.4a through Figure 29.4c indicate that the poorest shallow hydrologic condition in regard to GPR drainage pipe detection occurs with a wet soil surface and a static water table located

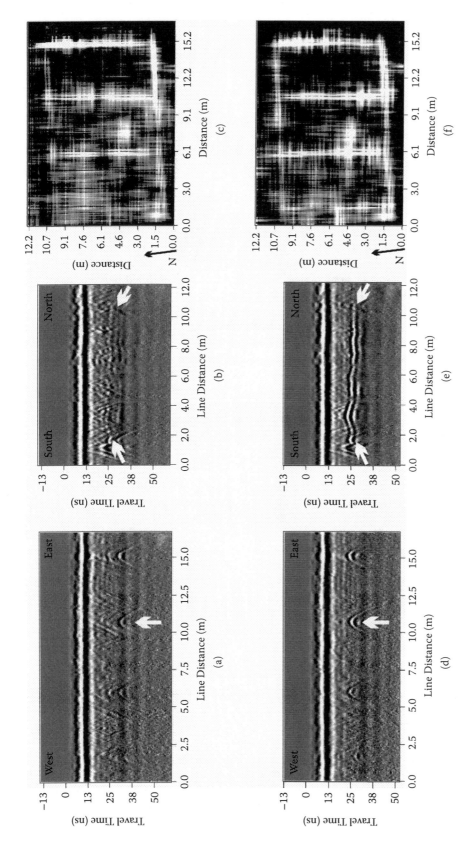

FIGURE 29.4 ElectroScience Laboratory (ESL) test plot ground-penetrating radar (GPR) profiles and amplitude maps showing the effect of shallow hydrologic conditions on drainage pipe detection; (a), (b), and (c) wet surface and water table raised 0.5 m above drainage pipes, and (d), (e), and (f) a very moist soil profile with pipes totally drained of water.

well above the position of the drainage pipes. The western-most of the four drainage pipe reflection hyperbolas is extremely subtle in Figure 29.4a. The banded linear feature representative of a buried drainage pipe oriented directly along trend of the GPR data collection line is almost absent in Figure 29.4b. Additionally, the subsurface drainage pipe system is difficult to discern on the western end of the Figure 29.4c GPR amplitude map. With the water table above the drainage pipes, the dielectric constant contrast between the saturated soil surrounding the pipe and the water inside the pipe is such that the effective radar reflection coefficient for the pipe is fairly low, and there is less reflected radar energy from the pipe that returns to the receiving antenna.

A better shallow hydrologic condition for GPR drainage pipe detection occurs with a very moist soil profile surrounding air filled drainage pipe (Figure 29.4d through Figure 29.4f). All four of the reflection hyperbolas are apparent in Figure 29.4d. The banded linear feature showing the complete length of a drain line is very clear in Figure 29.4e. The complete subsurface drainage system is well defined in Figure 29.4f. Consequently, the dielectric constant contrast between the wet soil surrounding the pipe and the air inside the pipe is such that the effective radar reflection coefficient for the pipe is fairly high and there is more reflected radar energy from the pipe that returns to the receiving antenna. Here it is worth noting that the strength of the 250 MHz antenna GPR drainage pipe response for a dry to moderately dry soil profile and air-filled pipes, as exhibited in Figure 29.3c and Figure 29.3d, falls in between the strengths of the GPR drainage pipe response due to a wet soil profile and water-filled pipes versus a wet soil profile and air-filled pipes.

The results for this portion of the research project provide some important guidelines as to when a GPR drainage pipe mapping survey should be conducted with regard to the shallow hydrologic conditions present. Clearly, GPR surveys should be avoided when the water table is above the elevation of the drainage pipes. This shallow hydrologic condition takes place most often in the hours (or day or two at most) directly following a substantial rainfall event and before much soil drainage has occurred. Although less typical, the water table is also elevated above the drainage pipes, usually for prolonged periods of time, at locations where controlled drainage and subirrigation methods are in use. Moderately dry to dry soil profiles with the water table below the drain lines (pipes are completely air filled) are an acceptable shallow hydrologic condition for the use of GPR to locate drainage pipes. The moderately dry to dry soil profile, low water table condition is fairly common during periods where there has been little rainfall, especially during the growing season when evapotranspiration rates are high. A better shallow hydrologic condition for GPR drainage pipe detection occurs with a very moist soil profile and a water table at or below the drain lines (pipes are completely or at least largely air filled). The very moist soil, low water table conditions often occur during wet periods, especially after a day or two following a significant rainfall event, during which most of the excess soil water has had a chance to drain. It is important to point out that substantially increased soil wetness and the corresponding increase in soil electrical conductivity, which reduces radar signal penetration depth beneath the ground surface, does not in itself preclude using GPR to find buried agricultural drainage pipes.

Although perhaps not strongly emphasized in the previous discussion, Figure 29.3 and Figure 29.4 provide important insight on the impacts of pipe construction material and drain line orientation with respect to GPR drainage pipe detection. The Figure 29.3a, Figure 29.3c, Figure 29.3e, 29.4a, and Figure 29.4d GPR profiles were generated from measurements obtained along a transect oriented perpendicular to the four clay tile and CPT north–south trending drainage pipes at the ESL test plot (see Figure 29.2). Importantly, none of these profiles when viewed separately depict any noticeable difference in GPR response between adjacent clay tile and CPT pipes. Appearances of the clay tile and CPT drain lines are likewise similar on the Figure 29.4c and Figure 29.4f GPR time-slice amplitude maps of the ESL test plot. Therefore, the pipe construction material, clay tile or CPT, seems to have very little influence on the GPR drainage pipe detection response.

Figure 29.3a, Figure 29.3c, Figure 29.4a and Figure 29.4d show that an upside-down U-shaped feature called a "reflection hyperbola" is the typical drainage pipe response depicted on a GPR profile generated from data collected along a transect oriented perpendicular to the drain lines. The reflection hyperbola apex coincides with the top of pipe position. A banded linear feature is the drainage pipe response most commonly found on a GPR profile generated from data collected along a transect oriented directly along trend over a buried drainage pipe (Figure 29.3b, Figure 29.3d, Figure 29.4b and Figure 29.4e). The position of the top of the banded feature corresponds to the top of the buried drain line.

Figure 29.5 displays 250 MHz antenna drainage pipe detection results for a test plot in northwest Ohio and provides a good example of the drainage pipe response shown on GPR profiles generated from measurements collected along transects that are not oriented 90° (perpendicular) or 0° (along trend) to the drain line. Figure 29.5a is a west-to-east GPR profile, Figure 29.5b is a south-to-north GPR profile, Figure 29.5c is a GPR time-slice amplitude map of the test plot, and Figure 29.5d is an interpreted map of the drainage pipe pattern based on the time-slice amplitude map. The Figure 29.5d interpreted map shows all the drain lines trending east–west except one that trends southeast–northwest. The measurement transect for the south-to-north GPR profile (Figure 29.5b) crosses the southeast–northwest drain line at an angle of 60°, and the measurement transect for the west-to-east GPR profile (Figure 29.5a) crosses the southeast–northwest drain line at an angle

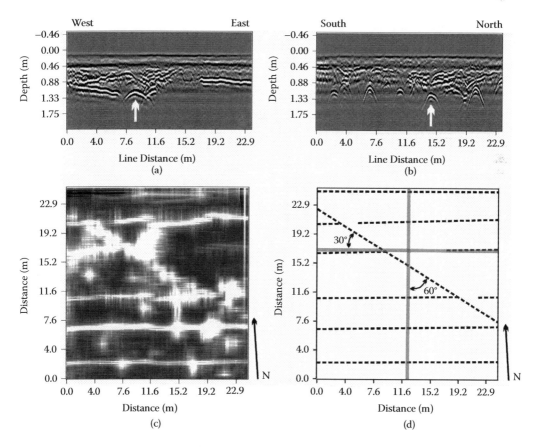

FIGURE 29.5 Ground-penetrating radar (GPR) results from a test plot in northwest Ohio emphasizing changes in the GPR drainage pipe response due to different orientations of the GPR measurement transect relative to the drain line, (a) west to east GPR profile, (b) south to north GPR profile, (c) 25 to 40 ns GPR time-slice amplitude map, and (d) interpreted map of drainage pipe pattern. (Dashed black lines represent drainage pipes, and solid gray lines show the position of the GPR measurement transects of profiles a and b.)

TABLE 29.1

Summary of GPR Drainage Pipe Detection Results for Fourteen Test Plots across Ohio

Test Plot	Region within Ohio (Closest City or Town)	Surface Soil Textural Class[1]	Average Soil Electrical Conductivity (mS/m)	Amount of Pipe Located (%)
1	Central Ohio (ESL–Columbus)	Silty clay	15.6	100
2	Central Ohio (Columbus)	Clay to silty clay	30.6	75
3	Central Ohio (Columbus)	Clay loam	20.5	50
4	Central Ohio (Columbus)	Clay loam	16.2	100
5	Central Ohio (Columbus)	Silty clay loam	9.6	50
6	Central Ohio (Columbus)	Silty clay	12.1	90
7	Southwest Ohio (Washington Courthouse)	Clay	28.0	0
8	Northwest Ohio (Defiance)	Clay	76.9	75
9	Northwest Ohio (Defiance)	Silty clay	76.8	100
10	Northwest Ohio (Defiance)	Sandy loam	15.9	100
11	Northwest Ohio (Delta)	Clay loam	22.6	100
12	Northwest Ohio (Delta)	Sandy clay loam to sandy loam	12.0	100
13	Northwest Ohio (Lima)	Silty clay to silty clay loam	—	100
14	Northwest Ohio (Hoytville)	Silty clay	31.6	0

of 30°. The GPR response to the southeast–northwest drain line shown on the Figure 29.5b south-to-north profile is highlighted with an upward pointing arrow and is represented by a reflection hyperbola that laterally is slightly stretched compared to other reflection hyperbolas within the profile representing drain lines oriented perpendicular to the measurement transect. The GPR response to the southeast–northwest drain line shown on the Figure 29.5a west-to-east profile is again highlighted with an upward pointing arrow and is represented by a reflection hyperbola that has been substantially stretched laterally. Consequently, as the orientation of the GPR measurement transect relative to the drain line changes from 90° (perpendicular) to 0° (along trend), the GPR profile drainage pipe response changes from a tight, narrow reflection hyperbola, to a laterally stretched reflection hyperbola, to a banded linear feature.

The overall effectiveness of GPR drainage pipe detection was assessed at fourteen test plots in central, southwest, and northwest Ohio using a Noggin[plus] unit with 250 MHz center frequency antennas. The location, surface soil texture, average soil electrical conductivity, and the GPR drainage pipe detection effectiveness for each test plot are provided in Table 29.1. The textural class of the surface soil (2.5 to 15 cm depth) was determined by particle size analysis (Wray, 1986). Average soil electrical conductivity was calculated from 14,610 Hz electromagnetic induction measurements taken at the test plots with a Geophex, Ltd. GEM-2 ground conductivity meter. Table 29.1 indicates that GPR was successful in finding on average 74 percent of the total amount of pipe present at the fourteen test plots. In seven test plots, 100 percent of the pipe was located, and in two test plots, none of the pipe was found. All in all, the GPR method worked quite well in finding clay tile and corrugated plastic tubing drainage pipe down to depths of around 1 m within a variety of different soil materials. GPR was even proven capable of locating buried drainage pipes in silty clay and clay soils having extremely high average electrical conductivity values above 75 mS/m, which would typically be expected to severely limit radar signal penetration depth (test plots 8 and 9). It is still unclear as to why GPR detected none of the drainage pipe in test plots 7 and 14.

29.4   SUMMARY

The major findings regarding the use of GPR to locate buried agricultural drainage pipes are as follows:

1. A GPR unit with 250 MHz center frequency antennas seemed to work best for detecting buried drainage pipe under conditions typical in Ohio.
2. Shallow hydrologic conditions with a wet soil surrounding a water-filled drainage pipe produce the poorest GPR drainage pipe response. Shallow hydrologic conditions with a wet soil surrounding an air-filled drainage pipe produce a much better GPR drainage pipe response. The GPR drainage pipe response for a dry to moderately dry soil profile and air-filled pipes falls somewhere in between.
3. The type of drainage pipe present, whether clay tile or CPT, does not seem to impact the GPR response.
4. The orientation of the GPR measurement transect with respect to the drain line, from perpendicular to directly along trend, governs the type of drainage pipe response found on a GPR profile, which ranges, respectively, from a tight, narrow reflection hyperbola, to a laterally stretched reflection hyperbola, to a banded linear feature.
5. Overall, the GPR method works quite well in finding clay tile and CPT drainage pipe down to depths of around 1 m within a variety of different soil materials.

REFERENCES

Allred, B.J., N.R. Fausey, L. Peters, Jr., C. Chen, J.J. Daniels, and H. Youn. 2004. Detection of buried agricultural drainage pipe with geophysical methods. *Appl. Eng. in Agric.* v. 20, pp. 307–318.

Allred, B.J., J.J. Daniels, N.R. Fausey, C. Chen, L. Peters, Jr., and H. Youn. 2005. Important considerations for locating buried agricultural drainage pipe using ground penetrating radar. *Appl. Eng. in Agric.* v. 21, pp. 71–87.

Boniak, R., S.K. Chong, S.J. Indorante, and J.A. Doolittle. 2002. Mapping golf course green drainage systems and subsurface features using ground penetrating radar. In *Proceedings of SPIE, Vol. 4758, Ninth International Conference on Ground Penetrating Radar.* pp. 477–481. S.K. Koppenjan and H. Lee, editors. April 29–May 2, 2002. Santa Barbara, CA. SPIE. Bellingham, WA.

Chow, T.L. and H.W. Rees. 1989. Identification of subsurface drain locations with ground-penetrating radar. *Can. J. Soil Sci.* v. 69, pp. 223–234.

Conyers, L.B. and D. Goodman. 1997. *Ground-Penetrating Radar: An Introduction for Archaeologists.* Alta-Mira Press. Walnut Creek, CA.

Hayakawa, H. and A. Kawanaka. 1998. Radar imaging of underground pipes by automated estimation of velocity distribution versus depth. *J. Appl. Geophysics.* v. 40, pp. 37–48.

LaFaleche, P.T. J.P. Todoeschuck, O.G. Jensen, and A.S. Judge. 1991. Analysis of ground probing radar data: Predictive deconvolution. *Canadian Geotechnical J.* v. 28, pp. 134–139.

Reynolds, J.M. 1997. *An Introduction to Applied and Environmental Geophysics.* John Wiley & Sons. Chichester, UK.

Schwab, G.O., R.K. Frevert, T.W. Edminster, and K.K. Barnes. 1981. *Soil and Water Conservation Engineering,* 3rd Edition. John Wiley & Sons. New York, NY.

Sharma, P.V. 1997. *Environmental and Engineering Geophysics.* Cambridge University Press. Cambridge, UK.

Sutinen, R. 1992. *Glacial Deposits, Their Electrical Properties and Surveying by Image Interpretation and Ground Penetrating Radar.* Bulletin 359. Geological Survey of Finland. Rovaniemi, Finland.

USDA Economic Research Service. 1987. *Farm Drainage in the United States: History Status, and Prospects.* USDA Miscellaneous Publication 1455. U.S. Dept. of Agriculture. Washington, DC.

Wensink, W.A., J. Hofman, and J.K. Van Deen. 1991. Measured reflection strengths of underwater pipes irradiated by pulsed horizontal dipole in air: Comparison with continuous plane-wave scattering theory. *Geophysical Prospecting.* v. 39, pp. 543–566.

Wray, W.K. 1986. *Measuring Engineering Properties of Soil*. Prentice Hall. Englewood Cliffs, NJ.
Zucker, L.A. and L.C. Brown, Editors. 1998. *Agricultural Drainage: Water Quality Impacts and Subsurface Drainage Studies in the Midwest*. Ohio State University Extension Bulletin 871. The Ohio State University, Columbus, OH.

30 Using Ground-Penetrating Radar to Estimate Tree Root Mass
Comparing Results from Two Florida Surveys

John R. Butnor, Daniel B. Stover, Brian E. Roth,
Kurt H. Johnsen, Frank P. Day, and Daniel McInnis

CONTENTS

30.1 INTRODUCTION

Roots make up 5 to 45 percent of tree biomass in upland forests worldwide (Cairns et al., 1997), but estimates from urban settings are limited. Roots of field-grown trees are difficult to assess and delineate without laborious and destructive excavations. For this reason, relative to studies on above-ground biomass, root systems have rarely been studied; often more is assumed about their nature than is really known. Foresters and ecologists are interested to learn how different management techniques (site preparation, cultural practices, and harvesting methods) and natural disturbances affect tree productivity and carbon sequestration. Tree roots play a dynamic role in sustainable forest productivity and serve as a conduit for atmospheric carbon to enter the rhizosphere. After a harvest, fire, or other disturbance, the majority of the recalcitrant carbon is retained in roots for a time and slowly released by oxidative processes, and a small fraction becomes a stable constituent of the soil. Research interests aside, trees play an important role in urban environments, providing shade, reducing temperatures, and providing aesthetics. Municipalities, arborists, and property owners are interested in mapping roots to establish protection zones during construction (Jim, 2003), assessing tree health and taking proactive steps to help urban trees thrive.

Ground-penetrating radar (GPR) can be used to detect and monitor roots if there is sufficient electromagnetic contrast with the surrounding soil matrix. This methodology is commonly used in archeology and civil engineering for non-destructive testing of concrete as well as road and bridge surfaces. GPR is ideal for these applications because the electrical properties of concrete and rebar (steel) are very different. Under amenable conditions (i.e., electrically resistive, sandy soils are ideal), tree roots are detectable and can be quantified. GPR has been used to resolve roots and buried organic debris, assess root size, map root distribution, and estimate root biomass (Butnor et al., 2001, 2003; Stover et al., 2007). Being non-invasive and non-destructive, GPR allows repeated measurements that facilitate the study of root system development. Roots as small as 0.5 cm in diameter have been detected at depths of less than 50 cm with a 1500 MHz antenna in well drained, coarse-textured soils (Butnor et al., 2003). There has been considerable interest in mapping tree root systems to understand root architecture and soil volume utilization (Cermak et al., 2000; Hruska et al., 1999; Stokes et al., 2002). However, without intensively detailed, methodical scanning of small grids, it is not possible to separate roots by size class or depth (Wielopolski et al., 2000). Orientations of roots, geometry of root reflective surfaces, and proximity of other adjacent roots presently confound attempts to delineate root size classes in forest soils. In addition, this methodology cannot determine differences in species within a mixed stand.

We present the results of two case studies conducted in Florida which were aimed at determining tree root biomass with GPR in a pine plantation and a native scrub-oak shrubland ecosystem to highlight the technical considerations, successes, and examples of where clutter confounded meaningful interpretation.

30.2 MATERIALS AND METHODS

The fundamentals of GPR are described in Chapter 7 and practical considerations are emphasized in Case History 12.7. In the studies presented here, tree roots were the target of interest, and GPR was applied to detect electrical discontinuities between roots and the surrounding soil. A variety of frequencies (400 to 1500 MHz) have been used to locate coarse roots; 400 MHz antenna can resolve discontinuities to a depth of 2 m or more, however the smallest detectable root is 4 to 5 cm in diameter (Butnor et al., 2001; Hruska et al., 1999); 900 MHz antennas consistently resolve roots >2 to 3 cm; and under optimal conditions, while 1500 MHz center frequencies can detect roots as small as 0.5 cm, penetration is usually limited to ~50 cm (Butnor et al., 2001). Using the aforementioned frequencies, it is not possible to resolve fine roots that are commonly classified as <0.2 cm diameter. Because most tree roots are located at relatively shallow soil depths <0.5 m (Hruska et al., 1999; Jim, 2003) the high-frequency antenna is a good choice to maximize root detection. A trade-off exists where low-frequency antennas penetrate deeper in the soil, but provide lower data resolution, and higher frequencies provide greater resolution with decreased depth penetration. Ground-coupled antennas need to be in close contact with the soil surface; this is especially important with higher-frequency antennas (e.g., 1500 MHz) where air-gaps greater than a couple of centimeters will cause a deleterious loss in resolution. This presents problems in forested terrains where the presence of herbaceous vegetation, fallen trees limbs, and irregular soil surfaces impedes the travel of an antenna, requiring additional site preparation before data acquisition (Butnor et al., 2001).

The goals of this study were to test the feasibility of using GPR to quantify root mass and root distribution at two sites that share similar soils but markedly different vegetation types. One site was a scrub-oak ecosystem located on a subtropical barrier island in the northern part of the Kennedy Space Center (KSC), Merritt Island, Brevard County, Florida. Several oak species (*Quercus* sp.) dominate the system because of a prescribed fire ten years earlier, and they represent most of the aboveground biomass (Stover et al., 2007). The other was a 5-year-old loblolly pine (*Pinus taeda*) plantation in northern Florida near Sanderson. Both sites exhibited sandy soils in the near surface, the KSC being excessively well drained and the Sanderson site being moderately well drained.

In 2005, a SIR-2000 GPR system, manufactured by Geophysical Survey Systems, Inc. (GSSI; Salem, NH) was fitted with a custom-designed sampling rig that steadied the high-frequency antenna (model 5100, 1500 MHz antenna) and incorporated a survey wheel to meter electromagnetic pulses (Figure 30.1) to make measures at Sanderson. Measures at KSC were made with a SIR-3000, using a model 5100, 1500 MHz antenna fitted with a much smaller survey wheel than depicted in Figure 30.1. Before spatial distribution can be assessed, it is necessary to sample test transects to be able to correlate radar data with destructively sampled soil cores. With both systems, measures

FIGURE 30.1 SIR 2000 ground-penetrating radar (GPR) system connected to a 1500 MHz antenna mounted on a skateboard deck equipped with a survey wheel encoder.

were made by slowly drawing the survey rig along a measurement transect (200 scans per m) while ensuring that the antenna remained in contact with the soil surface. The locations of the verification cores were electronically marked on the data file during collection. The resulting scan was a two-dimensional profile (transect length by depth) where electromagnetic anomalies are located.

Postcollection processing was accomplished using RADAN 6.5 software (GSSI) with the following steps (Butnor et al., 2003; Cox et al., 2005; Stover et al., 2007):

1. *Crop.* Each radargram was cropped to the diameter of a 15 cm area soil core.
2. *Background Removal.* Root structures appear as hyperbolic reflectors, and parallel bands represent plane reflectors such as ground surface and soil layers. Parallel bands were removed with a horizontal finite impulse response filter (FIR) method called background removal.
3. *Migration.* Kirchoff migration was applied to correct the position of objects and collapse hyperbolic diffractions based on signal geometry (Daniels, 2004).
4. *Hilbert Transformation.* Hilbert transformation was applied to express the relationship between magnitude and the phase of the signal, allowing the phase of the signal to be reconstructed from its amplitude, allowing subtle discontinuities to be detected (Oppenheim and Schafer, 1975).

Root mass was quantified by converting radargrams to bitmap images and using Sigma Scan Pro Image Analysis software (Systat) with the following steps (Butnor et al., 2003; Cox et al., 2005; Stover et al., 2007):

1. *Gray Scale.* Each image was converted to an 8-bit grayscale image.
2. *Intensity Threshold.* Root mass was quantified by using pixel intensity (proxy for amplitude derived from Hilbert transform). Intensity is a relative measure ranging from 0 (black) to 255 (white). An intensity threshold range of 60 to 255 pixels, which was able to delineate roots as small as 0.5 cm was applied.
3. *Statistical Analysis.* Linear regression was performed to quantify the relationship between total root biomass from the soil cores and the GPR-derived index (pixels within the threshold range).

Once transects were scanned, extreme care was used to extract the cores in the correct location to correspond to the radargram. The cores were screened, and the biomass was separated into live roots and dead organic debris, dried and weighed.

30.3   RESULTS AND DISCUSSION

Pine roots are usually found in clusters where it can be difficult to separate individual roots, especially when their orientation is unknown and may be confused with plane reflectors (Figure 30.2A). Background removal removes most of the clutter associated with surface reflections, soil horizons, and moisture gradients (Figure 30.2B). Kirchoff migration serves to collapse the hyperbolas closer to their representative size and location (Figure 30.2C). The geometry of reflected signals may be very useful for extracting information about the reflective surface of the root, including diameter if the root orientation is known (Barton and Montagu, 2004). When a cylindrical root is scanned at a 90° angle it is possible to calculate diameter, but in the field roots present themselves in a variety of shapes and orientations. The Hilbert transform is a useful means to illustrate and quantify the reflection amplitude apart from reflector geometry (Figure 30.2D).

At Sanderson, the correlation between live root mass collected with cores and GPR data was less than desirable (r = 0.51). The roots were almost exclusively pine, and the dead organic debris was composed of decaying palmetto roots, residual slash, and dead roots. Debris from the previous pine plantation had been plowed into the soil and had not decomposed completely after 5 years. From forty cores, more than half of the biomass was classified as dead organic debris. GPR was not able separate out live and dead material. A compelling illustration of this problem is shown in Figure 30.3. Two soil core locations were scanned in a region of high reflectivity, despite having almost the same amount of biomass and a similar number of pixels in the analysis threshold (GPR index), core A contained 17 percent live root mass and core B contained 73 percent (Figure 30.3).

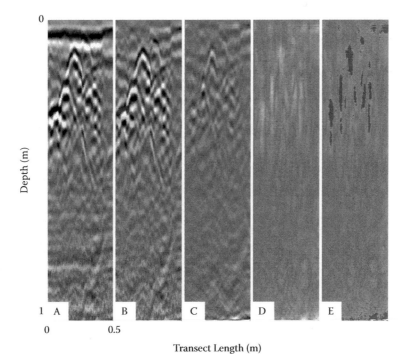

FIGURE 30.2   Example of a radargram collected at the Sanderson loblolly pine plantation (A) and processed using background removal (B), Kirchoff migration (C), Hilbert transformation (D) with RADAN software. Amplitude intensity was visualized and quantified with Sigma Scan Pro software (E).

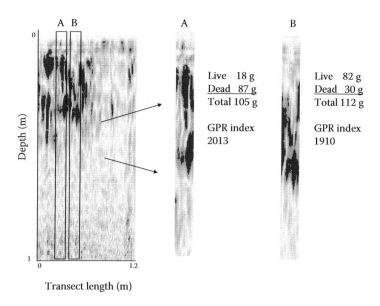

Live   18 g
Dead   87 g
Total 105 g

GPR index
2013

Live   82 g
Dead   30 g
Total 112 g

GPR index
1910

Transect length (m)

FIGURE 30.3   Two locations (A and B) were scanned in an area of high reflectivity at Sanderson, Florida, and destructively sampled with a 15 cm diameter soil core. Core A had low root density and a large mass of dead material, the opposite ratio was observed in core B. Ground-penetrating radar (GPR) could not discriminate between roots and dead organic material.

When live and dead root mass were combined, a strong positive relationship was observed with the GPR data (r = 0.80). The high proportion of buried organic debris at Sanderson limits the utility of the technique to detect live roots. At KSC the correlation between GPR and live roots was much better (r = 0.84). The Sanderson findings highlight the fact that presently, it is not possible to separate live and dead buried biomass. At KSC, dead biomass represented only 13 percent of the total buried biomass which clearly enhanced the accuracy of root biomass using GPR. When the live and dead root mass were combined at KSC and compared to GPR-derived predictions the correlation was excellent (r = 0.82). Previous forest studies minimized the importance of understanding the reflectivity of organic debris; it was thought that dead roots rapidly took on dielectric properties of the surrounding soil (Butnor et al., 2001). The issue of organic debris never came up at a site in Bainbridge, Georgia, which had been used and plowed for decades in traditional agricultural systems, exposing any residual debris to oxidation (Butnor et al., 2003). GPR has subsequently been used to detect dead root fragments in peach orchards (*Prunus persica*) to specifically identify and quantify organic debris that harbors harmful fungi (*Armillaria* sp.) that could be detrimental to new plantings (Cox et al., 2005). At Sanderson, the problem of interfering organic debris resulting from site preparation was likely exacerbated by the preponderance of saw palmetto which is decay resistant compared to other forest slash.

When monitoring root development and accretion over time, it is important for small changes to be detected. The sandy soils at KSC drain rapidly, producing little change in radargram analysis, except in the case of saturated field moisture conditions. GPR data interpretation and predictive root biomass equations are not limited to just conditions present at the time of core collection on very coarse soils, but this should be verified for specific sites. The Sanderson site was remeasured in 2003, 2004, and 2005, and a very interesting phenomenon was observed. The primary research goal was to determine the effect of genetic improvement, tree spacing, and fertilizer application on tree productivity (above- and belowground). Between 2004 and 2005, there was an apparent and unexpected drop in belowground biomass as estimated with GPR. Approximately 10 percent of the survey was remeasured with litter in place and again with litter removed to quantify the effects of litter on root resolution. Scanning through the dry, air-filled leaf litter served to defocus the

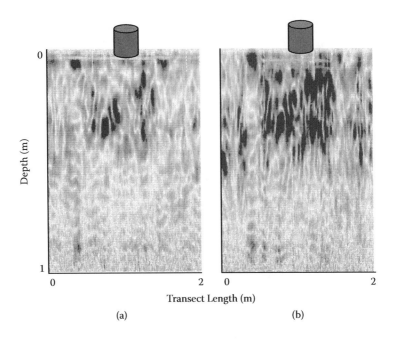

FIGURE 30.4 Example of antenna defocusing near the trunk of a loblolly pine caused by thick, dry pine straw, 20 cm deep on pine root discrimination (a) compared with enhanced discrimination when the pine straw is raked away (b).

1500 MHz antenna, it created an offset between the antenna and the soil surface which was largely air filled and degraded the ability to detect roots (Figure 30.4). If litter depth was uniform across all treatments, the defocusing effect could have been mitigated. However, because the experimental manipulations of tree genetics and fertilizer application resulted in differential litter depth (5 cm versus nearly 20 cm of dry newly deposited pine straw), the differences in root mass could not be assessed.

The Sanderson site was expected to be ideal for radar studies, but several unforeseen complications negated the quantitative value of the GPR-derived estimates of root mass. The study was important to highlight potential sources of data clutter in forests. The soil core calibrations at KSC became very useful for analyzing the effect of elevated atmospheric CO_2 concentrations on belowground biomass accumulation in the scrub system. Elevated atmospheric CO_2 concentration has been experimentally increased at the KSC site using open-top chambers (OTC) since 1996 (Day et al., 2006). CO_2 enters the chambers through air blowers and circulates throughout the chamber and exits at the open chamber top. During the ninth year of the CO_2 study, sixteen OTCs (eight elevated and eight ambient CO_2) were surveyed with a 1500 MHz antenna. The soil core calibrations at KSC (described earlier) were used to convert GPR index values from the OTC scans to mass of belowground biomass per unit area g m^2. Significantly greater coarse root biomass was present in plots treated with elevated compared to ambient CO_2 ($p = 0.049$) (Stover et al., 2007).

The scrub ecosystem is dominated by large belowground structures such as stems, lignotubers, and burls that serve as carbohydrate reserves that are critical during regrowth following a disturbance in this fire-controlled system. Approximately 86 percent of total biomass was allocated belowground in the scrub system, suggesting greater carbon sequestration in larger belowground structures under elevated CO_2 conditions. Similar to Sanderson, the KSC site was unable to adequately distinguish and separate live and dead root mass within the data.

The spatial distribution of root systems is important when examining their role in resource acquisition, storage, and structural support. Multiple plots (0.25 m^2) were intensively scanned (2 cm scan width) in an intersection grid pattern with a 1500 MHz antenna and later excavated at the KSC

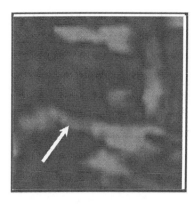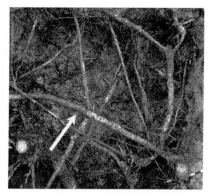

FIGURE 30.5 Visual correlation between ground-penetrating radar (GPR) data and coarse root excavation pits. Large roots spanning the entire length of the validation pit were most likely to be observed in the GPR data. Each GPR data image (left) represents a "slice" of the three-dimensional profile at the appropriately correlated depth with a viewing depth of 15 to 20 cm.

scrub-oak site. The data were processed with RADAN software which permits intersections of the X-Y scans to be interpreted as three-dimensional volumetric data. Resulting three-dimensional models were compared to photographs from the excavation for verification of spatial orientation and distribution of roots (Figure 30.5). Smaller objects that were parallel to one another or that were "shadowed" by neighboring root were typically masked or grouped together as a single object. Larger roots having sharp or unique angles were easiest to identify in the data-image comparisons (Figure 30.5). Initial determination of root architecture was limited by lack of adequate sample resolution (i.e., less than 1 cm width) and accurate methods to quantify the object size in the three-dimensional models. However, these results indicate with further refinement that spatial distribution can be quantified under the sandy soil conditions, thus elucidating environmental changes in root deployment, carbon storage and sequestration, and resource foraging in the scrub ecosystem.

30.4 SUMMARY

Tree root mass can be accurately estimated with GPR when there is good electromagnetic contrast between soil and roots. Calibrating and scaling GPR data to estimate belowground biomass using destructive sampling with 15.24 cm diameter soil cores can be successful, but sites differ in the type of targets that are resolved. Detailed site surveys should be conducted to determine if non-target reflectors (slash, rocks, buried organic matter, rodent tunnels, and surface discontinuities) will degrade the potential to quantify live tree roots. Natural ecosystems and managed forests present obstacles that may interfere with acceptable antenna and ground coupling; scanning over leaf litter, especially in treatments that can affect litter depth should be avoided. Overall, the application of ground radar technology to root and forest ecology studies provides unique, nondestructive means to sample root biomass and spatial distribution at least in some ecosystems.

REFERENCES

Barton, C.V.M. and Montagu, K.D., Detection and determination of root diameter by ground penetrating radar under optimal condition, *Tree Physiol.*, 24, 1323, 2004.

Butnor, J.R., Doolittle, J.A., Kress, L., Cohen, S., and Johnsen, K.H., Use of ground penetrating radar to study tree roots in the southeastern United States, *Tree Physiol.*, 21, 1269, 2001.

Butnor, J.R., Doolittle, J.A., Johnsen, K.H., Samuelson, L., Stokes, T., and Kress, L., Utility of ground penetrating radar as a root biomass survey tool in forest systems, *Soil Sci. Soc. Am. J.,* 67, 1607, 2003.

Cairns, M.A., Brown, S., Helmer, E.H., and Baumgardner, G.A., Root biomass allocation in the world's upland forests, *Oecologia*, 111, 1, 1997.

Cermak, J., Hruska, J., Martinkova, M., and Prax, A., Urban tree root systems and their survival near houses analyzed using ground penetrating radar and sap flow techniques, *Plant Soil*, 103, 2000.

Cox, K.D., Scherm, H., and Serman, N., Ground penetrating radar to detect and quantify residual root fragments following peach orchard clearing. *HortTechnology*, 15, 600, 2005.

Daniels, D.J., *Ground penetrating radar*, Second edition. Institution of Electrical Engineers, London, U., 2004.

Day, F.P, Stover, D.B., Pagel, A.L., Hungate, B.A., Dilustro, J.J., Herbert, B.T., Drake, B.G., and Hinkle, C.R., Rapid root closure after fire limits fine root responses to elevated atmospheric CO_2 in a scrub oak ecosystem in central Florida, USA, *Glob. Change Biol.*, 12, 143, 2006.

Hruska, J., Cermak, J., and Sustek, S., Mapping tree root systems with ground-penetrating radar, *Tree Physiol.*, 19, 125, 1999.

Jim, C.Y., Protection of urban trees from trenching damage in compact city environments, *Cities*, 20, 87, 2003.

Oppenheim, A.V. and Schafer, R. W., *Digital signal processing*, Prentice Hall, Englewood Cliffs, NJ, 1975.

Stokes, A., Fourcaud, T., Hruska, J., Cermak, J., Nadyezdhina, N., Nadyezhdin, V., and Praus, L., An evaluation of different methods to investigate root system architecture of urban trees in situ: 1. Ground penetrating radar, *J. Arboriculture*, 28, 2, 2002.

Stover, D.B., Day, F.P., Butnor, J.R., and Drake, B.G., Effect of elevated CO_2 on coarse-root biomass in Florida scrub detected by ground-penetrating radar, *Ecology* 88, 1328, 2007.

Wielopolski, L., Hendrey, G., and McGuigan, M., Imaging tree root systems in situ. In Noon, D.A., Stickley, G.F., and Longstaff, D. (Eds), *Proceedings of the Eighth International Conference on Ground Penetrating Radar*, The International Society of Optical Engineering, Bellingham, WA, 642, 2000.

Glossary

Agricultural Geophysics Agricultural application of geophysical methods to investigate the shallow subsurface (within a typical depth range of 0 to 2 m).

Agrigeophysics Contraction of the term "agricultural geophysics."

Apparent Soil Electrical Conductivity (Inverse of *apparent soil electrical resistivity.*) The electrical conductivity measured for a bulk volume of soil using resistivity and electromagnetic induction geophysical methods. The measured electrical conductivity represents the true value for the entire bulk soil volume only if soil electrical conductivity is homogeneous. However, soil electrical conductivity is more likely to be heterogeneous; hence, the use of the term "apparent."

Apparent Soil Electrical Resistivity (Inverse of *apparent soil electrical conductivity.*) The electrical resistivity measured for a bulk volume of soil using resistivity and electromagnetic induction geophysical methods. The measured electrical resistivity represents the true value for the entire bulk soil volume only if soil electrical resistivity is homogeneous. However, soil electrical resistivity is more likely to be heterogeneous; hence, the use of the term "apparent."

Attenuation Reduction in the amplitude or energy of a wave as it propagates through a medium. (a) Seismic wave amplitude reduction results from geometric spreading, frictional dissipation of elastic energy into heat, and mode conversion. (b) Generally, the reduction in amplitude of a plane electromagnetic wave traveling through soil or rock depends on the magnetic permeability, dielectric constant, and electrical conductivity of the soil or rock along with the frequency of the electromagnetic wave.

Azimuthal Rotation Resistivity Measurement A field data collection mode for the resistivity method employed to quantify the horizontal directional components of resistivity (horizontal resistivity anisotropy), which may provide useful information on the presence and trends of aligned features in the subsurface. Azimuthal rotation resistivity data collection is commonly carried out with either the electrode array midpoint or one of the array's end points kept stationary, while successive resistivity measurements are made as a linear electrode array is pivoted about this stationary point in increments of $10°$ to $20°$ through a complete sweep of $180°$ or $360°$. The spacing distances between electrodes remain constant as the array is pivoted about a point.

Bulk Modulus A measured constant that defines the stress–strain behavior of an elastic material due to pressure change. The bulk modulus along with the rigidity modulus and density determine the seismic P-wave velocity in soil or rock materials.

Capacitively Coupled Resistivity Methods Resistivity geophysical methods that utilize a capacitive coupling approach to introduce electric current into the subsurface. This approach typically employs coaxial cables that are laid out on the ground surface. The coaxial cable and the soil surface essentially form a large capacitor. The metal shield of the coaxial cable is one of the capacitor plates, and the soil surface is the other capacitor plate, with the outer insulation of the coaxial cable acting as the dielectric material separating the two plates. The system transmitter applies an alternating current to the coaxial cable side of the capacitor, in turn generating alternating current in the soil on the other side of the capacitor.

Cation Exchange Capacity (CEC) The amount of cations per unit dry soil weight electrostatically adsorbed at soil particle surfaces. The measured value is commonly reported in meq/100 g.

Clay Soil particles with an equivalent diameter less than 0.002 mm. The clay size fraction of a soil is composed primarily of clay minerals and organic matter.

Clay-pan A dense, compact, high clay content layer within the soil profile. The presence of a clay-pan will tend to impede root growth and the vertical movement of water.

Common Midpoint GPR Data Collection A field data collection mode for ground-penetrating radar normally used for determining subsurface radar velocities. With this mode of data collection, the midpoint between the transmitting and receiving antennas is kept stationary, and the transmitting and receiving antennas are successively moved farther and farther apart.

Common Offset GPR Data Collection The field data collection mode typically employed with ground-penetrating radar. Transmitting and receiving antennas for this mode have a relatively narrow, but constant, separation distance, and measurements are collected along single transects or a grid set of transects that cover the study area.

Constant Separation Traversing Resistivity Measurement A field data collection mode for the resistivity method in which the spacing distance between electrodes remains constant as the electrode array is moved along a transect. Constant separation traversing measurements are usually collected along a series of transects forming a grid that covers a study area, thereby making this data collection mode ideal for highlighting horizontal variations in apparent soil resistivity (or apparent soil electrical conductivity).

Critical Angle Typically refers to the incident angle (referenced to an imaginary line perpendicular to the interface) for a seismic wave that upon encountering a subsurface interface produces a refracted seismic wave (or head wave) that travels along the interface.

Crop Yield The amount of crop production per unit farm field area. The spatial variation in crop yield across a field is measured using devices called yield monitors.

Degree of Saturation The fraction of the soil or rock porosity that is filled with water.

Density The mass per unit volume of a material.

Dielectric Constant (Also called *relative permittivity*.) A dimensionless constant indicative of a material's ability to store charge in the presence of an electric field, or stated in a different manner, a material's ability to become polarized within an electric field (e.g., the alignment of polar molecules with the vector orientation of the electric field). Dielectric constant values for Earth materials typically range from 1 to 80, with 1 being the value for air and 80 being the value for water. The dielectric constant is an important material property for geophysics, especially with regard to ground-penetrating radar methods. The velocity, v_m, with which a radar wave travels through a particular soil or rock Earth material is governed by the dielectric constant of that material and can be calculated with the following equation:

$$v_m = \frac{v_0}{\sqrt{\varepsilon_r}}$$

where v_0 is the speed of light in air, and ε_r is the material's dielectric constant (or relative permittivity). Given a radar wave normally incident at a subsurface interface between two soil or rock layers, the reflected and transmitted radar wave amplitude or energy is likewise controlled by dielectric constant values, in this case, those for the materials on both sides of the interface. The equations for the radar wave reflection coefficient, R, and the transmission coefficient, T (see Glossary definitions) are provided below, and as shown, these equations are solely functions of dielectric constant.

$$R = \frac{\sqrt{\varepsilon_{r2}} - \sqrt{\varepsilon_{r1}}}{\sqrt{\varepsilon_{r2}} + \sqrt{\varepsilon_{r1}}}$$

and

$$T = 1 - \frac{\sqrt{\varepsilon_{r2}} - \sqrt{\varepsilon_{r1}}}{\sqrt{\varepsilon_{r2}} + \sqrt{\varepsilon_{r1}}}$$

where ε_{r1} is the dielectric constant of the first layer through which the radar wave is traveling before it encounters the subsurface interface, and ε_{r2} is the dielectric constant of the second layer on the opposite side of the interface.

Electrical Conductivity The capability of a material to transmit electric current.

Electrical Resistivity The capability of a material to oppose the transmission of electric current.

Electric Current Flow of electric charge. Charge carriers can be electrons or dissolved electrolytes (ions). Electric current is generally either unidirectional (direct current) or alternates in direction with a given frequency (alternating current).

Electric Field (Also referred to as the *electric potential gradient*.) A vector quantity having a magnitude at any point defined by the change in electric potential per unit distance.

Electric Potential Potential energy for a unit electric charge due to its position at some point within an electric field.

Electric Potential Difference The difference in electric potential between two points in an electric field.

Electrolyte A chemical compound capable of dissociating into ions when dissolved in solution. These ions can then act as electric charge carriers for the transmission of electric current through the solution. The common dissolved ions found in soil solution are SO_4^{2-}, Cl^-, HCO_3^-, NO_3^-, PO_4^{3-}, Ca^{2+}, Mg^{2+}, K^+, Na^+, and NH_4^+.

Electromagnetic Induction (EMI) Methods Geophysical investigation methods used to measure subsurface electrical conductivity (or its inverse, electrical resistivity). An instrument called a ground conductivity meter is commonly employed for the relatively shallow investigations needed for agricultural purposes. In regard to operation, an alternating electrical current is passed through one of two small electric wire coils spaced a set distance apart and housed within the ground conductivity meter, which is positioned at, or a short distance above, the ground surface. The applied current produces an electromagnetic field around the "transmitting" coil, with a portion of the electromagnetic field extending into the subsurface. This electromagnetic field, called the primary field, induces an alternating electrical current within the ground, in turn producing a secondary electromagnetic field. Part of the secondary field spreads back to the surface and the air above. The second wire coil acts as a receiver measuring the resultant amplitude and phase components of both the primary and secondary fields. The amplitude and phase differences between the primary and resultant fields are then used, along with the intercoil spacing, to calculate an "apparent" value for soil electrical conductivity (or resistivity).

Electromagnetic Wave Skin Depth The depth beneath the ground surface at which a plane electromagnetic wave's amplitude has been reduced to $1/e$ (approximately 37 percent) of its surface value.

Field Capacity An initially saturated or near-saturated soil that is freely capable of being drained reaches a volumetric water content called "field capacity" when all the possible gravity drainable water has been removed from the soil.

Field-Scale Experiments (With respect to agriculture.) Experiments often conducted on farms, where the land area being investigated is of a field size generally associated with traditional agricultural practices. Field-scale experiments will, when needed, utilize typical farm equipment.

Fragipan A high bulk density layer within the soil profile that can impede root growth and the vertical movement of water.

Frequency Wave cycle repetition rate.

Galvanic Contact Resistivity Methods Resistivity geophysical methods that employ electrodes to directly apply electric current into the subsurface.

Geographic Coordinate System A spatial reference system using latitude and longitude to describe locations on the Earth's surface.

Geographic Information System (GIS) A computer-based information management system capable of storing, editing, integrating, analyzing, and displaying spatially referenced data.

GIS: Layer A group of geographic features with thematic similarities having the same geographic extent, coordinate system, and attributes.

GIS: Raster Data A data format in which attribute values are referenced to individual cells within a spatially defined grid.

GIS: Vector Data A data format in which features and associated attributes are spatially referenced to points, lines, or areas enclosed by polygons.

Geophone A device for recording artificially generated seismic waves, accomplished by converting ground vibrations into voltage measurements.

Geophysics There are several definitions for geophysics, but in the context of this book, geophysics is defined as the application of physical quantity measurement techniques to provide in situ information on conditions or features beneath the Earth's surface.

Global Positioning System (GPS) A global satellite navigation system employing twenty-four satellites in medium Earth orbit, which transmit precise microwave signals, thereby enabling signal receivers to be accurately located.

GPS: 2D Operating Mode A two-dimensional GPS position fix that includes only horizontal coordinates (no GPS elevation). It requires a minimum of three visible satellites.

GPS: 3D Operating Mode A three-dimensional GPS position fix that includes horizontal coordinates, plus elevation. It requires a minimum of four visible satellites.

GPS: Accuracy A measure of how close an estimate of a GPS position is to the true location (how close a fix comes to the actual position).

GPS: Almanac A set of parameters included in the GPS satellite navigation message that a receiver uses to predict the approximate location of a satellite. The almanac contains information about all of the satellites in the constellation.

GPS: Ambiguity The initial bias in a carrier-phase observation of an arbitrary number of cycles. The initial phase measurement made when a GPS receiver first locks onto a GPS signal is ambiguous by an integer number of cycles because the receiver has no way of knowing the exact number of carrier wave cycles between the satellite and the receiver. This ambiguity, which remains constant as long as the receiver remains locked on the signal, is established when the carrier-phase data are processed.

GPS: Anti-Spoofing Encryption of the P-code to protect the P-signals from being "spoofed" through the transmission of false GPS signals by an adversary.

GPS: Carrier Frequency A radio wave that conveys or carries some kind of modulation. The frequency of an unmodulated output of a radio transmitter. The GPS L1 carrier frequency is 1575.42 MHz and L2 carrier frequency is 1227.6 MHz.

GPS: Carrier Phase The accumulated phase of either the L1 or L2 carrier of a GPS signal, measured by a GPS receiver since locking onto the signal (also called integrated Doppler).

GPS: Coarse/Acquisition Code (C/A Code) The standard positioning signal the GPS satellite transmits to the civilian user. It contains the information the GPS receiver uses to fix its position and time, and is accurate to 100 meters or better.

GPS: Coordinates A set of numbers that describes your location on or above the Earth. Coordinates are typically based on latitude and longitude lines of reference or a global or regional grid projection.

GPS: Cycle Slip A discontinuity in GPS carrier-phase observations, usually of an integer number of cycles, caused by temporary signal loss. If a GPS receiver loses a signal temporarily, due to obstructions for example, when the signal is reacquired there may be a jump in the integer part of the carrier-phase measurement due to the receiver incorrectly predicting the elapsed number of cycles between signal loss and reacquisition.

GPS: Data Message A message included in the GPS signal that reports the satellite location, clock corrections, and health. Included is approximate information about the other satellites in the system as well.

GPS: Datum The coordinate system used to define position on the Earth surface.

GPS: Differential GPS (DGPS) An extension of the GPS system that uses land-based radio beacons to transmit position corrections to GPS receivers. DGPS reduces the effect of selective availability, propagation delay, and so forth, and can improve position accuracy to better than 10 meters.

GPS: Dilution of Precision (DOP) A measure of the GPS receiver and satellite geometry, A dimensionless number that accounts for the contribution of relative satellite geometry to errors in position determination. A low DOP value indicates better relative geometry and higher corresponding accuracy. The DOP indicators are GDOP (geometric DOP), PDOP (position DOP), HDOP (horizontal DOP), VDOP (vertical DOP), and TDOP (time clock offset).

GPS: Doppler Effect The shift in the frequency of a received radio signal due to the relative motion of the transmitter and receiver.

GPS: P-Code The precise code of the GPS signal typically used only by the U.S. military. It is encrypted and reset every seven days to prevent use from unauthorized persons. The code consists of about 2.35×1014 chips and is sent at a rate of 10.23 megabits per second. At this rate, it would take 266 days to transmit the complete code. Each satellite is assigned a unique one-week segment of the code that is reset at Saturday/Sunday midnight. The P-code is currently transmitted on both the L1 and L2 frequencies.

GPS: Pseudo-Random Code Pseudorandom Noise (PRN) Code. Deterministic binary sequences with noise-like properties. These codes are used in spread-spectrum communications systems and in ranging systems such as GPS. Two PRN codes are transmitted by GPS satellites: the C/A-code and P-code. The identifying signature signal transmitted by each GPS satellite and mirrored by the GPS receiver in order to separate and retrieve the signal from background noise.

GPS: Pseudorange The measured distance between the GPS receiver and the GPS satellite using uncorrected time comparisons from satellite-transmitted code and the local receiver's reference code.

GPS: Real-Time Kinematic (RTK) The DGPS procedure whereby carrier-phase corrections are transmitted in real time from a reference receiver to the user receiver. RTK is often used for the carrier-phase integer ambiguity resolution approach.

GPS: Relative Accuracy The accuracy with which a user can measure position relative to that of another user on the same navigation system at the same time.

GPS: Selective Availability (SA) The random error, which the government can intentionally add to GPS signals, so that their accuracy for civilian use is degraded. SA is not currently in use.

GPS: Wide Area Augmentation System (WAAS) A system of satellites and ground stations that provide GPS signal corrections for better position accuracy. A form of DGPS in which the user's GPS receiver receives corrections determined from a network of reference stations distributed over a wide geographical area. A WAAS-capable receiver can give a position accuracy of better than three meters, 95 percent of the time. WAAS consists of approximately twenty-five ground reference stations positioned across the United States that monitor GPS satellite data. Two master stations, located on either coast, collect data from the reference stations and create a GPS correction message.

GPS: WGS-84 World Geodetic System, 1984. The primary map datum used by GPS. Secondary datums are computed as differences from the WGS 84 standard.

GPS: Y-Code The encrypted P-Code.

Gradiometer A device composed of two magnetometer sensors mounted a set distance apart and used to measure the magnetic field gradient.

Ground-Penetrating Radar Methods (GPR) Geophysical methods that employ radar signals to investigate the subsurface. The basic principle of operation for GPR involves using a transmitting antenna to direct radar signal into the subsurface. The radar signal travels through the ground via various direct, reflected, and refracted pathways from the transmitting antenna to a receiving antenna, which records the radar signal travel times and amplitudes. The radar signal information collected at the receiving antenna is then used to gain insight on below-ground conditions such as soil water content or to provide details on the positions and character of subsurface features including soil layers, utility lines, or buried drainage pipes.

Hydraulic Conductivity A porous media (soil or rock) proportionality constant relating hydraulic gradient to specific discharge. Hydraulic conductivity is a measure of the soil or rock ability to allow internal flow of water.

Inversion A computer analysis computational process for generating a spatial model of a soil or rock material physical property based on geophysical measurements.

Love Wave A type of seismic wave that travels along the surface. Love waves are essentially horizontal S-waves transmitted via multiple reflections between the top and bottom of a low seismic velocity surface layer.

Magnetic Field Strength A vector quantity representing the magnetic potential gradient at a point in space.

Magnetic Permeability A material property having a value that equals the ratio of the magnetic field strength to the magnetizing force.

Magnetic Susceptability A dimensionless material property representing the level to which a material can become magnetized.

Magnetometry Methods These geophysical investigation methods employ a sensor, called a magnetometer, to detect and quantify anomalies in the Earth's magnetic field that are indicative of various features beneath the surface. An anomaly is produced when a subsurface feature has a remanent magnetism or magnetic susceptibility that differs from its surroundings.

Map Projection Method A technique used to portray the Earth's curved surface on a flat map.

Map Scale The ratio between the distance on a map to the actual distance on the ground.

Non-Polarizing Electrode A specialized electrode used with resistivity and self-potential geophysical methods. These specialized electrodes minimize unwanted electric potentials that would occur when using metal stake electrodes that are in contact with the ground. A non-polarizing electrode is typically composed of a copper rod inserted through the lid of a container that is porous at its base and filled with a saturated, aqueous, copper sulfate solution.

Pesticide Partition Coefficient An equilibrium proportionality constant relating the soil solution pesticide concentration to the mass of adsorbed pesticide per unit soil particle mass.

Plane Coordinate System (Also called *rectangular coordinate system.*) A spatial reference system where locations are described on a flat map representation of the Earth's curved surface. This flat map representation is produced using a map projection method.

Pore Water Pressure Potential Component of the total pore water energy potential due to water pressure. Pore water pressure is measured relative to atmospheric pressure, and as a consequence, the pore water pressure under unsaturated conditions has a negative value because the unsaturated pore water pressure is less than the atmospheric pressure.

Porosity The volume amount of pore space in a soil or rock per unit volume of soil or rock.

Precision Agriculture (Also referred to as *precision farming.*) Precision agriculture is a growing trend combining geospatial data sets, state-of-the-art farm equipment technology, geographic information systems (GISs), and Global Positioning System (GPS) receivers to support spatially variable field application of fertilizer, soil amendments, pesticides, irrigation, tillage effort, and so forth. Geospatial data sets can include soil property maps derived from geophysical surveys and crop production maps generated by yield monitors. The benefits of precision agriculture to farmers are maximized crop yields and reduced input costs. Precision agriculture can additionally provide important environmental benefits, because with optimal amounts of fertilizer, soil amendments, pesticides, irrigation, and tillage effort applied on different parts of the field, there are potentially less agrochemicals and sediment released offsite via subsurface drainage and surface runoff. Reducing the offsite discharge of agrochemicals and sediment, in turn, diminishes adverse environmental impacts on local waterways. In essence, precision agriculture techniques allow a farm field to be divided into different management zones for the overall purpose of optimizing economic benefits and environmental protection.

P-Wave (Also called a *primary wave*, a *compressional wave*, or a *longitudinal wave.*) A type of seismic body wave having an elastic back-and-forth particle motion orientation that coincides with the direction of wave propagation. P-waves can be transmitted through solid, liquid, and gas materials. P-waves are the fastest seismic waves, and their velocity, V_P, within a soil or rock material is given by the following equation:

$$V_P = \sqrt{\frac{k + \frac{4}{3}\mu}{\rho}}$$

where k is the bulk modulus, μ is the rigidity modulus (or shear modulus), and ρ is density. As indicated by the equation directly above, the P-wave velocity in soil or rock depends only on elastic moduli and density of these Earth materials.

Rayleigh Wave A type of seismic wave that travels along the surface. Rayleigh wave particle motion is elliptical retrograde in a vertical plane oriented coincident with the direction of wave propagation. The amplitude of a Rayleigh wave decreases exponentially with depth.

Reflection Coefficient Given a subsurface interface separating two materials having different properties, the reflection coefficient is the ratio of the reflected radar or seismic wave amplitude at the interface to the incident radar or seismic wave amplitude at the interface.

Remanent Magnetism Magnetism that exists in a material regardless of whether an external magnetic field is present.

Resistivity Methods Geophysical investigation methods used to measure subsurface electrical resistivity (or its inverse, electrical conductivity). With the conventional resistivity method, an electric current is supplied between two electrodes inserted at the ground surface, and voltage is concurrently measured between a separate pair of electrodes also inserted at the surface. The current, voltage, electrode spacing, and electrode configuration are then used to calculate a bulk soil electrical resistivity (or conductivity) value. There are a variety of electrode configurations, called electrode arrays, most of which are linear, with the Wenner, Schlumberger, and dipole-dipole arrays the ones employed the majority of the time (see Chapter 5). The development of continuous galvanic contact and capacitively coupled resistivity measurement techniques in the late 1980s and early 1990s has transformed the conventional resistivity method into an effective and efficient tool to assess soil conditions and properties in large agricultural fields.

Rigidity Modulus (Also called *shear modulus.*) A measured constant that defines the stress–strain behavior of an elastic material due to the application of shear stress. The rigidity modulus plays an important role in determining seismic wave velocities in soil and rock materials.

Runoff Water flow over the soil surface due to a rainfall or irrigation event.

Salinity The total soluble salt (dissolved electrolyte) concentration in a soil. High salinity levels in soil can adversely affect plant growth and land use. Salinity is typically determined by the electrical conductivity measurement of a solution extracted from a saturated soil paste.

Sand Soil particles with an equivalent diameter ranging from 0.05 to 2.0 mm.

Self-Potential Methods From an operational standpoint, self-potential methods are probably the simplest geophysical methods, requiring only the measurement of a naturally occurring electric potential difference between two locations on the ground surface. Essentially, a voltmeter is used to measure the electric potential difference between two nonpolarizing electrodes placed at the ground surface that are connected to the voltmeter via electric cable. Naturally occurring electric potential differences in the subsurface are often caused by electrokinetic, electrochemical, or ore body oxidation and reduction processes.

Seismic Adsorption Coefficient A constant value used in the calculation of seismic wave attenuation that accounts for elastic energy loss due to frictional dissipation.

Seismic Methods Geophysical subsurface investigation methods employing explosive, impact, vibratory, or acoustic energy sources to introduce elastic (or seismic) waves into the ground. The seismic waves are timed as they travel via direct, reflected, or refracted paths through the subsurface from the source to the sensors, which are called geophones. Information on the timed arrivals and amplitudes of the seismic waves measured by the geophones is then used to determine below-ground conditions or to locate and characterize subsurface features.

Seismic Waves Elastic vibrations capable of propagating through Earth materials such as soil and rock.

Silt Soil particles with an equivalent diameter ranging from 0.002 to 0.05 mm.

Site-Specific Management The use of spatially governed agricultural practices to manage crops, soil, pests, and so forth, based on information obtained regarding variable conditions within a farm field.

Sodium Adsorption Ratio (SAR) A quantity used as an indicator of the exchangeable sodium percentage (ESP), which is the percentage of the soil cation exchange capacity occupied by sodium ions (Na^+). Both SAR and ESP are used to gauge soil sodicity that reflects not only the amount of sodium present but also its impact on plant growth. The value of SAR is commonly determined from the calcium ion (Ca^{2+}), magnesium ion (Mg^{2+}), and sodium ion (Na^+) concentrations in a solution extracted from a saturated soil paste, and it is calculated using the following equation:

$$SAR = \frac{\left[Na^+\right]}{\sqrt{\dfrac{\left[Ca^{2+}\right] + \left[Mg^{2+}\right]}{2}}}$$

where the values in brackets represent ion concentrations in meq/L.

Soil Drainage Class A soil can be assigned to one of seven drainage classes based on the frequency and duration for periods of water saturation or partial water saturation.

Soil Gravimetric Water Content Weight amount of water per unit weight of dry soil.

Soil Organic Carbon (SOC) The amount of carbon contained in organic matter that is present within a soil. Soil organic carbon is usually expressed as a fraction of total soil weight.

Soil Organic Matter (SOM) The amount of plant and animal material at various stages of decomposition that is present within a soil. Soil organic matter is usually expressed as a fraction of a total soil weight.

Soil Particle Size Distribution (Also referred to as *soil texture*.) The relative proportions of a soil's sand-, silt-, and clay-sized particles.

Soil pH A measure of acidity or alkalinity of a soil. The soil pH is measured on a scale of 0 to 14, with 0 being the extreme for acidity and 14 being the extreme for alkalinity. A soil pH of 7 reflects neutral conditions in which the soil is neither acidic nor alkaline. Plants and micro-organisms do not tolerate extremes in soil pH.

Soil Productivity Zones Areas within a farm field having similar crop production potential.

Soil Profile Sequence of soil horizons (layers) from the surface down to the parent material. The soil profile is derived from the parent material via weathering processes.

Soil Solution The aqueous phase of a soil containing dissolved ions, other solutes, and suspended colloids.

Soil Volumetric Water Content Volume amount of water per unit volume of soil.

Specific Surface Total soil particle surface area per unit dry weight of soil.

Subsurface Drainage As applied to agriculture, refers to the use of jointed or perforated buried pipes to remove excess water from the soil for the purpose of enhancing crop growth and improving field trafficability.

S-Wave (Also called a *secondary wave, shear wave,* or *traverse wave.*) A type of seismic body wave having an elastic particle motion that is perpendicular to the direction of wave propagation. There are two kinds of S-waves: the SV-wave and the SH-wave. The particle motion for an SV-wave has a vertical component. SH-waves, on the other hand, have a particle motion that is completely horizontal. S-waves are only capable of traveling through solid material, and not liquids or gases. S-waves are slower than P-waves and have a velocity, V_S, given by:

$$V_S = \sqrt{\frac{\mu}{\rho}}$$

where μ is the rigidity modulus and ρ is density. The S-wave velocity, as indicated by the equation above, is governed strictly by shear stress elastic behavior and density of the soil or rock through which the S-wave travels.

Thematic Map A map that provides information on a single topic or theme. Examples include maps of crop yield and soil volumetric water content.

Time Domain Reflectometry (TDR) A technique used primarily for in situ measurement of volu-metric water content for a small volume of soil. The technique is also employed to measure soil electrical conductivity. The probe used with the technique usually has two metal rod waveguides that are inserted into the soil. The probe generates a high-frequency radar pulse that travels along the waveguides, reflects at the ends of the waveguides, and then returns along the original path. The travel time of the radar pulse is used to calculate the radar veloc-ity in the soil. The radar velocity determines the soil dielectric constant, which, in turn, is strongly correlated with the volumetric water content present in the soil.

Tomography As applied to geophysics, refers to techniques used to determine the spatial distribu-tion of a soil or rock material physical property within a defined boundary. These techniques are based on measurements of some form of energy transmitted through the area or volume enclosed within the boundary. Furthermore, measurements are obtained for multiple energy source and sensor positionings along the periphery of the area or volume being investigated. Measurements are then input into image reconstruction computer software employing inver-sion techniques to determine the physical property spatial distribution within the bounded soil or rock material.

Tortuosity The ratio of the average path length a water or solute molecule travels in a porous media from one point to another compared to the straight-line path length between the two points. Tortuosity is an indicator of the amount of connectivity between pores within a soil or rock.

Transmission Coefficient Given a subsurface interface separating two materials having different properties, the transmission coefficient is the ratio of the amplitude for a radar or seismic wave transmitted through the interface to the amplitude of the radar or seismic wave incident at the interface.

Vertical Electric Sounding Resistivity Measurement A resistivity field data collection mode used for obtaining information on resistivity (or electrical conductivity) variations with depth. The procedure for vertical electric sounding involves keeping the midpoint of the resistivity array stationary, while the overall array length is successively increased. As the electrode array is lengthened, electric current penetrates farther into the subsurface, providing a greater depth of investigation.

Wavelength The beginning to end length of a wave cycle.

Wet Aggregate Stability A measure of the degree to which soil retains its structure when wet. Wet aggregate stability is expressed as a percentage of soil dry weight after wet sieving to the soil dry weight before wet sieving.

Wide-Offset Reflection GPR Data Collection A ground-penetrating radar data collection mode in which the transmitting and receiving antennas are kept at a constant but much wider separation distance than is normal for a more typical ground-penetrating radar survey. This type of data collection mode is commonly used for mapping near-surface soil volumetric water content.

Wilting Point The volumetric water content level below which plants can no longer extract water from soil. When the soil volumetric water content falls below this level, plants begin to wilt.

Index

Note: Page numbers in **bold** refer to figures or tables.